PLANT STRESS TOLERANCE
Physiological & Molecular Strategies

PLANT STRESS TOLERANCE
Physiological & Molecular Strategies

Editor

A. Hemantaranjan

Department of Plant Physiology
Institute of Agricultural Sciences
Banaras Hindu University,
Varanasi – 221 005 INDIA

Published by:

SCIENTIFIC PUBLISHERS (INDIA)

Jodhpur

–

5 A, New Pali Road
P.O. Box 91
Jodhpur - 342 001
INDIA

Delhi

–

4806/24, Ansari Road
Daryaganj
New Delhi - 110 002
INDIA

Germany

–

Landstr, 240
Frankfurt - 60388
GERMANY

E-mail: info@scientificpub.com
Website: www.scientificpub.com

Print : 2023

ISBN: 978-81-7233-985-2
eISBN: 978-93-86102-19-5

Printed in India

PREFACE

The book entitled *"**Plant Stress Tolerance – Physiological & Molecular Strategies**"* has been especially edited for holistic development of the science of agriculture and crop production under distinctly changing environment. Resource utilization is always overlooked; hence a brief focus on sustainability has been remarkably presented to prove the meaningfulness of this publication. This book brings ingenious applied researches highlighting the major environmental factors coupled with scrupulous strategies in solving abiotic stresses in varied micro and macro agro-climatic conditions , in general, and unfolding the basis for tolerance mechanisms in plant systems, in particular.

This revealing book begins by escalating our knowledge on ultra techniques in molecular plant physiology precisely, which is extremely needed in the present scenario. Liquid Chromatography Mass Spectrometry (LC-MS) is a combination of physical separation property of chromatography and mass analysis property of spectrometry and is a useful tool for molecular biology research, hence, presents a broad aspect of its applications in "Omics" research. Correspondingly, recent developments in the molecular biology techniques offer the potential to increase the rate of genetic improvement. Translating our research under controlled conditions into field conditions is the future target. This is needed to build up more understanding of the heterogeneous conditions prevalent in the field for linking the knowledge generated through the phenotyping studies in controlled and semi controlled environment in to natural environment. The successful application of genomics for crop improvement is possible only by the precise phenotyping and have been exhaustively embracing the pertinent features like high-throughput evaluation in controlled environment, deeply lacking in the present breeding system. The exercise of imaging techniques in the high-throughput phenotyping has opened up new horizons for abiotic stress tolerance. The details of plant phenotyping while increasing the cost-effectiveness of phenotyping for drought in several crops have been focused. Additionally, morpho-physiological and molecular modulations in plants in response to drought and focuses on the physiological impacts of high temperature stress and various tolerance mechanisms adapted by the plants to mitigate the high temperature stress followed by improving grain yield in wheat (*Triticum aestivum* L.) by stem reserve remobilization under heat stress have been the major segments of this book emphasizing the major current challenge, which is to break the yield barriers.

Plants can be genetically modified to enhance their tolerance, uptake, translocation, accumulation and volatilization capabilities for Hg phyto-remediation. In view of this noticeable problem, the short term responses of crops under mercury contamination at hazardous waste sites have been highlighted in this book. Further, the molecular genetics and physiological mechanisms of submergence tolerance in rice have been minutely described by covering vital aspects judiciously. Subsequently, this carries reviews on crucial aspects of paclobutrazole and its mitigating effect on the harmful effects caused by a number of abiotic stresses including UV-radiation on plant species carried out in the last two decades provide new dimension opening wider avenues for research especially for crop plants cultivated under varying environments.

Scientific elucidations on soil microbial diversity and nutrient transformation as influenced by introduction of transgenic plants underlining future environmental evaluation of the impact of the diverse transgenic crops leading to an improved understanding of soil biological functions and processes could be also obtained coupled with the worldwide crucial stress like iron-chlorosis in calcareous soil, noticeably of a major concern for better yields and productivity, explained satisfactorily on physiological and molecular bases with possible management strategies in this exceptionally designed book. In the post-genomic era this particular endeavour is accomplishing knowledge for the strategic move toward elite research in plant molecular physiology too, which precisely explains the significance of post-translational modifications for proteomic research, since post-translational modification provides a dynamic mechanism for regulation of protein function. Successively, *in vitro* regeneration and acclimatization studies of important endangered medicinal plants of Asia reveal that the cultured cells and tissue can take several pathways for production of true-to-type plants in large numbers are the preferred ones for commercial multiplication have been illustrated. Afterwards, this book presents a bioassay-directed discovery of natural products for their antimicrobial potential and development of medicine from bioassay-directed separation of compounds followed by a detailed account on *Swertia chirayita*, an endangered medicinal herb of temperate Himalaya to make this section highly exciting from academic and applied points of view. Nevertheless, two stirring chapters highlight the physiological and biochemical properties of a desirable plant *Gliricidia* and its cultivation develops scope for remunerative venture for farmers, on the one hand, besides the significance of thiourea for crop improvement and performance of crops that has been judiciously compiled with genuine and pragmatic attention towards sustainability of agriculture, on the other hand, which is indeed a vital issue for each and every discipline of agriculture for over two decades. Finally, overview on the role of allelopathy in sustainable agriculture-with special reference to *Parthenium* has been exhaustively brought together, which focuses upon conspicuous and indispensable aspects of allelopathy from the past to the present scenario.

In this enthusiastic endeavour, I am delighted to state my genuine respect to all distinguished and talented contributors from well known institutions for bringing up this unrivalled, sensible, considerate and far-reaching book up to the international standard created on the **Centennial Year** of the **Banaras Hindu University**. My heartfelt thankfulness is very much due to the Hon'ble Vice-Chancellor, Professor G.C. Tripathi, Banaras Hindu University, the Director, Institute of Agricultural Sciences, the Dean, Faculty of Agriculture and my admired colleagues all over the globe for their consistent moral support in the fulfillment of the devoted but enormous academic task.

Last but not the least, I am beholden to my family for their blessings and good wishes in this mission. Besides these, I am extremely thankful to the excellent and talented staff members of the Scientific Publishers, Jodhpur, India, for their real competence in the perfect printing of international standard.

Professor A. Hemantaranjan
Ph.D., Fellow,
Indian Society for Plant Physiology

October 29, 2015

Department of Plant Physiology
Institute of Agricultural Sciences
Banaras Hindu University
Varanasi 221 005, INDIA
E-mail: hemantaranjan@gmail.com

CONTRIBUTORS

A. Hemantaranjan
Department of Plant Physiology, Institute of Agricultural Sciences, Banaras Hindu University, Varanasi – 221 005, INDIA

A. K. Trivedi
National Bureau of Plant Genetic Resources, Regional Station Bhowali, Nainital– 263132, INDIA

Ancy Fernandes
Department of Botany, University School of Sciences, Gujarat University, Ahmedabad-380009 INDIA

B. Basak
ICAR-Directorate of Medicinal and Aromatic Plants Research, Boriavi, Anand 387 310, INDIA

Bandana Bose, Department of Plant Physiology, Institute of Agricultural Sciences, Banaras Hindu University, Varanasi-221005, INDIA

Bhawna Pant
Department of Plant Physiology, Institute of Agricultural Sciences, Banaras Hindu University, Varanasi-221005, INDIA

Binayak Chakraborty
College of Agriculture, Navasri Agricultural University, Waghai – 394 730, The Dangs, Gujarat, INDIA

D. Vijayalakshmi
Department of Crop Physiology, Tamil Nadu Agricultural University, Coimbatore, INDIA

Debarati Bhaduri
ICAR- Directorate of Groundnut Research, Junagadh-362 001, INDIA

Dhiman Mukherjee
Regional Research Station, Uttar Banga Krishi Viswavidyalaya, Kalimpong (Darjelling)-734301 (W.B), INDIA

Dinesh K. Yadav
Department of Plant Physiology, Institute of Agricultural Sciences, Banaras Hindu University, Varanasi - 221 005, INDIA

Jayanti Srivastava
FIST-DST PG Dept., of Botany, S.M.M.T. (PG) College, Ballia- 277001

Jyoti Kumari
Germplasm Evaluation Division, National Bureau of Plant Genetic Resources, New Delhi 110012, INDIA

K. Vanitha
Tamilnadu Rice Research Institute (TRRI), Tamil Nadu Agricultural University, Aduthurai, Thanjavur 612 101, INDIA

Koushik Chakraborty
ICAR-Directorate of Groundnut Research, Ivnagar Road, Junagadh – 362 001, INDIA

Kuldeep Yadav
Plant Tissue Culture Laboratory, Department of Botany, Kurukshetra University, Kurukshetra -136 119, INDIA

Lina Singh
FIST-DST PG Dept., of Botany,
S.M.M.T. (PG) College,
Ballia- 277001, INDIA

M.Raveendran
Department of Plant
Biotechnology, Tamil Nadu
Agricultural University,
Coimbatore, INDIA

Md. Afjal Ahmad
Department of Plant Physiology,
Institute of Agricultural Sciences,
Banaras Hindu University,
Varanasi - 221 005, INDIA

Narender Singh
Plant Tissue Culture Laboratory,
Department of Botany,
Kurukshetra University,
Kurukshetra -136 119, INDIA

P. Vivitha
Department of Crop Physiology,
Tamil Nadu Agricultural
University, Coimbatore, INDIA

Padmanabh Dwivedi
Department of Plant Physiology,
Institute of Agricultural Sciences,
Banaras Hindu University,
Varanasi - 221 005, INDIA

Prasann Kumar
Department of Plant Physiology,
Institute of Agricultural Sciences,
Banaras Hindu University,
Varanasi - 221 005, INDIA

Pravin Prakash
Department of Plant Physiology,
Institute of Agricultural Sciences,
Banaras Hindu University,
Varanasi - 221 005, INDIA

Preeti Singh
Department of Plant Physiology,
Institute of Agricultural Sciences,
Banaras Hindu University,
Varanasi - 221 005, INDIA

R.S. Meena
Department of Agronomy,
Institute of Agricultural Sciences,
Banaras Hindu University,
Varanasi - 221 005, INDIA

Nisha Raghava
FIST-DST PG
Dept., of Botany,
S.M.M.T. (PG) College,
Ballia- 277001, INDIA

Ravindra P Raghav
FIST-DST PG Dept., of Botany,
S.M.M.T. (PG) College,
Ballia- 277001, INDIA

Renu Singh
Department of Botany,
University School of Sciences,
Gujarat University,
Ahmedabad-380009, INDIA

Ruchi Bansal
Germplasm Evaluation Division,
National Bureau of Plant Genetic
Resources,
New Delhi 110012, INDIA

Sananda Mondal
Department of ASEPAN, Institute
of Agriculture, Visva-Bharati,
Sriniketan-731236, INDIA

Sharmistha Pal, ICAR-Central Soil
and Water Conservation Research
and Training Institute, Research
Centre, Madhya Marg,
Chandigarh 160 019, INDIA

Sundeep Kumar
Division of Genomic Resources,
National Bureau of Plant Genetic
Resources,
New Delhi 110012, INDIA

T. Parthasarathi
e - Extension Centre, Directorate
of Extension Education, Tamil
Nadu Agricultural University,
Coimbatore 641 003, INDIA

V. Ravichandran
Department of Rice, Tamil Nadu Agricultural University,
Coimbatore 641 003, INDIA

V.P. Singh
College of Agriculture,
JNKVV, Ganj Basoda,
Vidisha, M.P. 464221, INDIA

Vijai P.
Department of Plant Physiology,
Institute of Agricultural Sciences,
Banaras Hindu University,
Varanasi - 221 005, INDIA

Vikender Kaur
Germplasm Evaluation Division,
National Bureau of Plant Genetic Resources,
New Delhi 110012, INDIA

CONTENTS

Section III:
Microbial Diversity and Molecular Strategies in Plant Nutrition

Section IV:
Proteomic Research

Section V:
Medicinal Plants, In Vitro Regeneration & Natural Products

Section VI:
Plant Physiology in Sustainability of Agriculture

Section VII:
Comprehensive Review

Section - I

Ultra Techniques in Plant Physiology

1

APPLICATIONS OF LIQUID CHOMATOGRAPHY MASS SPECTROMETRY FOR PROTEOMICS

A.K. Trivedi[1]* and A. Hemantaranjan[2]

[1]*National Bureau of Plant Genetic Resources, Regional Station Bhowali, Nainital (Uttarakhand) – 263 132 INDIA*
[2]*Department of Plant Physiology, Institute of Agricultural Sciences, Banaras Hindu University, Varanasi – 221 005 INDIA*

INTRODUCTION

Liquid Chromatography Mass Spectrometry (LC-MS) is a combination of physical separation property of chromatography and mass analysis property of spectrometry. LC-MS is a useful tool for molecular biology research. Peptide and protein can be quantified by LC-MS using several strategies. It is frequently used to analyze difference between samples at the protein expression level, in post- translational modification (PTMs), in the components of protein complexes and in intracellular protein localization. With application of this technique thousand of proteins can now be compared in a single experiment.

LC-MS based techniques are preferred method for high-content quantitative analysis of metabolites. Amount of these metabolites can be used to study pathway activity. Moreover, LC-MS based techniques are widely used to quantify the activities of biochemical pathways. Such as activity of protein kinase/phosphatase reaction pairs is encoded in the phosphoproteome, this is a reflection of the activity of key signalling pathways linking activation of surface receptors to outputs. Therefore, quantitative methods for these omics such as metabolomics, phospho-proteomics, lipidomics, glycomics, etc., all have in common that these techniques provide a indication of enzyme or in turn pathway activity. Application of these new techniques is collectively termed as 'activitomics'. Activomics gives in depth under standing of enzymes and pathways activity and ultimately establish biological relationship. Proteomics and activitomics can be judiciously applied for study of the regulation of cellular function through signaling. These may also be applied for the study of the response to various stresses. In addition, this can be used for the study of cellular process as well as protein changes associated with normal as well as in a particular experimental condition like biotic and abiotic stress condition of plants.

APPLICATIONS OF LC-MS

As a powerful analytical technique LC-MS has many applications in molecular biology, some important applications are as following:

Proteomics

Protein Expression

One of the first uses of proteomics was as an alternative to cDNA microarrays for the analysis of gene expression (Mann, *et al.,* 2001). The rationale of using proteomics rather than transcriptomics for measuring gene expression is that functional expression of a gene requires not only its transcription to an mRNA molecule but also the translations to protein. Initial studies using 2D gels demonstrated that mRNA levels are often a poor indication of protein levels (Gygi *et al.*, 1999). Later using LC-MS/MS, it was suggested that there was a significant correlation between mRNA levels and protein expression in approximately 70% of cases, thus implying that 30% of expression is controlled at the level of translation. Thus measuring protein expression rather than mRNA levels, would be a more direct and relevant measure of gene expressions as this would account for the contribution of both translation and transcription. Proteomics also has applicability to the analysis of proteins in different cellular compartments.

It is often argued that gene expression microarrays offer a more comprehensive picture of gene expression than do approaches based on protein quantification. This view is based on the fact that probes for whole genomes can now be printed on microarray chips. However, not all genes are expressed at a given time or tissue and head to head comparison of LC-MS-based proteomic and microarray data has shown that the two techniques offer a similar depth of analysis (Cox *et al.,* 2007).

However, although analysis of gene expression at the protein level is conceptually more powerful than is analysis of gene expression at the mRNA level, performing LC-MS- based proteomics is much more technically demanding, time consuming and expensive than is performing microarray analysis. For these reasons, the latter is still the method of choice for the analysis of gene expression for the most researchers. This trend may change in the near future if mass spectrometers become less expensive, with greater duty cycle and dynamic range, and novel quantification methods and computer programs for their implementation become available.

Organelle Proteomics

LC-MS can be used for the precise study of organelle proteome. This technique can be used to reveal the protein composition of intracellular compartments and organelles (Burnet *et al.,* 2003; Dreger 2003; Robinson *et al.,* 2007). It can be used to annotate the proteome by assigning protein their respective intracellular location which gives clue about the specific function of protein as well as function of the organelle. All the organelle proteomic experiments require fractionation step to isolate the organelle under investigation; this is usually carried out by differential or sucrose gradient centrifugation.

High sensitivity and dynamic range of modern mass spectrometer detects even minor contaminants in these biochemical preparations of subcellular structures. Therefore, quantitative analysis by LC-MS/MS does not allow discriminating true organelle components to contaminants. Now strategies have been developed to quantum the enrichment of validated markers of these fractions. This is the basis of protein correlation profiling (PCP) (Andersen *et al.,* 2003) as well as localization of organelle proteins by isotope tagging (LOPIT) (Dunkley *et al.,* 2004). Main difference between these two approaches is that PCP uses label free LC-MS to quantify the relative enrichment of components in the subcellular fractions, whereas LOPIT relies on the labelling of components before quantification. PCP has been used to investigate the composition and dynamic of nucleoler proteins (Andersen *et al.,* 2003) and for comprehensive characterization of subcellular proteomes (Foster *et al.,* 2006). LOPIT has been applied to the characterization of membrane proteins (Sadowski *et al.,* 2006) and to identify the protein composition of organelles.

Protein Interaction

Proteins do not work in isolation but instead have to interact with other molecules to function. Thus, protein-protein, protein-lipid and protein-nucleic acid interaction has fundamental roles in biology. LC-MS is routinely used to detect novel protein-protein interactions (Abu-Farha *et al.,* 2008; Kocher *et al.,* 2007; and Yarmush *et al.,* 2002). A recent example is the identification of pyruvate kinase M2 as a phosphotyrosine-binding protein (Christofk *et al.,* 2008). This study used LC-MS to quantify the enrichment of binding of proteins to tyrosine-phosphorylated peptides relative to binding of their non-phosphorylated peptide counterparts.

As with the case of organelle protein analysis, detecting a protein in a protein complex by LC-MS/MS dose not necessarily mean that this protein is part of the complex. Protein complexes are difficult, possibly impractical, to isolate to homogeneity and the precence of non-specific contaminants is often impossible to avoid. An approach to overcome this problem is to over express protein inducement with several tags for tandem affinity purification (Rigaut *et al.,* 1999). This approach can certainly result in the identification of novel protein binders but because it needs to over express the protein inducement, it may result in the introduction of false positive without biological relevance. An alternative approach is to use quantitative LC-MS to identify dynamic protein complexes that form after cells are stimulated with specific agents. LC-MS can then be used to quantify proteins present in complexes after cell stimulation, thus identifying dynamic members of these complexes that may have a functional role in controlling the molecular biology of the complex under study. This strategy should have wide applicability not only for the identification of protein-binding partners but also for mapping protein-lipid and protein-nucleic acid interaction.

Activitomics

Applications of LC-MS for quantitative analysis of enzymatic activity is known as activitomics. This approach consists of quantifying product of *in vitro* enzymatic reaction by LC-MS, a technique that allows multiplexing enzymatic measurements with unstitched specificity (Greis, 2007). This technique can be used to quantify the activity of many enzymes that provide a read out of pathway activity (Cutillas *et al.*, 2006) The advantage of measuring enzymatic reaction, rather than the amounts of enzyme in cells, is that the modulation of enzymatic activity is a complex process, and protein amounts are often a poor indication of enzymatic and pathway activities. This concept is well understood for metabolic pathway where for metabolic control analyses (MCA) are well developed (Casecante *et al.*, 2002). It is through the application of MCA that biochemist and biotechnologists have realized that measuring protein amount is uninformative for the prediction of biological out put as a result of specific perturbation. System biology and modeling thus need to take into account the share of control that enzyme have in their pathway for a full understanding of how biochemical pathways are regulated (Cascant *et al.,* 2002). Advance in systems biology research thus need method that can be used to quantify enzymatic activity as comprehensively as possible, as this is an essential requirement for the construction of reasonably accurate model of biochemical pathways, the ultimate aim of many biological fields including metabolism and cell signaling research.

Phosphoproteomics

Quantitative analysis of phosphorylation may be regarded as an activitomic approach for comprehenshiv quantification for pathways controlled by protein kinase. LC-MS allows the quantification of thousand of phosphorylation sites on proteins and these approaches are increasingly used in several biological fields including cell signaling and cancer research (Kruger *et al.*, 2008; Olsen *et al.*, 2006; Trinidad *et al.*, 2006; Villen *et al.,* 2008). For example, quantitative LC-MS has been applied to the analysis of growth factor signalling (Olsen *et al.,* 2006). It has been observed that kinase pathways downstream of growth factors are more complex than previously thought. With the use of these techniques now it is possible to 'routinely' identify and quantify thousand of phosphopeptides per experiment.

A question that needs to be addressed is what is the value of these analyses (Lienhard, 2008)? Indeed, the significance of phosphoprotein quantification is some time not fully under stood. A misconception often found in the literature is that phosphorylation activates the enzymatic activity of proteins. Although this is case for many enzymes but it is not the case for many other. It has also been proposed that much of the phosphorylation occurring in eukaryotes is non-functional (Lienhard, 2008), with functional events occurring in a background of phosphorylation 'noise' for which there is no selection against.

The hypothesis of the existence of nonfunctional phosphorylation is difficult to prove or disprove, because phosphorylation with no apparent function under determined conditions may have function under other conditions. It should also be noted that activation of enzymatic activity is only one of the properties that phosphorylation may attribute to a protein. Indeed, not all proteins are enzymes and phosphorylation can also affect proteins localization and their ability to interact with other molecules in cells, which may be as important in controlling the biological properties of proteins as the modulation of enzymatic activity. Anyway of whether or not non-functional phosphorylation exists, the value of phosphoproteomics lies in its potential ability to serve as a comprehensive read-out of kinase activity in cells.

Phosphoprotien amounts are also the result of their expression in addition to the activity of kinase and phosphatases acting upon them, but since substrate level control enzyme activity, this view still holds. When a phosphoprotein substrate is not expressed, that kinase pathway cannot be active regardless of the activity of the kinase. Thus since protein kinase are key component in the controlling of signaling pathways, in-depth analysis of phosphorylation by LC-MS may be regarded as an activitomic approach that offers the unique prospect or comprehensive quantification of signaling at the system level. Thousand of phosphorylation sites can now be quantified with relatively good precision, accuracy and throughput. The challenge now is to link each of the thousands of phosphorylation sites quantifiable by LC-MS to the kinase responsible for their phosphorylation.

Metabolomics

As a result of the activities of metabolic enzymes different metabolites are formed. The metabolic status of cells and whole organisms is reflected in the composition of metabolites. Measuring metabolic concentration may therefore be regarded as an activitomic approache that quantifies the activity of metabolic enzyme and pathway. Routinely, preferred method of choice of researcher is NMR because of greater throughput and coverage of the analysed metabolome (Mertz *et al.,* 2008).

Metabolomic techniques based on LC-MS are still in their infancy. To derive metabolite profiles by MS and LC-MS is relatively easy but obtaining the identity of these metabolites is not simple. This is because of the many different chemical structures that form the backbone of these small molecules, which makes deriving general rules for their identification by MS/ MS, analogous to those derived for peptides, which is not easy task. Metabolomics will require the construction of wide-ranging spectral libraries of metabolite CID MS/ MS data, similar to those available for electron impact and chemical ionization spectra. In spite of these technical limitations, LC-MS is increasingly used for metabolic profiling and to quantify the fluxes of metabolic pathways. For example several metabolic pathways may be altered in diseased cells. Identifying these different routs and how these are regulated could be exploited by using this technique.

Lipidomics

LC-MS is used to the quantitative analysis of lipids. Levels of these lipids reflect the activity of enzymes involved in their production, modification and degradation. Hence, lipidomics can be regarded as an activitomic approach to quantify the activity of enzymes with roles in lipid metabolism, including lipid kinases and phosphatase. As with protein kinases and protein phosphatases, lipid kinases and phosphatases have important roles in controlling signalling pathways with roles in fundamental biological processes. Therefore, quantifying the activity of lipid kinase/phosphatase pair has several applications in basic and applied research.

In phosphoinositide (PI) research LC-MS could have a key role which is an example of lipidomic/ activitomic application (Wakelame *et al.,* 2007). PI signaling controls fundamental biological processes including growth, apoptosis, cell cycle progression and proliferation. Understanding the roles of different PI-modifying enzymes requires methods to quantify their activity status. Several studies have reported the use of LC-MS to precisely quantify PI species. The advantage of using LC-MS for phospholipid analysis is that by using normal phase LC, one dose not needs to deacetylate lipids prior to the analysis (Pettitt *et al.*, 2006). Therefore, the analysis can also consider the preference of enzymes for the phosphorylation of lipids with different acyl composition. By using ESI-MS, It has been confirmed that there at least eight different Phosphoinositide-3,4,5 triphosphate (PIP3) species that differ in their acyl composition (Milen *et al.,* 2005). This type of information cannot be obtained by other methods for detecting PI species like those based on HPLC. Moreover, LC-MS could in principle be used to profile all PI species in a single analysis, thus allowing the quantitative estimation of PI-kinase and phosphatase activities in a single run.

LC-MS has many applications in biomolecular research. In addition to the quantification of gene expression at the protein level, as an alternative to microarray analysis, LC-MS can be used to comprehensively profile the protein composition of subcellular organelles and protein complexes and to provide a picture of the dynamic protein composition of macromolecular structures. Another important and gradually increasing application of LC-MS is for in depth quantification of enzymatic activities in cells. Enzymes are important mediators of biological function, but since their level may not reflected their activity status, standard proteomic approaches cannot be used to infer the activity status of biochemical pathways. Therefore, methods that can be used to quantify enzymatic activities in a comprehensive manner will have many applications in biological research. It is expected that quantitative LC-MS will continue to have important roles in these workflows. Several techniques for quantitative LC-MS have been developed over the last decade and some important ones are described here as following.

QUANTITATIVE LC-MS TECHNIQUES FOR PROTEOMICS:

LC-MS has become a useful molecular biology tool and multiple strategies for peptide, protein and enzyme activity quantitation by LC-MS have been developed and applied to address a wide range of biological questions. LC-MS protein quantification strategies fall into two broad categories:

1. Labeling Methods
2. Label free quantification

These strategies are mostly used for relative quantification of proteins/peptides between two or more samples of interest. Nevertheless, these methods have also been adapted for absolute quantitation, where the introduction of a known amount of internal standard (often an isotopically labeled version of the peptide of interest) into the test samples permits the direct quantitation based on the peak intensity (ion current) of the standard versus endogenous peptide. In general terms, label free quantitation methods involve comparison of peptide ion currents between samples or the comparison of spectral counts for peptide ions from particular proteins across the samples. In terms of quantitative accuracy and coverage, the label-free approaches are more reliant on chromatographic reproducibility and the ability to detect the same ions from LC-MS run to run. Quite the opposite, labeling strategies permit the mixing of samples prior to LC-MS, and in some cases upstream of any fabrication. Thus, multiple specimens can be run simultaneously with the same peptides (or protein) beings identically separated and co-eluted into the mass spectrometer with ion intensities being directly compared in the same MS or MS/MS scans. The labeling strategies thus improve throughput and quantitative accuracy.

Labelling Methods for Quantitative LC-MS in Proteomics

Differential labeling approaches for quantitative LC-MS in proteomics fall into the two main categories: (i) those which use chemical derivatisation or enzymatic modification of proteins or peptides after sample collection and those which use incorporation of isotope-labeled amino acids in vivo. The chemical labeling approaches make use of tags with the same (isobaric labeling) or different masses (isotopic labeling). Isobaric labeling is represented by isobaric tags for relative and absolute quantitation (iTRAQ) (Ross *et al.,* 2004), consists of peptide tags which generate specific fragment ions by MS/ MS. Samples are differentially labeled and then combined and concurrently analyzed by LC-MS with relative quantitation performed by comparison of intensities of the 'reporter' fragments in the MS/MS spectra. On the contrary, isotopic labeling methods, such as isotope- coded affinity tags (ICATs) (Gygi *et al.,* 1999) or proteolytic ^{18}O labeling (Yao *et al.*, 2001), generate pairs (or more) of peptides with a mass difference introduced by the label. The ion intensities of the isotopic forms of the labeled peptides which should have identical LC elution profiles are then compared to give a peptide ratio of the 'heavy' – labeled versus 'light' –labeled peptide. Likewise, differential isotopic labeling in vivo allows quantification of peptide following incorporation of 'light' (^{12}C, ^{14}N, ^{1}H) and 'heavy' stable isotopic- labeled (^{13}C, ^{15}N, ^{2}H) amino acids and is illustrated by the stable isotope labeling of amino acids in culture (SILAC) strategy (Ong *et al.,* 2002).

In all of these methods the ratio of detected 'reporter' fragments or isotopically labeled peptides are computed and integrated in to protein ratios which can then be evaluated statistically. Multiple software solution for the analysis of quantitative information using these labeling strategies is available (Mueller *et al.,* 2008). Although many of these solutions are instrument, data or tag dependent, they all work on the same principle where by isotopically labeled peptide pairs (or reporter ions) are extracted on the basis of their characteristic mass difference and successful MS/MS peptide assignments. Ratios of the extracted isotopic pairs are then computed for performing statistical evaluation. If mass difference of the tags is smaller then interpretation of the data and accurate quantification becomes more difficult because the isotope envelopes of the differentially labeled peptides may overlap.

When using any labeling approach for LC-MS, the, labels are best introduced at the earliest point in the workflow to minimize differences introduced into the samples by handing or quantitative difference between LC-MS runs. In this way, the in vivo labeling strategies do better than the chemical and enzymatic labeling strategies in terms of accuracy of quantitation; though, multiplexing using in vivo labelling is presently more limited than chemical labeling, where 12-plex strategies have been reported (Zeng *et al.,* 2009).

Isotope-Coded Affinity Tags (ICAT) and Variations

Even though stable isotope labeling for protein quantitation were previously reported, ICAT was the first strong and universal differential labeling strategy to be developed for quantitative LC-MS and is based on cysteine thiol group modification using iodoacetamide tags. Initially, ICAT was used to examine the expression profiles of yeast growing on either galactose or ethanol in single analysis (Gygi *et al.,* 1999). Stable isotopes were incorporated into intact proteins after lysis by selective alkylation of cystienes with either a heavy (deuterium D_8) or alight (deuterium D_0) reagent bearing a biotin tag. Prior to LC-ESI-MS analysis, the protein mixture was digested with trypsin and the ICAT labeled (cysteine-containing) peptides enriched on monomeric avidin-agarose. This has the advantage of simplifying the peptide mixture for downstream LC-MS/MS-based identification. Griffin *et al.* combined ICAT labeling with LC-MALDI-MS/MS for targeted protein identification for better sample utilization and reduce MS instrument time (Griffin *et al.,* 2001).

Several technical problems were reported with the original deuterated ICAT reagents. Differential elution of ICAT labeled peptide pairs in reversed phase HPLC (RP-LC) leading to errors in the quantification (Zhang *et al.,* 2002; Zhang *et al.,* 2001) was a major problem. Moreover, relatively low efficiencies of labeled peptide collision-induced dissociation (CID), which was contemplated to be due to the relatively large size of the ICAT moiety (Borisov *et al.*, 2002) was also a problem. Poor recovery of tagged peptides from avidin may also contribute to reduced proteomic coverage to improve these issues. A new generation of ICAT reagents were developed where the deuterium atoms were substituted for nine ^{12}C (light) or ^{13}C (heavy) atoms and an acid – cleavable linker was also added. These new generation cleavable ICAT (cICAT) reagents gave more precise co-migration of the light and heavy – tagged peptides in RP–LC, whilst the cleavage strategy eliminated undesired residual fragmentation of biotinylated peptides and improved recovery (Yu *et al.,* 2004). Nevertheless, an additional clean-up step was required for removal of the cleaved biotin moiety prior to LC-MS analysis. a subsequent large –scale investigation of the cICT strategy, which measured protein changes in *Escherichia coli* treated with an inhibitor of fatty acid biosynthesis, quantified more than 24,000 peptides in four independent runs using an ion-trap mass spectrometer (Molloy *et al.,* 2006).

Variations to the original ICAT strategy have also been reported. These include fluorescent-labelled ICAT (FCAT), allowing absolute quantification by fluorescence measurement and peptide enrichment using an anti-FITC antibody or iminodiacetic acid coated beads (Rivera-Monroy *et al.,* 2009); visible isotope-coded affinity tags (VICAT) that contain a visible probe for monitoring chromatographic behavior and a photo releasable biotin affinity tags for selective capture and release of labelled peptides (Bottari *et al.,* 2004; Lu *et al.,* 2007); acid-labile isotope-coded extractants (ALICE), a class of chemically modified reins that contain a thiol-reactive group to capture cysteine-containing peptides and a heavy or a light isotope coded, acid labile linker (Qiu *et al.,* 2002); and metal coded affinity tags (MeCATs), where different element- coded metal chelates are used for cysteine labelling which can be purified on metal chelate-specific affinity resins (Whetstone *et al.,* 2004). The compatibility and robustness of the MeCAT technology for the relative quantification was recently shown using standard LC-MS techniques and offered the unique advantage of absolute quantification via inductively coupled plasma mass spectrometer (Ahrends *et al.,* 2009).

Amino and Carboxyl Group Differential Isotope Labeling

As the majority of protease-generated peptides as well as some proteins do not contain cysteine, the ICAT strategy has limitation in terms of proteomic coverage, particularly for peptidomic analyses. Thus, alternative methods have been described where the amino groups of the N terminus and lysine residues or the carboxylic acids of the C terminus and aspartic and glumatic acids residues are labeled with either deuterium free (light) or deuterium containing (heavy) tags.

Although relatively inexpensive, formaldehyde, the acetic and succinic anhydrides and PIC convert positively charged amines into neutral (formldehyde/acetyl/PIC) or negatively charged sites (succinyl), with the effect some peptides give weak MS signals in the positive ions mode (Che *et al.,* 2005). Weak signals may also occur due to partial labeling of lysine ε-amino groups (Lemmel *et al.,* 2004). Labeling with N-nicotinoyloxy- succinimide for nicotinylation or [3-(2,5-dioxopyrrolidin- 1-yl oxycarbonyl) propyl] trimethylammonnium chloride to give the 4-trimethlylammoniumbutyryl (TMAB) modification avoids this problem since the positive charge of the amino group is maintained (Zhang *et al.,* 2002; Morano *et al.,* 2008). However, it has been reported that TMAB labelled peptides can partially decompose during CID, making data interpretation and quantification more complex (Che *et al.*, 2005; Morano *et al.*, 2008). As with the cysteine-labelling isotopic tags, pairs of peptides labeled with the light and heavy deuterated forms of these tags do not always co-elute in RP-LC (Che *et al.*, 2005). Amino-labeling reagents have also been used for N-terminal labeling alone, where lysine side chains are first blocked by guanidination (Zappacosta *et al.,* 2004). Finally, a method for labelling the amino groups of intact proteins using isotope coded tags has been reported (Schmidt *et al.,* 2009). Termed isotope coded protein labelling (ICPL), the method used D_O- and D_4-N-nicotinoylloxy succinimide or D_O-, D_3 and D7- butyric acid-2,5- dixopyrrolidin-1-yl-ester to label proteins prior to digestion with trypsin or Glu-C. The method was shown to be accurate and reproducible for quantification and circum vents technical variations introduce at the digestion step.

Differential Proteolytic $^{16}O/^{18}O$ Labeling

Differential stable isotope labeling can also be achieved enzymatically. The major method relies on the oxygen atom exchange that takes place at the C-terminal carboxyl group of peptides during proteolytic digestion. Here, one or two ^{16} O atoms are replaced by one or two ^{18} O atoms through enzyme-catalysed exchange in the presence of $H_2\,^{18}O$. Since any variability in labeling relies solely on the digestion step, this method is expected to give smaller technical variations than do the two- step chemical labeling methods, where both labeling and digestion are a source of introduced variations. The method was first suggested as a protein quantification tool when ^{18}O-labelled internal standard were generated for absolute quantification by MALDI-MS (Mirgorodskaya *et al.,* 2000). This was followed by the reporting of conditions for protein labeling (Stewart *et al.,* 2001) and then the first proteomic applications of the method, where trypsin was used to incorporate two ^{18}O atoms into the C termini of the tryptic peptides and was applied to compare proteins from two serotypes of adenovirus (Yao *et al.,* 2001). Subsequent work by the same group showed that Glu-C could also be used for labeling, that ^{18}O-labelled and unlabelled (^{16}O) peptide pairs co eluted in RP-LC and measu-rement of isotope ratios by LC-MS were accurate and precise (Reynolds *et al.,* 2002). Applications include combining with 2D-LC-FT-ICR and an accurate mass and time (AMT) tag strategy to identify and quantify 429 distinct plasma proteins from an individual prior to and after lipopoly-saccharide administration (Qian *et al.,* 2005); combining with $^{16}O/^{18}O$-menthanol esterification, immobilised metal-ion affinity chroma-tography and RP-LC followed by neutral loss-dependent MS^3 for phospho-peptide identification in the study of lysophosphatidic acid-induced chemotaxis (Ding *et al.,* 2008); the differential analysis of NF- $_k$B transcription factor complexes following TNF-α stimulation (Bantscheff *et al.,* 2004); the labeling of a 'universal' reference sample of food plasma for spiking into individual unlabelled samples for quantitative analysis across clinical samples (Qian *et al.,* 2009).

Several drawbacks of the proteolytic labeling strategy are apartment. Only two samples can be compared simultaneously, C-terminal peptides of proteins can not be quantified and variable incorporation of ^{18}O into peptides can occur (Julka *et al.,* 2004). There is also a lack of computational tools for accurate quantification of peptide differences and this is exacerbated by the overlap of isotopic envelopes for ^{16}O and $^{18}O_1$- and $^{18}O_2$ –labelled peptides. However, some methods have been reported for correcting $^{16}O/^{18}O$ ratios overlapping isotopic multiplets (Johnson *et al.*, 2004; Rao *et al.*, 2005).

Isobaric Tagging (ITRAQ, TMT, CILAT)

Isobaric tagging is a multiplex peptide labeling method that relies on the introduction of stable isotope tags that are chemically identical but distinguishable by MS/ MS due to their fragmentation into reporter ions of different masses. The most commonly used method of this type has been the 4-plex iTRAQ reagents which are N-hydroxysuccinimidyl esters for the labeling of Mary amino groups. Specific reporter group in each tag based on N-methylpiperazine generates ions of 114, 115, 116 and 117 m/z upon CID fragmentation. These appear in an ion-sparse region of MS/MS spectra and their relative intensities provide the relative abundance of labeled peptides between the samples. The reporter groups in each tag are mass-balanced with a linker group making the tags isobaric. The major advantage of this MS/MS-dependent strategy is that the multiplex labeling does not increase the mass complexity of the samples and only peptides subjected to CID fragmentation are quantified. In addition, higher signal-to-noise ratios can be achieved with MS/MS-based detection versus MS-mode measurements.

Stable Isotope Labelling of Amino Acids in Culture (SILAC)

SILAC is an in vivo stable isotope labeling method which uses heavy and light versions of essential amino acids that are added to the growth media of metabolically active cells. The idea behind the strategy originates from comparative proteomic experiments of simple model organisms (Oda *et al.,* 1999; Washburn *et al.*, 2003) and plants (Bindschedler *et al.,* 2008) which can be grown in either medium containing ^{14}N at natural abundance (99.6%) or the same medium enriched in ISN.

The SILAC method has mostly been used for the comparion of protein expression and PTMs between two or more samples. Internal standard SILAC approaches have also been applied, where equal amounts of individual heavy-lebelled samples are mixed and compared with each of the light-labelled samples, thereby improving cross-sample quantitative accuracy (Yan *et al.,* 2005). Absolute SILAC for protein quantitation is another variation, where recombinant proteins are SILAC labeled during their production and then used as internal standards by mixing directly into cell or tissue lysates (Hanke *et al.,* 2008). An adaptation of the SILAC protein profiling method called heavy methyl SILAC has also been reported, where cells convert (13C, D3] Met to (13C, D3] S-adenosyl methionine, the sole biological methyl donor. The heavy methyl groups are incorporated into proteins during in vivo methylation for the labelling and relative quantification of this PTM (Ong *et al.,* 2004).

The major drawback of the SILAC method is that it can be used only for samples where in vivo labelling is possible. This has restricted its use to simple organisms or cultured cell models, and so the quantitative comparison of proteins derived from tissues is not possible.

Label-Free Quantification

The protein labeling techniques described above can quantify the relative abundance of proteins and peptides across samples. Even though useful for the investigation of cell biological processes, their main drawback is that labelling methods can be used to compare only a limited number of specimens. This feature means that it is often difficult to obtain sufficient data to perform statistical analysis and to infer biological significance of differences observed with these proteomic experiments. The proteomics community has therefore, shown great interest in the development and use of label free quantitative LC-MS as an alternative (Schmidt *et al.*, 2009; Wang *et al.*, 2008; America *et al.*, 2008; Braisted *et al.*, 2008; Cheng *et al.*, 2009). Particular advantage of label-free quantitative MS are their wide applicability, low expense and their ability to compare unlimited numbers of samples. In addition to its uses for relative quantification, techniques based on the label free concept may also be used for absolute quantification on a global scale.

Label free quantitative methods are those techniques that do not require chemical derivatisation step or incorporation or stables isotopes by proteins in metabolically active cells. They can be classified into two groups: (i) methods based on the evaluation of ion currents and (ii) methods based on spectral counting.

Methods Based on Ion Currents

LC/MS and LC-MS/MS techniques are routinely used to quantify small molecules like natural metabolites and pesticides. Results of such analyses are so accurate that these are accepted by regulatory and official bodies and in court. Hence, it is proper to say that LC-MS is a quantitative technique beyond doubt. Quantification by LC-MS involves measuring the elution profile of molecules and correlating this with their concentration in samples. This can be done by calculating peak areas or heights of extracted ion chromatograms (XICs) of the m/z of the molecule to be quantified. The specificity of the quantification is enhanced when this is done by MS/MS, which involves measuring the elution profile on one or more fragment ions produced as a result of fragmentation of the molecular ion by, for example, collision-induced-dissociation (CID). This technique thus measure the transition of parent to fragment ion and is commonly referred to as selected reaction monitoring (SRM) or multiple reaction monitoring (MRM), depending on the number of molecules being analyzed in an LC-MS run; these are normally carried out in triple quadruple instruments.

In principle the same concepts and techniques used to quantify small molecules may be used to quantify peptides and proteins. However, until relatively recently, the consensus in the proteomics community was that labeling approaches were needed to accurately quantify proteins present in complex mixtures (Gygi *et al.,* 1999; Mann, 1999). The reason for the initial reluctance of the proteomics community to make use of these more straightforward, non-labeling modes of quantification was probably due to the belief that, because of the extreme complexity of the samples, signal suppression was going to adversely affect protein quantification. Although ion suppression is commonly observed in the LC-MS analysis of small molecules, there is in fact very little experimental evidence to support the notion that ion suppression occurs under the experimental conditions normally encountered in LC-MS/MS settings for proteomics.

Chelius and Bondarenko (Chelius *et al.,* 2002) was one of the first groups to show that the ratios of peak areas in XICs could be used to quantify relative abundance of proteins across samples. Later it was shown that protein quantification with this technique was also suitable for the analysis of proteins in complex mixtures (Silva *et al.,* 2006; Wiener *et al.,* 2004; Chelius *et al.,* 2002).

It was also demonstrated that by spiking the samples with non-labelled internal standards at known concentration, the accuracy of quantification could be improved by normalising for protein losses during handling and for differences in the analytical behaviour of the LC-MS/MS instrumentation (Cutillas *et al.,* 2005). It was later proposed that introduction of an internal standard (IS) could also be used to quantify proteins in absolute units (Silva *et al.,* 2006). This technique considers the peak areas of the three most intense peptides derived from a protein. It was found that the intensities of these three peaks arc the same regardless of protein identity; thus by comparing the intensities of a known amount of IS with those of the protein to be quantified, one can provide estimates of protein amounts in absolute units with a certain degree of accuracy.

A common strategy for comparing LC-MS runs is to align and compare chromatograms resulting from the analysis of the samples to be evaluated (Mueller *et al.,* 2008). After alignment the intensities of mass features identified are compared using statistical tests. The identities of differentially expressed peptide ions are then obtained by including their masses and retention times in an inclusion list for targeted LC-MS/MS analysis.

Spectral Counting Techniques

In complex mixtures, such as a peptide mixture obtained by the digestion of a total cell lysate or the proteome of a biofluid, there is a staggering number of peptides (perhaps in excess of 100,000 different components), and therefore, even fast scanning tandem mass spectrometers can not possibly select all peptide ions for MS/MS in data-dependent acquisition (DDA) experiments. In this type of experiment, peptide ions are selected for MS/MS as they elute from the LC column when their intensities reach a predefined threshold value in the MS scan. Peptides producing larger ionic intensities are preferentially selected over peptides with less intense signals. Since proteins produce several peptides when digested with a protease, peptides derived from more abundant proteins will be selected for MS/MS more often than peptides derived from less abundant proteins. It has been found that this spectral count (essentially number of times peptides from a given protein are selected for MS/MS) is roughly proportional to protein concentration and has therefore, been used as the basis for several protein quantification strategies "which differ in the way spectral counts are normalized and in their scope (Ishihama *et al.,* 2005; Lu *et al.,* 2007). Normalization is essential since large proteins produce more peptides than smaller proteins. There may also be differences in the quality of peptides produced from the digestion of different proteins.

Conclusion

Several methods now exist for the quantification of peptides, proteins and activities. Quantifying sites of modification and metabolites can be used as a measure of enzymatic and pathway activity. This is an exciting realization because this implies that it may soon be possible to quantify pathway fluxes with the same depth of analysis as that achieved for the quantification of proteins. LC-MS is likely to continue playing a central role in this regard. An understanding of the pros and cons of the different techniques will assist in the most appropriate use of quantitative LC-MS for proteomic research.

References

Abu-Farha, M., Elisma, F. and Figeys, D. (2008) Identification of protein- protein interaction by mass spectrometry coupled techniques. *Adv. Biochem. Eng. Biotechnol.* **110,** 67-80.

Ahrends, R., Pieper, *S.,* Neumann, B., Scheler, C. and Linscheid, M. W. (2009) Metal-coded affinity tag labeling: a demonstration of analytical robustness and suitability for biological applications. *Anal. Chem.* **81**, 2176-2184.

America, A. H. and Cordewener, J. H. (2008) Comparative LC-MS: a landscape of peaks and valleys. *Proteomics* **8**, 731-749.

Andersen, J.S., Wilkinson, C.J., Mayor, T., Mortensen, P., Nigg, E.A. and Mann, M.,(2003) Proteomic Characterization of the human centrosome by protein correlation profiling. *Nature* **426,** 570-574.

Bantscheff, M., Dumpelfeld, B. and Kuster, B. (2004) Femtomol sensitivity postdigest (18) O labeling for relative quantification of differential protein complex composition. *Rapid Commun. Mass Spectrom.* **18**, 869-876.

Bindschedler, L.V., Palmblade, M. and Cramer, R. (2008) Hydroponic isotope labelling of entire plants (HILEP) for quantitative plant proteomics; an oxidative stress case study. *Phytochemistry* **69**, 1962-1972.

Borisov, O.V., Goshe, M.B., Conrads, T.P., Lukov, V.S., Veenstra, T.D. and Smith, R.D. (2002) Low-energy collision- induced dissociation fragmentation analysis of cysteinyl-modified peptides. *Anal. Chem.* **74**, 2284-2292.

Bottari, P., Aebersold, R., Turecek, F. and Gelb, M. H. (2004) Design and synthesis of visible isotope-coded affinity tags for the absolute quantification of specific proteins in complex mixtures. *Bioconjug. Chem.* **15**, 380-388.

Braisted, J.C., Kuntumalla, S., Vogel, C., Marcottc, E.M., Rodrigues, A.R., Wang, R., Huang, S.T., Ferlanti, E.S., Saeed, A.I., Fleischmann, R.D., Peterson, S.N. and Pieper, R. (2008) The APEX quantitative proteomics tool: genera-ting protein quantitation estimates from LC-MS/MS proteomics results. *BMC Bioformatics* **9**, 529.

Burner, S., Thibault, P., Gagnon, E., Kearney, P., Bergeron, J.J. and Desjardins, M. (2003) Organelles proteomics: looking at less to see more. *Trends Cell. Bio.* **13,** 629-638.

Cascante, M., Boros, L.G., Comin-Anduix, B., de Atuari, P., Centteless, J.J. and Lee, P.W. (2002) Meatbolic control analysis in drug discovery and disease. *Nat. Biotechnol.* **20,** 243-249.

Che, P. Y. and Pricker, L.D. (2005) Quantitative peptidomics of mouse pituitary: comparison of different stable isotopic tags. J. *Mass Spectrom.* **40**, 238-249.

Chelius, D. and Bondarenko, P.V. (2002) Quantitative profiling of proteins in complex mixtures using liquid chromatography and mass spcctrometry. *J. Prteomic. Res.* **1**, 317-323.

Cheng, F.Y., Blackburn, K., Lin, Y.M., Goshe, M.B. and Williamson, I.D. (2009) Absolute protein quantification by LC/MS(E) for global analysis of salicylic acid-induced plant protein secretion responses. *J. Proteome Res.* **8**, 82-93.

Christofk, H.R., Vander, Heiden, M.G., Harris, M.H., Ramanathan, A., Gerszten, R.E., Wei, R., Fleming, M.D., Schreiber, S.L. and Cantley, L.C. (2008) The M2 splice isoform of pyruvate kinase is important for cancer metabolism and tumor growth. *Nature* **452,** 230-233.

Christofk, H.R., Vander, Heiden, M.G., Wu, N., Asara, J.M. and Cantley, L.C., (2008) Pyruvate kinase M2 is a phosphotyrosine binding protein. *Nature* **452,** 181-186.

Cox, J. and Mann, M. (2007) Is proteomics the new genomics? *Cell* **130,** 395-398.

Cutillas, P. R., Biber, I., Marks, J.,]acob, R., Stieger, B., Cramer, R., Waterfield, M., Burlingame, A.L. and Unwin, R.J .(2005) Proteomic analysis of plasma membrane vesicles isolated from the rat renal cortex. *Proteomics* **5**, 101-112.

Cutillas, P.R., Khwaja, A., Graupera, M., Pearce, W., Gharbi., S., Waterfield, M. and Vanhaesebroeck, B. (2006) Ultrasensitive and absolute quantification of the phophoinositide 3-kinase/Akt signal transduction pathways by mass spectrometry. *Proc. Natl. Acad. Sci. USA* **103,** 8959-8964.

Ding, S.J., Wang, Y., Jacobs, J.M., Qian, W. J., Yang, F., Tolmachev, A.V., Du, X., Wang, W., Moore, R.J., Monroe, M.E., Purvine, S.O., Waters, K., Heibeck, T. H., Adkins, J.N., Camp, D.G., Klemke, R.L. and Smith, R.D. (2008) Quantitative phosphoproteome analysis of Iysophosphatidic acid induced chemotaxis applying dual step (18) O labeling coupled with immobilized metaion affinity chromatography. *J. Proteome Res.* **7**, 4215-4224.

Dreger, M. (2003) Subcellular proteomics. *Mass Spectrom. Rev.* **22,** 27-56.

Foster, L.J., de Hoog, C.L., Zhang, Y., Xei, X, Mootha, V.K. and Mann, M. (2006) A mammalian orgenell map by protein correlation profiling. *Cell.* **125,** 187-199.

Greis, K.D. (2007) mass spectrometry for enzyme assays and inhibitor screening: an emerging application in pharmaceuticals research. *Mass Spectrom. Rev.* **26,** 324-339.

Griffin, T.J., Gygi, S.P., Rist, B., Aebersold, R., Loboda, A., Jilkine, A., Ens, W. and Standing, K.G. (2001) Quantitative proteomic analysis using a MALDI quadrupole time-of-flight mass spectrometer. Anal. Chem. **73**, 978-986.

Gygi, S.P., Rist, B., Gerber, S.A., Turecek, F., Gelb, M.H. and Aebersold, R. (1999) Quantitative analysis of complex protein mixtures using isotope-coded affinity tags. *Nat. Bjotechnol.* **17**, 994-999.

Gygi, S.P., Rochon, Y., Franza, B.R. and Aebersold, R. (1999) Correlation between protein and mRNA abundance in yeast. *Mol. Cell. Bio.* **19,** 1720-1730.

Hanke, *S.,* Besir, H., Oesterhelt, D. and Mann, M. (2008) Absolute SILAC for accurate quantitation of proteins in complex mixtures down to the attomole level. *J. Proteome Res.* **7**, 1118-1130.

Ishihama, Y., Oda, Y., Tabata, T., Sato, T., Nagasu, T., Rappsilber, J. and Mann, M. (2005) Exponentially modified protein abundance index (emPAI) tor estimation of absolute protein amount in proteomics by the number of sequenced peptides per protein. *Mol. Cell. Proteomics* **4,** 1265-1272.

Johnson, K.L. and Muddiman, D.C. (2004) A method for calculating 160/180 peptide ion ratios for the relative quantification of proteomes. J. *Am. Soc. Mass Spec- from.* **15**,437-445.

Julka, *S.* and Regnier, F. (2004) Quantification in proteomics through stable isotope coding: a review. *J. Proteome Res.* **3**, 350-363.

Kocher, T. and Superti-Funga, G. (2007) Mass spectrometry- based functional proteomics: form molecular machines to protein networks. *Nat. Methods* **4,** 807-815.

Kruger, M., Kratchmarova, I., Blagoev, B., Tseng, Y.H., Khan, C.R. and Mann, M. (2008) Dissection of the insulin signalling pathways via quantitative phosphoproteomics. *Proc. Natl. Acad. Sci. USA* **105,** 2451-2456.

Lemmel, C., Weik, *S.,* Eberle, U., Deng-jel, J., Kratt, T., Becker, H.D., Rammensee, H.G. and Stevanovic, S. (2004) Differential quantitative analysis of MHC ligands by mass spectrometry using stable isotope labelling. *Nat. Biotechnol.* **22**, 450-454.

Lienhard, G.E. (2008) Non-functional phosphorylations *Trends Biochem. Sci.* **33,** 351-352.

Lu, P., Vogel, C., Wang, R., Yao, X. and Marcotte, E.M. (2007) Absolute protein expression profiling estimates the relative contributions of transcriptional and translational regulation. *Nat. Biotechnol.* **25**, 117-124

Lu, Y., Bottari, P., Aebersold, R., Turecek, F. and Gelb, M.H. (2007) Absolute quantification of specific proteins in complex mixtures using visible isotope-coded affinity tags. *Methods Mol. Biol.* **359**, 159-176.

Mann, M. (1999) Quantitativc proteomics? *Nat. Biotechnol.* **17**, 954-955.

Mann, M., Hendrickson, R.C. and Pandey, A. (2007) Analysis of proteins and proteomes by mass spectrometry. *Annu. Rev. Biochem.* **70,** 437-473.

Metz, T.O., Page, J.S., Baker, E.S., Tang, K., Ding. J., Sheh, Y. and Smith, R.D. (2008) High resolution separations and improved ion production and transmission in metabolomics. *Trends Anal. Chem.* **27,** 205-214.

Milne, S.B., Ivanova, P.T., DeCamp, D., Hsueh, R.C. and Brown, H.A. (2005) A targeted mass spectrometric analysis of phosphatidylinositol phosphate species. *J. Lipid Res.* **46,** 1796-1802.

Mirgorodskaya, O.A., Kozmin, Y.P., Titov, M.I., Korner, R., Sonksen, C.P. and Roepstorff, P. (2000) Quantitation of peptides and proteins by matrix-assisted laser desorption/ionization mass spectrometry using (18)0-labeled internal standards. *Rapid Commun. Mass Spectrom.* **14**, 1226-1232.

Molloy, M. P., Donohoe, S., Brzezinski, E. E., Kilby, G. W., Stevenson, T.I., Baker, J.D., Goodlett, D.R. and Gage, D.A. (2005) Large-scale evaluation of quantitative reproducibility and proteome coverage using acid cleavable isotope coded affinity tag mass spectrometry for protecomic profiling. *Proteomics* **5**, 1204-1208.

Morano, C., Zhang, X. and Fricker, L.D. (2008) Multiple isotopic labels for quantitative mass spectrometry. *Anal. Chem.* **80**, 9298-9309.

Mueller, L.N., Brusniak, M.Y., Mani, D. and Aebersold, R. (2008) An assessment of software solutions for the analysis of mass spectrometry based quantitative proteomics data. *J. Proteome Res.* **7**, 51-61.

Oda, Y., Huang, K., Cross, F.R, Cow-burn, D. and Chait, B.T. (1999) Accurate quantitation of protein expression and site specific phosphorylation. *Proc. Natl. Acad. Sci. USA* **96,** 6591-6596.

Olsen, J.V., Blagoev, B., Gnad, F., Macek, B., Kumar, C., Mortensen, P. and Mann, M. (2006) Global, in vivo, and site-specific phosphorylation dynamic in signalling networks. *Cell* **127,** 635-648.

Ong, S. E., Blagoev, B., Kratchmarova, I,, Kristensen, D. B., Steen, H., Pandey, A. and Mann, M. (2002) Stable isotope labeling by amino acids in cell culture, SILAC, as a simple and accurate approach to expression proteomics. *Mol. Cell. Proteomics* **1**, 376-386.

Ong, S. E., Mittler, G. and Mann, M. (2004) Identifying and quantifying in vivo methylation sites by heavy methyl SILAC. *Nat. Methods* **1**, 119-126.

Pettitt, T.R., Dove, S.K., Lubben, A., Calaminus, S.D. and Wakelame, M.J. (2006) Analysis of intact phosphoinositides in biological samples. *J. Lipid Res.* **47,** 1588-1596.

Qian, W.J ., Monroe, M.E., Liu, T., Jacobs, J.M., Anderson, G.A., Shen, Y., Moore, R.J., Anderson, D.J., Zhang, R., Calvano, S.E., Lowry, S.F., Xiao, W., Moldawer, L.L., Davis, R.W., Tompkins, R.G., Camp, D.G. and Smith, R.D. (2005) Quantitative proteome analysis of human plasma following in vivo lipopolysaccharide administration using 160/180 labeling and the accurate mass and time tag approach. *Mol. Cell. Proteomics* **4,** 700-709.

Qian, W. J., Liu, T., Petyuk, V.A., Gritsenko, M. A., Petritis, B. O., Polpitiya, A.D., Kaushal, A., Xiao, W., Finnerty, C.C., Ieschke, M. G., Iaitly, N., Monroe, M. E., Moore, R. J., Moldawer, L. L., Davis, R. W., Tompkins, R. G., Hern-don,

D.N., Camp, D.G. and Smith, R.D. (2009) Large-scale multiplexed quantitative discovery proteornics enabled by the use of an (18)0-labeled "universal" reference sample. *J. Proteome Res.* **8**, 290-299.

Qiu, Y., Sousa, E.A., Hewick, R.M. and Wang, I.H. (2002) Acid-labile isotope- coded extractants: a class of reagents for quantitative mass spectrometric analysis of complex protein mixtures. *Anal. Chem.* **74**, 4969-4979.

Rao, K.C., Carruth, R.T. and Miyagi, M. (2005) Proteolytic 180 labeling by peptidyl-Lys metalloendopeptidase tor comparative proteomics. *J. Proteome Res.* **4**, 507-514.

Reynolds, K. J., Yao, X. and Fenselau, C. (2002) Proteolytic 180 labeling for comparative proteomics: evaluation of endoprotease Glu-C as the catalytic agent. *J. Proteome Res.* **1**, 27-33.

Rigaut, G., Shevchenko, A., Rutz, B., Wilm, M., Mann, M. and Seraphin, B., (1999) A generic protein purification method for protein complex characterization and proteome exploration. *Nat. Biotechnol.* **17,** 1030-1032.

Rivera-Monroy, Z., Bonn, G. K, and Guttman, A. (2009) Fluorescent isotope-coded affinity tag 2: peptide labelling and affinity capture. *Electrophoresis* 30, 1111-1118.

Robinson, C.V., Sali, A. and Baumeister, W. (2007) The molecular sociology of the cell. *Nature* **450,** 973-982.

Ross, P. L., Huang, Y.N., Marchese, J. N., Williamson, B., Parker, K., Hattan*S.,* Khainovski, N., Pillai, *S.,* Dey, *S.,* Daniels, *S.,* Purkayastha, *S.,* Juhasz, P., Martin, *S.,* Bartlet-Jones, M., He, F., Jacobson, A. and Pappin, D. J. (2004) Multiplexed proteinquantitation in *Saccharomyces cerepjsjae* using amine-reactive isobaric tagging reagents. *Mol. CeJJ. Proteomics* **3**, 1154-1169.

Sadowski, P.G., Dunkley, T.P., Shadforth, I.P., Dupree, P., Bessant, C., Griffin, J.L. and Lilley, K.S., (2006) Quantitative proteomic approach to study subcellular localization of membrane proteins. *Nat. Proto.* **1,** 1778-1789.

Schmidt, *A.,* Bisle, B. and Kislinger, T. (2009) Quantitative peptide and protein profiling by mass spectrometry. *Methods Mol. Biol.* **492**, 21-38.

Silva, J. C., Denny, R., Dorschel, C., Goren- stein, M. V., Li, G. Z., Richardson, K., Wall, D. and Geromanos, S. J. (2006) Simultanous qualitative and quantitative analysis of the *Escherichia coil* proteome: a sweet tale. *Mol. Cell. Proteomics* **5**, 589-607.

Stewart, II, Thomson, T. and Figeys, D. (2001) 180 labeling: a tool for proteomics. *Rapid Commun. Mass Spectrom.* **15**, 2456-2465.

Trinidad, J.C., Specht, C.G., Thalhammer, A., Schoepfer, R. and Burlingame A.L. (2006) Comprehensive identification of phosphorylation sites in postsynaptic densite preparations. *Mol. Cell. Proteomics* **5,** 914-922.

Villen, J. and Gygi, S.P. (2008) The SCX/IMAC enrichment approach for global phosphorylation analysis by mass spectrometry. *Nat. Protoc.* **3,** 1630-1638.

Wakelam, M.J., Pettitt, T.R. and postle, A.D., (2007) Lipidomic analysis of signaling pathways. *Methods Enzymol.* **432,** 233-246.

Wang, M., You, J., Bemis, K. G., Tegeler, T. J. and Brown, D. P. (2008) Label- free mass spectrometry-based protein quantification technologies in proteomic analysis. *Brief Funct. Genomic Proteomic* **7**, 329-339.

Washburn, M.P., Koller, A., Oshiro, G., Ulaszek, R.R., Plouffe, D., Deciu, C., Winzeler, E. and Yastes, J.R. (2003) Protein pathway and complex clustering of correlated mRNA and protein expression analysis in *Saccharomyces cerevisiae*. *Proc. Natl. Acad. Sci. USA* **100,** 3107-3112.

Whetstone, P. A., Butlin, N. G., Corneillie, T.M. and Meares, C.F. (2004) Element-coded affinity tags for peptides and proteins. *Bioconjug. Chem.* **15**, 3-6.

Wiener, M.C., Sachs, J., Oeyama, E.G. and Yates, N.A. (2004) Oincrential mass spectrometry: a label-tree LC-MS mcthod tor tinding signiticant dmcrences in complex pcptidc and protcin mixtures. *Anal. Chem.* **76**, 6085-6096.

Yan, Y., Weaver, V.M. and Blair, I.A. (2005) Analysis of protein expression during oxidative stress in breast epithelial cells using a stable isotope labeled proteome internal standard. *J. Proteome Res.* **4**, 2007-2014.

Yao, X., Freas, A., Ramirez, J., Demirev, P.A. and Fenselau, C. (2001) Proteolytic 180 labeling for comparative proteomics: model studies with two serotypes of adenovirus. *Anal. Chem.* **73**, 2836-2842.

Yarmush, M.L. and Jayaraman, A. (2002) Advances in proteomic technologies. *Annu. Rev. Biomed. Eng.* **4,** 349-373.

Yu, L.R., Conrads, T.P., Uo, T., Issaq, H.J., Morrison, R.S. and Veenstra, T.D. (2004) Evaluation of the acid-cleavable isotope-coded affinity tag reagents: application to camptothecintreated cortical neurons. *J. Proteome Res.* 3, 469-477.

Zappacosta, F. and Annan, R.S. (2004) N-terminal isotope tagging strategy for quantitative proteomics: results-driven analysis of protein abundance changes. *Anal. Chem.* 76, 6618 - 6627.

Zeng, D. and Li, S. (2009) Improved ClLAT reagents for quantitative proteomics. *Bioorg. Med Chem. Lett* **19,** 2059-2061.

Zhang, R., Sioma, C.S., Thompson, R.A., Xiong, L. and Regnier, F.E. (2002) Controlling deuterium isotope effects in comparative proteomics. *Anal. Chem.* 74, 3662-3669.

Zhang, R., Sioma, C.S., Wang, S. and Regnier, F.E. (2001) Fractionation of isotopically labelled peptides in quantitative proteomics. *Anal. Chem.* 73, 5142-5149.

Zhang, R. and Regnier, F.E. (2002) Minimizing resolution of isotopically coded peptides in comparative proteomics. J. *Proteome Res. J*, 139-147.

2

HIGH-THROUGHPUT PHENOTYPING FOR ABIOTIC STRESS TOLERANCE

***Ruchi Bansal[1], Jyoti Kumari[1], Vikender Kaur[1], Sundeep Kumar[2] and V.P. Singh[3]**

[1] Germplasm Evaluation Division,
National Bureau of Plant Genetic Resources, New Delhi 110012
[2]Division of Genomic Resources,
National Bureau of Plant Genetic Resources, New Delhi 110012
[3]College of Agriculture, JNKVV, Ganj Basoda, Vidisha, M.P. 464221
**Corresponding author: ruchibansal06@gmail.com*

INTRODUCTION

Our population is increasing exponentially and since 1950 to 2010, population has increased from 2.5 billion to 6.9 billion. The United Nations population projections have indicated that our population could reach 9.15 billion in 2050. Demands for the agricultural products is expected to increase at 1.1 per cent per year from 2005 to 2050 (Alexandratos and Bruinsma, 2012). We require doubling our food production to fulfil the increasing demands of the growing population, while being resilient to the changing climatic conditions with limited land, water and nutrient resources. To achieve the goal, we need to increase the yield at the rate of 2.4% per year (Ray *et al.,* 2012; 2013). Extensive breeding and agronomic efforts increased world cereal production by 1,225 million tonnes at 1.9 per cent per year from 1961 to 2007 (Alexandratos and Bruinsma, 2012). With the limiting resources of land and water, now the target is to achieve the yield with climate resilient crops. In the present scenario of climate change, abiotic stresses (drought, salinity and extreme temperature) are the major factors responsible for yield losses up to 50%

(Qin *et al.,* 2011). To integrate the resilience in to our crops, we require new sources of resistance against different abiotic and biotic stresses. Genetic resources for food and agriculture can play an important role in adaptation to the consequences of climate change. They may provide us the new sources of resistances for particular stresses. Therefore, we need to evaluate a large number of germplasms collected from different agro-climatic zones to search for the new source of resistance to different abiotic and biotic stresses. Most of the phenotyping methods being used till date are destructive, time consuming and labour intensive. These methods can provide accurate information with a limited number of samples. Since, we need to screen a large number of population, high-throughput phenotyping can serve the purpose (Kumar *et al.*, 2015).

Recent developments in the molecular biology techniques offer the potential to increase the rate of genetic improvement. Genomics has the potential to facilitate the dissection of complex traits such as abiotic stress tolerance. But successful application of genomics for crop improvement is possible only by the precise phenotyping (Arvidsson *et al.,* 2011). Screening of large population for different traits require high-throughput and fast screening techniques. Conventional phenotyping techniques are manual, labour intensive and destructive, which is the main limitation. High-throughput evaluation in controlled environment is sorely lacking in the present breeding system. It is the need of the hour to correlate the gene function and plant phenotype precisely in the target environment (Kumar *et al.*, 2015).

HIGH-THROUGHPUT PHENOTYPING

The phenotype is the resultant of the complex interactions between genotype and the environmental factors. Plant phenotyping is the comprehensive evaluation of complex traits as well as individual parameters which form the basis of complex traits. We study the complex traits by measurement of different quantifiable parameters such as biomass, photosynthetic efficiency, root architecture, canopy characteristics, yield attributing traits. Phenotyping may be destructive as well as non destructive. The term high through put phenotyping refers to noninvasive phenotyping on a large scale with in a fraction of time. It is an upcoming approach, which enable us in tracing the links between genotype, environment and phenotype. Conventional methods of phenotyping growth involve destructive analysis for measuring different parameters, but now the recent developments in imaging technologies have enabled the comprehensive and non destructive analysis of plants. The combined application of robotics and image analysis under has led to a revolutionary change in the phenotyping (Arvidsson *et al.,* 2011). Bioinfornatics approaches enable us in analyzing the very high dimensional data rapidly. The application of image analysis has been known for decades in the field of remote sensing in agriculture and forestry; however its recent application in phenotyping plants for stress tolerance in a high-throughput manner has opened new opportunities for plant biologists.

Tremendous growth in technological interventions has reduced the cost of genotyping, but we lack behind in our ability of accurate phenotyping. It is especially true in the case of very complex traits such as abiotic stress tolerance. Several physiological traits are being used for the phenotyping of population under stress, but inaccuracy in the measurement of a trait at a specific growth stage has been a problem since a very long time. Due to bottlenecks in the present phenotyping techniques, high-throughput phenotyping has been emerged as a powerful tool to characterize large number of lines/individual plants. It requires only a small fraction of time and labor (Furbank, 2009). To dissect the complex traits, we need to evaluate thousands of lines (Myles *et al.,* 2009). Molecular breeding through QTL mapping or genome wide association studies also involve the phenotyping of a large number of population across the year and the locations (Myles *et al.,* 2009; Jannink *et al.,* 2010). Nowadays high-throughput phenotyping platforms are being used in the controlled conditions as well as in the field (White *et al.,* 2012; Petrozza *et al.,* 2014).

Imaging Techniques in High-Throughput Phenotyping

Use of image analysis algorithm is the base of phenotyping studies. Different image analysis software tools are being developed. These input driven soft wares are manual, semi automated and automated. Imaging basically relies on the interaction between light and plants reflecting in the form of wavelength specific absorbance, reflectance and transmittance (Li *et al.,* 2014). Plant phenotyping involves imaging at different wavelengths (Fig. 1). Visible cameras detect light in the visible range (400-700 nm). Infrared camera use near infra red light and near infrared cameras detect near infrared and short wave infra red light. Thermal cameras sense long wave infra red radiations, while, hyper spectral cameras detect spectral bands ranging from 350-2500 nm.

Development in automation, robotics, sensor and imaging technologies has provided lot of choice for lab research, screening systems. These imaging techniques are widely used in the measurement of different physiological traits (Table 1). In the last decade, automated phenotyping systems have been deployed in the controlled environments. These systems are designed for large scale phenotyping of model plant Arabidopsis to cereals (Jansen *et al.,* 2009; Golzarian *et al.,* 2011). There are two type of imaging systems depending upon the movement. They may be categorized as Sensor to plant: systems, in which the sensor moves at different positions to reach the plant (Granier *et al.,* 2006; Arvidsson *et al.,* 2011) and Plant to sensor, when the position of sensor is fixed and the Plants moves to reach the sensor (Golzarian *et al.,* 2011). High-throughput phenotyping is extensively being used under abiotic stress as well as biotic stress conditions. Some of the examples of automated phenotyping techniques used for abiotic stress tolerance are given in Table 2. Most of these phenotyping platforms are suitable for the water deficit environments, where the target environment is to be defined in terms of moisture content (Granier *et al.,* 2006; Berger *et al.,* 2010). High-throughput phenotyping has also been done for tolerance to salt stress, cold stress and nutrient deficiency (Rajendran *et al.,* 2009; Humplik *et al.,* 2015; Neilson *et al.,* 2015).

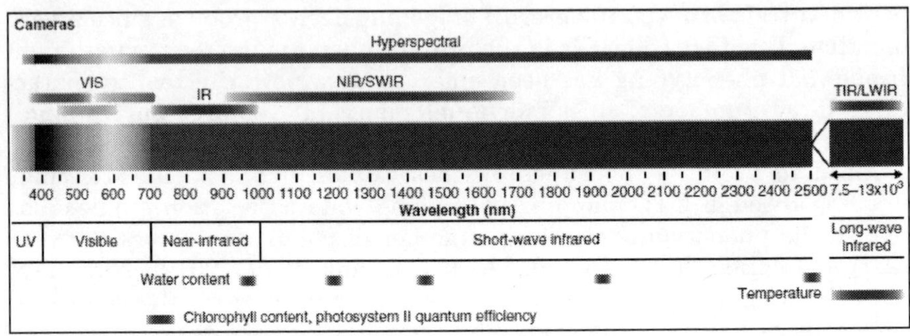

Fig. 1. Light spectrum being used to capture images for high-throughput phenotyping (Source: Fahlgren *et al.,* 2015)

Table 1: Application of imaging techniques in non invasive phenotyping in plants

S.No.	Imaging Technique	Parameters	References
1.	Visible light imaging	Projected area, growth dynamics, yield traits, root architecture, Germination rate, early embryonic axis growth, height, size, morphology, flowering time	Granier *et al.,*2006; Skiryex *et al.,* 2011; Tisne *et al.,* 2013; Bresson *et al.,* 2013; Vasseur *et al.,* 2014; Bresson *et al.,* 2014
2.	Fluorescence imaging	Photosynthetic status, Quantum yield, non photochemical quenching, shoot architecture	Woo *et al.,*2008; Jansen *et al.,*2009; Munns *et al.,* 2010; Feher-Juhasez *et al.* 2014; Richard *et al.,* 2015
3.	Thermal imaging	Canopy temperature	Jones *et al.,* 2009; Rajendran *et al.,* 2009; Harris *et al.,* 2010; Munns *et al.,* 2010
4.	Near infra red imaging	Water content composition parameters for seeds, leaf area index	Cseri *et al.,* 2013; Honsdorf *et al.,* 2014; Chen *et al.,* 2014; Neumann *et al.,* 2015

S.No.	Imaging Technique	Parameters	References
5.	Hyperspectral imaging	Leaf water status, leaf growth, coverage density	Golzarian *et al.* ,2011; Schilling *et al.,* 2014
6.	3D imaging	Shoot structure, leaf angle, root architecture	Ivanov *et al.,* 1995; Omasa *et al.,* 2007; Eietel *et al.,* 2014; Metzner *et al.,* 2015
7.	Laser imaging	Shoot structure, biomass, leaf angle, canopy structure, root architecture	Tilly *et al.,* 2012
8.	Magnetic resonance images	Morphometry, water content	Moradi *et al.,* 2010
9	Positron emission tomography	Transport partitioning, flow, velocity	Buhler *et al.,* 2011
10	Computed tomography	Tillers, morphometric parameters, grain quality	Gregory *et al.,* 2003; Hargreaves *et al.,* 2009

Table 2: High-throughput phenotyping in response to abiotic stress

S.No.	Plant	Nature of stress	Imaging	Reference
1.	Arabidopsis	Drought	RGB (top view), Fluorescence imaging	Granier *et al.,* 2006; Jansen *et al.,* 2009; Skiryex *et al.,* 2011; Tisne *et al.,* 2013; Bresson *et al.,* 2013; Vasseur *et al.,* 2014; Bresson *et al.,* 2014
2.	Wheat	Drought	RGB (Multiple view), Thermo-imaging, Fluorescence imaging	Seelig *et al.,* 2008; Suzuki Pinto *et al.,* 2010; Burling *et al.,* 2013 Feher-Juhasez *et al.,*2014; Richard *et al.,* 2015
	Wheat	Salt	RGB (Multiple view) Thermoimaging	Rajendran *et al.,* 2009; Sirault *et al.,* 2009; Harris *et al.,* 2010; Suzuki Pinto *et al.,* 2010
3.	Barley	Drought	RGB (Multiple view), hyperspectral NIR, Fluorescence imaging	Munns *et al.,* 2010; Cseri *et al.,* 2013; Honsdorf *et al.* ,2014; Chen *et al.,* 2014, Neumann *et al.,* 2015
	Barley	Salt	RGB (multiple view)	Harris *et al.,* 2010; Golzarian *et al.,* 2011; Schilling *et al.,* 2014

S.No.	Plant	Nature of stress	Imaging	Reference
4.	Rice	Salt	RGB (multiple views); Fluore-scence imaging, Thermoimaging	Hairmansis *et al.*, 2014; Siddiqui *et al.*, 2014
5.	Cotton	Drought	RGB (Multiple view) Thermoimaging	Andrade-Sanchez *et al.*, 2014
6.	Sorghum	Nutrient deficiency	RGB (multiple views); hyper-spectral NIR	Neilson *et al.*, 2015
7.	Tomato	Drought	RGB (multiple views); hyper-spectral NIR; Fluorescence imaging	Petrozza *et al.*, 2014
8.	Pea	Cold stress	RGB (mutiple views); Fluorescence imaging	Humplik *et al.*, 2015

RGB Imaging

The most commonly used imaging technique is RGB (Red, Green, Blue) imaging. It is the simplest method of imaging, which is based on silicon sensors (CCD and CMOS arrays). RGB phenotyping use the spectrum of visible range (400-700 nm) and allow the two dimensional imaging. Visible band cameras are either conventional digital cameras or RGB/CIR cameras. The captured image is pre-processed and segmented for very fine colour discrimination with out losing the sharpness. RGB imaging provides the most affordable solutions for phenotyping. RGB imaging has been widely used in phenotyping studies for assessing the projected area, growth dynamics, yield traits, root architecture, germination rate, early embryonic axis growth, height, size, morphology and flowering time. Studies have been done by capturing top view or the multiple angle images to derive a mathematical equation for estimation of biomass/area. In cereals, multiple images with different angles have been used in reducing the image occlusions. Studies on detached leaves for leaf shape have also been done by RGB imaging and have been applied in association mapping. In Arabidopsis, barley and tobacco, highly significant linear/polynomial correlations were observed while calibrating the shoot area based on total leaf area, fresh and dry shoot weight (Walter *et al.*, 2007; Golzarian *et al.*, 2011).

The limitations of RGB imaging are the background noise due to soil and overlapping of leaves during growth and development. Therefore, the set up for the imaging studies should be the growth stage specific in a particular species. Precise measurement of biomass additionally requires precise measurement of leaf thickness, so that the growth rates can be predicted with higher degree of confidence.

Fluorescence Imaging

Fluorescence is the emission of light of shorter wavelength after absorption of light. Fluorescence imaging is the imaging of fluorescence signals and is generally carried out by the CCD cameras which capture the fluorescence signals. Samples are illuminated with the visible or UV light by using pulsed lasers, pulsed flashlight lamps and light emitting diodes (LED). UV illumination (340-360 nm) generates two types of signals (either in red to far red region or blue to green region). In the imaging technique, fluorescence emission is captured from blue (440 nm), green (520 nm), red (690 nm) and far red (740 nm) region of the spectrum.

Fluorescence imaging has been widely used in the measurement of chlorophyll fluorescence. Chlorophyll fluorescence is an important parameter, which is being extensively used in the phenotyping under controlled as well as field conditions. Chlorophyll molecules emit the fluorescence in red and far red spectral bands. Chlorophyll fluorescence provides an estimate of quantum efficiency of photosystem II (Fv/Fm) for early detection of abiotic stress. Pulse amplitude modulation based instruments measure quantum efficiency, electron transport rate, non-photochemical quenching (Maxwell and Johnson, 2000). Chlorophyll fluorescence has been used as a screening criterion in many cereal and horticultural crops. In Arabidopsis, measurement of chlorophyll fluorescence under drought conditions in ecotype C24 and PARP mutants. Fluorescence imaging was used to study the chlorophyll fluorescence as well as for estimation of leaf area to calculate the relative growth rate. In Arabidopsis, chlorophyll fluorescence was not much affected by water limited conditions (Jansen *et al.,* 2009). Under Saline conditions also, variation in chlorophyll fluorescence was not significant in wheat (James *et al.,* 2002). Chlorophyll fluorescence was found to be more important in case of heat, cold and uv stress. Under these stress conditions, light harvesting complex in the photosystems are more prone to damage due to changes in photosynthesis and membrane properties. Semi automated systems have used fluorescence cameras and validation has been carried out (Woo *et al.,* 2008; Jansen *et al.,*2009). In a recent study in wheat, chlorophyll fluorescence kinetics has been used as early stress detection parameter and it has been used to map QTL under heat stress at seedling stress. A very large population of mutants was characterized by using the fluorescence meters and fluorescence cameras (Lu *et al.,* 2011), but the application of active fluorescence using 2D fluorescence is limited to few plant species (Jansen *et al.,* 2009).

Most of the studies have been carried out at the seedling stage. Imaging of mature plants is complex because it requires the construction of a three dimensional model. Recent developments for measuring photosynthesis at the canopy level are laser induced fluorescence and sun induced fluorescence, but they need to be validated for reproducibility.

Thermal Imaging

Thermal imaging is used for visualization of the infrared radiations. The most commonly used wavelengths for thermal imaging is 300-500 nm and 700-1400 nm. Thermal imaging is used to estimate the leaf surface temperature, which depends on the evaporation/transpiration from leaf surface. Estimation of canopy temperature by thermal imaging provides is used to detect the changes in plant water relations in response to stress. Thermal imaging has vast application in the lab and field based phenotyping. Canopy temperature depression (canopy temperature differences with respect to the surrounding air) is currently being used as one of the most important selection trait for drought stress. Stomatal conductance, transpiration rate and canopy temperature are directly correlated to each other (Jones *et al.,* 2002; 2009). Drought tolerant genotypes maintained higher stomatal conductance and photosynthetic rate (Medrano *et al.,* 2002). Measurement of canopy temperature by infra red thermometer is generally carried out in the breeding programs for drought prone environment. Handheld thermometers and guns are now cheaply available and can be used conveniently. But manual phenotyping of very large populations by hand held gadgets require time and labour for walking in the plots measuring canopy temperature. Changes in the environmental conditions during data recording over long distance also affect the data. But phenotyping by the imaging sensors can acquire data in a large number of plots at the same time. Furthermore, thermal cameras allow the more precise measurements in a single image in a fraction of time (Jones *et al.,* 2009). Phenotyping by automated imaging set ups in the controlled environment provides us an opportunity to evaluate the material for target environments (Sirault *et al.,* 2009). Thermal imaging was used at individual plant level in wheat seedlings to capture the differences in canopy temperature in response to salt stress with a differential of 1° in temperature (Fig. 1a) (Sirault *et al.,* 2009). Thermography was also used in a wheat trial imaged from cherry picker (Fig. 1b) (Munns *et al.,* 2010). Studies have also been performed in barley and Arabidopsis in two dimensional structures, but study of plants with three dimensional morphology should take care of the orientation of leaves with respect to light and angle at which different images are to be taken (Merlot *et al.,* 2002; Xie *et al.,* 2006; Grant *et al.,* 2006; Jones *et al.,* 2009).

Limitations of thermal imaging include the background noise due to soil and effect of wind and cloudiness (Jones *et al.,* 2009, Matsuda *et al.,* 2012). Though, the use of thermal cameras provide a fast screening strategy, but the understanding of the basic principles of heat fluxes in the natural environment is necessary before phenotyping. Surface temperature measurements require extensive calibrations. Orientation of leaf towards the incident radiation and camera angle used to capture the images should be taken into consideration during data analysis.

Imaging Spectroscopy

Use of multispectral and hyperspectral cameras over a very wide range of wave length has made it possible to apply the imaging spectroscopy in phenotyping. Solar radiations mostly lie in the range of visible and near infrared region. In the visible range, canopy reflectance is low, while, in the near infrared wavelengths, reflectance is very high and it is called 'red edge'. Reflectance pattern is also governed by leaf thickness and canopy architecture. Above 1300 nm, reflectance reduces due to increase in water absorption at 1450 nm, 1930 nm and 2500 nm (Knipling, 1970). These imaging systems can scan the wavebands of green reflectance, near infrared to mid infrared regions. The spectral measurements have been used in deriving the vegetation reflectance indices. These indices have been used for estimation of the characters particularly correlated with green colour as greenness, pigments and nitrogen in the large scale phenotyping in the breeding programs. The color information can be used to estimate the degree of senescence. Biomass accumulation and greenness has been estimated in the field using reflectral measurements (Campos *et al.,* 2004; Harris *et al.,* 2010). SPAD measurements and NDVI (Normalized difference vegetation index) are the most commonly used indices for phenotyping in the field (Ruan *et al.,* 2001; Richardson *et al.,* 2002). Multispectral and hyperspectral measurements are used to estimate the canopy water content. It uses the absorption bands in the infra red range. Since the measurements are highly influenced by the atmospheric vapor, few studies have reported a very good correlation between stress indices and the relative water content of leaves under severe drought (Eitel *et al.,* 2006; Seelig *et al.,* 2008). Hyperspectral imaging has been used in tomato (Petrozza *et al.,* 2014), Arabidopsis (Harshvardhan *et al.,* 2014), barley (Chen *et al.,* 2014) under drought stress.

It generates very large volume of data and the cost of multispectral and hyperspectral cameras are very high. Therefore, for large scale phenotyping, reductions in cost and data analysis are major issues, which need to be addressed.

THREE DIMENSIONAL MAPPING

The imaging techniques used for the three dimensional reconstruction of canopy are light detection and ranging (LIDAR), stereovision, photon mixer devices, time of flight cameras and Microsoft kinect. The most commonly used sensor is LIDAR. It is a remote sensing technology, which uses that uses liar sensors for measurement of 3D distribution of plant canopy. It is a robust and accurate technique with high resolution. In wheat, laser scanners have been used to map the leaf area index and in estimation of plant area density profile (Hosoi *et al.,* 2009). 3D data acquired from LIDAR provides accurate estimates of canopy structure, biomass and nitrogen (Omasa *et al.*, 2007; Tilly *et al.*, 2012; Eitel *et al.*, 2014).

Stereovision has also been used to construct 3D model of plants. It provides the estimates of plant height, leaf area and shape and was used for estimation of growth (Ivanov *et al.*, 1995; Rovira *et al.*, 2005; Takizama *et al.*, 2005). Surface texture of the object and sun light are the important factors affecting stereovision performance. Time of flight cameras have also been used in the phenotyping studies (Song *et al.*, 2011; Nakarni *et al.*, 2012), but low resolution and sensitivity to sunlight are the major limitations in wide applications. Microsoft Kinect has also been used in phenotyping by Azzari *et al.*, 2013. But sensitivity to light is the main limitation. Presently 3D imaging has been used in the study of growth and developmeny in the plant species, but its potential for abiotic stress is yet to be explored.

TOMOGRAPHIC IMAGING

Tomographic imaging is done by nuclear magnetic resonance imaging (MRI), positron emission tomography (PET) and X-ray computed tomography (CT). MRI detects the nuclear resonance signals originating from ^{1}H, ^{13}C, ^{14}N and ^{15}N. This technique provides 3D data on plant structure, root systems, estimates on water content and water transport (Moradi *et al.*, 2010; Hillnhutter *et al.*, 2012). Positron emission tomography is nuclear imaging technique. It produces images by detecting the pair of gamma rays based on the distribution of labeled compounds such as ^{11}C, ^{13}N or ^{52}Fe. It is used in estimating the transport velocity and the lateral losses along the paths (Buhler *et al.*, 2011). CT has been widely used in the study of root architecture in wheat and barley (Gregory *et al.*, 2003; Hargreaves *et al.*, 2009). Comparative study of 3D root imaging by MRI and CT in bean plants revealed that both the techniques performed well for small sized pots, while for medium sized pots, CT was better as compared to MRI. For large pots, MRI delivered more information than CT (Metzner *et al.*, 2015). These imaging are being used mainly in the root phenotyping and have not been used in phenotyping for abiotic stress tolerance.

PHENOTYPING PLATFORMS

Recent developments in sensing technologies, automation, aeronautics and computing are being used to develop controlled environment based and field based phenotyping platforms (Table 3). Controlled environment based phenotyping platforms are developed in public domain and are also being sold commercially (Lemna Tec, Plantscreen, PHENOPSIS, WIWAM, GROWSCREEN). These platforms are special structures designed for large scale phenotyping of selected species ranging from arabidopsis to cereals. These controlled environment systems are deployed in growth chambers and green houses. Studies in the controlled environment are generally performed to minimize the variability between plants in statistically significant way. Accurate measurement of the environmental conditions either in controlled can help in finding out the principal environmental factor affecting the phenotype. But controlled environment has its own limitations and can not truly represent the field conditions. Space availability in the growth chambers affect the pattern of growth and development of plants and therefore, it is not feasible to assess the effect of stress on reproductive growth of plant. Real characterization of plant's response to stress is problematic in controlled environment systems. Therefore, field based phenotyping platforms are getting more attention as compared to controlled platforms (White *et al.,* 2012).

Table 3: Comparative account of different phenotyping platforms

S.No.	Phenotyping platforms		Advantages	Limitations
1.	Controlled environment based		Automated operations, High reproducibility	Cost is high and limited number of plots can be monitored.
2.	Field based	Ground	High spatial resolution GPS/GIS tagging	Variability due to time taken in covering the whole experimental area
		Aerial	Cover a very large area with in a fraction of time	Spatial resolution depends on the speed and height, Maximum payload

Field based phenotyping platforms may be ground or aerial based platforms depending on the vehicle used to carry the sensor (Fig. 2 A, B). Ground based options include high clearence tractors, linear move or central pivot irrigation systems, while, aerial based platforms use manned fixed and rotatory wing aircraft, unmanned aircraft and tethered aerostats.

Fig. 2. Different types of field based phenotyping platforms (A) High clearance prototype tractor (White *et al.,* 2012) **(B)** Zeppelin NT Aircraft (Liebisch *et al.,* 2015)

Phenomics Facilities

Australian Plant Phenomics Facility (APPF) has two facilities, one the high resolution plant phenomics centre at Canberra and another the plant accelerator at Adelaide. This facility was developed for high-throughput phenotyping of plants across a range of growth conditions (glass houses and growth chambers). Others highthroughput phenotyping facilities are Crop Design in Gatersleben, Germany and phenoArch in France. In Germany, the Institute for Phytosphere Research (IPR) in Jülich established the Jülich Phenomics Centre in 2007. Recently, Indian Council of Agricultural Research, New Delhi under National Initiative on Climate Resilient Agriculture (NICRA) project has established managed stress environment phenomics facilities with visual, IR thermal and NIR root and shoot imaging facilities at Indian Agricultural Research Institute (IARI), New Delhi and Central Research Institure for Dryland Agriculture, Hyderabad, while these facilities are being established at Nationa Institute for Abiotic Stress Management, Baramati and Indian Institute of Horticultural Research, Bengaluru. Further, Indian Council of Agriculture Research is establishing a National Phenomics Facility for high-throughput non-destructive automated controlled environment phenotyping facility at IARI, New Delhi.

Conclusions

High-throughput phenotyping is an emerging research area. It is a combined application of plant biology, sensor technology and automated engineering. It has significant potential to systematically address the challenges in the phenotypic studies. Use of imaging techniques in the high-throughput phenotyping has opened the new horizons. Application of noninvasive phenotyping is increasing day by day for screening in controlled as well as field conditions for abiotic stress tolerance. Canopy structure, growth analysis and photosynthetic performance can be measured non-destructively in model plant and cereal crops. Study of root architecture in the field is still a problem to be addressed. Mechanistic understanding and deep phenotyping play a key role in identification and evaluation of useful parameters. Success of these technologies depends on precise environmental monitoring and use of information technology for data management. Achieving high-throughput screening requires comprehensive facilities with respect to set up for different environments. For large scale phenotyping, reduction in cost and gain in time are the future targets. Successful application of high-throughput phenotyping can be achieved only through the collective efforts of plant biologist, physicist, mathematician and engineer.

Future Prospects

Translating our research under controlled conditions into field conditions is the future target. We need to develop more understanding of the heterogenous conditions prevalent in the field, so that we can link the knowledge generated through the phenotyping studies in controlled and semi controlled environment in to natural environment. Cheaper phenotyping solutions with in acceptable limits are needed to be developed for field as well controlled conditions. Quick and robust root phenotyping techniques are required to characterize the root architecture.

References

Alexandratos N and Bruinsma J 2012 World agriculture towards 2030/2050: the 2012 revision. ESA working paper no. 12-03, Agriculture Development Economics Division. www.fao.org/economic/esa

Andrade-Sanchej P, Gore MA, Heun JT, Thorp KR, Carmo-Silva AE *et al* 2014 Development and evaluation of a field-based high throughput phenotyping platform. *Functional Plant Biology* **41** 68-79

Arvidsson S, Perez-Rodriguez P and Mueller-Roeber B 2011 A growth phenotyping pipeline for *Arabidopsis thaliana* integrating image analysis and rosette area modeling for robust quantification of genotype effects. *New Phytologist* **191** 895–907

Australian plant phenomics facility. Available online: http://www.plant-phenomics.org.au/

Azzari G, Goulden ML and Rusu RB 2013 Rapid characterization of vegetation structure with a microsoft kinect sensor. *Sensors* **13** 2384–2398

Berger B, Parent B and Tester M 2010 High-throughput shoot imaging to study drought responses. *Journal of Experimental Botany* **61** 3519-28

Berger B. de Regt B and Tester M 2012 Trait dissection of salinity tolerance with plant phenomics. In Shabala s, Cuin TA editors Plant salt Tolerance. New york city Humana Press pp. 399-413

Bresson J, Varoquaux F, Bontpart T, Touraine B and Vile D 2013 The PGPR strain *Phyllobacterium brassicacearum* STM 196 induces a reproductive delay and physiological changes that result in improved drought tolerance in Arabidopsis. *New Phytologist* **200** 558-69

Bresson J, Vasseur F, Dauzat M, Labadie M, Varoquaux F *et al* 2014 Interact to survive : *Phyllobacterium brassicacearum* improves arabidopsis tolerance to severe water deficit and growth recovery. *PLoS One* **9** e107607

Bühler J, Huber G, Schmid F and Blümler P 2011 Analytical model for long-distance tracer-transport in plants. *Journal of Theoretical Biology* **270** 70–79

Burling K, Cerovic ZG, Cornic G, Ducruet JM, Noga G *et al.* 2013 Fluorescence-based sensing of drought induced stress in the vegetative phase of four contrasting wheat genotypes. *Environmental and Experimental Botany* **89** 51-59

Campos H, Cooper M, Habben JE, Edmeades, GO and Schussler JR 2004 Improving drought tolerance in maize: a view from industry. *Field Crops Research* **90** 19-34

Chen D, Neumann K, Friedel S, Kilian B, Chen M *et al.* 2014 Dissecting the phenotypic components of crop plant growth and drought responses based on High-throughput image analysis. *Plant and Cell* **26** 4636-55

Cseri A, Sass L, Torjek O, Pauk J, Vass I *et al* 2013 Monitoring drought responses of barley genotypes with semi robotic phenotyping platform and association analysis between recorded traits and allelic variants of some stress genes. *Australian Journal of Crop Sciences* **7** 1560-70

Eitel JU, Magney TS, Vierling LA, Brown TT and Huggins DR 2014 Lidar based biomass and crop nitrogen estimates for rapid, non-destructive assess-ment of wheat nitrogen status. *Field Crops Research* **159** 21–32

Eitel JUH, Gessler PE, Smith AMS and Robberecht 2006 Suitability of existing and novel spectral indices to remotely detect water stress in *Populus* spp. *Forest Ecology and Management* **229** 170-182

Fahlgren N, Gehan MA and Baxter I 2015 Lights, camera, action: high-throughput plant phenotyping is ready for a close up. *Current Opinion in Plant Biology* **24** 93-99

Feher-Juhasez E, Majer P, Sass L, Lantos C, Csiszar J *et al.* 2014 Phenotyping shows improved physiological traits and seed yield of transgenic wheat plants expressing the alfalfa aldose reductase under permanent drought stress. *Acta Physiologiae Plantarum* **36** 663-73

Furbank RT 2009 Plant phenomics: from gene to form and function. *Functional Plant Biology* **36** 5

Golzarian M, Frick R, Rajendran K, Berger B, Roy S *et al.* 2011 Accurate inference of shoot biomass from high-throughput images of cereal plants. *Plant Methods* **7** 2

Granier C, Aguirrezabal L, Chenu K, Cookson SJ, Dauzat M *et al.* 2006 PHENOPSIS, an automated platform for reproducible phenotyping of plant responses to soil water deficit in *Arabidopsis thaliana* permitted the identification of an accession with low sensitivity to soil water deficit. *New Phytologist* **169** 623–635

Grant OM, Chaves MM and Jones HG 2006 Optimizing thermal imaging as a technique for detecting stomatal closure induced by drought stress under greenhouse conditions. *Physiologiae Plantarum* **127** 507-518

Gregory PJ, Hutchison DJ, Read DB, Jenneson PM and Gilboy WB *et al* 2003 Non-invasive imaging of roots with high resolution X-ray micro-tomography. *Plant and Soil* **255** 251–59

Hairmansis A, Berger B, Tester M and Roy SJ 2014 Image-based phenotyping for non-destructive screening of different salinity tolerance traits in rice. *Rice* **7** 16

Hargreaves C, Gregory P and Bengough AG 2009 Measuring root traits in barley (*Hordeum vulgare* ssp. Vulgare and ssp. spontaneum) seedlings using gel chambers, soil sacs and X-ray microtomography. *Plant and Soil* **316** 285-97

Harris BN, Sadras VO and Tester M 2010 A water-centred framework to assess the effects of salinity on the growth and yield of wheat and barley. *Plant and Soil* **336** 377-89

Harshavardhan VT, Vanson L, Seiler C, Junker A, Weigelt-Fischer K *et al* 2014 AtRD22 and AtUSPL1, members of the plant specific BURP domain family involved in *Arabidopsis thaliana. PLos One* **9** e110065

Hillnhu¨ tter C, Sikora RA, Oerke E-C and van Busschoten, D 2012 Nuclear magnetic resonance: a tool for imaging below ground damage caused by *Heterodera schachtii* and *Rhizoctonia solani* on sugar beet. *Journal of Experimental Botany* **63** 319–327

Honsdorf N, March TJ, Berger B, Tester M and Pillen K 2014 High-throughput phenotyping to detect drought tolerance QTL in wild barley introgression lines. *PLoS One* **9** e97047

Hosoi F, Omasa K 2009 Estimating vertical plant area density profile and growth parameters of a wheat canopy at different growth stages using three-dimensional portable lidar imaging. *ISPRS Journal of Photogrammetry and Remote Sensing* **64** 151–158

Humplik JF, Lazar D, Furst T, Husickova A and Hybl M *et al* 2015 Automated integrative high-throughput phenotyping of plant shoots: a case study of the cold-tolerance of pea (*Pisum sativum* L.). *Plant Methods* **11** 1-11

Ivanov N, Boissard P, Chapron M and Andrieu B 1995 Computer stereo plotting for 3-D reconstruction of a maize canopy. *Agricultural and Forest Meteorology* **75** 85–102

James RA, Rivelli AR, Munns R and von Caemmerer S 2002 Factors affecting CO_2 assimilation, leaf injury and growth in salt-stressed durum wheat. *Functional Plant Biology* 29 1393-1403

Jannink JL, Lorenz AJ and Iwata H 2010 Genomic selection in plant breeding: From theory to practice. *Briefings in Functional Genomics* **9** 166–177

Jansen M, Gilmer F, Biskup B, Nagel KA, Rascher U *et al.* 2009 Simultaneous phenotyping of leaf growth and chlorophyll fluorescence via GROWS-CREEN FLUORO allows detection of stress tolerance in *Arabidopsis thaliana* and other rosette plants. *Functional Plant Biology* **36** 902–914

Jones HG, Serraj R, Loveys BR, Xiong L, Wheaton A *et al* 2009 Thermal infrared imaging of crop canopies for the remote diagnosis and quantification of plant responses to water stress in the field. *Functional Plant Biology* **36** 978–89

Jones HG, Stoll M, Santos T, de Sousa C, Chaves MM *et al* 2002 Use of infrared thermography for monitoring stamatal closure in the field: application to grapevine. *Journal of Experimental Botany* **53** 2249-2260

Knipling EB 1970 Physical and physiological basis for the reflectance of visible and near IR radiation from vegetation. *Remote Sensing of Environment* **1** 155–59

Kumar J, Pratap A and Kumar S (2015) Plant Phenomics: an overview. In: Phenomics in crop plants: trends, options and limitations. Springer, New Delhi, India pp. 1-10

Li L, Zhang Q and Huang D 2014 A review of imaging techniques for plant phenotyping. *Sensors* **14** 20078-20111

Liebisch F, Kirchgessner N, Schneider D, Walter A and Hund A 2015 Remote, aerial phenotyping of maize traits with a mobile multi-sensor approach. *Plant Methods* **11** 9

Lu Y, Savage LJ and Last RL 2011 Chloroplast phenomics: systematic phenotypic screening of chloroplast protein mutants in *Arabidopsis*. *Methods in Molecular Biology* **775** 161–85

Matsuda O, Tanaka A, FujitaT, Iba K 2012 Hyperspectral imaging techniques for rapid identification of Arabidopsis mutants with altered leaf pigment status. *Plant and Cell Physiology* **53** 1154-70

Maxwell K and Johnson GN 2000 Chlorophyll fluorescence-a practical guide. *Journal of Experimental Botany* **51** 659-68

Medrano H, Escalona JM, Bota J, Gulias J and Flexas J 2002 Regulation of photosynthesis of C3 plants in response to progressive drought: stomatal conductance as a reference parameter. *Annals of Botany* **89** 895-905

Merlot S, Mustilli AC, Genty B, North H, Lefebvre V *et al* 2002. Use of infrared thermal imaging to isolate Arabidopsis mutants defective in stomatal regulation. *The Plant Journal* 30:601–9

Metzner R, Eggert A, Dusschoten D, Pflugfelder D, Gerth S S *et al.* 2015 Direct comparison of 3D imaging of root systems in soil: potential and challenges for root trait quantification. *Plant Methods* **11** 17

Moradi A, Oswald S, Nordmeyer-Massner J, Pruessmann K, Robinson B *et al* 2010 Analysis of nickel concentration profiles around the roots of the hyper-accumulator plant berkheya coddii using mri and numerical simulations. *Plant and Soil* **328** 291–302

Munns R, James RA, Sirault XRR, Furbank RT and Jones HG 2010 New phenotyping methods for screening wheat and barley for beneficial responses to water deficit. *Journal of Experimental Botany* **61** 3499–507

Myles S, Peiffer J, Brown PJ, Ersoz ES, Zhang Z *et al.* 2009 Association mapping: critical considerations shift from genotyping to experimental design. *Plant and Cell* **21** 2194-2202

Nakarmi A and Tang L 2012 Automatic inter-plant spacing sensing at early growth stages using a 3D vision sensor. *Computers and Electronics in Agriculture* **82** 23–31

Neilson EH, Edwards AM, Blomstedt CK, Berger CK, Berger B *et al* 2015 Utilization of a high-throughput shoot imaging system to examine the dynamic phenotypic responses of a C4 cereal crop plant to nitrogen and water deficiency over time. *Journal of Experimental Botany* **66** 1817-32

Neumann K, Klukas C, Friedel S, Rischbeck P, Chen D *et al* 2015 Dissecting spatio-temporal biomass accumulation in barley under different water regimes using high-throughput image analysis. *Plant Cell and Environment* Doi: 10.1111/pce.12516. in press.

Omasa K, Hosoi F and Konishi A 2007 3D lidar imaging for detecting and understanding plant responses and canopy structure. *Journal of Experimental Botany* **58** 881–98

Petrozza A, Santaniello A, Summerer S, Di Tommaso G, Di Tommaso D *et al* 2014 Physiological responses to Megafol treatments in tomato plants under drought stress: A phenomic and molecular approach. *Scientia Horticulturae* **174** 185-92

Qin F, Shinozaki K and Yamaguchi-Shinozaki K 2011 Achievements and challenges in understanding plant abiotic stress responses and tolerance. *Plant and Cell Physiology* **52(9)** 1569-1582

Rajendran K, Tester M and Roy SJ 2009 Quantifying the three main components of salinity tolerance in cereals. *Plant Cell and Environment* **32** 237–49

Ruan WR, Solie JB, Johnson GV, Stone GV, Stone ML *et al* 2001 In-season prediction of potential grain yield in winter wheat using canopy reflectance. *Agronomy Journal* **93** 131-138

Ray DK, Ramankutty N, Mueller ND, West PC and Foley JA 2012 Recent patterns of crop yield growth and stagnation. *Nature Communications* **3** 1293

Ray DK, Mueller ND, West PC and Foley JA 2013 Yield trends are insufficient to double global crop production by 2050. *PLoS ONE* **8** e66428

Richard CA, Hickey LT, Fletcher S, Jennings R, Chenu K *et al* 2015 High-throughput phenotyping of seminal root traits in wheat. *Plant Methods* **11** 13

Richards RA, Rebetzke GJ, Watt M, Condon AT, Spielmeyer W *et al* 2010 Breeding for improved water productivity in temperate cereals: Phenotyping, quantitative trait loci, markers and the selection environment. *Functional Plant Biology* **37** 85–97

Richardson AD, Duigan SP and Berlyn GP 2002 An evaluation of noninvasive methods to estimate foliar chlorophyll content. *New Phytologist* **153** 185-194

Rovira-Más F, Zhang Q and Reid J 2005 Creation of three-dimensional crop maps based on aerial stereoimages. *Biosystems Engineering* **90** 251–259

Schilling RK, Marschner P, Shavrukov Y, Berger B, Tester M *et al.* 2014 Expression of the Arabidopsis vacuolar H+ -pyrophosphatase gene (AVP1) improves the shoot biomass of transgenic barley and inceased\s grain yield in a saline field. *Plant Biotechnology Journal* **12** 378-86.

Seelig HD, Hoehn A, Stodieck L, Klaus D, Adams IW *et al* 2008 The assessment of leaf water content using leaf reflectance ratios in the visible, near-, and short-wave-infrared. *International Journal of Remote Sensing* **29** 3701–3713

Siddiqui ZS, Cho JI, Park SN, Kwon TR, Ahn BO *et al* 2014 Phenotyping in rice in salt stress environment using high-throughput infrared imaging. *Acta Botanica Croatia* **73(1)** 149-158

Sirault XRR, James RA and Furbank RT 2009 A new screening method for osmotic component of salinity tolerance in cereals using infrared thermography. *Functional Plant Biology* **36** 970-977

Skirycz A, Vandenbroucke K, Clauw P, Maleux K, De Meyer B *et al* 2011 Survival and growth of *Arabidopsis* plants given limited water are not equal. *Nature Biotechnology* **29** 212–14

Song Y, Glasbey CA, van der Heijden GW, Polder G and Dieleman JA 2011 Combining stereo and time-of-flight images with application to automatic plant phenotyping. In *Image Analysis*; Springer: Berlin/Heidelberg, Germany, pp. 467–478

Suzuki Pinto R, Reynolds MP, Mathews KL, McIntyre CL, Olivares-Villegas JJ *et al* 2010 Heat and drought adaptive QTL in a wheat population designed to minimize confounding agronomic efforts. *Theoritical and Applied Genetics* **121** 1001-1021

Takizawa H, Ezaki N, Mizuno S and Yamamoto S 2005 Plant recognition by integrating color and range data obtained through stereo vision. *Journal of Advanced Computer Intelligence and Intelligent Informatics* **9** 630–636

Tilly N, Hoffmeister D, Liang H, Cao Q, Liu Y *et al* 2012 Evaluation of terrestrial laser scanning for rice growth monitoring. *The International Archives of Photogrammetry,. Remote Sensing and Spatial Information Sciences.* **39** B7

Tisne A, Serrand Y, Bach L, Gilbault E, Ben Ameur R *et al.* 2013 Phenoscope: an automated large scale phenotyping platform offering high spatial homogeneity. *The Plant Journal* **74** 534-44

Vasseur F, Bontpart T, Dauzat M, Granier C and Vile D 2014 Multivariate genetic analysis of plant responses to water deficit and high temperature revealed contrasting adaptive strategies. *Journal of Experimental Biology* **65** 6457-69

Walter A, Scharr H, Gilmer F, Zierer R, Nagel KA *et al.* 2007 Dynamics of seedling growth acclimation towards altered light conditions can be quantified via GROWSCREEN: a set up and procedure designed for rapid optical phenotyping of different plant species. *New Phytologist* **174** 447-55

White JW, Andrade-Sanchez P, Gore MA, Bronson KF, Coffelt TA *et al* 2012 Field-based phenomics for plant genetics research. *Field Crops Research* **133** 101–112

Woo NS, Badger MR, Pogson BJ. 2008. A rapid, non-invasive procedure for quantitative assessment of drought survival using chlorophyll fluorescence. *Plant Methods* **4** 27

Xie XD, Wang YB, Williamson L, Holroyd GH, Taglivia C *et al* 2006 The identification of genes involved in the stomatal response to reduced atmospheric relative humidity. *Current Biology* **16** 882-887.

Section II:

Abiotic Stresses - Physiological and Molecular Implications

3

PLANT PHENOTYPING UNDER DROUGHT

T. Parthasarathi[1], K. Vanitha[2] and V. Ravichandran[3]

[1]e - Extension Centre, Directorate of Extension Education,
Tamil Nadu Agricultural University, Coimbatore, India - 641 003.
[2]Tamilnadu Rice Research Institute (TRRI),
Tamil Nadu Agricultural University, Aduthurai, Thanjavur, India - 612 101.
[3]Department of Rice, Tamil Nadu Agricultural University,
Coimbatore, India - 641 003.
Email: tparthasarathicrp@gmail.com

INTRODUCTION

Plant derived products are at the center of grand challenges posed by increasing requirements for food, feed and raw materials. Integrating approaches across all scales from molecular to field applications are necessary to develop sustainable plant production with higher yield and using limited resources. While significant progress has been made in molecular and genetic approaches in recent years, the quantitative analysis of plant phenotypes, structure and function of plant has become the major bottleneck.

Plant phenotyping is an emerging science that links genomics with plant ecophysiology and agronomy. The functional plant body (PHENOTYPE) is formed during plant growth and development from the dynamic interaction between the genetic background (GENOTYPE) and the physical world in which plants develop (ENVIRONMENT) (Fig. 1). These interactions determine plant performance, productivity measured as accumulated biomass and commercial yield and resource use efficiency.

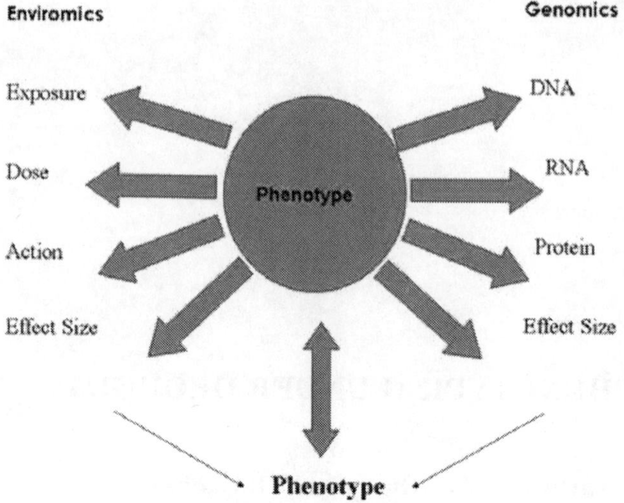

Fig. 1. Climate change and scarcity of arable land constitute additional challenges for future scenarios of sustainable agricultural production. It is necessary and urgent to increase the impact of plant sciences through practical breeding for varieties with improved performance in agricultural environments. The understanding of the link between genotype and phenotype is currently hampered by insufficient capacity (both technical and conceptual) of the plant science community to analyze the existing genetic resources for their interaction with the environment. Advances in plant phenotyping are therefore, a key factor for success in modern breeding and basic plant research.

PHENOTYPING

Phenotype *(from Greek phainein, 'to show' + typos, 'type')* is composite of an organism's observable characteristics of traits: morphology, development, biochemical or physiological properties and phenology. Genotype of an organism is the inherited instructions it carries within its genetic code. Phenotype is the physical manifestation of an organism's allellic combination (genotype). Phenotype is the observable/ qualifiable/ quantifiable representation of a trait. Phenotypic plasticity is the range of phenotypes a single genotype can express as a function of its environment. Phenotypes result from the expression of an organism's genes as well as the influence of environmental factors and the interactions between the two.

Net Result of Expression and Interaction of Many Genes with the Environmental Factors (GXE)

Plant phenotyping is the comprehensive assessment of plant complex traits such as growth, development, tolerance, resistance, architecture, physiology, ecology, yield, and the basic measurement of individual quantitative parameters that form the basis for the more complex traits. Examples for such direct measurement parameters are image-based projected leaf area, chlorophyll fluorescence, stem diameter, plant height/ width, compactness, stress pigment concentration, tip burn, internode length, colour, leaf angle, leaf rolling, leaf elongation, seed number, seed size, tiller number, flowering time and germination time etc. In simple, phenotyping is characterizing plant performance and structure-function dynamics.

After all, plant phenotyping has been performed by farmers and breeders for the few more thousand years, essentially since the days humans started to carefully select grasses to increase yield or enhance other desirable traits. In the past, phenotyping was mostly based on experience and intuition, in a process where measurement and interpretation were not separated. This highly integrated approach compensated for the human deficiencies in performing reproducible, objective measurements and dealt with the individual subjectivity factor of the phenotyping person. A large set of different aspects led to the development of the ever-increasing new field of highly automated, non-destructive *plant phenotyping*

Plant phenotyping used for the comprehensive assessment of plant complex traits such as growth, development, tolerance, resistance, architecture, physiology, ecology, yield, and the basic measurement of individual quantitative parameters that form the basis for the more complex traits. Phenotyping need to generate more quantitative data for genetic analysis and modelling of plant development.

Phenotyping is essential for

- Functional analysis of specific genes
- Forward and Reverse genetic analysis
- Production of new plants with beneficial characteristics

WHAT DOES 'GOOD PHENOTYPING' MEAN?

Good phenotyping must reduce the genotype-phenotype gap (Quantitative traits determinants of drought resistance). Basic attributes of good phenotyping carried out with appropriate genetic materials are accuracy and precision of measurements. Accuracy involves the degree of closeness of a measured or calculated quantity to its actual (true) value. Complexity of phenotyping a large number of genotypes (e.g., a mapping population) for drought-adaptive features is exemplified by those traits such as stomatal conductance and tissue water potential. The value of which can vary considerably within a rather short timeframe due to changing environmental conditions.

The need for Good Phenotyping Practice: Standards in Plant Phenotyping (European Plant Phenotyping Network consortium)

The ground breaking discoveries in molecular biology have allowed us a detailed characterization and understanding of complex genetic networks. However, the plant phenotype is the unit of importance in particular for practical issues such as crop production (Fiorani and Schurr, 2013; Pieruschka and Poorter, 2012). Advances in phenotyping as the process of quantitative characterizing the phenotype has become the major limiting step. The gap between the knowledge about genes and phenotypes is particularly large in analyses of plant-environment interactions that are urgently needed for research and application to sustainable and resource-efficient crop production in the context of climate change and varying agricultural production conditions (Houle *et al.,* 2010).

One of the key elements in the phenotyping is the establishment of common standards and good phenotyping practice integrates the phenotyping activities. The designing and performing "reference experiment" with certain number of the same genotypes of a monocot and a dicot screened at different locations. The important initial step in this experiment is the definition of standards for both environmental monitoring and trait assessment. Plant structure and function of selected genotypes will be analyzed under a range of environmental conditions defined by the diverse geographical location of the participating labs. Thus, standardization of environmental monitoring, trait assessment and data management is the key to elaborate and test methods to evaluate the responses of different genotypes to environmental conditions, by exploiting the contrasts of environmental conditions within different platforms. While in previous studies (Massonnet *et al.,* 2010) the experimental conditions were approximately constant in growth chambers, this cannot be maintained under greenhouse conditions at different locations. This requires a high level of standardization of environmental and phenotypic measurements. This provides thus an important step towards standardization of phenotyping experiments, which goes beyond the interests of the scientists but addresses plant phenotyping community.

GOOD PHENOTYPING PRACTICE

The main objective of this work is to define essential elements of phenotyping procedures able to:

- Optimize protocols aimed at answering biological questions, from experiment design to data interpretation
- Compare results and performing meta-analyses of data from different platforms and experiments.

The guidelines of Good Phenotyping Practice based on experience from integrated phenotyping platforms, technologies, Information and Technology. Test the reproducibility of data in phenotyping platforms and facilitate the transfer of information from controlled-conditions platforms towards the performance of genotypes in the field. At present this has not been systematically tested. The good phenotyping will address this point with a set of experiments comparing different platforms at different stringency of environmental conditions with respect to the ranking of genotypes on key traits. This effort aims at identifying the sensitive parameters that need to be controlled for robust and reliable results, such as accessing the phenotyping protocols including the assessment of environmental data, selected traits and data analysis tools.

Collecting Good Phenotypic Data

First is to simplify 'the system' in order to facilitate elucidation of the function of the relevant loci for target traits (i.e., the reductionist approach). Second is to evaluate the broader value of findings in a breeding, agronomically sound context (i.e., the holistic approach), physiology, epistatic interactions, pleiotropic effects of complex traits inevitably limit and blur the identification of the main factors leading to specific phenotypes (e.g., drought-resistant vs drought-susceptible). Good phenotyping is a critical issue for any kind of experimental activity. Challenges faced by investigating the effects of drought on crops are difficulties in standardizing, controlling and monitoring environmental conditions (especially in field). Collecting accurate phenotypic data are relevant to Target Prone Environment (TPE) has always been a major challenge for the improvement of quantitative traits.

The success of this endeavour is intimately connected with the heritability of the trait, namely the portion of the phenotypic variability accounted for by additive genetic effects that can be inherited through sexually propagated generations (Falconer, 1981). Trait heritability varies according to: (i) the genetic makeup of the materials under investigation; (ii) the conditions under which the materials are investigated; and (iii) the accuracy and precision of the phenotypic data. With only a few notable exceptions (e.g., flowering time and carbon-isotope discrimination), most of the traits determining the performance of crops under drought conditions usually have low (ca 0.30 - 0.40) or, at best, intermediate (ca 0.40 - 0.70) heritability. This impairs our capacity to dissect their genetic basis properly and, more importantly, reduces the effectiveness of phenotypic selection (Falconer, 1981; Melchinger *et al.*, 2004).

Influence of growth stages on phenotyping

Phenotyping occurs at flowering or shortly after additional bias is introduced if the tested genotypes differ considerably in flowering time and / or maturity. Phenotyping all accessions on the same date provides data collected from plants at different physiological stages, a circumstance that could introduce significant bias in the interpretation of cause-effect relationships between secondary traits and yield. In maize some biochemical factors, such as concentration of sucrose in the placental-chalazal area of the kernel, exert a strong and timely effect on reproductive fertility around flowering but not a week earlier or later.

Types of Phenotyping

Phenotyping in field condition

Following general factors should be evaluated carefully to ensure collection of meaningful phenotypic data in field experiments under water-limited conditions:

- • Experimental design
- • Heterogeneity of experimental conditions between and within experimental units
- • Size of the experimental unit and number of replicates
- • Number of sampled plants within each experimental unit
- • Genotype-by-environment-by-management interaction

Relative impact of each factor on the quality of the phenotypic data to be collected will vary greatly according to each experiment.

Phenotyping in controlled environment

Grain yield and its components are best phenotyped in field trials, measuring secondary traits in plants grown in controlled environment facilities (greenhouse, growth chamber, rain out shelters, etc.) (Fig. 2) takes advantage of an accurate control of main environmental parameters of moisture stress, air humidity, temperature, light, etc., that vary greatly in field experiments. Plants grown under controlled conditions allows for collection of phenotypic data and samples at same growth stage and under similar conditions. This allows for more accurate assessment of constitutive capacity of different genotypes to accumulate drought-adaptive compounds.

Fig. 2. Phenotyping in controlled environment facilities (Source: Controlled Environment Chambers and Rain Out Shelters (ROS) of TNAU, Coimbatore, by Authors)

PHENOTYPING STRATEGIES

Phenotyping strategy covers the preliminary and advanced phenotyping, study of drought resistance mechanisms, and evaluation of hybrids in addition to segregating and advanced populations.

Preliminary phenotyping

Preliminary phenotyping includes local and international cultivars and elite lines with high genetic diversity. About 30 days after emergence for rice and 20 days for beans, two drought treatments are applied, namely maintenance of adequate soil moisture as in the early plant growth stage and induced drought up to the final growth stage. Irrigation is applied using a linear moving sprinkler irrigation system with electronic control of the irrigation water supply.

The discrimination of lines is based on the drought susceptibility index (DSI) defined by Fisher and Maurer (1978) as:

$$DSI = (Yw/ods - Yw/ds) / (Yw/ods \cdot D)$$

where, D = 1 - YMcdh / YMw/ods and Yw/ds and Ycdh are the yields of a given line without and with drought, respectively, and YMw/ds and YMw/ods are the yield averages of the experiments, without and with drought, respectively.

Advanced phenotyping

The preliminary evaluation provides a reduced number of improved lines for inclusion in advanced phenotyping. Phenotyping at this level can include lines from international cooperative programmes. The lines included in the phenotyping are maintained under good soil moisture conditions during the vegetative stage. Then, three water regimes are applied: severe water stress, moderate water stress and no water stress.

In the third treatment, irrigation is controlled using a tensiometer. Irrigation of approximately 25mm depth is applied whenever necessary. The moderate water stress and severe water stress treatments receive 2/3 and 1/3 respectively of the water applied in the non water stressed treatment. Lines are evaluated for yield potential as well as for agronomic and morphological characteristics, such as shoot dry matter, plant height and 100 seed weight. The DSI and canopy temperature are determined. In addition, yield components such as percentage of spikelet sterility, number of grains per panicle and tiller fertility are evaluated for rice and flower abscission, number of pods per plant and number of seeds per pods for common beans.

Trial Planning - Definition of the Target Environments

Target population of environments (TPE)

Each breeding programme must clearly define the TPE for which it is developing varieties. TPE is the set of all environments, fields and seasons in which improved varieties are expected to do well. However, the environments must be sufficiently similar for one genotype to perform well in all of them.

Water relations

Water is the most abundant constituent of all physiologically active plant cells, ranging between 55-85% of the water content of leaves. Water is an important parameter for plant growth or yield representing a complex hydrodynamic system, which involves absorption of water from the soil, water translocation throughout the plant and, its loss to the environment by transpiration. The capacity of plants to meet the demand and thus to avoid water deficit depends on this hydrodynamic system. Since water relations of a plant are closely related and affected by the environment we need to closely monitor at least:

- Soil properties in particular the water content which determines the water that is potentially available for transpiration,
- Air humidity and temperature around the leaves as the two major factors which affect the evaporation of water into the air.

Monitoring water levels to characterize drought

Water supply can be monitored during crop growth to determine the timing and severity of drought to further define the TPE. The pattern of water level recorded over the season can be used to characterize three different types of drought:

- An early drought that occurs during vegetative growth
- An intermittent mid-season drought that occurs between tillering and mid-grain filling
- A late drought that occurs during flowering and grain filling

In addition to knowing the frequency, it is also important to know the severity. For this, it is necessary to compare the yields under the drought and irrigated conditions or, if irrigation is not possible, choose a well-watered site such as the bottom of the topo sequence. Then turn go for phenotyping screening.

MECHANISMS OF DROUGHT ADAPTATION CONCEPTS, NOMENCLATURE AND TRAITS

In agriculture, 'drought' refers to a condition in which the amount of water available through rainfall and/or irrigation is insufficient to meet the transpiration needs of the crop. Modern breeding activities, including phenotyping conditions, have predominantly targeted the latter levels of stress. Yield remains an elusive and neglected concept under water-limited conditions; it is an appropriate way to gauge the overall phenotypic value of any accession.

Phenotyping for drought performance is not just a matter of choosing the right combination of traits and measuring them at the right time. It is also necessary to cope with other sources of uncertainty relating to the need for suitable test sites with a drought cycle, irrigation system and trained staff. Efficient phenotyping (frequently termed 'precision phenotyping') implies meeting two requirements. The first requirement is proper stress management of the agronomic conditions (including irrigation management and agroclimatic record) in order to impose as closely as possible the desired stress in terms of severity and occurrence during the crop cycle. Ultimately, knowledge contributed by any drought related study should help address the yield issue either directly or indirectly. Strategies to characterize and evaluate the crop response to drought, under controlled environment conditions with different water regimes.

Timing of Measurement and Sample Collection

Morpho-physiological traits were fluctuate widely during the circadian cycle, such as plant's water status, ABA content, stomatal conductance, leaf rolling, canopy temperature, etc., choosing most appropriate time for measurement and/or sample collection is very critical. Balota *et al.* (2007) have investigated timing of measuring CTD on breeding selections of wheat in relation to growth stage, time of day and weather. Although under dry conditions long-term mean CTD at noon and yield were correlated in two growing seasons, the relation of short-term CTD readings to grain yield was highly variable (Balota *et al.*, 2007). Genotype effects on CTD were detected for all hours of day and night. Genotype-by-hour interaction was non-significant at night, suggesting that night-time measurements may provide more stable conditions for CTD comparison among genotypes.

Traits responsible under drought phenotyping

The morpho-physiological traits and the corresponding QTLs that affect drought tolerance can be categorized as constitutive (i.e., also expressed under well-watered conditions) or drought-responsive (i.e., expressed only under pronounced water shortage; Blum, 2006; Lafitte and Edmeades, 1995). While drought-responsive traits/ QTLs usually affect yield only under rather severe drought conditions, constitutive traits/ QTLs can affect yield at low and intermediate levels of drought stress as well. The response of QTLs for drought-adaptive traits (e.g., accumulation of osmolytes, relocation of water-soluble carbohydrates, etc.) to drought is probably due to regulation of the expression of the underlying structural genes in response to signalling cues such as abscisic acid (ABA) accumulation (Bray, 2002) that are reinforced by cellular dehydration. Under appropriate soil moisture conditions, the presence of QTLs for traits usually classified as constitutive but difficult to measure (e.g., root depth) can be revealed by the collocation of QTLs for traits indicative of the water status of the plant such as ABA content, stomatal conductance, canopy temperature depression, etc. Additionally, QTL alleles that constitutively accelerate flowering, e.g., in maize (Salvi *et al.*, 2007; Ducrocq *et al.*, 2008) will positively influence yield when drought occurs around flowering and / or during grain filling (Reynolds and Tuberosa, 2008).

The traits to be considered as potential selection targets for improving yield under water limited conditions must be genetically (i.e., causally) correlated with yield and should have a greater heritability than yield itself (Blum, 1988; Monneveux and Ribaut, 2006).

Drought Phenotyping Characters

The following phenotyping parameters were observed in screening tolerant genotypes. Such as early vigour, root architecture, flowering time, carbon isotope discrimination, stomatal conductance, canopy temperature depression, abscisic acid, osmotic adjustment, chlorophyll concentration, stay-green, delayed leaf senescence and remobilization of water-soluble carbohydrates.

Early vigour

Early and vigorous establishment under conditions of low evapotranspiration may allow annual crops to optimize WUE and limit the loss of water due to direct evaporation from the soil surface. This leaves more stored water available for later developmental stages when soil moisture becomes progressively exhausted and increasingly limiting for yield (Rebetzke *et al.*, 2007; Reynolds and Tuberosa, 2008; Richards, 2006; Richards *et al.*, 2002; 2007; Slafer *et al.*, 2005). Early establishment also reduces the occurrence of inhibition of stomatal conductance as a consequence of root-borne signalling such as from ABA through the xylem flow (Davies *et al.*, 2000; Ren *et al.*, 2007), caused by an excessively shallow and superficial root system (Blum, 1996; Giuliani *et al.*, 2005b).

Root architecture

Roots exhibit an astounding level of morphological plasticity in response to soil physical conditions (Passioura, 1983; Siopongco *et al.*, 2009). This peculiarity allows plants to adapt better to the chemical and physical properties of the soil, particularly under drought conditions (Bacon *et al.*, 2002; Maggio *et al.*, 2001; Yu *et al.*, 2007). The concept of root ideotype should be elaborated only after gaining a detailed understanding of: (i) the factors that limit the availability of soil moisture to the crop; and (ii) the metabolic cost sustained by the plant to develop and maintain a more vigorous root system.

In rice, an enhanced capacity to penetrate a soil hardpan is considered an essential feature for the development of deeper roots under rainfed lowland conditions (Fukai and Cooper, 1995) and is a key factor in drought adaptation in areas where water supply is limited (Siopongco *et al.*, 2009). Therefore, roots able to proliferate quickly in the subsurface layers of the soil prior to and during the early stages of drought should allow more prolonged water extraction and consequently, better maintenance of leaf water potential and photosynthesis. Gel-filled chambers, soil sacs and X-ray microtomography have also been used to investigate bi- and tri-dimensional root architecture (Bengough *et al.*, 2004; Hargreaves *et al.*, 2009; Norton and Price, 2009; Sanguineti *et al.*, 2007). These experiments are particularly suited to the discovery of QTLs that are prevalently expressed in a constitutive fashion and which, as such, are more likely to influence root architecture across different soil conditions.

Flowering time

Flowering time is recognized as the most critical factor to optimize adaptation, hence yield, in environments differing in water availability and distribution during the growing season (Richards, 1991; 2006). Positive associations between plasticity of yield and flowering time across different levels of water availability have been reported in different crops (Sadras *et al.*, 2009). Therefore, in addition to phenology per se (i.e., mean time to a phonological stage), plasticity of phenological development merits consideration as a distinct trait influencing crop adaptation. In annual crops, the genetic basis of flowering time is more complex in temperate species (e.g., barley, wheat, rye, etc.) as compared to tropical species (e.g., rice, sorghum, maize, etc.), due to the presence in the former group of vernalization genes influencing flowering time in response to low temperatures.

Carbon isotope discrimination

Carbon isotope discrimination ($\Delta^{13}C$) measures the ratio of stable carbon isotopes ($^{13}C/^{12}C$) in the plant dry matter compared to the ratio in the atmosphere (Condon *et al.*, 1990). Because of differences in leaf anatomy and the mechanisms of carbon fixation in species with the C_3 or C_4 pathway, studies on $\Delta^{13}C$ have wider implications for C_3 species where the variation in $\Delta^{13}C$ is larger than in C_4 species and has a greater impact on crop yield (Condon *et al.*, 1990; 2006). Commonly, but not always (Turner *et al.*, 2007), $\Delta^{13}C$ is negatively associated with WUE over the period of dry mass accumulation (Araus *et al.*, 2002; Condon *et al.*, 1990; 2004; Rebetzke *et al.*, 2002; Royo *et al.*, 2008; Xu *et al.*, 2007). Under drought stress, $\Delta^{13}C$ is a good predictor of stomatal conductance (Condon *et al.*, 2002) and WUE in different crops (Tambussi *et al.*, 2007; Turner, 1997).

$\Delta^{13}C$ measured in grains correlates positively with growth cycle duration (Araus *et al.*, 1997) and negatively with leaf temperature (Richards *et al.*, 2002). Therefore, the relationship between $\Delta^{13}C$ and grain yield depends on the environmental conditions, the phenology of the crop and the plant organ (e.g., leaf or grain) from which the samples are collected (Araus *et al.*, 1997; Condon *et al.*, 2004; Merah *et al.*, 2001). For these characteristics, $\Delta^{13}C$ is an attractive breeding target for improving WUE and yield, while the high cost required measuring each sample makes it an interesting candidate for Marker Assisted Selection (Fig. 3.).

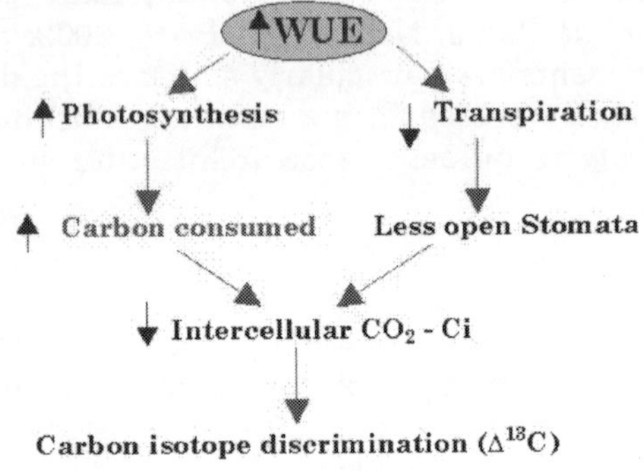

Fig. 3. Images reproduced (Source: Araus *et al.,* 1997)

Stomatal conductance

Stomatal conductance plays a pivotal role in determining $\Delta^{13}C$ and WUE (Condon *et al.*, 2002; Richards *et al.*, 2002; 2007). In cotton, high yields have been associated with high stomatal conductance and reduced leaf temperature (Radin *et al.*, 1994; Singh *et al.*, 2007; Ulloa *et al.*, 2000). Altogether, this indicate the possibility of raising the yield potential, hence the amount of water used by the crop, through an indirect selection for stomatal conductance and/or leaf temperature.

A more attractive and integrative way to indirectly monitor stomatal conductance through an extended time-period is based on the measurement of the natural oxygen isotope composition ($d^{18}O$) in leaf and grain materials (Barbour *et al.*, 2000; Ferrio *et al.*, 2007). Compared with stomatal conductance, measuring $d^{18}O$ in plant material offers four advantages: (i) it provides an integrated measure of stomatal conductance and leaf temperature over the period that the analysed tissue was formed; (ii) it avoids a number of experimental problems typical of measuring stomatal conductance; (iii) it allows for the collection of a large number of samples and (iv) requires very little labour in the field. However, the authors (Barbour *et al.*, 2000) caution that $d^{18}O$ is a questionable yield predictor when stomatal conductance and grain yield are not strongly correlated.

Canopy temperature depression

Canopy temperature depression (CTD) as measured by thermal imaging is the difference in temperature between the canopy surface and the surrounding air. CTD is a highly integrating trait resulting from the effects of several biochemical and morphophysiological features acting at the stomata, leaf and canopy levels. In the field, genotypes with a cooler canopy temperature under drought stress, or a higher CTD, use more of the available water in the soil to avoid excessive dehydration (Blum, 1988; Ferrio *et al.*, 2001; Ludlow and Muchow, 1990; Reynolds *et al.*, 2007). Infrared thermometry can report subtle differences in leaf temperature in both field and controlled conditions (Ayeneh *et al.*, 2002; Blum *et al.*, 1982; Chaerle *et al.*, 2007; Jones *et al.*, 2003). "Canopy temperature epitomises a mechanism of dehydration avoidance expressed throughout the cycle and across latitudes, which can be utilized as a selection criterion to identify high yielding wheat genotypes or as an important predictor of yield performance under drought" (Olivares-Villegas *et al.*, 2007). Thermal imaging successfully distinguished between irrigated and non-irrigated plants of different species, with strong correlations between thermal indices and stomatal conductance as measured with a leaf porometer.

Abscisic acid

One of the main factors influencing leaf temperature via an effect on transpiration through stomatal conductance is the concentration of ABA in the leaf tissue and, ultimately, in guard cells (Sirichandra *et al.*, 2009; Wasilewska *et al.*, 2008). Therefore, ABA is a fundamental component of the mechanisms allowing the plant to match the water demand with the water supply and to optimize growth and survival in response to both daily and more long-term environmental fluctuations (Xiong *et al.*, 2007; Zhang and Davies, 1990). Indeed, an increase in ABA concentration is a universal response observed in plants subjected to drought and other abiotic stresses (Quarrie, 1991; Setter, 2006). ABA has been shown to affect many of the traits that influence the water balance of the plant through both dehydration avoidance and dehydration tolerance (Thompson *et al.*, 2007). It has also been shown that ABA facilitates water uptake into roots as the soil begins to dry, particularly under non-transpiring conditions, when the apoplastic path of water transport is largely excluded (Hose *et al.*, 2001). Additionally, ABA modulates the expression of a large number of genes whose products protect the cell from the harmful effects of dehydration (Bray, 2002; Seki *et al.*, 2007). In rice, selection for reduced ABA root signalling has been advocated as a means for better exploitation of subsoil water under mild or transient water deficit (Siopongco *et al.*, 2008; 2009). Due to the availability of ABA-specific monoclonal antibodies (Quarrie *et al.*, 1988; Walker-Simmons *et al.*, 1991) that allows for the cost-effective measurement of a large number of samples, several studies have been devoted to the identification of QTLs for ABA concentration and the analysis of their associated effects on other drought-related traits and yield (Giuliani *et al.*, 2005b; Landi *et al.*, 2005; 2007; Lebreton *et al.*, 1995; Pelleschi *et al.*, 2006; Reymond *et al.*, 2003; Sanguineti *et al.*, 1999; Tuberosa *et al.*, 1998; 2002a).

Osmotic adjustment (OA)

Osmatic Adjustment is a metabolic process entailing a net increase in intercellular solutes in response to water stress (Morgan, 1984; Serraj and Sinclair, 2002; Zhang *et al.*, 1999). As soil moisture declines, OA favours turgor maintenance, and hence the integrity of metabolic functions. Importantly, OA can bias estimates of the value of relative water content, as has been shown in wheat and barley (Boyer *et al.*, 2008). Under conditions of severe dehydration, a higher capacity to accumulate osmolytes may help plants withstand a prolonged drought spell and undergo a more prompt and complete recovery upon rehydration.

Chlorophyll concentration, stay-green and delayed leaf senescence

A well-sustained source capacity is a key factor to maximize yield potential during both vegetative and reproductive phases, particularly under source-limiting conditions that commonly characterize drought stressed crops. Therefore, delaying leaf senescence maintains transpiration and increases cumulative photosynthesis over the crop life cycle (Borrell *et al.*, 2001; Jiang *et al.*, 2004). This is a strategy that is adequate for soils with appreciable water reserves but may otherwise cause severe stress at the end of the growth season due to increased transpiration.

Remobilisation of water-soluble carbohydrates (WSC)

Remobilization of water-soluble carbohydrates (WSC) from the stem and leaves can mitigate the negative effects on grain filling caused by post anthesis drought tolerance (Araus *et al.*, 2002; Blum, 1988; 1998; Rebetzke *et al.*, 2008b; Reynolds and Tuberosa, 2008; Reynolds *et al.*, 2007). Rebetzke *et al.* (2008b) phenotyped three wheat mapping populations for WSC concentration (WSC-C) and for WSC mass per unit area (WSC-A). Genotypes with high WSC-C were commonly shorter, flowered earlier and produced significantly fewer tillers than those of low WSC-C. Drought-adaptive traits, far from being exhaustive, indicates that genetic variability in drought tolerance and WUE can be traced to the interaction of a multitude of quantitatively inherited morphophysiological features, whose effects on yield can vary greatly both in terms of magnitude and direction according to the prevailing drought scenario and other yield constraints.

Alternative Approaches to Phenotyping for Water-Use Efficiency

Specific leaf area

The specific leaf area (SLA) is the ratio of leaf area to leaf dry weight, and is an indirect measure of leaf expansion. Higher SLA values represent a larger surface area for transpiration, Hence, SLA and WUE would be inversely related. SLA can be used as an alternative for rapid estimation of genetic variability in WUE.

Specific leaf nitrogen (SLN)

Expressed as the amount of nitrogen present per unit leaf area has, as a result, emerged as a useful alternative approach to assessing differences in WUE. The latter depends significantly on the biochemical efficiency of carbon fixation in the chloroplasts, protein content and, hence, the leaf nitrogen status is an important determinant of photosynthesis

Osmotic Adjustment -Pressure Chamber (Leaf Ψ)

The Ψw of a plant cell is governed by the equation: $\Psi w = \Psi s + \Psi p$, where Ψs is the osmotic potential and Ψp is the pressure potential. A higher Ψp can be achieved at a given Ψw by accumulating solutes inside the cell, thus lowering Ψp. The accumulation of additional solutes in response to a low Ψs is termed 'osmotic adjustment' (OA). Thus, turgor maintenance, and hence the integrity of metabolic functions.

Chlorophyll Fluorescence- (Plant Physiologists Stethoscope)

The following fluorescence parameters are very useful in monitoring and judging the physiological state of plants under drought. ΦPSII = Quantum yield of PSII. Fv/Fm=Maximum potential quantum yield of PSII (Indicator of photosynthetic efficiency). Fv = Fm-Fo. qP = Photochemical quenching (proportion of open PSII) (PQ). qN= non photochemical quenching (NPQ). ETR (J) =Electron Transport Chain.

Stay-green trait

This explains stay-green as the non senescence or delayed senescence of leaves. More green leaf area after and during grain filling is an important trait component for sustaining yield potential under stress during grain filling. Measuring of SLN, SPAD value, SLW, green leaf area and senescence scoring are the grain filling stage-trait. Non-senescence is under relatively simple genetic control and can readily be improved by conventional or molecular breeding.

Epicuticular wax load

Epicuticular wax load (bloom) is the covering of the aerial portions of the plant with a thick, amorphous layer of wax, as occurs in most sorghum. Epicuticular wax has been implicated as a feature of sorghum that enhances its drought resistance. Thus epicuticular wax may enhance WUE in sorghum by regulating nighttime water loss.

Leaf water retention capacity

Leaf water retention capacity (%) = (fresh weight – dry weight)/fresh weight x 100. The principle behind this technique is that drought-stressed or hardened plants retain more water than unstressed plants and, when the stress is imposed across many different genotypes, the tolerant genotypes exhibit a greater capacity for water retention in the leaf tissue.

The difference in leaf water retention capacity (LWRC) may be due to the tightness of stomatal closure or to other causes such as cuticular resistance to water loss.

Leaf rolling score in rice (0 to 9) standard evaluation system for rice (ses) by IRRI

Simple scoring methods such as leaf drying and rolling, scored around flowering stage can be a useful secondary traits for differences in plant water status. Scores and symptoms for leaf rolling and drought resistance at vegetative stage

Scores	Reaction	Leaf rolling
0	Highly	No symptoms of stress
1	Resistant	No rolling
3	Moderately resistant	Partially rolled, unrolled in evening
5	Intermediate	Partially; unrolling at late evening and early morning
7	Susceptible	Complete, unrolling in morning
9	Highly susceptible	Like tube; no unrolling in morning

Phenotyping - overcoming the bottleneck

Genotyping is relatively fast, genome sequencing advancing techniques. Phenotyping is still time consuming and labor intensive. To solve this, technological advances essential for high throughput phenotyping by using robotics, non-destructive image analysis and powerful computers. Conventional methods for phenotyping plants are frequently laborious, destructive and involve the removal of plant biomass for analysis. Recent developments in high-throughput, non-destructive imaging technologies allow a researcher to obtain multiple images of the same plant at different time points and at different wavelengths, thereby offering new non-destructive methods for acquiring quantitative data on plant growth, health and water use under drought. Phenmoics is the study of how the genetic makeup of an organism determines its appearance, function and performance. Phenomics techniques can cut phenotyping time down from weeks to minutes. Phenome is the interaction between gene and environment or the expression of the genome as traits in a given environment.

HIGH THROUGHPUT PHENOTYPING

The high throughput phenotyping is the answer for the gap in accurate phenotyping method and variation in data set. This will reduce errors and increase the precision in recording the data of large genotype population.

Essentiality of high throughput in different growing conditions (e.g., under abiotic and biotic stress condition) on many different lines (to discover desirable lines)

- Mutant populations
- Mapping populations
- Breeding populations
- Germplasm populations

Useful approach for complex and multigenic traits for drought tolerance

- Infra-red imaging of transpiration
- Hyperspectral imaging of C, N, phenolics
- FTIR imaging at cellular level
- Chlorophyll fluorescence imaging of photosynthesis
- Hyperspectral sensing of stress tolerance

High Throughput Phenotyping Ways

- Digital imagery
- Thermal imagery
- Spectral reflectance
- Stable isotopes

Digital imagery

Digital imagery used for estimating leaf N content, early biomass and response to water-limited conditions (Fig. 4). Digital picture properties (picture-derived vegetation indices, picVIs) as a phenotyping tool for crop breeding under well-watered and rainfed conditions.

Fig. 4. Source: Feher *et al.*, manuscript; Denes Dudits Institute of Plant Biology, Biological Research Centre, H.A.S., Sazeged.

Pictures taken approximately two weeks prior to anthesis showed hue and brightness were highly correlated with measurements of NDVI. The view of digital imagery procedure was expressed in Fig. 5.

Fig. 5. Images reproduced from Fincher *et al.* (2011). Australia Plant Phenomics Facility.

Estimation of shoot biomass from digital image

Improved estimate of biomass when age of the plant is taken into account

$$Y = a0 + a1\times(G+B+Y) + a2\times(G+B+Y)\times H$$

(H = number of days after seed preparation date); (Correction for leaf colour did not greatly improve weight estimates); (Cross validation run 10x).

Thermal Imagery

Water deficiency, an immediate plant response is a reduction in transpiration to reduce water loss, increasing leaf (and canopy) temperature. Infrared (IR) thermography used to determine leaf and canopy temperature. It is an indirect estimate of plant water status (Garrity and O'Toole, 1994). IR thermometers have been successfully deployed in wheat breeding for both drought and heat screening (Brennan *et al.*, 2007). In maize crop, there is association between leaf temperature differences and relative biomass accumulation in response to drought at the seedling stage. Infrared camera used to take images at canopy level of maize under reproductive stage drought stress between anthesis and blister stage (Fig. 6) (Romano *et al.*, 2011). IR thermal imaging used as a potential tool accelerates phenotyping and screening in maize water stress (drought) breeding programs. Leaf or canopy temperature can be used directly as the absolute value or, more frequently as the difference from air temperature (canopy temperature depression, CTD), which takes into consideration daily changes in air temperature. Plant temperature also allows the calculation of the crop water stress index (CSWI), which gives an indication of the degree of stress in a crop (Romano *et al.*, 2011).

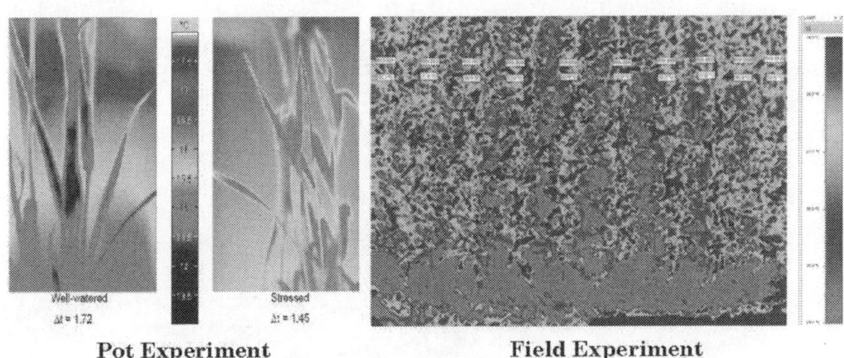

Fig. 6. Images reproduced from Golzarian *et al.* (2010) and Romanio *et al.* (2011) from Wheat and Maize drought phenotyping study.

Thermal images indicate respiratory function in plants under low water content. Thermal images of tropical and sub-tropical maize of 5 genotypes in 2-row plots expo sed to reproductive stage drought stress at CIMMYT experimental station (Tlaltizapan, Mexico). Emitted long-wave IR radiation is processed into a color digital image display of different temperatures in target area (Romano *et al.*, 2011).

Fig. 7. Thermal images are reproduced from A. Wheaton, B.R. Loveys and H.G. Jones (unpublished research)

A visible image of a grape vineyard taken from a balloon at a height of ~80 m. Mean canopy temperatures of three areas of crop to three irrigation treatments (T2, T3 & T4 with decreasing amounts of irrigation applied) are indicated (A. Wheaton, B.R. Loveys and H.G. Jones, unpubl. data). Thermal images acquired over the corn field at 0.4 m pixel resolution sowing the Tc-Ta changes at four different times of day. The greatest thermal variability between corn variety plots is obtained at midday, continuing during afternoon (Fig. 7.). Thermal seedlings treated with drought (3day old seedling). Thermal image of a wheat trial imaged al images of durum wheat from a cherry picker (Fig. 8.)

Fig. 8. Images reproduced from TRENDS in Plant Sciences (Furbank and Tester, 2011)

Stable isotopes

Stable isotope is a powerful selection trait may integrate in time (i.e., along the crop cycle) and level or organization (whole plant-canopy level) the behavior of the crop under stress. Carbon Isotopic Discrimination (CID) (^{13}C) indirect analysis (i.e., based in a physiological trait other that of yield, the primary or target trait) which has been successfully deployed in breeding program (Richards, 2006). Oxygen isotope enrichment (^{18}O) or oxygen isotope composition (δ18O) is a "time-integrative" indicator of transpiration and effective use of water in different plant species (Cabrera - Bosquet *et al.*, 2011a) including maize (Araus *et al.*, 2010). Effective use of water is related to genotypic capacity to use available water to sustain transpiration under unfavorable environmental conditions (Blum, 2009). Natural variation in plant N isotope composition (δ15N) is potentially useful for genotypic screening under drought might linked to N metabolism (Yousfi *et al.*, 2012). Large-scale phenotyping of genotypes, the marker assisted selection, isotopic signatures estimated in a fast and low-cost manner by near-infrared reflectance spectroscopy (NIRS) which was applied to select genotypes with better performance under drought (Cabrera-Bosquet *et al.*, 2011b).

Thermal infrared imaging to access transpiration rate

Thermal infrared imaging or infrared thermography (IRT) to measure leaf or canopy temperature, coloured images were used to identify sunlight and shaded leaves in a canopy, which determine the respective leaf temperature (Leinonen and Jones, 2004; Moller *et al.*, 2007). Thermal imaging or 'thermography' can be used to extract leaf or canopy temperature at the single plant level or plot level. Visible, NIR imaging used to know the growth and water status of plants. Hyperspectral reflectance measurements used for identification of waveband signatures (spectral indices) indicative of plant stress levels.

Visible and NIR imaging in assessing drought response of plants

Tran and Grishko (2004) used an NIR camera fitted with an acousto-optic tuneable filter to monitor changes in leaf water content of detached olive leaves over time. Increasing wavelengths, beyond 1300 nm, reflecta-nce decreases gradually due to an increased absorption by water present in the leaves with characteristic water absorption bands at 1450 nm, 1930 nm, and 2500 nm (Knipling, 1970): this is the basis for using NIR imaging to study water stress of plants. Leaf, canopy architecture (leaf thickness, growth), Greenness and biomass accumulation were measured by using NIR image (Harris *et al.*, 2007).

Imaging techniques relevant for drought research

- Colour imaging - Biomass, structure, phenology, Leaf health (chlorosis, necrosis)

- Near infrared imaging - Tissue water content, Soil water content
- Far infrared imaging - Canopy/leaf temperature
- Fluorescence imaging - Physiological state of photosynthetic machinery
- Automated weighing and watering - Water usage, imposing drought/salinity conditions

Traits should be Measured in High Throughput

Colour imaging

- Plant area, volume, mass, Biomass, structure, phenology
- Senescence, relative chlorophyll content
- Leaf health (chlorosis, necrosis) (Fig. 9)
- Seed yield, Agronomic traits

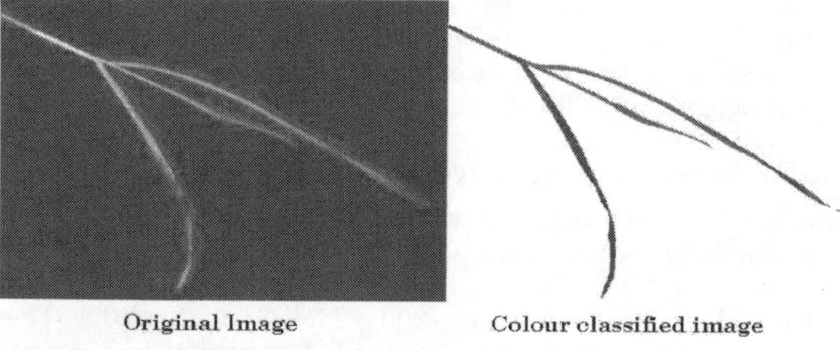

Fig. 9. Use of colour information, e.g., leaf rolling screen in barley (Source: Julie Hayes, Margie Pallotta and Tin Sutton, ACPEG, Australians Plant Phenomics Facility)

Near infrared imaging

- Tissue water content
- Soil water content

Far infrared imaging

- Canopy/leaf temperature/water use/salt tolerance

Fluorescence imaging

- Physiological state of photosynthetic machinery

Hyperspectral imaging

- Carbohydrates, pigments & proteins

FTIR imaging spectoscopy

- Cellular localization metabolites (sugar, protein 7 aromatics)

Carbon isotopic ratio

- Transpiration efficiency, phototsynthetic efficiency

Automated weighing and watering

- Water usage, imposing drought/salinity conditions

Summary of non-destructive imaging methods to study Physiological changes occurring under drought

Physiological changes occurring under drought	Non-destructive measurement method	Technical challenges	References
Decreased stomatal conductance	Thermal infrared	Influence of environmental conditions (air temperature, wind speed etc.); Orientation of leaves to light source and camera	Jones *et al.*, 2009
Decreasing leaf water content	Near infrared	Low sensitivity; Absolute measurements of possible at this stage, only monitoring of changes over time	Seelig *et al.*, 2009 Seelig *et al.*, 2008
Early onset of senescence	RGB / visible (or fluorescence if available if applicable)	RGB /visible	Rajendran *et al.*, 2009
Decreased growth rate and biomass accumulation			Rajendran *et al.*, 2009
Decrease in photosynthetic activity (only under severe stress conditions)	Fluorescence	Pulsed imaging setups currently limited to about 100 cm^2 and only optimized for planophyll plants	Baker, 2008; Woo *et al.*, 2008; Jansen *et al.*, 2009

Evaluation of different field devices for genotype study in response to drought (Kamoshita and Fukai, Unpublished observations)

Field devices for drought study	Cost	Strengths	Limitations	Climate and soils	References
1. Late planting with drainage in rainy season trial	Large uniform field management	High chance of reproductive terminal drought	Photoperiod and non-sensitive	Semi-arid tropics	Pantuwan *et al.*, 2002a

2. Dry season trial	Large uniform field management	High chance of drought, vegetative drought	Photoperiod non- sensitive, genotype-by season interaction	Semi-arid tropics	Pantuwan *et al.*, 2004
3. Line source sprinkler	Equipment, water source, monitoring	Different water regimes	Wind, space	Semi-arid to arid climate	Garrity and O'Toole, 1994
4. Rainout shelter	Construction	All types of drought	Space	Cost	Lilley and Fukai, 1994
5. Green-house	Construction	All types of drought	Space and rhiz-osphere differ-ences (small and loose)	Cost	Yadav *et al.*, 1997 Wade *et al.*, 2000
6. Root restriction	Rhizosphere manipulation	Evaluation of non-root traits	Space	Hardpan, simulated lowland	Kato *et al.*, 2007
7. Raised bed	Rhizosphere manipu-lation	Dry surface soil (interr-upt capillary water)	Space	Sub-humid climate	Kato *et al.*, 2007

Non-Contacting High Throughput Sensing of Drought Stressed Plants

Classical measurements for estimating water status in plants using oven drying or pressure chambers are tedious and time consuming. Simplified, rapid assessment of the plant water status or related properties such methods are not only useful for irrigation management purposes, but would also allow for the efficient screening of large populations of plants as part of a high-throughput system to precisely evaluate the phenotype for breeding purposes.

Non-contacting high throughput sensing methods:

1. Passive reflectance sensors
2. Active reflectance sensors
3. Laser-induced chlorophyll fluorescence sensing
4. Thermographic camera
5. Thermal near infrared sensor
6. GPS unit

Drought phenotyping on crop basis

The drought phenotyping experiments on crop growth and yield experiments were detailed in these following aspects. So, far there are many crops were phenotyped under drought. Few of the examples were groundnut, wheat, sorghum, common bean, chickpea, pigeonpea, cowpea banana, maize, sweet potato and rice.

Drought Phenotyping of Rice

One of the major limitations to the improvement of rice for drought-prone areas has been the lack of appropriate methods to impose drought routinely and reliably in order to select more tolerant lines. Many methods have been used to impose drought in order to have a better understanding of the mechanisms that lead to higher yields and the traits that can be selected for drought tolerance.

Putative traits for drought tolerance in rice

Leaf-rolling score

Leaf rolling score is an stress indicator used to reduce transpiration. Used during vegetative stress; high heritability (ca 0.8), but low/no association with yield (Courtois *et al.*, 2000).

Osmotic adjustment (OA)

To allow turgor maintenance at low plant water potential. *Indica* types have high OA, *japonica* types have low OA. This trait has been associated with a yield advantage in wheat, especially in terminal stress environments (Lilley *et al.*, 1996).

Deeper, thicker roots

The function of deeper and thicker roots was to explore a greater soil volume. There is evidence from MAS that increasing root mass below 30cm results in greater yield under stress. No evidence on root thickness per se. Large-scale screening is difficult. (Yadav *et al.*, 1997).

Root-pulling resistance

For root penetration into deeper soil layers which Is correlated with larger root system (Pantuwan *et al.*, 2002b).

Greater root penetration ability

Root penetration ability is to explore a larger soil volume. Most studies use artificial barriers with known mechanical resistance. There is some controversy regarding how well this mimics the soil situation (Ali *et al.*, 2000; Clark *et al.*, 2000).

Membrane stability

The membrane stability of plant is to allow leaves to continue functioning at high temperature. Genotypic differences are clear and has been duly linked to heat tolerance in several species. Link to drought tolerance is less evident (Tripathy *et al.*, 2000).

Leaf relative water content (RWC)

The leaf relative water content indicates the maintenance of favourable plant water status. Trait has rather low heritability; QTLs not repeatable. (Courtois *et al.*, 2000).

Water-use efficiency (WUE)

The water use efficiency indicates greater carbon gain per unit of water lost by transpiration. Carbon isotope discrimination (Δ13C) provides an integrated measure of WUE over the season. It has been used successfully for crops in more arid climates but has not been applied to rice (Specht *et al.*, 2001).

Grain yield under stress is the primary trait for selection in breeding programmes for drought-prone environments. However, it is sometimes useful to screen for secondary traits as well. These traits are plant characteristics that are associated with yield under stress, and they can provide additional information for scientists. Secondary traits expected to be of value in some drought-tolerance breeding programmes (Source: Lafitte *et al.*, 2003). The secondary traits are flowering/maturity date, flowering delay, per cent fertile spikelets, leaf rolling score, leaf death score and canopy temperature.

Drought Phenotyping of Wheat

Wheat faces different drought scenarios worldwide; consequently, the physiological traits that confer drought resistance in specific environments may be very distinct. The potential value of each trait needs to be considered with respect to the type of drought environment in which a cultivar is to be adapted. Secondary traits may be particularly suited to improving the selection response for stress conditions if they avoid any confounding effects of stress timing on yield (e.g., drought and flowering dates), and allow the selection to be focused on a specific type of drought.

Primary putative traits for drought tolerance

Grain yield under stress is the primary trait for selection in breeding programmes for drought-prone environments.

Traits related to pre-anthesis growth

Total biomass was calculated by spectroradiometer, which was to measure spectra of light reflected by canopy (Aparicio *et al.*, 2002; Royo *et al.*, 2003). Green biomass was calculated from digital imagery (Casadesús *et al.*, 2005).

Traits relating to access to water

Stomatal conductance (gs) and Canopy temperature depression (CTD) can give indications of water extraction patterns. Abscisic acid (ABA) accumulation in stomatal regulation under drought. Higher transpiration is a positive trait selecting for higher yield potential or better adaptation to moderate drought stress. Carbon isotope discrimination (Δ13C) estimation of Δ13C is done through the near-infrared spectroscopy (NIRS) technique (Clark *et al.*, 1995; Ferrio *et al.*, 2001).

Traits relating to WUE

Chlorophyll fluorescence used for screening for spike photosynthetic capacity under stress (P Horton, personal communication). Stay-green trait may indicate the presence of drought avoidance mechanisms. Senescence as a secondary trait is considered for performance under drought (Rharrabti *et al.*, 2001).

Traits relating to photoprotection

Effects of photo-inhibition are alleviated by antioxidants (SOD and POX) to increase in quantity in response to drought stress (Mittler and Zilinskas, 1994).

Thermal dissipation

The xanthophyll cycle is another protective mechanism which dissipate 75% of absorbed light energy (Niyogi, 1999)

(i) Rapid xanthophyll cycling

(ii) Up to 50% less leaf chlorophyll (a passive reduction of light absorbance) (Havaux and Tardy, 1999).

Epicuticular wax load (bloom) is covering of the aerial portions of plant with a thick, amorphous layer of wax. Epicuticular wax enhances the WUE by regulating night time water loss (Burow *et al.*, 2008).

Most drought physiology research in wheat has been conducted in controlled environments and has been poorly integrated into wheat breeding programmes. Secondary traits expected to be of value in some drought-tolerance wheat breeding programmes (Reynolds *et al.*, 2001b). The secondary traits were large seed size, long coleoptile, early ground cover, specific leaf dry weight, growth habit, tiller survival, earliness, number of grains per spike, vegetation indices, stomatal conductance, canopy temperature depression, CID, ash content, spike photosynthetic capacity, SPAD reading, leaf waxiness, leaf pubescence, leaf rolling, glume pubescence, leaf thickness, delayed senescence, fructanes content in stem, solute accumulation and ABA content.

Drought Phenotyping of Maize

Spectroradiometrical techniques used for evaluation of chlorophyll content and related traits (N content, green area), chlorophyll fluorescence, thermal imaging or digital imaging, abscisic acid (ABA), Carbohydrates and compatible solutes may play a role in osmotic adjustment (OA), maintaining turgor pressure in cells (in leaves as well as in reproductive organs) during water stress.

Secondary traits

1. Anthesis - silking interval (days)
2. Leaf rolling: 2,3 weeks after stop irrigation (Score 1-5) - before re-irrigation
3. Senescence: 3 weeks after stop irrigation (Score 1-10) - before re-irrigation (30 days after silking)
4. NDVI (Normalized Differential Vegetation Index): The green seeker: NDVI values can be used to predict green biomass.
5. Vegetative stage: weekly measurements are done from 3 leaves until 8 leaves stage.
6. Grain filling stage: weekly from two weeks after male flowering till one week prior maturity.
7. Stomatal conductance (mmol H_2O m^{-2} s^{-1}) is measured by leaf porometer
8. Leaf temperature (°C) is measured by Infrared thermometer
9. Vegetative stage: 4-5 weeks after planting and again 2 weeks later.
10. Reproductive stage: 2 days after stopped irrigation and then every weekly until re-irrigation. Measurements taken 3 plants/ plot. One leaf per plant at the 4th leaf below flag leaf during 10-15 h sunny days.
11. Chlorophyll content measured by using SPAD meter
12. Flowering stage: 1 time
13. Grain filling: every week during grain filling for 5 weeks.

Different devices to evaluate plant growth, phenology and water status

(a) Spectroradiometer with active sensor, which is used to measure normalized difference vegetation index (NDVI)

(b) Leaf chlorophyll meter

(c) Porometer used to measure the stomatal conductance (gs)

Transpiration may be broken down into two components. One is the leaf conductance (mostly determined by how open the stomata are, i.e., the stomatal conductance, gs) which really depends on the water status of the plant. The other is the evapotranspirative demand, which depends on environmental variables such as temperature, relative humidity and wind. Thus, gs may be used to screen for water status in maize (Sanguineti *et al.*, 1999), and the current generation of relatively low-cost and easy-to handle porometers such as the Decagon Leaf Porometer SC-1 or the Delta-T AP4 allow rapid (20–30s).

Canopy temperature

Depression of the canopy temperature reflects evaporative cooling of the leaf surface due to transpiration. Measurements are performed from a distance using infrared thermometers, which are inexpensive devices.

Oxygen isotope composition

The stable $_{13}C/_{12}C$ isotope composition (□$_{13}C$) measured in plant matter has been used to help breeding for drought adaptation in wheat and other small grain cereals. However, in maize, its C_4 metabolism prevents the use of □^{13}C as a tool for screening (Cabrera-Bosquet *et al.*, 2009b). The stable $_{18}O/_{16}O$ isotope composition of plant organic material ($_{18}O$) has been shown to reflect the isotope composition of soil water, evaporative $_{18}O$ enrichment in transpiring leaves, and isotopic exchange between oxygen atoms in organic molecules and local water in the cells in which the organic molecules are formed (Barbour, 2007). As plant material has been shown to record leaf evaporative conditions, measurement of $_{18}O$ enrichment of the plant matter compared with the source water may provide a powerful tool for plant breeders (Cabrera-Bosquet *et al.*, 2009a).

Other more novel phenotyping techniques based on chlorophyll fluorescence (Chaerle *et al.*, 2007), and thermal imaging (Chaerle *et al.*, 2007; Grant *et al.*, 2007) or digital imaging (Casadesus *et al.*, 2007), or even the use of spectroradiometers covering the region of 2-3 µm wavelength are promising, although still very expensive and at an early phase of their development.

Drought Phenotyping of Groundnut

Despite significant progress in understanding drought stress responses in plants, improving crop productivity under water-limited conditions has still remained an elusive goal.

Phenotyping for water-use efficiency (WUE)

The WUE can be determined at the whole plant level by gravimetry and at a single leaf level by gas exchange methods. More recent developments in the use of stable isotopes of carbon and oxygen as surrogates for WUE (Sheshshayee *et al.*, 2006a) and transpiration rate (Sheshshayee *et al.*, 2005) have provided a very accurate and high throughput phenotyping option for these traits.

Gravimetric approach for determining water use efficiency at the whole plant level

The accurate determination of the total water transpired by a plant over a specific period of crop growth and the total biomass the plant accumulated over the same period. At the whole plant level, water use and

WUE can be determined during a period from 35 to 65 days after sowing. This approach relies on the accurate determination of water content in the soil by weighing the containers at regular intervals, at least once every day. Initially, it is necessary to assess the weight of the container when empty (WE) and then after filling with either dry soil or a specific rooting mixture (WES). The amount of dry soil (WS) in each container would then be calculated from: WS = WES˜ WE. The pots with plants and the controls are weighed at least once daily and sufficient water added to bring the soil back to 100 percent FC (Udayakumar *et al.*, 1998a).

The water added daily over the entire experimental period is summed to arrive at the cumulative water added (CWAP) to the pots with plants. The total transpiration can be determined by subtracting the cumulative water added to the control containers (CWAc) from CWAP. Biomass and leaf area are recorded again for the plants in the remaining containers at the end of the experiment. Assuming linear growth during the experimental period, WUE and other physiological parameters are calculated as follows: WUE (g kg^{-1}) = (DMend - DMstart) / CWT where DMend and DMstart are the total dry matter (g pot^{-1}) measured at the end and start of the experiment, respectively.

Gas exchange approach for determining water use efficiency at the single leaf level

At the single leaf level, WUE is the ratio of carbon assimilation (A) to transpiration rate (T; Caemerrer and Farquhar, 1981; Farquhar and Sharkey, 1982) and is expressed as: A / T = gc (Ca - Ci) / gw (ei - ea) where gc and gw are diffusive conductance to CO_2 and water, respectively, Ca and Ci are the CO_2 concentration in ambient air and leaf intercellular spaces respectively, and ea and ei are the water vapour pressure in ambient air and leaf intercellular spaces respectively. A portable infrared gas analysers (IRGA) is routinely used to determine gas exchange traits such as carbon assimilation rate, stomatal conductance, transpiration rate and intercellular CO_2 concentration.

Stable isotope ratio - a powerful time-averaged option for estimating physiological traits

Isotope composition of carbon and oxygen provides very useful time-averaged information on several physiological traits such as WUE, transpiration rate, carboxylation efficiency and root traits. Further, determination of stable isotope ratios in a continuous flow mass spectrometer is high-throughput. Hence, stable isotope ratios provide a very powerful option to estimate physiological traits in large number of accessions.

Carbon isotopes

During photosynthesis, plants discriminate against the heavy isotope of carbon (13C), resulting in the depletion of $_{13}C$ content in biomass compared with atmospheric air (O'Leary, 1981). The extent of this depletion in the carbon isotope ratio ($_{13}C/_{12}C$), called 'carbon isotope discrimination' ($\Delta_{13}C$) has been shown to be related to the ratio of the partial pressures of CO_2 inside the leaf to that in ambient air (Pi/Pa), as follows (Farquhar *et al.*, 1989a): $\Delta 13C = a + (b - a)$ Pi/Pa where, a and b are isotope fractionations that occur during diffusion through stomata and carboxylation by Rubisco respectively (Farquhar *et al.*, 1982). Stomatal diffusion and carboxylation processes also regulate transpiration and photosynthesis, and hence WUE. With the ever increasing importance of breeding for improved productivity, especially under water-limited conditions, several experiments were initiated to select for low $\Delta_{13}C$ to improve WUE. However, improvement in productivity was not always noticed when selection was made for low $\Delta_{13}C$ (Condon *et al.*, 2004).

The use of stable oxygen isotopes has generated considerable interest in plant carbon and water relations in recent years. It has been fairly well documented that, during evaporation of water from lakes and oceans, water gets enriched with the heavy isotope of oxygen ($_{18}O$) because the water molecules containing the lighter isotopes diffuse relatively faster and have a higher vapour pressure than those with the heavy oxygen isotope (Craig and Gordon, 1965).

Measurement of root traits

Determination of genetic variability in root traits represents the most difficult challenge in crop improvement programmes. Several techniques have been developed ranging from hydroponics to growing plants in pipes (Venuprasad *et al.*, 2002). Such approaches have the inherent disadvantage that root growth is constrained by the space available. These disadvantages can largely be overcome by raising plants in specially constructed 'root structures'.

Alternative approaches to phenotyping for groundnut WUE

Although stable isotope ratios are accurate surrogates, progress in breeding for physiological traits can be hindered severely by the cost of a mass spectrometer and limited accessibility. Thus, simpler yet accurate alternatives are essential, at least as initial screening strategies. Such as, specific leaf area (SLA), specific leaf nitrogen (SLN), mineral ash content, SPAD chlorophyll meter reading (SCMR), intrinsic tolerance at the cellular level, membrane permeability based on leakage of solutes from leaf samples, oxidative damage and lipid peroxidation.

Traits associated with water mining from deeper soil profiles (root traits), efficiency of water use for biomass production (WUE), water conservation strategies associated with epicuticular waxes and regulation of stomatal movement, and superior intrinsic tolerance at the cellular level are a few of the most relevant traits that deserve exploitation.

Drought Phenotyping of Sorghum

Drought can occur at both pre-flowering and post-flowering stages of development, and has the most adverse effects on yield during and after anthesis (Kebede *et al.*, 2001). Recent studies by Prasad *et al.* (2008) have shown that heat stress occurring at the pre-flowering, flowering and post-flowering stages can also affect sorghum plants.

Canopy temperature measurements

Remotely-sensed infrared canopy temperature measurements provide an efficient method for rapid, non-destructive monitoring of the whole-plant response to water stress. With the advent of portable Infra Red Thermometers, the concept of using canopy temperature measurements to infer water stress and transpirational rates of genotypes has been advanced and received some interest as a drought-tolerance screening technique for research purposes.

Thermal infrared measurements of Tc have been used to screen for drought tolerance in many crops. Relationship between Tc and air temperature (Ta) differences (Tc – Ta = ΔT) and air VPD has been used by some researchers to evaluate drought response.

i. Identifying genotypes with relatively cooler temperatures under low soil moisture conditions
ii. Identifying relatively warmer genotypes under well-watered conditions

There are some limitations in using Tc measurements from drought-stressed environments. As soil water deficits increase, variability in soil texture and water holding capacity become an increasingly important source of variation that may mask genotypic differences. Another limitation is that remote Tc measurements require a dense canopy completely covering the soil surface (Jackson, 1982) in order to avoid errors resulting from viewing exposed soil surface.

Epicuticular wax load

Epicuticular wax load (bloom) is the covering of the aerial portions of the plant with a thick, amorphous layer of wax, as occurs in most sorghums. Burow *et al.* (2008) reported that correlation analysis showed significant negative associations between leaf epicuticular wax load with epidermal permeability and night-time conductance, which indicates the important role of epicuticular wax in these traits. These results suggest that epicuticular wax may enhance WUE in sorghum by regulating night time water loss.

Stay-green

Stay-green (non senescence or delayed senescence) under moisture stress is considered an important trait component for sustaining yield potential and, in some cases, also for sustaining yield under stress during grain filling (Borrell and Hammer, 2000; Sanchez *et al.*, 2002). Additionally, stay-green is an important factor in sustaining a positive nitrogen balance in sorghum (Borrell and Hammer, 2000).

On the other hand, there is an obvious advantage to stay-green in plants that are devoid of any effective capacity for stem-reserve utilization for grain filling under stress, such as in the normal 3-dwarf sorghum plant that is in wide use (Blum *et al.*, 1997).

DROUGHT PHENOTYPING OF COMMON BEAN

Many drought adaptation traits, such as phenology, root size and depth, hydraulic conductivity and storage of reserves are associated with plant development and structure, and are constitutive rather than stress induced (Chaves *et al.*, 2003).

Shoot traits

1. At mid-podfilling: dry weights of leaf biomass, stem biomass, pods plus reproductive structure biomass, total shoot biomass and leaf to stem ratio of dry weight.
2. At harvest: dry weights of stem biomass, pod biomass and seed biomass, number of pods per plant, dry weight of pod wall biomass and proportion of pod wall biomass to pod biomass, seed number per pod, 100 seed dry weight, seed number per area and pod number per area.
3. Seed yield: the two central rows of each plot are used to determine seed yield stem biomass reduction (%): (stem biomass dry weight at mid-podfilling - stem biomass dry weight at harvest) / stem biomass dry weight at mid-podfilling x 100
4. Grain filling index (GFI): the GFI for each genotype can be estimated from 100 seed dry weight under rainfed conditions / 100 seed dry weight under irrigated conditions x 100.
5. Yield production efficiency (g g^{-1}): seed biomass dry weight at harvest / total shoot biomass dry weight at mid-podfilling (adapted from Board and Maricherla, 2008).
6. Drought intensity index (DII): the DII for each growing season can be calculated as DII = 1 - Xds/Xns, where Xds and Xns are the mean of all genotypes under drought stress and no stress treatments, respectively.
7. Drought susceptibility index (DSI) for seed yield: the DSI for each genotype is calculated as follows: DSI = (1 - Yds/Yns) / DII, where Yds and Yns are mean yields of a given genotype in drought stress and no stress environments, respectively (Fisher and Maurer, 1978).

The nondestructive measurements that are related to physiological processes such as photosynthetic efficiency, total chlorophyll content (Soil–Plant Analyses Development or SPAD measurement), stomatal conductance, transpiration rate, leaf temperature (in both the morning and afternoon) and leaf water potential. The destructive measurements that are related to growth and metabolism include LAI, canopy dry weight per plant (leaf, stem and pod biomass), shoot nutrient (nitrogen and phosphorus) uptake, shoot and seed ash content, and shoot and seed total non-structural carbohydrates (TNC). Seed nitrogen, phosphorus, ash content and TNC can be measured at the time of harvest.

Target root traits in the field

Root traits associated with drought tolerance can be measured either in the field or in the greenhouse and these include root depth and root architectural traits. Rooting depth and root distribution under field conditions can be quantified using soil cores taken at different soil depths followed by root washing, scanning and weighing as described below for greenhouse root phenotyping. Deep rooting has been positively correlated with seed yield, crop growth, cooler canopy temperature and soil water extraction in common beans (Sponchiado *et al.*, 1989).

A number of shoot physiological characteristics are measured in a soil tube screening system assay. These include photosynthetic efficiency, total chlorophyll content (SPAD), stomatal conductance and transpiration rate, leaf temperature (both in the morning and afternoon), and leaf water potential. At the time of harvest (ca 35–45 days after planting and 3 weeks of drought stress), leaf area, shoot biomass distribution (leaf, stem, pod and root biomass), leaf TNC content, and root characteristics are determined. The soil tube is sliced into 5 layers (0-5, 5-10, 10-20, 20-40 and 40-75cm). Roots in each soil layer are washed free of soil and length, diameter, specific root length, and dry weight are determined. Root length and diameter are measured with an image analysis system (WinRHIZO, Regent Instruments Inc). Rao *et al.* (2006b) used the above soil tube screening system to evaluate the impact of drought on different genotypes of root growth and root distribution.

DROUGHT PHENOTYPING OF COWPEA

The developmental and morphological traits associated with these uses also influence the adaptation of the cowpea cultivars to drought. With respect to drought adaptation, in addition to careful consideration of optimal phenology and morphology, it should be ensured that the available vegetative-stage drought resistance is present. Increasing water-use efficiency (WUE), i.e, biomass production / crop water use, has been proposed as a mechanism to enhance the adaptation of crops to drought. By measuring the stable carbon isotope composition of cowpea tissues, genotypic differences in transpiration efficiency (TE) i.e., photosynthetic rate / transpiration rate, that were associated with genotypic differences in WUE have been detected (Hall *et al.*, 1994a; Ismail and Hall, 1992). Adaptation to drought may be enhanced by breeding cowpeas with deeper rooting for a target population of environments (TPE) where the crop is grown under limited rainfall with substantial water available deep in the soil. A method has been developed for selecting stable lines of cowpeas for differences in rate of rooting (Robertson *et al.*, 1985).

The seedling screen involving a shallow soil layer in a box (Singh and Matsui, 2002) can be used to screen parents and, if necessary, segregating generations. Singh and Matsui (2002) and Muchero *et al.* (2008) have described some variation in this trait but it is not known whether it is worth pursuing, since some cowpea cultivars already have very strong resistance to vegetative stage drought. The robustness of cowpeas under hot dry soil conditions during the seedling and vegetative stages could be enhanced considerably by incorporating resistance to ashy stem blight that has recently been discovered (JD Ehlers, unpublished observations).

DROUGHT PHENOTYPING OF BANANA

Drought tolerance in bananas is the ability to survive under water scarcity during various stages of crop growth, without significant yield reductions. Plant growth is the real measurement because the whole plant is involved in the comparison among treatments. Besides plant height, pseudostem girth, phyllochron and leaf emergence rate, the following growth parameters should be measured. Leaf area (L) can be measured according to Summerville (1944), as $L = \Sigma\ 0.83\ lb$, where l refers to length and b to the width (measured at the widest portion) of the jth leaf, j being the order of the leaf.

Growth analysis

Dry matter content

Data on dry matter content can be recorded only at the end of the experiment.

Relative growth rate

The relative growth rate (RGR) is normally analysed by destructive sampling. The above mentioned constraints also apply to this parameter (Turner and Lahav, 1983).

RGR is calculated as follows: RGR of $L = (\ln L_2 - \ln L_1) / (t_2 - t_1)$

RGR of plant dry matter $= (\ln W_2 - \ln W_1) / (t_2 - t_1)$

where, L_1, W_1, L_2 and W_2 are total leaf area and total plant dry weights at two time intervals, t_1 and t_2 respectively.

Leaf geometry

The influence of water stress treatment on leaf geometry can be assessed at the end of the experiment for controls and treatments by measuring the phyllotaxy, the angles between the midribs and horizontal plane, leaf folding and 'choking'.

Relative leaf folding

Relative leaf folding (Ft) is calculated from: Ft = 100 (1 - c / b), where c is the apparent leaf width from the outer edge of one lamina half to the other, and b is the leaf width when the laminae are horizontal (no folding). The Ft represents a proportional reduction in the exposed leaf area to a vertical beam of natural light.

Choking

Choking is the shortening of the distance between the petioles of alternate leaves and is a feature commonly associated with cool weather (Cann, 1964) and drought (Kuhne and Green, 1970). Choking can be quantitatively expressed using the ratio of H/n, where H is the pseudostem height and n is the number of leaves produced. Plants are choked where there is a low H/n ratio over control.

Leaf emergence rate

The leaf emergence rate (LER) is a useful index of the vegetative development rate of a banana plant and is closely related to temperature. The leaves emerged during the experimental period are noted in both control and treated plants.

Plant water status

Relative water content

The relative water content (RWC) is the ratio of water present in the leaf disc at the time of sampling to that present in the excised disc after it has been fully dehydrated.

Leaf water retention capacity

Assessment of the rate of water loss from excised leaves or plants has shown some promise for differentiating drought resistance of wheat cultivars (Bayles *et al.*, 1937; Clarke and McCaig, 1982; Dedio, 1975; Salim *et al.*, 1969; Sandhu and Laude, 1958). The difference in leaf water retention capacity (LWRC) may be due to the tightness of stomatal closure (Kirkham *et al.*, 1980) or to other causes such as cuticular resistance to water loss (Clarke and McCaig, 1982).

Plant function

Leaf gas exchange

Leaf gas exchange or the rate of extrusion of the leaf is a more sensitive method for determining the response of banana plants to water deficit. This parameter can be measured along with other photosynthetic parameters in any portable photosynthesis measuring system, e.g., an infra-red gas analyser (IRGA). The third youngest leaf is used, with a minimum of three measurements for each leaf.

Chlorophyll content

The chlorophyll content is measured from the third youngest leaf lamina, sampling from both sides of the middle portion of the leaf from three plants, using three replications for each treatment. Leaf discs of 40-50mg in weight can be extracted in 10 mL of dimethyl sulphoxide (DMSO) in a glass test tube covered with aluminium foil and kept in an oven at 65°C for 4h (Hiscox and Israelstam, 1979). The optical density (OD) of the extract is read in a spectrophotometer at 645 nm and 663 nm. Chlorophyll a (Chl a) and chlorophyll b (Chl b) concentrations (in µg ml^{-1}) are calculated using the formula given by Arnon (1949):

$$\text{Chl } a = 12.7 \text{ OD663} - 2.69 \text{ OD645}$$

$$\text{Chl } b = 22.9 \text{ OD645} - 4.68 \text{ OD663}$$

Chlorophyll stability index

The chlorophyll stability index (CSI) is an indication of the stress tolerance capacity of a plant. A high CSI value means that the stress did not have much effect on the chlorophyll content of the plant and the chlorophyll available for photosynthesis, on increased dry matter production and on higher productivity. CSI is measured by following the method of Koleyoreas (1958).

Drought susceptibility index

The drought susceptibility index (DSI) is measured as: DSI = [1˜ (Ys / Yw)] /DII where DII = (1˜ Xd / Xp), and Xd and Xp are the mean experiment yield of all genotypes grown under drought stress and well watered regimes, respectively. In addition to yield and yield parameters, the traits measured in the preliminary evaluation also have to be taken into consideration for correlating with economic yield.

Yield and yield parameters

Any stress effect ultimately has to be evaluated in terms of economic yield. Therefore, the water stress imposed at the third and fifth month after planting and at flowering for a period of four weeks has to be evaluated in terms of yield. Yield components such as the number of hands, the number of fingers, and the rate of finger growth (in terms of length and circumference) must be compared in stressed and unstressed plants. At harvest, the individual finger weight difference can be calculated for the stressed and unstressed plants as: geometric mean yield (GM) = (Ys x Yw) / 2 where, Ys = genotypic performance under stress and Yw = genotypic performance under well-watered conditions.

Great potential exists for improving drought resistance in crop plants. Exploiting this potential will be enhanced by more systematic application of physiological and genomic tools and continued genetic and mechanistic analysis of a range of diverse germplasm both from within the species and from close relatives. At present, the most important traits appear to be those associated with rooting depth and photosynthate remobilisation, but other traits may emerge in the future.

Conclusions and Future Perspectives

Long before the genomic era, techniques for assessing plant phenotypes were needed as new cultivars and/or varieties were being developed through conventional breeding approaches. Today, however, tools specific to the high-throughput screening of plant phenotypes offers to promote understanding of gene function and will soon become a valuable component of the plant genomic toolbox. Although the scientific community currently relies heavily on visual evaluation of mutants, imaging techniques that allow immediate and non-invasive detection of altered plant characteristics, before visual symptoms appear must be developed for monitoring changes in phenotypic characteristics of transgenic plants expressing those genes. The utilization of techniques/approaches that allow for a precise control of the water regime (e.g., irrigated trials in dry regions, rainout shelters, etc.) and a reduction of the experimental noise (e.g., precision application of nutrients and water, remote sensing for simultaneous phenotyping, etc.) coupled with the adoption of high-throughput platforms will streamline the collection of good phenotypic data while increasing the cost-effectiveness of phenotyping for drought.

References

Ali ML, Pathan MS, Zhang J, Bai G, Sarkarung S and Nguyen HT 2000. Mapping QTLs for root traits in a recombinant inbred population from two indica ecotypes of rice. *Theoretical and Applied Genetics*. 101:756–766.

Aparicio N, Villegas D, Araus JL, Blanco R and Royo C 2002. Seedling development and biomass as affected by seed size and morphology in durum wheat. *Journal of Agricultural Science* 139:143–150.

Araus JL, Amaro T, Zuhair Y and Nachit MM 1997. Effect of leaf structure and water status on carbon isotope discrimination in field-grown durum wheat. *Plant, Cell & Environment* 20:1484–1494.

Araus JL, Cabrera-Bosquet LL and S′anchez C 2010 Is heterosis in maize mediated through better water use? *New Phytol*. 187, 392– 406.

Araus JL, Slafer GA, Reynolds MP and Royo C (2002). Plant breeding and drought in C3 cereals: What should we breed for? *Annals of Botany* 89:925–940.

Arnon DL (1949). Copper enzyme in isolated chloroplasts: polyphenol oxidase in Beta vulgaris. *Plant Physiology* 24:1–5.

Bacon MA, Davies WJ, Mingo D and Wilkinson S (2002). Root signals. In: Plant roots: The hidden half, Third edition (Waisel Y, Eshel A and Kafkafi U, eds). Marcel Dekker, Inc, Monticello, New York, USA, pp 461–470.

Baker NR. 2008. Chlorophyll fluorescence: a probe of photosynthesis in vivo. *Annual Review of Plant Biology* 59, 89–113.

Balota M, Payne WA, Evett SR and Lazar MD (2007). Canopy temperature depression sampling to assess grain yield and genotypic differentiation in winter wheat. *Crop Science* 47:1518–1529.

Barbour MM (2007). Stable oxygen isotope composition of plant tissue: a review. *Functional Plant Biology* 34:83–94.

Barbour MM, Fischer RA, Sayre KD and Farquhar GD (2000). Oxygen isotope ratio of leaf and grain material correlates with stomatal conductance and grain yield in irrigated wheat. *Australian Journal of Plant Physiology* 27:625–637.

Bayles BB, Taylor JW and Bartel AT (1937) Rate of water loss in wheat varieties and resistance to artifi cial drought. *Journal of the American Society for Agronomy* 29: 40–52.

Bengough AG, Gordon DC, Al-Menaie H, Ellis RP, Allan D, Keith R, Thomas WTB and Forster BP (2004). Gel observation chamber for rapid screening of root traits in cereal seedlings. *Plant and Soil* 262:63–70.

Blum A (1988). Breeding for stress environments, CRC Press, Boca Raton, USA, 245 pp.

Blum A (1996). Crop responses to drought and the interpretation of adaptation. *Plant Growth Regulation* 20:135–148.

Blum A (2006). Drought adaptation in cereal crops: a prologue. In: Drought adaptation in cereals (Ribaut J-M, ed). The Haworth Press Inc, Binghamton NY, USA, pp 3–15.

Blum A (2009) Effective use of water (EUW) and not water-use efficiency (WUE) is the target of crop yield improvement under drought stress. *Field Crop Res.* 112, 119–123.

Board JE and Maricherla D (2008). Explanations for decreased harvest index with increased yield in soybean. *Crop Science* 48:1995-2002.

Borrell AK and Hammer GL (2000). Nitrogen dynamics and the physiological basis of stay-green in sorghum. *Crop Science* 40:1295–1307.

Blum A, Golan G, Mayer J and Sinmena B (1997). The effect of dwarfi ng genes on sorghum grain fi lling from remobilized stem reserves under stress. *Field Crops Research* 52:43–54.

Boyer JS, James RA, Munns R, Condon TAG and Passioura JB (2008). Osmotic adjustment leads to anomalously low estimates of relative water content in wheat and barley. *Functional Plant Biology* 35:1172–1182.

Bray EA (2002). Abscisic acid regulation of gene expression during water-deficit stress in the era of the Arabidopsis genome. *Plant, Cell & Environment* 25:153–161.

Brennan JP, Condon AG, Van Ginkel M and Reynolds MP (2007). An economic assessment of the use of physiological selection for stomatal aperture-related traits in the CIMMYT wheat breeding programme. *Journal of Agricultural Science* 145:187–194.

Burow GB, Franks CD and Xin Z (2008). Genetic and physiological analysis of an irradiated bloomless mutant (epicuticular wax mutant) of sorghum. *Crop Science* 48:41–48.

Cabrera-Bosquet L, Albrizio R, Nogu´es S, Araus JL (2011a) Dual ?13C/δ18O response to water and nitrogen availability and its rela- tionship with yield in field-grown durum wheat. *Plant Cell Environ.* 34, 418–433.

Cabrera-Bosquet L, Molero G, Nogués S and Araus JL (2009a). Water and nitrogen conditions affect the relationships of Δ13C and Δ18O with gas exchange and growth in durum wheat. *Journal Experimental Botany* 60:1633–1644.

Cabrera-Bosquet L, S´anchez C, Rosales A, Palacios-Rojas N, Araus JL (2011b) NIRS-assessment of δ18O, nitrogen and ash content for improved yield potential and drought adaptation in maize. *J. Agric. Food Chem.* 59, 467–474.

Caemerrer S and Farquhar GD (1981). Some relations between the biochemistry of photosynthetic and gas exchange of leaves. *Planta* 53:376–387.

Cann HJ (1964). How cold weather affects banana growing in New South Wales. *Agricultural Gazette of New South Wales* 75:1012–1019.

Casadesús J, Biel C and Savé R (2005). Turf color measurement with conventional digital cameras. In: EFITA/WCCA Joint Congress in Agriculture (Boaventura Cunha J and Morais R, eds). Universidade de Trás- os-Montes e Alto Douro, Vila Real, Portugal, pp 804–811.

Casadesús J, Kaya Y, Bort J, Nachit MM, Araus JL, Amor S, Ferrazzano G, Maalouf F, Maccaferri M, Martos V, Ouabbou H and Villegas D (2007). Using vegetation indices derived from conventional digital cameras as selection criteria for wheat breeding in water-limited environments. *Annals of Applied* Biology 150:1–10.

Chaerle L, Leinonen I, Jones HG and Van der Straeten D (2007). Monitoring and screening plant populations with combined thermal and chlorophyll fluorescence imaging. *Journal of Experimental Botany* 58:773–784.

Chaves MM, Maroco JP and Pereira JS (2003). Understanding plant responses to drought – from genes to the whole plant. *Functional Plant Biology* 30:239–264.

Clark DH, Johnson DA, Kephart KD and Jackson NA (1995). Near infrared refl ectance spectroscopy estimation of 13C discrimination in forages. *Journal of Range Management* 48:132–136.

Clark LJ, Aphale SL and Barraclough PB (2000). Screening the ability of rice roots to overcome the mechanical impedance of wax layers: importance of test conditions and measurement criteria. *Plant and Soil* 219:187–196.

Clarke JM and McCaig TN (1982). Evaluation techniques for screening for drought resistance in wheat. *Crop Science* 22:503–506.

Condon AG, Farquar GD and Richards RA (1990). Genotypic variation in carbon isotope discrimination and transpiration efficiency in wheat. Leaf gas exchange and whole plant studies. *Australian Journal of Plant Physiology* 17:9–22.

Condon AG, Farquar GD, Rebetzke GJ and Richards RA (2006). The application of carbon isotope discrimination in cereal improvement for water-limited environments. In: Drought adaptation in cereals (Ribaut J M, ed). The Haworth Press, Inc, Binghamton, New York, USA, pp 171–219.

Condon AG, Richards RA, Rebetzke GJ and Farquhar GD (2002). Improving intrinsic water-use efficiency and crop yield. *Crop Science* 42:122–131.

Condon AG, Richards RA, Rebetzke GJ and Farquhar GD (2004). Breeding for high water-use effi ciency. *Journal of Experimental Botany* 55:2447–2460.

Courtois B, McLaren G, Sinha PK, Prasad K, Yadav R and Shen L (2000). Mapping QTLs associated with drought avoidance in upland rice. *Molecular Breeding* 6:55–66.

Craig L and Gordon LI (1965) Deuterium and oxygen-18 variations in the ocean and the marine atmosphere. In: Proceedings of a Conference on Stable Isotopes in Oceanographic Studies and Paleotemperatures (Tongiorgi E, ed). Spoleto, Italy, pp 9–130.

Davies WJ, Bacon MA, Thompson DS, Sobeih W and Rodriguez LG (2000). Regulation of leaf and fruit growth in plants growing in drying soil: exploitation of the plants' chemical signalling system and hydraulic architecture to increase the efficiency of water use in agriculture. *Journal of Experimental Botany* 51:1617–1626.

Dedio W (1975). Water relations in wheat leaves as screening tests for drought resistance. *Canadian Journal of Plant Sciences* 55: 369–378.

Ducrocq S, Madur D, Veyrieras JB, Camus- Kulandaivelu L, Kloiber-Maitz M, Presterl T, Ouzunova M, Manicacci D and Charcosset A (2008). Key impact of Vgt1 on flowering time adaptation in maize: Evidence from association mapping and ecogeographical information. *Genetics* 178:2433–2437.

Farquhar GD and Sharkey TD (1982). Stomatal conductance and photosynthesis. *Annual Review of Plant Physiology* 33:317–345.

Farquhar GD, Ehleringer JR and Hubick KT (1989a). Carbon isotope discrimination and photosynthesis. *Annual Review of Plant Physiology and Plant Molecular Biology* 40:503–537.

Farquhar GD, O'Leary MH and Berry JA (1982). On the relationship between carbon isotope discrimination and the intercellular CO_2 concentration in leaves. *Australian Journal of Plant Physiology* 9:121–131.

Ferrio JP, Bertran E, Nachit MM, Royo C and Araus JL (2001). Near infrared refl ectance spectroscopy as a new surrogate analysis for ?13C in mature kernels of durum wheat. *Australian Journal of Agricultural Research* 52:809–816.

Ferrio JP, Mateo MA, Bort J, Abdalla O, Voltas J and Araus JL (2007). Relationships of grain D13C and D18O with wheat phenology and yield under water-limited conditions. *Annals of Applied Biology* 150:207–215.

Fiorani F, Schurr U (2012) Future Scenarios for Plant Phenotyping. *Annual Review of Plant Biology* **64,** 1-17.

Fischer RA and Maurer R (1978). Drought resistance in spring wheat cultivars. I. Grain yield responses. *Australian Journal of Agricultural Research* 29:897–912.

Fukai S and Cooper M (1995). Development of drought-resistant cultivars using physio- morphological traits in rice. *Field Crops Research* 40:67–86.

Garrity DP and O'Toole JC (1994) Screening rice for drought resistance at the reproductive phase. *Field Crops Research* 39:99–110.

Giuliani S, Sanguineti MC, Tuberosa R, Bellotti M, Salvi S and Landi P (2005b). Root-ABA1, a major constitutive QTL, affects maize root architecture and leaf ABA concentration at different water regimes. *Journal of Experimental Botany* 56:3061–3070.

Grant OM, Tronina L, Jones HG and Chaves MM (2007). Exploring thermal imaging variables for the detection of stress responses in grapevine under different irrigation regimes. Journal of Experimental Botany 58:815–825.

Hall AE, Richards RA, Condon AG, Wright GC and Farquhar GD (1994a). Carbon isotope discrimination and plant breeding. Plant Breeding Reviews, Volume 12, Chapter 4, John Wiley and Sons, New York, USA, pp 81–113.

Hargreaves CE, Gregory PJ and Bengough AG (2009). Measuring root traits in barley (*Hordeum vulgare* ssp *vulgare* and ssp *spontaneum*) seedlings using gel chambers, soil sacs and X-ray microtomography. Plant and Soil 316:285–297.

Harris K, Subudhi PK, Borrell A, Jordan D, Rosenow D, Nguyen H, Klein P, Klein R and Mullet J (2007). Sorghum stay-green QTL individually reduce post-flowering drought-induced leaf senescence. Journal of Experimental Botany 58:327–338.

Havaux M and Tardy F (1999). Loss of chlorophyll with limited reduction of photosynthesis as an adaptive response of Syrian barley landraces to high-light and heat stress. Australian Journal of Plant Physiology 26:569–578.

Hiscox JD and Israelstam GF (1979). A method for the extraction of chlorophyll from leaf tissue without maceration. Canadian Journal of Botany 57:1332–1334.

Houle D, Govindaraju DR and Omholt S (2010) Phenomics: the next challenge. *Nature Reviews Genetics* **11,** 855-866.

Ismail AM and Hall AE (1992). Correlation between water-use efficiency and carbon isotope discrimination in diverse cowpea genotypes and isogenic lines. Crop Science 32:7-12.

Jackson RD (1982). Canopy temperature and crop water stress. Advanced Irrigation 1:43–85.

Jansen M, Gilmer F, Biskup B, Nagel KA, *et al.* 2009. Simultaneous phenotyping of leaf growth and chlorophyll fluorescence via GROWSCREEN FLUORO allows detection of stress tolerance in Arabidopsis thaliana and other rosette plants. Functional Plant Biology 36, 902–914.

Jiang GH, He YQ, Xu CG, Li XH and Zhang Q (2004). The genetic basis of stay-green in rice analyzed in a population of doubled haploid lines derived from an indica by japonica cross. Theoretical and Applied Genetics 108:688–698.

Jones HG, Archer N, Rotenberg E and Casa R (2003). Radiation measurement for plant ecophysiology. Journal of Experimental Botany 54:879–889.

Jones HG, Serraj R, Loveys BR, Xiong LZ, Wheaton A, Price AH. 2009. Thermal infrared imaging of crop canopies for the remote diagnosis and quantification of plant responses to water stress in the field. Functional Plant Biology 36, 978–989.

Kato Y, Kamoshita A and Yamagishi J (2007). Evaluating the resistance of six rice cultivars to drought: root restriction and the use of raised beds. Plant and Soil 300:149–161.

Kebede H, Subudhi PK, Rosenow DT and Nguyen HT (2001). Quantitative trait loci infl uencing drought tolerance in sorghum (*Sorghum bicolor* L Moench). Theoretical and Applied Genetics 103:266–276.

Kirkham MB, Smith EL, Dhanasobhon C and Drake TI (1980). Resistance to water loss of winter wheat flag leaves. Cereal Research. Communication 8:393–399.

Knipling EB. 1970. Physical and physiological basis for the reflectance of visible and near IR radiation from vegetation. Remote Sensing of Environment 1, 155–159.

Koleyoreas SA (1958). A new method for determining drought resistance. Plant Physiology 33:232–233.

Kuhne FA and Green GC (1970). Bananas choked by drought. Farm South Africa No 410: (5)1–22.

Lafitte HR, Blum A and Atlin G (2003). Using secondary traits to help identify drought tolerant genotypes. In: Breeding rice for drought-prone environments (Fischer KS, Lafitte R, Fukai S, Atlin G and Hardy B, eds). IRRI, Los Baños, The Philippines, pp 38–39

Lafitte HR and Edmeades GO (1995). Stress tolerance in tropical maize is linked to constitutive changes in ear growth characteristics. Crop Science 35:820–826.

Leinonen I, Jones HG. 2004. Combining thermal and visible imagery for estimating canopy temperature and identifying plant stress. Journal of Experimental Botany 55, 1423–1431.

Lilley JM and Fukai S (1994). Effect of timing and severity of water deficit on four diverse rice cultivars: 1. Rooting pattern and soil water extraction. Field Crops Research 37:205–213.

Lilley JM, Ludlow MM, McCouch SR and O'Toole JC (1996). Locating QTL for osmotic adjustment and dehydration tolerance in rice. Journal of Experimental Botany 47:1427–1436.

Lorens GF, Bennett JM and Loggale LB (1987). Differences in drought resistance between two corn hybrids. I. Water relations and root length density. Agronomy Journal 79:802–807.

Ludlow MM and Muchow RC (1990). A critical evaluation of traits for improving crop yields in water-limited environments. Advances in Agronomy 43:107–153.

Maggio A, Hasegawa PM, Bressan RA, Consiglio MF and Joly RJ (2001). Unravelling the functional relationship between root anatomy and stress tolerance. Australian Journal of Plant Physiology 28:999–1004.

Massonnet C, Vile D, Fabre J, Hannah MA, Caldana C, Lisec J, Beemster GTS, Meyer RC, Messerli G, Gronlund JT, Perkovic J, Wigmore E, May S, Bevan MW, Meyer C, Rubio-Di¡az S, Weigel D, Micol JL, Buchanan-Wollaston V, Fiorani F, Walsh S, Rinn B, Gruissem W, Hilson P, Hennig L, Willmitzer L, Granier C (2010) Probing the Reproducibility of Leaf Growth and Molecular Phenotypes: A Comparison of Three Arabidopsis Accessions Cultivated in Ten Laboratories. *Plant Physiology* **152,** 2142-2157.

Melchinger AE, Utz HF and Schon GC (2004). QTL analyses of complex traits with cross validation, bootstrapping and other biometric methods. Euphytica 137:1–11

Mittler R and Zilinskas BA (1994). Regulation of pea cytosolic ascorbate peroxidase and other antioxidant enzymes during the progression of drought stress and following recovery from drought. Plant Journal 5:397–405.

Moller M, Alchanatis V, Cohen Y, Meron M, Tsipris J, Naor A, Ostrovsky V, Sprintsin M, Cohen S. 2007. Use of thermal and visible imagery for estimating crop water status of irrigated grapevine. Journal of Experimental Botany 58, 827–838.

Monneveux P and Ribaut J M (2006). Secondary traits for drought tolerance improvement in cereals. In: Drought adaptation in cereals (Ribaut J M, ed). The Haworth Press Inc, Binghamton, New York, USA, pp 97–143.

Morgan JW (1984). Osmoregulation and water stress in higher plants. Annual Review of Plant Physiology 35:299–319.

Muchero W, Ehlers JD and Roberts PA (2008). Seedling stage drought-induced phenotypes and drought-responsive genes in diverse cowpea genotypes. Crop Science 48:541–552.

Niyogi KK (1999). Photoprotection revisited: genetic and molecular approaches. Annual Review of Plant Physiology and Plant Molecular Biology 50:333–359.

Norton GJ and Price AH (2009). Mapping of quantitative trait loci for seminal root morphology and gravitropic response in rice. Euphytica 166:229–237.

O'Leary MH (1981). Carbon isotope fractionation in plants. Phytochemistry 20:53–567.

Olivares-Villegas JJ, Reynolds MP and McDonald GK (2007). Drought-adaptive attributes in the Seri/Babax hexaploid wheat population. Functional Plant Biology 34:189–203.

Pantuwan G, Fukai S, Cooper M, Rajatasereekul S and O'Toole JC (2002a). Yield response of rice (*Oryza sativa* L) genotypes to different types of drought under rainfed lowlands. I. Grain yield and yield components. Field Crops Research 73:153–168.

Pantuwan G, Fukai S, Cooper M, Rajatasereekul S and O'Toole JC (2002b). Yield response of rice (*Oryza sativa* L) genotypes to different types of drought under rainfed lowlands. III. Plant factors contributing to drought resistance. Field Crops Research 73:181–200.

Pantuwan G, Fukai S, Cooper M, Rajatasereekul S, O'Toole JC and Basnayake J (2004). Yield response of rice (*Oryza sativa* L) genotypes to drought under rainfed lowlands. IV. Vegetative stage screening in the dry season. Field Crops Research 89:281–297.

Passioura JB (1983). Roots and drought resistance. Agricultural Water Management 7:265-280.

Pieruschka R, Poorter H (2012) Phenotyping plants: genes, phenes and machines. *Functional Plant Biology* **39,** 813-820.

Prasad PVV, Pisipati SR, Mutava RN and Tuinstra MR (2008). Sensitivity of grain sorghum to high temperature stress during reproductive development. Crop Science 48:1911–1917.

Quarrie SA (1991). Implications of genetic differences in ABA accumulation for crop production. In: Abscisic acid: physiology and biochemistry (Davies WJ and Jones HG, eds). Bios Scientific Publishers, Oxford, UK, pp 227–243.

Quarrie SA, Whitford PN, Appleford NEJ, Wang TL, Cook SK and Henson IE (1988). A monoclonal antibody to (S)-abscisic acid: its characterisation and use in a radioimmunoassay for measuring abscisic acid in crude extracts of cereals and lupin leaves. Planta 173:330–339.

Rajendran K, Tester M, Roy SJ. 2009. Quantifying the three main components of salinity tolerance in cereals. Plant, Cell and Environment 32, 237–249.

Rao IM, Polania J, Garcia R and Beebe S (2006b). Development of a greenhouse soil tube method to quantify phenotypic differences among advanced lines in root development and distribution under drought stress. Annual Report 2006. Project IP-1: Bean improvement for the tropics. CIAT, Cali, Colombia, pp 19–25.

Rebetzke GJ, Condon AG, Richards RA and Farquhar GD (2002). Selection for reduced carbon isotope discrimination increases aerial biomass and grain yield of rainfed bread wheat. Crop Science 42:739–745.

Rebetzke GJ, Ellis MH, Bonnett DG and Richards RA (2007). Molecular mapping of genes for coleoptile growth in bread wheat (*Triticum aestivum* L). Theoretical and Applied Genetics 114:1173–1183.

Rebetzke GJ, van Herwaarden AF, Jenkins C, Weiss M, Lewis D, Ruuska S, Tabe L, Fettell NA and Richards RA (2008b). Quantitative trait loci for water-soluble carbohydrates and associations with agronomic traits in wheat. Australian Journal of Agricultural Research 59:891–905.

Ren HB, Wei KF, Jia WS, Davies WJ and Zhang JH (2007). Modulation of root signals in relation to stomatal sensitivity to root-sourced abscisic acid in drought-affected plants. Journal of Integrative Plant Biology 49:1410–1420.

Reynolds M and Tuberosa R (2008). Translational research impacting on crop productivity in drought-prone environments. Current Opinion in Plant Biology 11:171–179.

Reynolds M, Dreccer F and Trethowan R (2007). Drought-adaptive traits derived from wheat wild relatives and landraces. Journal of Experimental Botany 58:177–186.

Reynolds MP, Trethowan R, van Ginkel M and Rajaram S (2001b). Application of physiology in wheat breeding. In: Application of physiology in wheat breeding (Reynolds MP, Ortiz-Monasterio JI and McNab A, eds). CIMMYT, Mexico DF, Mexico, pp 2–10.

Rharrabti Y, Villegas D, García del Moral LF, Aparicio N, El Hani S and Royo C (2001). Environmental and genetic determination of protein content and grain yield in durum wheat under Mediterranean conditions. Plant Breeding 120:381–388.

Richards R, Rebetzke G, Condon A and van Herwaarden A (2002). Breeding opportunities for increasing the efficiency of water use and crop yield in temperate cereals. Crop Science 42:111–121.

Richards RA (1991). Crop improvement for temperate Australia: future opportunities. Field Crops Research 96:141–169.

Richards RA (2006). Physiological traits used in the breeding of new cultivars for water-scarce environments. Agricultural Water Management 80:197–211.

Richards RA, Watt M and Rebetzke GJ (2007). Physiological traits and cereal germplasm for sustainable agricultural systems. Euphytica 154: 409–425.

Robertson BM, Hall AE and Foster KW (1985). A fi eld technique for screening for genotypic differences in root growth. Crop Science 25: 1084–1090.

Romano G, Zia S, Spreer W, Sanchez C, Cairns J, Araus JL, M¨uller J (2011) Use of thermography for screening genotypic water stress adaptation in tropical maize. Comp. Elect. Agric. 79, 61–74.

Royo C, Aparicio N, Villegas D, Casadesús J, Monneveux P and Araus JL (2003). Usefulness of spectral reflectance indices as durum wheat yield predictors under contrasting Mediterranean conditions. International Journal of Remote Sensing 24:4403–4419.

Royo C, Martos V, Ramdani A, Villegas D, Rharrabti Y and del Moral LFG (2008). Changes in yield and carbon isotope discrimination of Italian and Spanish durum wheat during the 20th century. Agronomy Journal 100:352–360.

Sadras VO, Reynolds MP, de la Vega AJ, Petrie PR and Robinson R (2009). Phenotypic plasticity of yield and phenology in wheat, sunflower and grapevine. Field Crops Research 110:242–250.

Salim HH, Todd GW and Stutte CA (1969) Evaluation of techniques for measuring drought avoidance in cereal seedlings. Agronomy Journal 61:182–185.

Salvi S, Sponza G, Morgante M, Tomes D, Niu X, Fengler KA, Meeley R, Ananiev EV, Svitashev S, Bruggemann E, Li B, Hainey CF, Radovic S, Zaina G, Rafalski JA, Tingey SV, Miao GH, Phillips RL and Tuberosa R (2007). Conserved noncoding genomic sequences associated with a flowering-time quantitative trait locus in maize. Proceedings of the National Academy of Sciences of the United States of America 104:11376–11381.

Sanchez AC, Subudhi PK, Rosenow DT and Nguyen HT. (2002) Mapping QTLs associated with drought resistance in sorghum (*Sorghum bicolor* L Moench). Plant Molecular Biology 48: 713–726.

Sandhu AS and Laude HH (1958) Tests of drought and heat hardiness of winter wheat. Agronomy Journal 50:78–81.

Sanguineti MC, Li S, Maccaferri M, Corneti S, Rotondo F, Chiari T and Tuberosa R (2007). Genetic dissection of seminal root architecture in elite durum wheat germplasm. Annals of Applied Biology 151:291–305.

Sanguineti MC, Tuberosa R, Landi P, Salvi S, Maccaferri M, Casarini E and Conti S (1999). QTL analysis of drought related traits and grain yield in relation to genetic variation for leaf abscisic acid concentration in fi eld-grown maize. Journal of Experimental Botany 50:1289–1297.

Seelig HD, Hoehn A, Stodieck LS, Klaus DM, Adams WW, Emery WJ. 2009. Plant water parameters and the remote sensing R (1300)/R (1450) leaf water index: controlled condition dynamics during the development of water deficit stress. Irrigation Science 27, 357–365.

Seki M, Umezawa T, Urano K and Shinozaki K (2007). Regulatory metabolic networks in drought stress responses. Current Opinion in Plant Biology 10:296–302.

Serraj R and Sinclair TR (2002). Osmolyte accumulation: can it really help increase crop yield under drought conditions? Plant, Cell & Environment 25:333–341.

Setter TL (2006). The role of abscisic acid under water-limited conditions. In: Drought adaptation in cereals (Ribaut J M, ed). The Haworth Press, Inc, Binghamton, New York, USA, pp 505–530.

Sheshshayee MS, Bindumadhava H, Rachaputi NR, Prasad TG, Udayakumar M, Wright GC and Nigam SN (2006a). Leaf chlorophyll concentration relates to transpiration effi ciency in peanut. Annals of Applied Biology 148:7–15.

Sheshshayee MS, Bindumadhava H, Ramesh R, Prasad TG, Lakshminarayana M and Udayakumar M (2005). Oxygen isotope enrichment (?18O) as a measure of time-averaged transpiration rate. Journal of Experimental Botany 56(422):3033– 3039.

Singh BB and Matsui T (2002). Cowpea varieties for drought tolerance. In: Challenges and opportunities for enhancing sustainable cowpea production (Fatokun CA, Tarawali SA, Singh BB, Kormawa PM and Tamò M, eds). Proceedings of the World Cowpea Conference III, IITA, Ibadan, Nigeria, 4–8 September 2000. IITA, Ibadan, Nigeria, pp 287–300.

Singh RP, Prasad PVV, Sunita K, Giri SN and Reddy KR (2007). Influence of high temperature and breeding for heat tolerance in cotton: A review. Advances in Agronomy 93:313–386.

Siopongco J, Sekiya K, Yamauchi A, Egdane J, Ismail AM and Wade LJ (2008). Stomatal responses in rainfed lowland rice to partial soil drying; Evidence for root signals. Plant Production Science 11:28–41.

Siopongco J, Sekiya K, Yamauchi A, Egdane J, Ismail AM and Wade LJ (2009). Stomatal responses in rainfed lowland rice to partial soil drying; comparison of two lines. Plant Production Science 12:17–28.

Sirichandra C, Wasilewska A, Vlad F, Valon C and Leung J (2009). The guard cell as a single-cell model towards understanding drought tolerance and abscisic acid action. Journal of Experimental Botany 60:1439–1463.

Slafer GA, Araus JL, Royo C and Garcia Del Moral LF (2005). Promising eco-physiological traits for genetic improvement of cereal yields in Mediterranean environments. Annals of Applied Biology 146:61–70.

Specht JE, Chase K, Macrander M, Graef GL, Chung J, Markwell JP, Germann M, Orf JH and Lark KG (2001). Soybean response to water: A QTL analysis of drought tolerance. Crop Science 41:493–509.

Sponchiado BN, White JW, Castillo JA and Jones PG (1989). Root growth of four common bean cultivars in relation to drought tolerance in environments with contrasting soil types. Experimental Agriculture 25:249–257.

Summerville WAT (1944). Studies on nutrition as qualified by development in *Musa cavendishii* Lamb. Journal of Agricultural Meterology 12:361–389.

Tambussi EA, Bort J and Araus JL (2007). Water use efficiency in C3 cereals under Mediterranean conditions: a review of physiological aspects. Annals of Applied Biology 150:307–321.

Thompson AJ, Andrews J, Mulholland BJ, McKee JMT, Hilton HW, Black CR and Taylor IB (2007). Over production of abscisic acid in tomato increases transpiration efficiency and root hydraulic conductivity and influences leaf expansion. Plant Physiology 143:1905-1917.

Tran CD, Grishko VI. 2004. Determination of water contents in leaves by a near-infrared multispectral imaging technique. Microchemical Journal 76, 91–94.

Tripathy JN, Zhang JX, Robin S, Nguyen TT and Nguyen HT (2000). Mapping quantitative trait loci for cell membrane stability in rice. Theoretical and Applied Genetics 100:1197–1202.

Tuberosa R, Salvi S, Sanguineti MC, Landi P, Maccaferri M and Conti S (2002a). Mapping QTLs regulating morpho-physiological traits and yield: Case studies, shortcomings and perspectives in drought-stressed maize. Annals of Botany 89:941–963.

Tuberosa R, Sanguineti MC, Landi P, Salvi S, Casarini E and Conti S (1998). RFLP mapping of quantitative trait loci controlling abscisic acid concentr-ation in leaves of drought- stressed maize (*Zea mays* L). Theoretical and Applied Genetics 97:744–755.

Tuberosa, R. (n.d.). I.1 Phenotyping drought-stressed crops: key concepts, issues and approaches, 1–36.

Turner DW and Lahav E (1983). The growth of banana plants in relation to temperature. Australian Journal of Plant Physiology 10:43 –53.

Turner NC (1997). Further progress in crop water relations. Advances in Agronomy 528:293–338.

Turner NC, Palta JA, Shrestha R, Ludwig C, Siddique KHM and Turner DW (2007). Carbon isotope discrimination is not correlated with transpiration efficiency in three cool-season grain legumes (Pulses). Journal of Integrative Plant Biology 49:1478– 1483.

Udayakumar M, Rao RCN, Wright GC, Ramaswamy GC, Ashok, Roy S, Gangadhar GC and Aftab Hussain IS (1998a). Measurement of transpiration effi ciency in fi eld condition. Journal of Plant Biology 1:69–75.

Venuprasad R, Shashidhar HE, Hittalmani S and Hemamalini GS (2002). Tagging quantitative trait loci associated with grain yield and root morphological traits in rice (*Oryza sativa* L) under contrasting moisture regimes. Euphytica 128:293–300.

Wade LJ, Kamoshita A, Yamauchi A and Azhiri-Sigari T (2000). Genotypic variation in response of rainfed lowland rice to drought and rewatering. I. Growth and water use. Plant Production Science 3:173–179.

Wasilewska A, Vlad F, Sirichandra C, Redko Y, Jammes F, Valon C, Frey NFD and Leung J (2008). An update on abscisic acid signaling in plants and more. Molecular Plant 1:198–217.

Woo NS, Badger MR, Pogson BJ. 2008. A rapid, non-invasive procedure for quantitative assessment of drought survival using chlorophyll fluorescence. Plant Methods 4, 14.

Xiong YC, Li FM, Zhang T and Xia C (2007). Evolution mechanism of non-hydraulic root-to-shoot signal during the anti-drought genetic breeding of spring wheat. Environmental and Experimental Botany 59:193–205.

Xu X, Yuan H, Li S and Monneveux P (2007). Relationship between carbon isotope discrimination and grain yield in spring wheat under different water regimes and under saline conditions in the Ningxia Province (North-west China). Journal of Agronomy and Crop Science 193:422–434.

Yadav R, Courtois B, Huang N and McLaren G (1997). Mapping genes controlling root morphology and root distribution in a doubled-haploid population of rice. Theoretical and Applied Genetics 94:619–632.

Yousfi S, Serret MD, M´arquez AJ, Voltas J, Araus JL (2012) Combined use of δ13C, δ18O and δ15N tracks nitrogen metabolism and genotypic adaptation of durum wheat to salinity and water deficit. New Phytol. (In press).

Yu GR, Zhuang J, Nakayama K and Jin Y (2007). Root water uptake and profile soil water as affected by vertical root distribution. Plant Ecology 189:15–30.

Zhang J and Davies WJ (1990). Does ABA in the xylem control the rate of leaf growth in soil dried maize and sunflower plants? Journal of Experimental Botany 41:1125–1132.

4

MORPHO-PHYSIOLOGICAL AND MOLECULAR BASIS OF DROUGHT RESPONSE IN PLANTS

Bhawna Pant[1], Sananda Mondal[2] and Bandana Bose[1*]

[1]*Department of Plant Physiology, Institute of Agricultural Sciences, Banaras Hindu University, Varanasi-221005*
[2]*Department of ASEPAN, Institute of Agriculture, Visva-Bharati, Sriniketan-731236* [1*]*Corresponding author: E-mail bbosebhu@gmail.com*

INTRODUCTION

Drought is a word simply meaning dry condition and the dryness occurs only in the absence of water. Drought in an area comes slowly with an extended rainless period. But the whole phenomenon of drought for a plant depends on the availability of water or it depends on the interplay of water and environment for specific crops and their growing seasons. For maize and wheat crop the effect of drought will differ due to their difference in growing season. The question arises, but why? This depends on the availability of water to that particular crop. Therefore, to enter in the plant's response to drought we must focus upon water a little. **Water** is a compound with the chemical formula H_2O. A water molecule contains one oxygen and two hydrogen atoms connected by covalent bonds. Water covers 71% of the Earth's surface (CIA- The world fact book), and is vital for all known forms of life (United Nations). However, water on earth moves recurrently through the water cycle of evaporation and transpiration (evapotranspiration), condensation, precipitation, and runoff, usually reaching the sea. Evaporation and transpiration contribute to the precipitation over land. It plays an important role in the world economy, as it functions as a solvent for a wide variety of chemical substances and facilitates industrial cooling and transportation. Approximately 70% of the fresh water used by humans goes to agriculture (Baroni *et al.*, 2007). The most important use of water in agriculture is irrigation, which is a key component to produce enough food. Irrigation takes up to 90% of water withdrawn in some developing countries ("WBCSD Water Facts and Trends) and significant proportions in more economically developed countries (United States, 30% of freshwater usage is for irrigation) (Water Use in the United States). It is found that a fifth of the world's population, more than 1.2 billion, lives in areas of physical water scarcity, where there is not enough water to meet all demands. Further, 1.6 billion people lives in areas experiencing economic water scarcity, where the lack of investment in water or insufficient human capacity makes it impossible for authorities to satisfy the demand for water. The report found that it would be possible to produce the food required in future, but that continuation of today's food production and environmental trends would lead to crises in many parts of the world. To avoid a global water crisis, farmers will have to strive to increase productivity to meet growing demands for food, while industry and cities find ways to use water more efficiently (Chartres and Varma, 2010).

On the earth life evolved in the medium of water. Water is an excellent solvent and has two other useful properties: the fact that ice floats enables aquatic organisms to survive beneath it in winter; and its molecules have electrically negative and positive ends, which enables it to form a wider range of compounds than other solvents can. Water has a pervading quality and ubiquitous and cannot be considered as simple diluents and performes lot of functions, viz., transports, structures' stabilisation, lubricant, reactants and partitions. Water also play an appreciable role in maintaining the structure, stability, dynamic, and function of biomolecules. Various plant processes such as photolysis of water in photosynthesis to hydrophobic bonding of macromolecules and to the maintenance of form in non woody tissue is balanced by water molecules (Hsiao, 1973). Whereas, water in plants can also maintain cell turgidity for structure and growth; transporting nutrients and organic compounds throughout the plant; comprising much of the living protoplasm in the cells; serving as a raw material for various chemical processes, including photosynthesis and through transpiration, buffering the plant against wide temperature fluctuations.

Among all the recognized abiotic stresses, water stress is most harmful stress in agriculture. It affects plant at different levels of their organization over space and time. Now a day various plant physiologists, agronomists, breeders and biotechnologists adapted many advanced management practices; breeding approaches (like Marker Assisted Selection, reverse breeding and double haploid) and biotechnological tools (Microarray, Transgenic, Gene Knockdown Techniques etc) to overcome the water stress and to improve resource use efficiency (nutrient and water) of plants.

Water stress condition occurs when the requirement of water exceeds than available amount during a definite phase or when its use is restricted due to poor quality/ quantity. The quantity (over exploitation, dry-river, etc) and quality (eutrophication, organic matter contamination, saline interruption etc.) of fresh water resources also decline which leads to water stress. Plants suffer due to water stress in two ways viz. (i) **Drought**: which is primarily caused by the water scarcity and (ii) **Waterlogging**: water exists in soil solution more than adequate amount but plants cannot uptake it – a situation commonly known as 'physiological drought' (Lisar *et al.*, 2012). With these points kept in mind the present review explores about the work related with drought (causing water scarcity) in respect to various types of modulations in plant's responses starting from morpho- physiological to molecular level.

Nevertheless, drought is a meteorological term and is usually defined as a phase lacking significant rainfall. Drought is a deficiency in precipitation over an extended period, usually a season or more, resulting in a water shortage causing adverse impacts on vegetation, animals, and/ or people. It is a normal, recurrent feature of climate that occurs in virtually all climatic zones, from very wet to very dry. Drought is a temporary aberration from normal climatic conditions, thus it can vary significantly from one region to another. Drought is different than aridity, which is a permanent feature of climate in regions where low precipitation is the norm, as in a desert (National Oceanic and Atmospheric Administration, 2003). It can have a substantial impact on the ecosystem and agriculture of the affected region. Although drought can persist for several years, even a short, intense drought can cause significant damage and harm to the local economy (Australian Drought and Climate Change). Drought stress tolerance is seen in almost all plants but its level varies from species to species and still within species. Dehydration of a living cell has great physical and chemical effects on cellular and macromolecular structures and usually causes irreversible changes, which finally lead to cell death. Hence, in terms of critical water level, two types of tolerance for living organisms are distinguished. "Drought tolerance" can be described as tolerance to moderate dehydration, with no bulk cytoplasmic water, which is present with 23% water on a fresh weight basis or 0.3 g H_2O/g dry weight. "Desiccation tolerance" usually refers to tolerance to further dehydration, when the hydration shell is gradually lost (Hoekstra *et al.*, 2001). It seems that plants cannot get away but acclimatize to the environment. Survival of plants in such stressful environment for so long is itself an indication that plants must have evolved with some mechanism such that they can sensitively distinguish the incoming stress and quickly regulates their physiologies to cope with it.

As the drought persists, the conditions surrounding it gradually worsen and its impact on the local population gradually increases. People tend to define droughts in three main ways (NOAA):

1. **Meteorological drought** is brought about when there is a prolonged period with less than average precipitation. Meteorological drought usually precedes the other kinds of drought.
2. **Agricultural droughts** are droughts that affect crop production or the ecology of the range. This condition can also arise independently from any change in precipitation levels when soil conditions and erosion are triggered by poorly planned agricultural endeavors causing a shortfall in water available to the crops. However, a traditional drought can be caused by an extended period of below average precipitation.
3. **Hydrological drought** is brought about when the water reserves available in sources such as aquifers, lakes and reservoirs fall below the statistical average. Hydrological drought tends to show up more slowly because it involves stored water that is used but not replenished. Like an agricultural drought, this can be triggered by more than just a loss of rainfall.

This article reviews important aspects of drought with special emphases on drought resistance mechanism of plants; physico-chemical and molecular mechanism of drought followed by drought induced proteins. In the end, a brief conclusive part elucidates the future prospects too.

DROUGHT RESISTANCE MECHANISM OF PLANTS

Water limited productivity of plants depends on the total amount of water available and on the water use efficiency of the plant. A plant that is capable of acquiring more water or that has higher water use efficiency will resist drought better. Some plants possess adaptations, such as the C_4 and CAM modes of metabolism, which allow them to exploit more arid environment. Indeterminate type of growth habit helps the plants to acclimatize in drought such as sorghum is more resistant to drought than corn because sorghum readily develops secondary shoots that can continue to grow and flower after the main shoot is fully mature. However, corn does not form secondary shoots, so if fruiting fails because of drought, the corn plant has no option to recoup the losses if rain occurs later (Eck, 1986).

An earliest response to stress is the biophysical events rather than chemical reaction due to dehydration. When the water content of the plant decreases, the cell shrinks and cell wall relaxed, the solutes in the cells become more concentrated. The plasma membrane becomes thicker and more compressed, as it covers a smaller area than before, because turgor-dependent activities such as leaf expansion and root elongation are the most sensitive to water deficit. So, reduction in leaf area is a first line of defense against drought. Plants mainly show a shallow root system when all soil layers are wetted and a loss of shallow root and proliferation of deep roots, resulted as water in top layer of the soil is depleted. Deeper root growth into wet soil can be considered a second line of defense against drought. The onset of stress is more rapid when the plant has reached to its full leaf area before initiation of stress, other responses protect the plant against immediate desiccation. Under these conditions closure of stomata can be considered a third line of defense against drought (Blum, 1996; De Leonardis, 2012).

Effect of Drought on Various Morpho-physiological Aspects

The response of plant for water scarcity is very complex involving changes and deleterious effects. Under field condition these responses can be synergistically or antagonistically adapted by the superimposition of other stress. Plant strategies to manage with drought normally involve mixture of stress avoidance and tolerance strategies that vary with genotype (Chaves *et al.*, 2002).

Photosynthetic response to drought stress is extremely complex. It involves the interplay of restrictions taking place at different sites of the cell/leaf and at different time scales in relation to plant development. The strength, period and rate of progression of the stress all influence plant responses to water scarcity and salinity, because these factors together dictate whether mitigation processes of stress in the system are to be coupled with acclimation or not. Acclimation responses under drought, which indirectly affect photosynthesis, include those related to growth inhibition or leaf shedding by restricting water expenses by source tissues, and help to maintain plant water status and therefore, plant carbon assimilation. Osmotic compounds that are synthesized in response to a slowly imposed dehydration also has a function in sustaining tissue metabolic activity. Acclimation responses to salinity also include synthesis of compatible solutes as well as adjustments in ion transport (such as uptake, extrusion and sequestration of ions). These responses eventually lead to restoration of cellular homeostasis, detoxification and therefore, results in survival of the system under stress (Chaves *et al.*, 2009).

Drought, as an abiotic stress which is multidimensional in environment, affects plants at various levels of their organization. In fact, under prolonged drought, many plants will dehydrate and die. Water stress in plants reduces the plant-cell's water potential and turgor, which elevate the solutes' concentrations in the cytosol and extracellular matrices. As a result, cell enlargement decreases leading to growth inhibition and reproductive failure. Drought not only affects plant water relations through the reduction of water content, turgor and total water, it also affects stomatal closure, limits gaseous exchange, reduces transpiration and stop carbon assimilation (photosynthesis) rates. Change in plants hydraulic conductivity has been evoked as playing a major role in short term stomatal regulation in woody plants (Saliendra *et al.,* 1995). The interaction between root chemical signaling and changes in plant hydraulic conductivity during drought is difficult to understand and need more attention (Jackson *et al.*, 2000). Negative effects on mineral nutrition (uptake and transport of nutrients) and metabolism leads to a decline in the leaf area and modification in assimilate partitioning among the organs. Change in plant cell wall elasticity and interruption of homeostasis and ion distribution in the cell has also been reported (Wu *et al.*, 1996).

One essential mechanism for reducing the shock of drought is early stomatal closure at the beginning of a period of water deficit. Stomatal closure reduces water loss, but also reduces the gas exchange between the plant and the ambient air. The reduced CO_2 intake then results in reduced photosynthesis (Chaves *et al.*, 2002). Nevertheless high yield requires high stomatal conductance to sustain a great CO_2 fixation. Some leaf traits, such as stomatal number/density and leaf mesophyll structure can be important in increasing the water use efficiency. In particular, studies carried out in wheat suggest that the high-yielding modern varieties are "opportunistic", that is they have high rates of stomatal conductance with optimal soil moisture, but markedly reduce stomatal conductance when soil moisture is limiting (Siddique *et al.*, 1990; Rizza *et al.*, 2011). Maximal rates of photosynthesis were also positively correlated with increased yields of advanced varieties, while leaf temperatures were negatively correlated (Fischer *et al.*, 1998). Leaf permeability is another crucial trait, as leaves can lose water through cuticle, increasing crop transpiration without an associated benefit in CO_2 fixation. Glaucousness, which is caused by the presence of epicuticular wax, can prevent these losses (Kerstiens, 2006). At cellular level, osmotic adjustment is an adaptive mechanism in which the accumulation of solutes helps to maintain a favorable gradient of water potential in the soil-plant-air system. It allow maintaining sufficient water absorption from a moderately dry soil for supporting photosynthetic and transpiration activity, and cell expansion for root growth (Mastrangelo *et al.*, 2011; Dichio *et al.*, 2006). Regarding the importance of this trait in improving grain yield in water stressed environment, a positive correlation between osmotic adjustment and yield increases has been found in particular in conditions of severe water stress (Serraj and Sinclair, 2002). Besides above-ground traits of plant, deep rooted cultivars have established a clear yield improvement under water stress conditions. An increased root development in presence of water stress represents a corresponding strategy to stomatal closure regulation. The influence of root structural design on yield and other agronomic traits, particularly under stress conditions, has been commonly reported in all major crops (Tuberosa *et al.*, 2002; de Dorlodot *et al.*, 2007). A deep and extended root system should allow exploring a greater soil volume and extracting more water.

Rice is a disreputably drought-susceptible crop due in part to its small root system, rapid stomatal closure and little circular wax during mild water stress (Hirasawa, 1999). Reduction of photosynthetic activity, accumulation of organic acids and osmolytes, and changes in carbohydrate metabolism, are typical physiological and biochemical responses to drought stress (Tabaeizadeh, 1998). Water deficit also increases the formation of reactive oxygen species (ROS) resulting in lipid peroxidation, protein denaturation and nucleic acid damage with severe consequences on overall metabolism (Hansen *et al.*, 2006). Drought tolerance is a cost-intensive phenomenon, as a considerable quantity of energy is spent to cope with it. The fraction of carbohydrate that is lost through respiration determines the overall metabolic efficiency of the plant (Davidson *et al.*, 2000). The root is a major consumer of carbon fixed in photosynthesis and uses it for growth and maintenance, as well as dry matter production (Lambers *et al.*, 1996). Plant growth and developmental processes as well as environmental conditions affect the size of this fraction (i.e., utilized in respiration). However, the rate of photosynthesis often limits plant growth when soil water availability is reduced (Huang and Fu, 2000). A negative carbon balance can occur as a result of diminished photosynthetic capacity during drought, unless simultaneous and proportionate reductions in growth and carbon consumption take place. The alternative pathway is not coupled with adenosine triphosphate synthesis, but can be induced in response to stress or inhibition of the main electron transfer pathway (Wagner and Moore, 1997). When plants are exposed to drought stress, they produce reactive oxygen species, which damage membrane components (Blokhina *et al.*, 2003). In this regard, alternative oxidase activity could be useful in maintaining normal levels of metabolites and reduce reactive oxygen species production during stress.

Role of Abscisic Acid (ABA) during Water Stress Conditions

Higher plants are sessile organisms that have evolved a high plasticity for adaptation to environmental challenges. The key role of abscisic acid (ABA), jasmonic acid (JA) and salicylic acid (SA) as primary signals in the regulation of plant defense has been well established (Bari and Jones 2009; Pieterse *et al.*, 2009). It appears that plants have evolved with two lines of responses to soil drying. Initial soil drying may be sensed by part of the root system and a root sourced ABA in the xylem may regulate the stomatal conductance such that water loss in the shoots may be reduced and leaf water deficit can be avoided (Zhang and Davies 1989a). When the soil drying is prolonged and getting more severe, the shoot water deficit becomes unavoidable and some older leaves may get wilted first because of their weak hydraulic association with the main stem. Such leaf wilting should report for the much accelerated ABA concentration in the xylem and much more severe stomatal inhibition that can be seen in the young leaves (Zhang and Davies 1989b).

These hormones generate a signal transduction network that leads to a cascade of events responsible for the physiological adaptation of the plant to stress. It should be noted that the degree of drought tolerance varies with developmental stages in most plant species (El-Far and Allan 1995; Reddy *et al.*, 2004; Rassaa *et al.*, 2008). Phytohormones are essential for the ability of plants to adapt to drought stress by mediating a wide range of adaptive responses. They often rapidly alter gene expression by inducing or preventing the degradation of transcriptional regulators via the ubiquitin–proteosome system. The phytohormone ABA serves as an endogenous messenger in biotic and abiotic stress responses of plants (Kuromori *et al.*, 2010).

ABA is defined as a stress hormone because of its rapid accumulation in response to stresses and its mediation of many stress responses that help plant survival over the stresses. A limiting water supply leads to an immediate hydraulic signal in plants that triggers ABA biosynthesis over long distances, whereas high humidity activates an ABA catabolizing P450 enzyme within minutes of perceiving the stress condition. Upon perception of a stress signal ABA formation is induced primarily in vascular tissues and ABA is exported from the site of biosynthesis and uptake is stimulated into other cells by specific ATP-dependent transporters. The mechanism allows the rapid distribution of ABA into neighboring tissues. Drought and high salinity result in strong increases of plant ABA levels, accompanied by a major change in gene expression and in adaptive physiological responses. One of the most considered topics in the response of plants to abiotic stress, particularly water stress, is ABA signaling and ABA-responsive genes. ABA synthesis is one of the fastest responses of plants to abiotic stress, triggering ABA-inducible gene expression and causing stomatal closure, thereby reducing water loss via transpiration and eventually restricting cellular growth. Numerous genes associated with ABA *de novo* biosynthesis and genes encoding ABA receptors and downstream signal relays have been characterized in *Arabidopsis thaliana.* Cytokinin (CK) is an antagonist to ABA, and the exposure of plants to water limiting conditions results in decreased levels of CK. Elevated CK levels promoted survival under water-stress conditions, inhibited leaf senescence and induced increased proline level (Tran *et al.*, 2007). Manipulation of endogenous CK levels was effective in delaying senescence. The IPT gene has been over-expressed in several plant species under different promoters and the transgenic plants were tested for tolerance to various environmental stresses.

PHYSICO-CHEMICAL AND MOLECULAR MECHANISM OF DROUGHT

Plant's adaptation to words drought involves many complex biological processes. It includes reduced growth, transcriptional activation or inactivation of specific genes, transient increases in the level of ABA, accumulation of compatible solutes and activation of scavenging enzymes, change in the level of ROS, activation in catabolism etc. Hence drought is associated with a number of modulations in plant's physico-chemical as well as molecular mechanisms, consequently to improve the adaptive commitments of the suffering system. Firstly onset of drought limits photosynthesis and cell growth, due to closure of stomata. The restricted photo-assimilation causes a reduction in leaf area but prolonged stress initiates the generation of stress relieving metabolites (Zhang *et al.* 1999) and then gene expression and protein synthesis (Lawlor and Tezara, 2009). Aranjuelo *et al* (2011) determined the physiological, metabolic and proteomic processes involved in photosynthetic inhibition in relation to decrease in nitrogenase activity during the deleterious effect of drought on alfalfa. Drought suffered plants invest a large quantity of C and N resources into the synthesis of osmoregulators like proline and pinitol; lower Rubisco activase content lowers the activation state of Rubisco protein activity. Proteomic characterization also showed the down regulation of Calvin cycle proteins involved in C regeneration. The down regulation of Rubisco protein content resulted in a decrease in nitrogenase activity in alfalfa nodule. The recognition and practical study of stress responsive genes explains the molecular mechanisms of the plant stress response and tolerance, and eventually leads to development of stress tolerance in crops. Abiotic stresses such as drought and high salinity lead to wide range of biochemical, physiological and morphological modulations in plants in the process of adaptation in respect to these unfavorable conditions. These adaptations require a large number of changes in gene expression. Many of the differentially expressed gene products protect plant cells from damage, such as dehydrins, enzymes for the synthesis of osmolytes and enzymes for the scavenging of reactive oxygen species (ROS). The production of these functional proteins is widely regulated by specific transcription factors. Transcription factors present the first level of regulation of mRNA metabolism, which controls the synthesis of pre mRNA. However, drought is one of the main factors limiting crop production. Research on the biology and genetics of drought resistance and breeding of drought resistant crops have, thus, become important field of current research in molecular biology and molecular propagation. During the regulation of drought stress responses, ABA generation occurs via osmotic stress, the latter is also regulated by various genes viz. ZEP, NCED, ABA2, LOSS/ABA3 and AAO responsible for the biosynthesis of ABA (Xiong *et al.*, 2001). Evidence also suggests during stress the catabolism of ABA is inhibited hence a pool of ABA is formed in system. Signal transduction pathway during drought stress may be dependent or independent by the ABA hormone in plants (Hirayama and Shinozaki, 2010). In *Arabidopsis* bZIP, a large family of transcription factors with 75 members has been reported by Choi *et al.* (2000). However MYC and MYB proteins activate accumulation of ABA during later stages of drought stress (Cominelli *et al.*, 2005). As regard to stomatal movement AtMY B60 and AtMYB 44 gens are found to be instrumental for transcriptional responses (Jung *et al.*, 2008). Gene families like DRE/CRT binding protein (CBF/DREB1 and DREB2) were described as a distinct target of drought ABA- independent signaling transduction pathways in *Arabidopsis* (Shinozahi and Yamaguchi-Shinozaki, 2000). Ramamoorthy *et al.* (2008) established that one or two WRKY domains, having 60 amino acid regions, sequenced as WRKYGQK and zinc finger like motif constitutes the WRKY protein, regulates a number of biotic and abiotic stress factors (Fig. 1).

ABA may interrelate with jasmonic acid (JA) and nitric oxide (NO) to stimulate stomatal closure, while its regulation of gene expression includes the induction of genes associated with response to ethylene, cytokinin, or auxin. ABA-mediated regulation of signal transduction pathways during drought stress is presented here as an example to illustrate the signal transduction elements that may play roles in cross-talk associated with the response to drought stress.

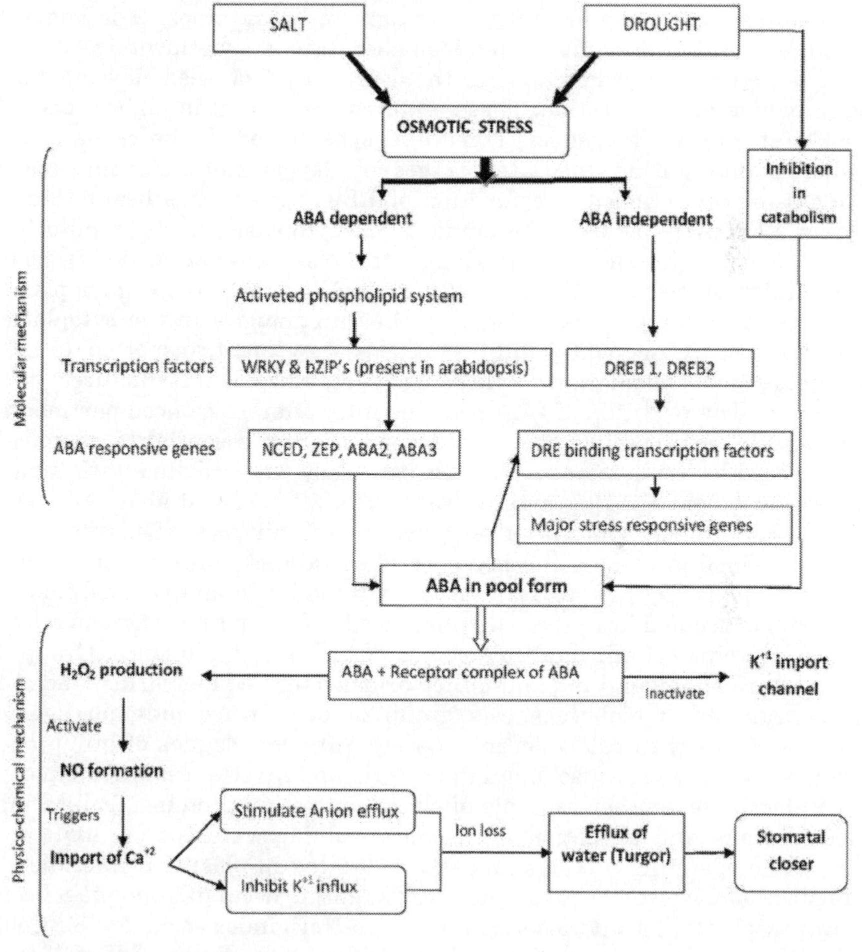

Fig. 1. Physico-chemical and molecular mechanism of ABA responses. DREB (dehydration responsive element binding transcription factors), NCED, ZEP, ABA2, ABA3(ABA responsive genes of different crops)

DROUGHT INDUCED PROTEINS

Synthesis of various stress induced proteins have direct functions in membrane and protection. They involved the acquisition of water and ions, as well as their transportation and homeostasis maintenance in basal cell functions. Late Embryogenesis Abundant protein (LEA proteins) family belongs to this group and was described as highly accumulated in plant embryos (Dure *et al.*, 1981; Galau *et al.*, 1986). The evolution of LEA proteins is one of these changes, which plays an important role in resistance to drought. This implies that studies on tolerant proteins, and the isolation, recognition and functional analysis of their genes will be of great advantage to the breeding of drought-resistant crops. It is general acquaintance that drought related food shortages is a worldwide problem. LEA proteins are formed during the late period of seed development accompanied by dehydration. These proteins are present in higher concentration at late embryogenesis stage and synthesized in the seeds of all higher plants during the later stages of development. As important drought-tolerance genes, the function of LEA proteins has been studied widely. LEA proteins are to be found in the cytoplasm, nucleus, mitochondria, vacuoles (Houde *et al.,* 1995, Egerton-Warburton *et al.,* 1997), near the cellular membrane (Danyluk *et al.,* 1998), as well as in amyloplasts (Rinne *et al.,* 1999), however, mainly they accumulate in the cytoplasm and nucleus. Subsequently after synthesis they are transported to cell organelles and membranes, where they stabilize cell structures and molecules. The partition of LEA proteins into different cell compartments determines their possible role in the regulation of essential biochemical processes, like replication or respiration. They are proteins with small molecular weight ranging mainly from 10 to 30 kDa and above 30 kDa. These LEA proteins comprise a very divergent family, classified into three loosely defined groups by the presence of certain sequence motifs. Three groups, one is defined to contain a hydrophilic 20-amino acid motif, members of group 2 have three distinct motifs (Y, S and K), whereas group 3 LEA proteins contain multiple copies of an 11-mer sequence. Group 2 proteins are only found in plants, they are also termed dehydrins, and are either truly late in embryogenesis (group 2a) or they are not embryogenic and associated with cold tolerance (group 2b). Homologues of group one and three proteins are also found in bacteria and invertebrates. Aquaporin (AQP) family is another example of stress-protecting proteins, facilitating water uptake and allocation by forming cellular water pores. AQP are membrane spanning proteins and can be located in plasma membrane or tonoplast (Johansson *et al.*, 2000). Differential accumulation has been shown in plant tolerant varieties (Montalvo-Hernandez *et al.*, 2008). Cold is another abiotic stress that induce AQP accumulation (Maurel and Chrispeels, 2001; Seki *et al.*, 2001) and also play an important role in plant recovery after the dehydration to resume cell growth and photosynthetic activity (Oono *et al.*, 2003).

Heat Shock Proteins (HSP), highly accumulated during stress and also known as molecular chaperones, are widely distributed in nature; they are involved in protein folding and assembly, as well as in the removal and disposal of non-functional proteins (Wang *et al.*, 2004). HSP are induced by drought and salinity (Alamillo *et al.*, 1995; Campalans *et al.*, 2001) and *in vivo* evidence suggests that HSPs prevent thermal aggregation (Lee *et al.*, 1995); thus facilitating the recovery of cell functions after abiotic stress.

Turnover of damaged macromolecules in stress is a basic process to maintain cell homeostasis. An increase in cysteine protease activity has also been observed under drought conditions (Koizumi *et al.*, 1993; Seki *et al.*, 2002); polyubiquitin involved in labeling and targeting for hydrolysis in the proteasome is also up-regulated in the stressed plants (Barrera-Figueroa *et al.*, 2007; Bartels and Sunkar, 2005).

Conclusion and Future Perspectives

The overall review concluded that the accessibility of water is one of the foremost important abiotic factors which shaped and keep on shaping plant's evolution. The drought stress can be evaded by plants through increasing water uptake (by deep root system) or decreasing water loss (by stomatal closer and small leaves). Guard cell signaling has a significant importance during drought stress responses as it is the main denominator of water resources within the plant. However, in near future the plants may suffer from water stress due to erratic rainfall, shifting of monsoon and an increase in drought condition resulting in more of evapo-transpiration thus a cause for fluctuation in temperature level towards higher side. Hence the plant scientists have to engage themselves more in physiological studies of those plants, surviving in drought condition naturally, to know that why and to what extent these systems modulate their inner mechanism to cope in this situation. Further these studies should be correlated with the molecular mechanism of the plants/crops, grown specifically for agricultural point of view by using modern technological knowhow which may open a channel to understand not only the genetical but the epigenetical domain of the plant as it may change their phenomics in response to various stresses specially in drought stress.

References

Alamillo J, Almogura C, Barteles D, Jordano J (1995) Constitutive expression of small heat shock proteins in vegetative tissues of the resurrection plant *Craterostigma plantagineum. Plant Mol. Biol.* 29:1093-1099.

Aranjuelo I, Molero G, Erica G, Avice JC, Nogues S (2011) Plant Physiology and proteomics reveals the leaf response to drought in alfalfa (*Medicago sativa* L.). *J. Exp. Bot.* 62:111-123 (DOI:10.1093/jxb/erg249.)

Bari R, Jones JD (2009) Role of plant hormones in plant defence responses. *Plant Mol. Biol.* 69:473-88.

Baroni L, Cenci L, Tettamanti M , Berati M (2007) Evaluating the environmental impact of various dietary patterns combined with different food production systems. *European Journal of Clinical Nutrition* 61: 279–286.

Barrera-Figueroa B, Pena-Castro J, Acosta-Gallegos JA, Ruiz-Medrano R, Xoeonostle-Cazares B (2007) Isolation of dehydration-responsive genes in a drought tolerant common bean cultivar and expression of a group 3 late embryogenesis abundant mRNA in tolerant and susceptible bean cultivar. *Functional Plant Biol.* 34:368-381.

Bartels D, Sunkar R (2005) Drought and salt tolerance in plants. *Crit. Rev. Plant Sci.* 24:23-58.

Blokhina O., Virolainen E., Fagerstedt K.V. (2003) Antioxidants, oxidative damage and oxygen deprivation stress: a review, *Ann. Bot.* 91,179–194.

Blum A (1996) Crop responses to drought and the interpretation of adaptation. *Plant Growth Regulation* 20: 135-148.

Campalans A, Pages M, Messeguer R (2001) Identification of differentially expressed genes by the cDNA-AFLP technique during dehydratation of almond (*Prunus amygdalus*). *Tree Physiol.* 21:633-643.

Chartres C, Varma S (2010) Out of water from abundance to scarcity and how to solve the world's water problems. *FT Press* (USA).

Chaves MM, Pereira JS, Maroco J, Rodrigues ML, Ricardo CPP, Oso´rio ML (2002) How plants cope with water stress in the field: photosynthesis and growth. *Annals of Botany* 89: 907–916.

Chaves MM, Flexas J, Pinheiro C (2009) Photosynthesis under drought and salt stress: regulation mechanisms from whole plant to cell. *Annals of Botany* 103:551–560.

Choi H, Hong J, Ha J, Kang J, Kim SY (2000) ABFs, a family of ABA-responsive element binding factors. *Journal of Biological Chemistry* 275:1723-1730.

CIA (2008) The world fact book .Central Intelligence Agency Retrieved 12-20.

Cominelli E, Galbiati M, Vavasseur A, Conti L, Sala T, Vuylsteke M, Leonhardt N, Dellaporta S, Tonelli C (2005) A guard-cell-specific MYB transcription factor regulates stomatal movements and plant drought tolerance. *Current Biology* 15:1196–1200.

Danyluk J, Perron A, Houde M, Limin A, Fowler B, Benhamou N, Sarhan F (1998) Accumulation of an acidic dehydrin in the vicinity of the plasma membrane during cold acclimation of wheat. *Plant Cell* 10: 623.638.

Davidson E.A., Verchot L.V., Cattanio J.H., Ackerman I.L., Carvalho H.M. (2000) Effects of soil water content on soil respiration in forests and cattle pastures of eastern Amazonian, *Biogeochemistry* 48, 53–69.

De Dorlodot S, Forster B, Pagè L, Price A, Tuberosa R, Draye X (2007) Root system architecture: opportunities and constraints for genetic improvement of crops. *Trends in Plant Science* 12: 474–481.

De Leonardis AM, Petrarulo M, De Vita P, Mastrangelo AM (2012) Genetic and molecular aspects of plant response to drought in annual crop species. Advances in Selected Plant Physiology Aspects, Dr. Giuseppe Montanaro (Ed.), ISBN: 978-953-51-0557-2.

Dichio B, Xiloyannis C, Sofo A, Montanaro G (2006) Osmotic regulation in leaves and roots of olive trees during a water deficit and rewatering. *Tree Physiol.* 26:179-85.

Dure L, Greenway SC, Galau GA (1981) Developmental biochemistry of cottonseed embryogenesis and germination: Changing messenger ribonucleic acid population as shown by in vitro and in vivo protein synthesis. *Biochemistry* 20:4162-4168.

Eck H V (1986) Effect of water deficits on yield, yield components, and water use efficiency of irrigation corn. *Agron. J.* 78: 1035 – 1040.

El-Far IA, Allan AY(1995) Responses of some wheat cultivars to sowing methods and drought at different stages of growth . *Aust. J. Agric. Sci.* 26: 267–277.

Egerton-Warburton LM, Balsamo RA, Close TJ (1997) Temporal accumulation and ultrastructural localization of dehydrins in Zea mays. *Physiol. Plant* 101: 545-555.

Fischer RA, Rees D, Sayre KD, Lu Z-M, Condon AG, Saavedra AL (1998) Wheat yield progress associated with higher stomatal conductance and photosynthetic rate and cooler canopies. *Crop Science* 38: 1467–1475.

Galau GW, Hughes DW, Dure L (1986) Abscisic acid induction of cloned cotton Late Embryogenesis Abundant (LEA) messenger RNAs. *Plant Mol. Biol.* 7:155-170.

Hansen JM, Go YM, Jones DP (2006) Nuclear and mitochondrial compartmentation of oxidative stress and redox signaling. *Annu. Rev. Pharmacol. Toxicol.* 46, 215–234.

Hirayama T, Kazuo Shinozaki K, (2010) Research on plant abiotic stress responses in the post-genome era: past, present and future, *Plant J.* 61: 1041–1052.

Hirasawa T (1999) Physiological characterization of rice plant for tolerance of water deficit. In: Ito O, O'Toole JC, Hardy B, editors. Genetic improvement of rice for water-limited environments. Los Ba˜nos, Philippines: International Rice Research Institute 89–98.

Hoekstra FA, Golovina EA, Buitink J (2001) Mechanisms of plant desiccation tolerance. *Trends Plant Sci.* 2001 6:431-8.

Houde M, Daniel C, Lachapelle M, Allard F, Laliberte S, Sarhan, F (1995) Immunolocalization of freezing-tolerance-associated proteins in the cytoplasm and nucleoplasm of wheat crown tissues. *Plant J.* 8: 583.593.

Hsiao CT (1973) Plant responses to water stress. *Ann Rev. Plant Physiol.* 24:519-70.

Huang B.R., Fu J. (2000) Photosynthesis, respiration, and carbon allocation of two cool-season perennial grasses in response to surface soil drying, *Plant Soil* 227, 17–26.

Jackson RB, Sperry JS, Dawson TE (2000) Root water uptake and transport: using physiological processes in global prediction, *Trends in Plant Science,* 5:482-488.

Johansson I, Karlsson M, Johanson U, Larsson C , Kjellbom P (2000) The role of aquaporins in cellular and whole plant water balance. *Biochim. Biophys. Acta,* 1465:324-342.

Jung C, Seo JS, Han SW, Koo YJ, Kim CH, Song SI, Nahm BH , Choi YD, Cheong JJ (2008) Over expression of AtMYB44 enhances stomatal closure to confer abiotic stress tolerance in transgenic Arabidopsis, *Plant Physiology* 146: 623–635.

Kerstiens G (2006) Water transport in plant cuticles: an update, *J. Expt. Bot.* 57: 2493-2499.

Koizumi M, Yamaguchi-Shinozaki K, Tsuji H, Shinozaki K (1993) Structure and expression of two genes that encode distinct drought- inducible cysteine proteinases in *Arabidopsis thaliana. Gene* 129:175-182.

Kuromori T, Miyaji T, Yabuuchi H, Shimizu H, Sugimoto E, Kamiya A (2010) ABC transporter AtABCG25 is involved in abscisic acid transport and responses. *Proc Natl Acad Sci USA* 107:2361–2366.

Lambers H., Atkin O.K., Scheureater I. (1996) Respiratory patterns in roots in relation to their function, in: Waisel Y. (Ed.), Plant Roots, The Hidden Half. Marcel Dekker, New York.

Lawlor DW, TezaraW (2009) Causes of decreased photosynthetic rate and metabolic capacity in water deficient leaf cells: a critical evaluations of mechanisms and integration of processes. *.Annals of Botany* 103:543-549.

Lee GJ, Pokala N, Vierling E (1995) Structure and in vitro molecular chaperone activity of cytosolic small heat shock proteins from pea. *J. Biol. Chem.* 270:10432-10438.

Lisar, Seyed YS, Motafakkerazad R, Hossain M M, Ismail M M, Rahman MM (2012) Water Stress in Plants: Causes, Effects and Responses, Water Stress (Ed.), ISBN: 978-953-307-963-9,

Maurel C, Chrispeels MJ (2001) Aquaporins a molecular entry into plant water relations. *Plant Physiol.* 125:135-138.

Mastrangelo AM, Mazzucotell E, Guerra D, De Vita P, Cattivelli L (2011) Improvement of drought resistance in crops: from conventional breeding to genomic selection In "Crop Stress and its Management: Perspectives and Strategies" Arun K. Shanker eds, DOI 10.1007/978-94-007-2220-0_7.

Montalvo- Hernandez L, Elias Piedra-Ibarra E, Gmez-silva L, Lira- Carmona R, Acosta- Gallegos JA (2008) Differential accumulation of mRNA in drought tolerant and susceptible common bean cultivars in response to water deficit.41 *New Phytol.* 177:102-113.

NOAA, (2003) All About droughts: National weather service. National Oceanic and Atmospheric Administration. http://www.nws.noaa.gov. NOAA factsheet, retrieved April 10, 2007

Oono Y, Seki M, Nanjo T, Narusaka M, Fujita M (2003) Monitoring expression profiles of Arabidopsis gene expression during rehydration process after dehydration using ca. 7000 full-length cDNA microarray. *Plant J.* 34:868-887.

Pieterse CMJ, Leon-Reyes A, Van der Ent S, Van Wees SCM (2009) Networking by small-molecule hormones in plant immunity. *Nat. Chem. Biol.* 5: 308–316.

Ramamoorthy R, Jiang SY, Kumar N, Venkatesh PN, Ramachandran S (2008) A comprehensive transcriptional profiling of the WRKY gene family in rice under various abiotic and phyto hormone treatments. *Plant Cell Physiology* 49: 865–879.

Rassaa N, Ben Haj, Salah H, Latiri K (2008) Thermal responses of Durum wheat *Triticum durum* to early water stress, Consequence on leaf and flower development. *C.R. Biologies* 331: 363–371.

Reddy AR, Chaitanya KV, Vivekanandan M (2004) Drought-induced responses of photosynthesis and antioxidant metabolism in higher plants. *J. Plant Physiol.* 161:1189–1202.

Rinne PL, Kaikuranta PL, Van der Plas LH, Van der Schoot C (1999) Dehydrins in cold-acclimated apices of birch (*Betula pubescens* Ehrh.): production, localization and potential role in rescuing enzyme function during dehydration. *Planta* 209: 377-388.

Rizza F, Ghashghaie, Meyer S, Matteu L, Mastrangelo AM, Badeck FW (2011) Constitutive differences in water use efficiency between two durum wheat cultivars. *Fields Crops Research* doi: 10.1016/J fcr.2011.09.001.

Saliendra NZ, Sperry JS, Comstock JP (1995) Influence of leaf water status on stomatal response to humidity, hydraulic conductance, and soil drought in *Betula occidentalis*. *Planta* 196:357-366.

Seki M, Narusaka M, Abe H, Kasuga M, Fujita M (2002) Monitoring the expression profiles of 7000 Arabidopsis genes under drought, cold and high-salinity stresses using a full length cDNA microarray. *Plant J.* 31:279-292.

Seki M, Narusaka M, Abe H, Kasuga M, Yamaguchi-Shinozaki, K. (2001) Monitoring the expression pattern of 1300 Arabidopsis genes under drought and cold stresses by using a full length cDNA microarray. *Plant Cell* 13:61-72.

Serraj R, Sinclair TR, (2002) Osmolyte accumulation: can it really help increase crop yield under drought conditions. *Plant Cell and Environ.* 25:333–41.

Siddique KHM, Tennant D, Perry MW, Belford RK (1990) Water use and water use efficiency of old and modern wheat cultivars in a Mediterranean-type environment. *Australian Journal of Agricultural Research* 41: 431 – 447.

Shinozahi ZK, Yamaguchi-Shinozaki K (2000) Molecular responses to dehydration and low temperature: differences and cross-talk between two stress signaling pathways, *Current Opinion in Plant Biology* 3: 217–223.

Tabaeizadeh Z (1998) Drought induced responses in plant cells. *International review of cytology* 182:193-247.

Tran LSP, Urao T, Qin F, Maruyama K, Kakimoto T, Shinozaki K, Yamaguchi-Shinozaki K (2007) Functional analysis of AHK1/ATHK1 and cytokinin receptor histidine kinases in response to abscisic acid, drought, and salt stress in Arabidopsis. *Proc Natl Acad Sci USA* 104:20623–20628.

Tuberosa R, Sanguineti MC, Landi P, Giuliani MM, Salvi S, Conti S (2002) Identification of QTLs for root characteristics in maize grown in hydroponics and analysis of their overlap with QTLs for grain yield in the field at two water regimes. *Plant Molecular Biology* 48: 697-712.

United Nations Un.org. 2005-03-22 Retrieved 2010-07-25.

Wagner AB, Moore AL (1997) Structure and function of the plant alternative oxidase: its putative role in the oxygen defence mechanism, *Bioscience Rep.* 17, 319–333

Wang W X, Vinocur B, Shoseyov O, Altman A (2004) Role of plant heat-shock proteins and molecular chaperones in the abiotic stress response. *Trends in Plant Sci.* 9:244-252.

WBCSD Water Facts and Trends Retrieved 2010-07-25.

Wu YJ, Sharp RE, Durachko DM (1996) Growth maintenance of the maize primary root at low water potentials involves increases in cell-wall extension properties, expansion activity, and wall susceptibility to expansions. *Plant Physiol* 111:765–772.

Xiong L, Ishitani M, Lee H, Zhu JK (2001) The Arabidopsis *LOS5/ABA3* locus encodes a molybdenum cofactor sulfurase and modulates cold and osmotic stress responsive gene expression. *Plant Cell* 13:2063–83.

Zhang J, Davies WJ (1989a) Abscisic acid produced in dehydrating roots may enable the plant to measure the water status of the soil . *Plant Cell Environ.* 12:73-81.

Zhang J, Davies WJ (1989b) Sequential responses of whole plant water relations towards prolonged soil drying and the mediation by xylem sap ABA concentrations in the regulation of stomatal behavior of sunflower plants. *New Phytol.* 113:167-17.

Zhang YL, Fan WH, Kinkema M, Li X, Dong X (1999) Interaction of NPR1 with basic leucine zipper protein transcription factors that bind sequences required for salicylic acid induction of the PR-1 gene. *Proc. Natl. Sci. USA* 96:6523-6528.

5

HIGH TEMPERATURE STRESS – PHYSIOLOGICAL, BIOCHEMICAL AND MOLECULAR IMPACTS AND TOLERANCE MECHANISMS

D. Vijayalakshmi[1*] and P.Vivitha[1]

[1]Department of Crop Physiology, Tamil Nadu Agricultural University, Coimbatore.
**Corresponding author: vijiphysiology@gmail.com*

INTRODUCTION

High Temperature Stress has become increasingly important as a yield-limiting factor, and the increased frequency of short-term but extremely high temperatures is particularly damaging (Intergovernmental Panel on Climate Change 2007). The global mean temperature increased by 0.6°C between 1990 to 2000, and is projected to increase by another 1.4 to over 5°C by 2100 (Houghton *et al.*, 2001; McCarthy *et al.*, 2001). Estimates range up to a 17% decrease in crop yield for each degree Celsius increase in average growing season temperature (Lobell and Asner, 2003). Further the climate change projections have shown an increase in maximum and minimum temperature by 3.1 to 3.7°C and 3.7 to 4.2°C, respectively at the end of the 21st century (Arthi Rani and Maragatham, 2013).

Plant life exists between temperatures of -8 to 9°C to +58°C. However, most of the plants are adapted to limited range of temperatures. If plants experience a temperature of above 45°C, plants are subjected to high temperature stress. During its growing season, temperature is the principal factor that determines rice growth, development and ultimately grain yield (Butler and Huybers, 2013). Increase of 15–20°C above normal temperature causes deeper modification of growth without being necessarily lethal *viz.*, protein denaturation, enzyme inactivation, reduction in chloroplast's photosynthetic activity. Heat sensitive cultivars, heat stress increases the costs of nitrogen reduction process, maintenance and continuance of production of chlorophyll in plant leaves. Therefore, these disorders severely decrease production and allocation of photo-assimilate to grains, which ultimately reduce grain yield. Therefore, replacement of heat-sensitive cultivars with heat-tolerant, adjustment of sowing time, choice of varieties with a growth duration allowing avoidance of peak stress periods, are some of the adaptive measures that will help the mitigation of forecast yield reduction due to global warming (Mombeini *et al.,* 2014). Hence, the review paper focuses on the physiological impacts of high temperature stress and various tolerance mechanisms adapted by the plants to mitigate the high temperature stress.

PHYSIOLOGICAL IMPACTS OF HEAT STRESS

Heat stress causes multifarious, and often adverse, alterations in plant growth, development, physiological processes and yield (Hasanuzzaman *et al.*, 2012; Hasanuzzaman *et al.*, 2013) (Fig. 1).

Heat Stress on Growth and Development

- Seedling establishment is hampered
- Reproductive (Pollen) development is affected
- Total biomass, grain and fruit development and quality is affected

Seedling establishment

Germination is an important growth stage affected by high temperature stress. Heat stress exerts negative impacts on various crops during seed germination though the ranges of temperatures vary largely on crop species (Johkan *et al.,* 2011; Kumar *et al.*, 2011). Reduced germination percentage, plant emergence, abnormal seedlings, poor seedling vigor, reduced radicle and plumule growth of geminated seedlings are major impacts caused by heat stress (Piramila *et al.,* 2012; Toh *et al.*, 2008). At very high temperatures (45°C) the rate of germination of most cereals was strictly prohibited and caused cell death and embryos for which seedling establishment rate was reduced (Cheng *et al.*, 2009). Long-term effects of heat stress on developing seeds may include delayed germination or loss of vigor, ultimately leading to reduced emergence and seedling establishment. Under diurnally varying temperatures, coleoptile growth in maize is reduced at 40°C and ceased at 45°C (Weaich *et al.*, 1996). ABA induced high temperature stress leads to inhibition of seed germination (Essemine *et al.,* 2010).

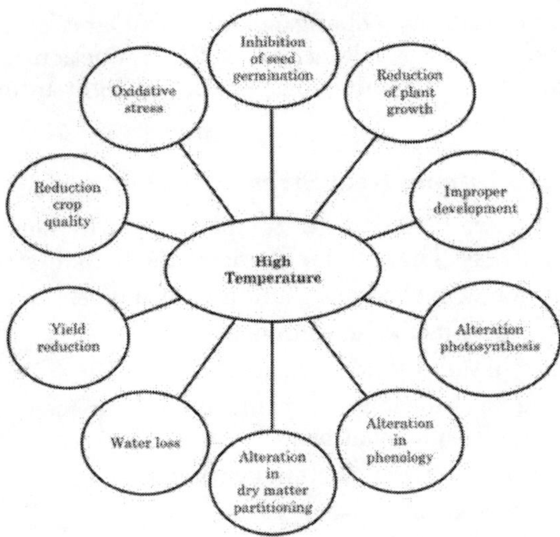

Fig. 1. Major effects of high temperature on plants (Source: Hasanuzzaman *et al.*, 2013)

Reproductive development

A few degrees elevation in temperature during flowering time can result in the loss of entire grain crop cycles (Lobell *et al.,* 2011). During reproduction, a short period of heat stress can cause significant decrease in floral buds and flowers abortion although great variations in sensitivity within and among plant species and variety exist (Sato *et al.,* 2006). The reasons for increasing sterility under high temperature stresses are impaired meiosis in both male and female organs, impaired pollen germination and pollen tube growth, reduced ovule viability, anomaly in stigmatic and style positions, reduced number of pollen grains retained by the stigma, disturbed fertilization processes, obstacle in growth of the endosperm, proembryo and unfertilized embryo (Cao *et al.,* 2008). The genetic variation in percent sterility among the selected backcross populations of rice grown in high temperature environments showed that large number of plants can be identified and selected with lower percent sterility (Manigbasi *et al.,* 2014).

Yield and Biomass

Elevated temperatures (1.5°C) increase in temperature has significant negative effects on crop yields (Warland *et al.,* 2006). Higher temperatures affect the grain yield mostly through affecting phenological development processes. It was demonstrated that increase of the average temperature 1°C decreased the grain yield of cereals by 4.1% to 10.0% (Wang *et al.,* 2012). Loss of productivity in heat stress is chiefly related to decreased assimilatory capacity which is due to reduced photosynthesis by altered membrane stability (Zhang *et al.*, 2006) and enhanced maintenance respiration costs (Reynolds *et al.*, 2007), reduction in radiation use efficiency (RUE), biomass production per unit of light intercepted by the canopy.

Cellular Changes During Heat Stress

When plants are exposed to temperatures higher than 45^{0}C it experiences heat stress. The cellular changes due to heat stress are

1. Disruption of cytoskeleton and microtubules.
2. Fragmentation of golgi complex
3. Increase in number of lysosomes
4. Swelling of mitochondria resulting in decreased respiration and oxidative phosphorylation
5. Disruption of normal protein synthesis
6. Disappearance of polysomes
7. Disruption of splicing of mRNA precursors
8. Cessation of pre-RNA processing
9. Decline in transcription by RNA polymerase I
10. Inhibition of chromatin assembly
11. Decline in DNA synthesis

Changes in membranes

Lipids are basic constituents of membranes. Physio-biochemical characteristics of lipid bilayers differ at different temperatures, with the species of lipid head group or esterified fatty acids. Trienoic fatty Acids (TAs) (poly unsaturated fatty acids having 3 cis-double bonds) increase in plants with high temperature tolerance. TAs is formed from increased activity of ώ– 3 fatty acid desaturase.

One of the typical heat stress symptoms is tissue senescence, characterized by membrane damage associated with increased fluidity of membrane lipids, lipid peroxidation, and protein degradation in various metabolic processes (Savchenko *et al.*, 2002). Membrane lipid saturation is therefore considered an important element in high temperature tolerance. A higher share of saturated fatty acids in membrane lipids increases the lipid melting temperature and prevents a heat-induced increase in the membrane fluidity. Therefore, to maintain membrane fluidity, plants increase the content of saturated and monounsaturated fatty acids, modulating their metabolism in response to increasing temperatures (Zhang *et al.*, 2005b). Thus, increasing the saturation level of fatty acids appears to be critical for maintaining membrane stability and enhancing heat tolerance (Larkindale and Huang, 2004).

Direct injuries due to high temperatures include protein denaturation and aggregation, and increased fluidity of membrane lipids. When plants are exposed to potentially harmful environmental conditions, such as strong light and/or elevated temperatures, the xanthophylls including violaxanthin, antheraxanthin and zeaxanthin partition between the light-harvesting complexes and the lipid phase of the thylakoid membranes. The resulting interaction of the xanthophyll molecules and the membrane lipids brings about a decreased fluidity (thermostability) of membrane and a lowered susceptibility to lipid peroxidation under high temperatures (Havaux, 1998).

Active Oxygen Species cause the autocatalytic peroxidation of membrane lipids and pigments thus leading to the loss of membrane semi-permeability and modifying its functions (Xu *et al.,* 2006). The OH− can damage chlorophyll, protein, DNA, lipids and other important macromolecules, thus fatally affecting plant metabolism and limiting growth and yield (Sairam and Tyagi, 2004). Heat stress induced oxidative stress also significantly increased the membrane peroxidation and reduced the membrane thermostability by 28% and 54% which surprisingly increased electrolyte leakage.

Membrane lipid saturation is considered as an important element in high temperature tolerance. Heat treatment increased relative quantities of linolenic acid (among galactolipids) and *trans*- 3-hexaldecanoic acid (among phospholipids) (Behl *et al.,* 1996). Currently, it is unknown whether a higher or a lower degree of membrane lipid saturation is beneficial for high-temperature tolerance (Klueva *et al.,* 2001). The contribution of lipid and protein components to membrane function under heat stress needs further investigation. Localization of LMW-HSPs with chloroplastic membranes upon heat stress suggests that they play a role in protecting photosynthetic electron transport (Heckathorn *et al.,* 1998).

Various steps in membrane lipid peroxidation

(a) Initiation of the peroxidation process by an oxidizing radical $X^{\bullet}$ by abstraction of a hydrogen atom, thereby forming a pentadienyl radical.

(b) Oxygenation to form a peroxyl radical and a conjugated diene.

(c) Peroxyl radical moiety partitions to the water-membrane interface where it is poised for repair by tocopherol.

(d) Peroxyl radical is converted to a lipid hydroperoxide, and the resulting tocopherol radical can be repaired by ascorbate.

(e) Tocopherol has been recycled by ascorbate; the resulting ascorbate radical can be recycled by enzyme systems. The enzymes phospholipase A2 (PLA2), phospholipid hydroperoxide glutathione peroxidase (PH-GPx), glutathione peroxidase (GPx) and fatty acyl-coenzyme A (FA-CoA) cooperate to detoxify and repair the oxidized fatty acid chain of the phospholipid. (Buettner, 1993).

Changes in Photosynthetic Apparatus

Changes in lipid and protein composition of thylakoid membrane

Level of free fatty acids increases 2-3 fold by the activation of endogenous lipases at high temperatures. Thylakoid membranes are rich in Monogalactosyl diacyl glycerol (MGDG), Digalacyosyl diacyl glycerol (DGDG), Sulfoquinovosyl diacyl glycerol and Phosphatidyl glycerol. Increase in temperature stress causes a decrease in MGDG / DGDG from 1.3 to 0.9. Increase in accumulation of DGDG – offers protection – capability to form lipid bilayers. High temperatures increase the fluidity and permeability of membranes resulting in decrease in proton gradient formation across the thylakoid membrane which is essential for ATP synthesis during light reaction. Synthesis of several thylakoid proteins are reduced which are given below.

(i) Apoprotein of reaction centre of PSII (P 680)

(ii) Subunits of alpha & beta ATP synthetase

(iii) Cytochrome f

(iv) Cytochrome b 559

(v) Apoprotein of core antenna complex cp47

All the above proteins are coded by the chloroplast genome.

Structural Changes of thylakoid

Important structural changes modify the thylakoid membrane integrity following exposure to elevated temperatures. The first changes appear in the 35-45^{0}C interval as a decreased membrane stacking and re-organisation of thylakoid membranes. Above 45^{0}C, a transformation of membranes into vesicles was found. Dissociation of major light harvesting complex (LHC) of PSII from the core complex was observed by freeze fracture experiments. Increased PSII / PSI ratio in the unstacked regions of the thylakoid was observed under heat stress. Analysis of protein composition of thylakoid membranes suggests that, during heat stress at 46^{0}C , PSII core complex dissociates from dimeric to monomeric form.

Inhibition of photosynthetic activity

Photosystem II is much less resistant to heat compared to PSI. Inactivation of PSII during heat stress is due to the extraction of divalent cations (Ca^{2+} and Mn^{2+}) from O2 evolving complex of PSII and dissociation of 33 kDa extrinsic polypeptide involved in the stabilisation of Mn cluster of O_2 evolving complex.

The antenna complex of PSII is also sensitive to heat stress. Thus chlorophyll fluorescence from chloroplasts decreases progressively with the elevated temperature from 20 to 48^{0}C. Thermal inhibition of PSII is followed by a decrease in the variable fluorescence (Fv) ie., photochemical efficiency of PSII but increases the Initial fluorescence (F_o). Increase in F_o

can be used as an indicator for irreversible damage in PSII, associated to LHCII dissociation and blocking the electron transference in the reductant side of PSII. All these events result in photochemical and carboxylative efficiency losses, and serious metabolic restrictions in the Calvin cycle, such as inactivation of Ribulose-1,5- bisphosphate carboxylase/ oxygenase (Rubisco) and variations in the metabolic pool, especially ATP and NADPH availability.

Photosystem I (P 700) is much resistant to heat stress. It retains the cyclic electron transport activity even if temperature rises above >45^{0}C because the dissociated LHC from PSII gets associated with PSI. Some other reasons believed to hamper photosynthesis under heat stress are reduction of soluble proteins, Rubisco binding proteins (RBP), large-subunits (LS), and small-subunits (SS) of Rubisco in darkness, and increases of those in light. High temperature also greatly affects starch and sucrose synthesis, by reduced activity of sucrose phosphate synthase, ADP-glucose pyrophosphorylase, and invertase.

Steps involved in inhibition of photosynthetic activity

1. PSII inhibition
2. Thylakoid permeability
3. PSI Stimulation
4. Cyclic electron transport stimulation
5. Stromal reductant use stimulation
6. State I to State II transition
7. Acidification of lumen
8. De-epoxidation of violaxanthin

Synthesis of Heat Shock Proteins

In response to sudden, 5 to 10°C rises in temperature, plants produce a unique set of proteins referred to as **heat shock proteins** (**HSPs**). This increase in expression is transcriptionally regulated. This dramatic upregulation of the heat shock proteins induced mostly by Heat Shock Factor (HSF) is a key part of the heat shock response. Production of high levels of heat shock proteins can also be triggered by exposure to different kinds of environmental stress conditions, such as infection, inflammation, exposure of the cell to toxins (ethanol, arsenic, trace metals and ultraviolet light, among many others), starvation, hypoxia (oxygen deprivation), nitrogen deficiency (in plants), or water deprivation. Consequently, the heat shock proteins are also referred to as stress proteins and their upregulation is sometimes described more generally as part of the stress response.

Most HSPs function to help cells withstand heat stress by acting as molecular chaperones. Heat stress causes many cell proteins that function as enzymes or structural components to become unfolded or misfolded, thereby leading to loss of proper enzyme structure and activity. Such misfolded proteins often aggregate and precipitate, creating serious problems within the cell. HSPs act as molecular chaperones and serve to attain a proper folding of misfolded, aggregated proteins and to prevent misfolding of proteins. This facilitates proper cell functioning at elevated, stressful temperatures.

Heat-shock proteins also occur under non-stressful conditions, simply "monitoring" the cell's proteins. Some examples of their role as "monitors" are that they carry old proteins to the cell's "recycling bin" and they help newly synthesised proteins fold properly. These activities are part of a cell's own repair system, called the "cellular stress response" or the "heat-shock response". The function of heat-shock proteins is similar in virtually all living organisms, from bacteria to humans.

Plants and most other organisms make HSPs of different sizes in response to temperature increases. The molecular masses of the HSPs range from 15 to 104 kDa (kilodaltons), and they can be grouped into five classes based on size. Different HSPs are localized to the nucleus, mitochondria, chloroplasts, endoplasmic reticulum, and cytosol. Members of the HSP60, HSP70, HSP90, and HSP100 groups act as molecular chaperones, involving ATP-dependent stabilization and folding of proteins, and the assembly of oligomeric proteins. Some HSPs assist in polypeptide transport across membranes into cellular compartments. HSP90s are associated with hormone receptors in animal cells and may be required for their activation, but there is no comparable information for plants. Heat Shock proteins 101 (HSP 101) and heat stress associated 32 KD protein (HSA32) at the post transcriptional level prolongs the effect of heat acclimation in rice seedlings (Meng-yi Lin, 2014).

Low-molecular-weight (15–30 kDa) HSPs are more abundant in higher plants than in other organisms. Whereas plants contain five to six classes of low-molecular- weight HSPs, other eukaryotes show only one class. The different classes of 15–30 kDa molecular-weight HSPs (smHSPs) in plants are distributed in the cytosol, chloroplasts, ER and mitochondria. The function of these small HSPs is not understood. Cells that have been induced to synthesize HSPs show improved thermal tolerance and can tolerate exposure to temperatures that are otherwise lethal. Some of the HSPs are not unique to high-temperature stress. They are also induced by widely different environmental stresses or conditions, including water deficit, ABA treatment, wounding, low temperature, and salinity. Thus, cells previously exposed to one stress may gain cross-protection against another stress. Such is the case with tomato fruits, in which heat shock (48 hours at 38°C) has been observed to promote HSP accumulation and to protect cells for 21 days from chilling at 2°C.

A transcription factor mediates HSP accumulation in response to heat shock. All cells seem to contain molecular chaperones that are constitutively expressed and function like HSPs. These chaperones are called **heat shock cognate proteins**. However, when cells are subjected to a stressful, but nonlethal heat episode, the synthesis of HSPs dramatically increases while the continuing translation of other proteins is dramatically lowered or ceases. This heat shock response appears to be mediated by a specific transcription factor (**HSF**) that acts on the transcription of HSP mRNAs. In the absence of heat stress, HSF exists as monomers that are incapable of binding to DNA and directing transcription (Fig. 2). Stress causes HSF monomers to associate into trimers that are then able to bind to specific sequence elements in DNA referred to as heat shock elements elements (HSEs). Once bound to the HSE, the trimeric HSF is phosphorylated and promotes the transcription of HSP mRNAs. HSP70 subsequently binds to HSF, leading to the dissociation of the HSF/HSE complex, and the HSF is subsequently recycled to the monomeric HSF form. Thus, by the action of HSF, HSPs accumulate until they become abundant enough to bind to HSF, leading to the cessation of HSP mRNA production.

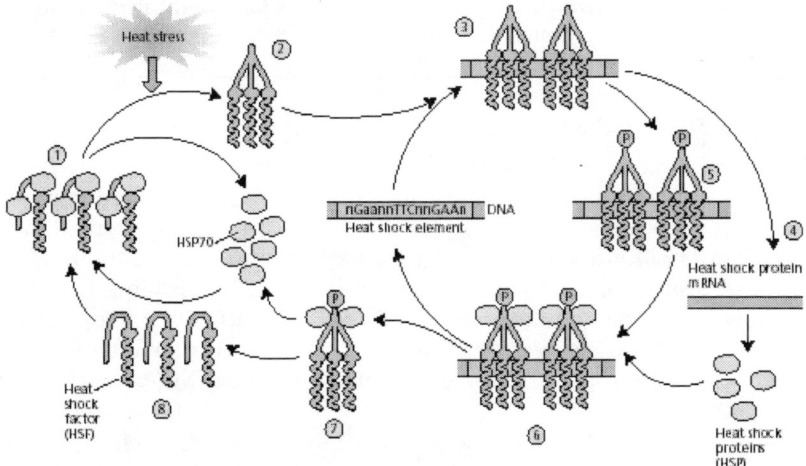

The heat shock factor (HSF) cycle activates the synthesis of heat shock protein in mRNAs. In non stressed cells, HSF normally exists in a monomeric state (1) associated with HSP70 proteins. Upon the onset of an episode of heat stress, HSP70 dissociates from HSF which subsequently trimerizes (2). Activate trimmers bind to heat shock elements (HSF) in the promoter of heat shock protein (HSP) genes (3), and activate the transcription of HSP mRNAs leading to the translation of HSPs among which are HSP70 (4). The HSF trimers associated with the HSE are phosphorylated (5) facilitating the bonong of HSP70 to the phosphorylated trimers (6). The HSP70 trimer complex (7) dissociates from the HSE and disassembles and dephosphorylates into HSF monomers (8), which subsequently bind HSP reforming the resting HSP70/HSF complex.

Fig. 2. Synthesis of Heat Shock Proteins from Heat Shock Factors (Source: Bray *et al.* 2000)

The HSPs are named according to their molecular weights, for example Hsp70 and Hsp90 each define families of chaperones. The major classes of heat shock proteins are tabulated below (Table 1).

Table 1: Major classes of heat shock proteins in plant system for heat stress tolerance.

Major Classes of Heat shock proteins	Functions
HSP100	ATP- dependent dissociation and degradation of aggregate protein
HSP90	Co-regulator of heat stress linked signal transduction complexes and manages protein folding. It requires ATP for its function
HSP70, HSP40	Primary stabilization of newly formed proteins, ATP-dependent binding and release
HSP60, HSP10	ATP- dependent specialized folding machinery
HSP 20 or smHSP	Formation of high molecular weight oligomeric complexes which serve as cellular matrix for stabilization of unfolded proteins. HSP100, HSP70, HSP40 are needed for its release

Synthesis of LEA Proteins

A large group of genes that are regulated by osmotic stress was discovered by examination of naturally desiccating embryos during seed maturation. These genes code for so-called **LEA proteins** (named for *late embryogenesis abundant*), and they are suspected to play a role in cellular membrane protection. Although the function of LEA proteins is not well understood they accumulate in vegetative tissues during episodes of osmotic stress. They are involved in putative activities such as ion sinks, stabilizers of membrane structure, buffers of hydrate water, antioxidants and/or chaperones.

The proteins encoded by these genes are typically hydrophilic, **Gly rich** with **amphipathic α-helices** and strongly bind water. Their protective role might be associated with an ability to retain water and to prevent crystallization of important cellular proteins and other molecules during desiccation. They might also contribute to membrane stabilization. LEA proteins are divided into different classes based on conserved sequence motifs. In general, LEA proteins are hydrophilic and have a biased amino acid composition, mostly lacking Cys and Trp. They are typically highly water soluble, and they often remain soluble after boiling. A protective function correlates with their cellular distribution: LEA polypeptides are found in all cell types, accumulating abundantly in cytoplasm or plastids. LEA proteins may be anchors in a structural network stabilizing cytoplasmic components during pollen drying.

Recent studies suggest that LEA proteins might behave as molecular chaperones, and the current study was undertaken to test this hypothesis. two group 2 LEA proteins, ERD (early response to dehydration) 10 and 14, are potent molecular chaperones. Recombinant forms of AavLEA1, a group 3 LEA proteins from the anhydrobiotic nematode Aphelenchus avenae, and Em, a group 1 LEA protein from wheat, have been subjected to functional analysis. Heat-stress experiments with citrate synthase, which is susceptible to aggregation at high temperatures, suggest that LEA proteins do not behave as classical molecular chaperones, but they do exhibit a protective, synergistic effect in the presence of the so-called chemical chaperone, trehalose. In contrast, both LEA proteins can independently protect citrate synthase from aggregation due to desiccation and freezing, in keeping with a role in water-stress tolerance; similar results were obtained with lactate dehydrogenase. This is the first evidence of anti-aggregation activity of LEA proteins due to water stress. Again, a synergistic effect of LEA and trehalose was observed, which is significant given that non-reducing disaccharides are known to accumulate during dehydration in plants and nematodes. There are five groups of LEA proteins which are given in Table 2.

Table 2: Classification of LEA proteins

Group (family name)	Proteins in the group	Structural characteristics and motifs	Functional information/ proposed function
Group 1 (D-19 family)	Cotton D-19 Wheat Em (early methionine labelled protein) Sunflower Hs cls 10 Barley B 19	Conformation is predominantly random coil with some predicted short α helices.Charged amino acids and glycine are abundant	Contains more water of hydration than typical globular proteins. Over expression confers water deficit tolerance on yeast cells.
Group 2 (D- 11 family) Also referred to dehydrins	Maize DHN1, M3, RAB17 Cotton D-11 Arabidopsis pRABT 1, ERD10, ERD14 Craterostigma pcC 27-04, Pc 6 -19 Tomato pLE4, TAS 14 Barley B8, B9, B17 Rice pRAB 16A Carrot pcEP40	Variable structure includes α helix – forming lysine rich regions.The consensus sequence for group 2 is EKKGIMDKIKELPG. The number of times this consensus repeates per protein varies. Often contains a poly (serine) region of variable length rich in polar residues and either Gly or Ala and Pro	Often localized to the cytoplasm or nucleus. More acidic members of the family are associated with the plasma membrane. May act to stabilize macro molecules at low water potential.
Group 3 (D-7 family)	Barley HVA 1 (ABA induced) Cotton D-7 Wheat pMA2005, pMA1949 Craterostigma pcC3-06	Eleven amino acid consensus sequence motif TAQAAKEKEXE is repeated in the protein. Contains α helices Dimeric protein	Transgenic plants expressing HVA1 demonstrate enhanced water deficit tolerance. D-7 is an abundant protein in cotton embryos. Each putative dimer of D-7 may bind as many as ten inorganic phosphates and their counterions.

Group 4 (D-95 family)	Soybean D-95 Craterostigma pcC27-45	Slightly hydrophobic N- terminal region is predicted to form amphipathic α helices	In tomato, a gene encoding a similar protein is expressed in response to nematode feeding
Group 5 (D-113 family)	Tomato LE 25 Sunflower hads11 Cotton D-113	Family members share sequence homology at the conserved N terminus. N terminal region is predicted to form α helices. C- terminal domain is predicted to be a random coil of variable length and sequence. Ala, Gly and Thr are abundant in the sequence	Binds to membrane and/or proteins to maintain structure during stress. Possibly functions in ion sequestration to protect cytosolic metabolism. When LE 25 is expressed in yeast, it confers salt and freezing tolerance D-113 is abundant in cotton seeds (up to 0.3 mM)

(Bray *et al.*, 2000)

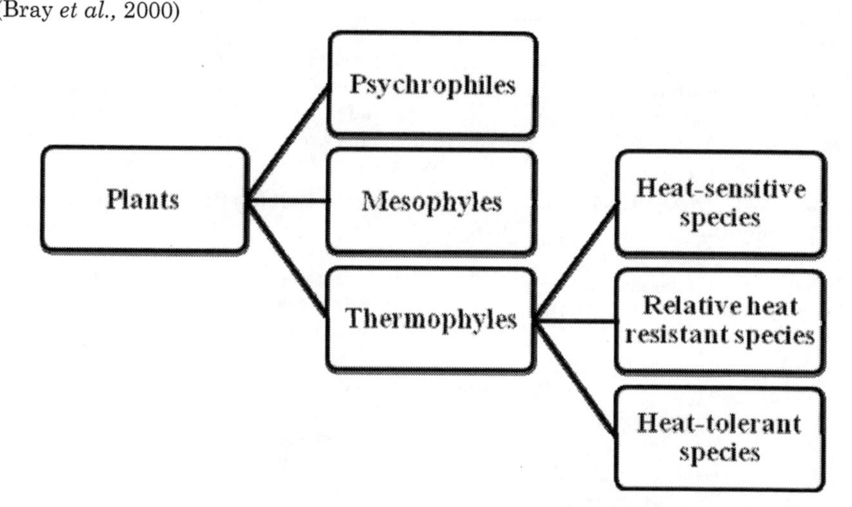

Fig. 3. Classification of plants on the basis of their heat tolerance. (Source: Larcher, 1995)

HEAT STRESS AND TOLERANCE MECHANISMS

Acclimation to High Temperature or Plant Adaptation to Heat Stress

Larcher, 1995 classified all the plant species into three groups on the basis of thermotolerance:

(a) Psychrophiles: which grow optimally at low temperature ranges between 0 and 10°C;

(b) Mesophyles: which favor moderate temperature and grow well between 10 and 30°C;

(c) Thermophyles: which grow well between 30 and 65°C or even higher (Fig. 3).

Plant adaptation to heat stress includes avoidance and tolerance mechanisms which employ a number of strategies:

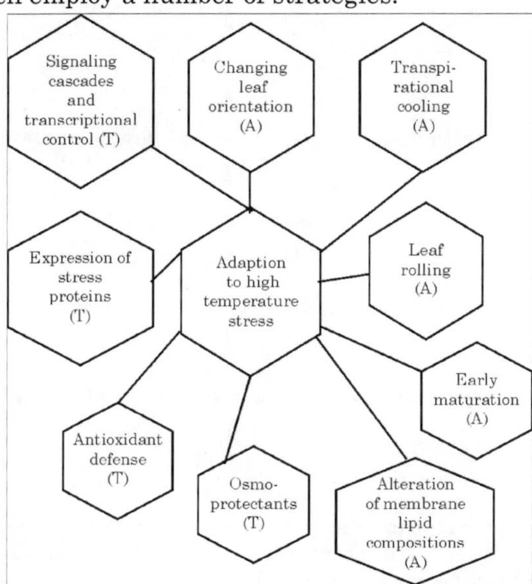

Fig. 4. Different adaptation mechanisms of plants to high temperature (A: Avoidance, T: Tolerance) (Source: Fitter and Hay, 2002)

Avoidance Mechanisms

Under HT conditions, plants exhibit various mechanisms for surviving which include long-term evolutionary phenological and morphological adaptations and short-term avoidance or acclimation mechanisms such as changing leaf orientation, transpirational cooling, or alteration of membrane lipid compositions. Closure of stomata and reduced water loss, increased stomatal and trichomatous densities, and larger xylem vessels are common heat induced features in plants (Srivastava *et al.,* 2012).

Tolerance Mechanisms

They include short term avoidance/acclimation mechanism or long term evolutionary adaptations. Some major tolerance mechanisms, including ion transporters, late embryogenesis abundant (LEA) proteins, osmoprotectants, antioxidant defense, and factors involved in signaling cascades and transcriptional control (Rodríguez *et al.*, 2005; Wang *et al.*, 2004). (Fig. 4). The induction and regulation of antioxidant defences are necessary for obtaining substantial tolerance against temperature stresses. The modification of the cellular redox state may be used to increase the yield and stress tolerance in plants and to improve agriculture (Awasthi *et al.*, 2015).

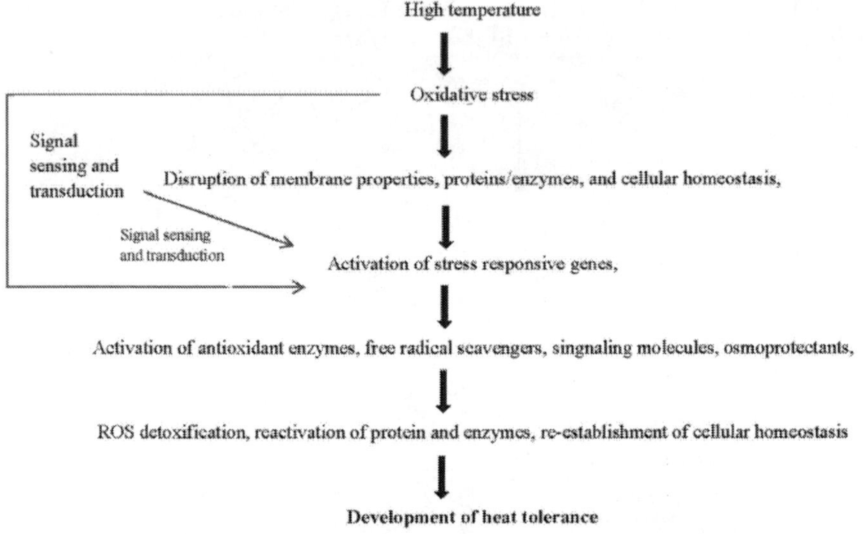

Fig. 5. Mechanism of Signal Transduction and Development of Heat Tolerance (Source: Kaur and Gupta, 2005)

Molecular and Biotechnological Strategies for Development of Heat Stress Tolerance in Plants

Heat Shock Proteins (HSPs)

Heat stress is responsible for the up-regulation of several heat inducible genes; commonly referred as "heat shock genes" (HSGs) which encode HSPs and these active products help in plant's survival under fatal HT (Fig. 5) (Chang *et al.*, 2007).

Synthesis of GABA under Heat Stress

One of the metabolic acclimations to heat stress is the accumulation of the nonprotein amino acid γ-aminobutyric acid (GABA). During episodes of heat stress, GABA accumulates to levels six- to tenfold higher than in unstressed plants. GABA is synthesized from the amino acid L-glutamate, in a single reaction catalyzed by the enzyme glutamate decarboxylase (GAD). GAD is one of several enzymes whose activity is modulated by the calcium-activated, regulatory protein calmodulin.

Calcium-activated calmodulin activates GAD and increases the biosynthesis rate of GABA. Possible functions of GABA in heat stress resistance are under investigation.

Heat stress causes a reduction in cytosolic pH from the normal slightly alkaline value, probably by inhibiting proton-pumping ATPases and pyrophosphatases that pump protons across the plasma membrane or into the vacuole. Additionally, heat stress effects a change in calcium homeostasis inside the cell by affecting the influx of calcium into the cytosol through either plasma membrane or vacuolar calcium channels or by action on efflux ATPases or proton cotransporters. This increase in cytosolic calcium leads to the activation of calmodulin (CaM), which binds to glutamate decarboxylase (GAD) converting it from the inactive to the active form. Glutamate conversion to √aminobutyric acid (GABA) is then accomplished consuming protons in the process and mediating an increase in cytosolic pH. CAX1 and CAX2 (Calcium exchangers) are transport proteins, ACA:Ca^{2+} ATPase.

Transgenic Approaches to Increase Heat Stress Tolerance

Utilisation of various heat shock proteins to develop thermotolerant plants: Transgenic tobacco expressing class I cytosolic small HSP gene, TLHS1 showed up to two times higher cotyledon opening rates in comparison to the transgenic tobacco seedlings carrying the TLHS1 gene in antisense orientation at high temperature stresses of 40°C and 45°C for 1 to 4 h. Expression of Athsp101 and sHSP17.7 into the basmati rice resulted in higher survival and better growth performance in the recovery phase following the heat stress. Overexpression of certain transcriptional regulators of HSP expression, like HSF1 and HSF3, triggered plants to constitutively express HSPs for higher basal thermotolerance. *Arabidopsis* plants engineered to overproduce glycine betaine, showed increased thermotolerance, suggesting that nitrogenous osmoprotectants can have a role in cellular protection against heat stress. Engineering tobacco plants to express bacterial aspartate decarboxylase, the enzyme catalyzing the decarboxylation of aspartate to betaalanine is another approach. The transgenic plants expressing the bacterial gene contained significantly increased levels of beta-alanine and total free amino acids and exhibited a thermotolerant phenotype, when compared to control plants. Dr. Anil Grover's group is working extensively on generating high temperature tolerant transgenic plants (Grover *et al.*, 2013).

Omics Approaches in Developing Heat Stress Tolerance

A large number of genes with potential roles in heat stress responses have been identified using genetic screens and genome wide expression studies (Yeh *et al.,* 2012). Transcriptory products (mRNAs), from such genes in the genome, have made their transcriptome (transcriptomics) and then proteome (proteomics) when they translate into the functional proteins (responsible for stress tolerance). Micro RNA plays an important functional role in such study and micromics will helps for the better understanding of tolerance. Several heat stress responsive miRNAs have been identified in plants and their role in osmolyte accumulation (osmoprotection) and nutrient starvation response have been established. Metabolic network of compatible solutes (Pro, GB, glucose, fructose, galactinol, raffinose, *etc.*) have important roles in tolerance to heat stress. After proper combination of transcript analysis and metabolite profile, integrative approaches have been developed to bring about heat tolerance (Fig. 6).

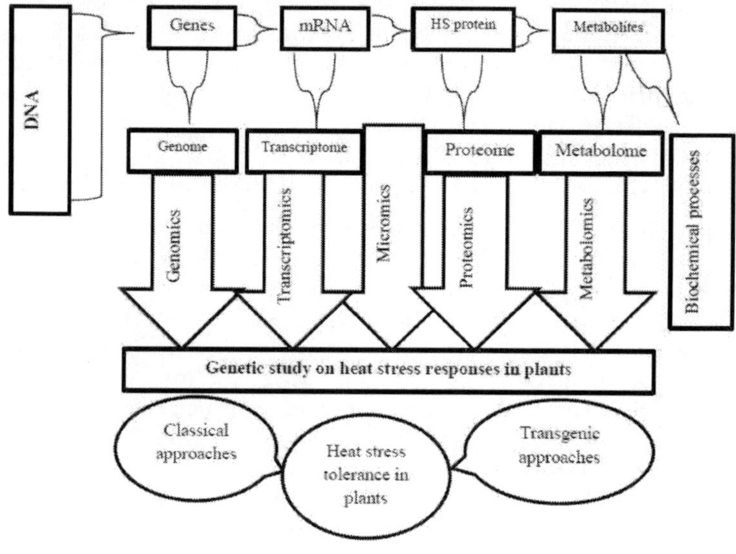

Fig. 6. Integrated circuit of different "omics" approaches that are connected to each other at molecular genetic level associated with heat stress tolerance in plants. Source: Hasanuzzaman *et al.,* 2013

Research initiatives at Tamil Nadu Agricultural University

TNAU is currently focusing on developing rice genotypes for heat stress. Large stocks of rice germplasm including Nagina 22 (N22) (donor for heat tolerance) are being maintained at TNAU. Crosses have already been made involving popular rice varieties of Tamil Nadu namely, Improved White Ponni x N22 and CO 51 x N22. Most significant heat-responsive QTL for spikelet fertility has been identified on chromosome 1 (qHTSF1.1) and chromosome 4 (qHTSF4.1). Hence, further studies will be targeted to introgress these QTLs, to evolved genotypes for high temperature stress tolerance.

Conclusion and Future Outlook

The diversity in Indian climate systems and the occurrence of four distinct seasons in most parts of the country make the climate change study in India challenging. The paper clearly shows that under high temperature conditions, plants accumulate different metabolites (such as antioxidants, osmoprotectants, heat shock proteins [HSPs], Lea proteins) and different metabolic pathways and processes are activated nor affected. These changes emphasize the importance of physiological, biochemical and molecular studies to reveal the mechanisms underlying stress responses. Molecular approaches that uncover the response and tolerance mechanisms will pave the way to engineering plants capable of tolerating High temperature stresses. The next challenging task of researchers will to utilize a genetic approach, especially with models combined with a better understood connection of the complimenting areas of physiology, biochemistry and molecular biology to the genetic component that controls them.

References

Arthi Rani and Maragatham, 2013. Effect of Elevated Temperature on Rice Phenology and yield. *Ind. J. Sci. and Tech.*, 5095-5097

Awasthi, R; Bhandari, K and Nayyar, H. 2015. Temperature Stress and Redox Homeostasis in Agricultural Crops. *Front. Environ. Sci.* 3:11. doi:10.3389/fenvs. 2015.00011

Behl, RK; Heise, KP and Moawad, AM. 1996. High temperature tolerance in relation to changes in lipids in mutant wheat. *Tropenlandwirt.,* 97: 131–135.

Bray, EA; Bailey-Serres, J and Weretilnyk, E. 2000. Responses to abiotic stresses. In W Gruissem, B Buchannan, R Jones, eds, Biochemistry and Molecular Biology of Plants. *American Society of Plant Physiologists., Rockville, MD, pp 1158–1249.*

Buettner, GR. 1993. The pecking order of free radicals and antioxidants: lipid peroxidation, α-tocopherol, and ascorbate. *Archives of Biochemistry and Biophysics.*, 300: 535–543

Butler, E and Huybers, P. 2013. Adaptation of US maize to temperature variations. *Nature Climate Change, 3:* 68-72.

Cao, YY; Duan, H; Yang, LN; Wang, ZQ; Zhou, SC and Yang, JC. 2008. Effect of heat stress during meiosis on grain yield of rice cultivars differing in heat tolerance and its physiological mechanism. *Acta Agron. Sin.*, *34*: 2134–2142.

Chang, Y; He, HS; Bishop, I; Hu, YM; Bu, RC; Xu, CG and Li, XZ. 2007. Long-term forest landscape responses to fire exclusion in the Great Xing'an Mountains, China. *Int. J. Wildl and Fire,* 16: 34–44.

Cheng, L; Zou, Y; Ding, S; Zhang, J; Yu, X; Cao, J and Lu, G. 2009. Polyamine accumulation in transgenic tomato enhances the tolerance to high temperature stress. *J. Integr. Plant Biol.*, *51*: 489–499.

Essemine, J; Ammar, S and Bouzid, S. 2010. Impact of heat stress on germination and growth in higher plants: Physiological, biochemical and molecular repercussions and mechanisms of defence. *J. Biol. Sci., 10*: 565–572.

Fitter, A.H.; Hay, R.K.M. Environmental Physiology of Plants, 3rd ed.; Academic Press: London, UK, 2002.

Grover, A; Mittal, D; Negi, M and Lavania, D. 2013. Generating high temperature tolerant transgenic plants: Achievements and challenges. *Plant Sci. 205–206:* 38–47.

Hasanuzzaman, M; Hossain, MA; da Silva, JAT and Fujita, M. 2012. Plant Responses and Tolerance to Abiotic Oxidative Stress: Antioxidant Defenses is a Key Factor. In *Crop Stress and Its Management: Perspectives and Strategies.*, Bandi, V., Shanker, A.K., Shanker, C., Mandapaka, M., Eds.; Springer: Berlin, Germany, pp. 261–316.

Hasanuzzaman, M; Nahar, K and Fujita, M. 2013. Extreme Temperatures, Oxidative Stress and Antioxidant Defense in Plants. In *Abiotic Stress—Plant Responses and Applications in Agriculture.*, Vahdati, K., Leslie, C., Eds.; InTech: Rijeka, Croatia, pp. 169–205.

Havaux, M. 1998. Carotenoids as membrane stabilizers in chloroplasts. *Trends in Plant Science.*, 3: 147–151.

Heckathorn, SA; Downs, CA; Sharkey, TD. and Coleman, J.S. 1998. The small, methionine-rich chloroplast heat-shock protein protects photosystem II electron transport during heat stress. *Plant Physiol.*, 116: 439–444.

Houghton, JT; Ding, Y; Griggs, DJ; Noguer, M; van der Linden, PJ; Dai, X; Maskell, K and Johnson, CA. 2001. *Climate Change 2001: The Scientific Basis, Cambridge University Press,* 787 pp.

Johkan, M; Oda, M; Maruo, T and Shinohara, Y. 2011. Crop production and global warming. In *Global Warming Impacts—Case Studies on the Economy, Human Health, and on Urban and Natural Environments*, Casalegno, S., Ed.; In Tech: Rijeka, Croatia, pp. 139–152.

Klueva, NY; Maestri, E; Marmiroli, N. and Nguyen, HT. 2001. Mechanisms of thermotolerance in crops. In: A.S. Basra (Ed.) *Crop Responses and Adaptations to Temperature Stress, Food Products Press, Binghamton, NY*, pp. 177–217

Kaur, N and Gupta, AK. 2005. Signal transduction pathways under abiotic stresses in plants. *Curr. Sci., 88*: 1771–1780.

Kumar, S; Kaur, R; Kaur, N; Bhandhari, K; Kaushal, N; Gupta, K; Bains, TS and Nayyar, H. 2011. Heat-stress induced inhibition in growth and chlorosis in mungbean (*Phaseolus aureus* Roxb.) is partly mitigated by ascorbic acid application and is related to reduction in oxidative stress. *Acta Physiol. Plant., 33*: 2091–2101.

Larcher, W. 1995. *Physiological Plant Ecology.*, Springer: Berlin, Germany.

Larkindale, J and Huang, B. 2004. Thermotolerance and antioxidant systems in Agrostis stolonifera: involvement of salicylic acid, abscisic acid, calcium, hydrogen peroxide, and ethylene. *J. Plant Physiol.,* 161: 405-413.

Lobell, DB and Asner, GP. 2003. Climate and management contributions to recent trends in U.S. agricultural yields. *Science., 299*, doi:10.1126/science.1078475.

Lobell, DB; Schlenker, W and Costa-Roberts, J. 2011. Climate trends and global crop production since 1980. *Science, 333:* 616–620.

Manigbasi, NL; Lambio, LAF; Madridi, LB and Cardenas, CC. 2014. Germplasm Innovation of Heat Tolerance in Rice for Irrigated Lowland Conditions in the Philippines. *Rice Science, 21(3):* 162–169

McCarthy, JJ; Canziani, OF; Leary, NA; Dokken, DJ and White, KS. 2001. *Climate Change 2001: Impacts, Adaptation, and Vulnerability.* Cambridge University Press, 1032 pp.

Meng-yi Lin; Kuo-hsing Chai; Swee-suak Ko; Lin-yun Kuang; Huu-Sheng Lur and Yee-yung Charng. 2014. A Positive Feedback Loop between heat shock protein 101 and heat stress associated 32 KD protein modulates long-term acquired thermotolerance illustrating diverse heat stress responses in rice varieties. *Plant Physiology, Vol. 164,* 2045–2053

Mombeini, M. Siadat, SA. Lack, S, Gilani, AA and Rasekh, AR. 2014. Physiological Responses to Heat Stress in Rice (*Oryza sativa.* L): I. Nitrogen Status, Chlorophyll Content and Cell Membrane Thermal Stability (CMTS) of Flag Leaf. *Adv. in Environ. Biol. 8(5)*: 1420-1430.

Piramila, BHM; Prabha, AL; Nandagopalan, V and Stanley, AL. 2012. Effect of heat treatment on germination, seedling growth and some biochemical parameters of dry seeds of black gram. *Int. J. Pharm. Phytopharmacol. Res., 1*: 194–202.

Reynolds, MP; Pierre, CS; Saad, ASI; Vargas, M and Condon, AG. 2007. Evaluating potential genetic gains in wheat associated with stress-adaptive trait expression in elite genetic resources under drought and heat stress. *Crop Sci., 47*: 172–189.

Rodríguez, M; Canales, E and Borrás-Hidalgo, O. 2005. Molecular aspects of abiotic stress in plants. *Biotechnol. Appl., 22*: 1–10.

Sairam, RK and Tyagi, A. 2004. Physiology and molecular biology of salinity stress tolerance in plants. *Curr. Sci.,* 86: 407–421.

Sato, S; Kamiyama, M; Iwata, T; Makita, N; Furukawa, H and Ikeda, H. 2006. Moderate increase of mean daily temperature adversely affects fruit set of *Lycopersicon esculentum* by disrupting specific physiological processes in male reproductive development. *Ann. Bot., 97*:731–738.

Savchenko, EV; Grigorashchenko, ON; Ogurtsov, AN; Rudenkov, VV; Gumenchuk, GB; Lorenz, M; Frankowski, M; Smith-Gicklhorn, AM and Bondybey, VE. 2002. Photo- and thermally assisted emission of electrons from rare gas solids. *Surf. Sci.,* 507-510, 754-761.

Srivastava, S; Pathak, AD; Gupta, PS; Shrivastava, AK and Srivastava, AK. 2012. Hydrogen peroxide-scavenging enzymes impart tolerance to high temperature induced oxidative stress in sugarcane. *J. Environ. Biol., 33*: 657–661.

Toh, S; Imamura, A; Watanabe, A; Okamot, M; Jikumaru, Y; Hanada, A; Aso, Y; Ishiyama, K; Tamura, N and Iuchi, S. 2008. High temperature-induced abscisic acid biosynthesis and its role in the inhibition of gibberellin action in *Arabidopsis* seeds. *Plant Physiol., 146*: 1368–1385.

Wang, W; Vinocur, B; Shoseyov, O and Altman, A. 2004. Role of plant heat-shock proteins and molecular chaperones in the abiotic stress response. *Trends Plant Sci., 9*: 244–252.

Wang, X; Cai, J; Liu, F; Jin, M; Yu, H; Jiang, D; Wollenweber, B; Dai, T and Cao, W. 2012. Pre-anthesis high temperature acclimation alleviates the negative effects of post-anthesis heat stress on stem stored carbohydrates remobilization and grain starch accumulation in wheat. *J. Cereal Sci., 55*: 331–336

Warland, JS; McDonald, MR and McKeown, AM. 2006. Annual yields of five crops in the family Brassicacae in southern Ontario in relation to weather and climate. *Can. J. Plant Sci., 86*: 1209–1215.

Weaich, K; Briston, KL and Cass, A.1996. Modeling pre-emergent maize shoot growth. II. High temperature stress conditions. *Agric. J.* 88: 398–403.

Xu, S; Li, J; Zhang, X; Wei, H and Cui, L. 2006. Effects of heat acclimation pretreatment on changes of membrane lipid peroxidation, antioxidant metabolites, and ultrastructure of chloroplasts in two cool-season turfgrass species under heat stress. *Environ. Exp. Bot., 56*: 274–285

Yeh, CH; Kaplinsky, NJ; Hu, C and Charng, YY. 2012. Some like it hot, some like it warm: Phenotyping to explore thermotolerance diversity. *Plant Sci., 195*: 10–23

Zhang, Q; Canagaratna, MC; Jayne, JT; Worsnop, DR and Jimenez, JL. 2005b. Time and size-resolved chemical composition of submicron particles in Pittsburgh – Implications for aerosol sources and processes, *J. Geophys. Res.,* 110, D07S09, doi:10.1029/2004JD004649,

Zhang, Y; Mian, MAR and Bouton, JH. 2006. Recent molecular and genomic studies on stress tolerance of forage and turf grasses. *Crop Sci., 46*: 497–511.

Źróbek-Sokolnik, A. Temperature stress and responses of plants. In Environmental Adaptations and Stress Tolerance of Plants in the Era of Climate Change; Ahmad, P., Prasad, M.N.V., Eds.; Springer: New York, NY, USA, 2012; pp. 113–134.

6

IMPROVING GRAIN YIELD IN WHEAT (*TRITICUM AESTIVUM* L.) BY STEM RESERVE REMOBILIZATION UNDER HEAT STRESS

Md. Afjal Ahmad[1] and Pravin Prakash[2*]

Department of Plant Physiology, Institute of Agricultural Sciences, Banaras Hindu University, Varanasi - 221 005
**E-mail: prakashpbhu@gmail.com*

INTRODUCTION

Wheat (*Triticum aestivum* L.) production has expanded even into the marginal areas that experience high temperatures because of the increasing demand for more wheat to feed the world. High temperature during the grain filling period of wheat shows many negative effects such as inhibition of photosynthetic process and acceleration of leaf senescence (Al-Khatib and Paulsen 1990). As a result, a great reduction occurs in nutrient supply for the developing grain from current assimilation.

It is one of the most important staple food crops of the world, occupying 17% of crop acreage worldwide, feeding about 40% of the global population and providing 20% (one fifth) of total food calories and protein in human nutrition (Gupta *et al.*, 2008). A massive increment from current 590 million metric tons per year to 840 million metric tons per year by 2025 to meet the global demand has been predicted (Murchie *et al.*, 2009). In India bread wheat accounts for approximately 95% of the wheat grown, while 4% is durum wheat and 1% is dicoccum wheat (Gupta, 2004). Around 312 wheat varieties were released during past 40 years in India's six wheat zones and have played a major role in enhancing productivity (Chatrath *et al.*, 2006). For India the demand for 2020 is estimated at 87.5

million tons, about 13 million tons more than the record production of 75 million tons in 1999-2000.

According to the Recent Assessment Report of the Intergovernmental Panel on Climate Change, global earth temperatures have increased by 0.74°C in the last century, and are likely to increase by 1.1 to 6.4°C by 2100 which will further aggravate the problem (IPCC, 2007). Water deficit is also a severe additional concern; the effects of heat and water deficits are difficult to separate under natural conditions since the minimal rainfall and the high temperature, which usually prevail during grain filling, enhance the rate of evapotranspiration, and result in extreme water deficits (Siddique *et al.*, 2000). Besides, current recommendations for agronomic practices that can reduce heat stress in crops rely heavily on additional inputs, especially irrigation water (Badaruddin *et al.*, 1999). Given the shrinkage of water resources around the globe and the very little scope for increasing acreage under wheat cultivation, the major current challenge is to break the yield barriers.

EFFECT OF HIGH TEMPERATURE STRESS ON WHEAT GROWTH

Wheat (*Triticum aestivum* L.) is grown under heat stress conditions during grain filling, pre- anthesis stored total non-structural carbohydrates (TNC) and nitrogen (N) could serve as alternative source of assimilates. Heat stress negatively affected grain yield, its components and N remobilization while it increased TNC remobilization because of the increasing demand for resources.

Temperature is an important determinate of all plant processes and is the dominant factor controlling wheat development (McMaster, 1997). Plant responses to high temperature vary with plant species, variety and phenological stages, and changes imposed may be direct on physiological processes or indirect on the developmental patterns (Rawson, 1992; Morison, 1996). Lillemo *et al.* (2005) reported an earlier survey estimate of around 9 million ha of wheat in tropical or subtropical areas experiencing yield losses due to high temperature stress. In the tropics, excess radiation and high temperatures are often the most limiting factors affecting plant growth and final crop yield (Wahid *et al.*, 2007). According to Larcher (1995) less heat for a prolonged period causes as much damage as more heat for a shorter duration. In the short term, the lethal temperature is likely to be higher because lethality requires direct damage to the integrity of the cell and may involve enzyme inactivation or destruction, inhibition of protein synthesis and degradation, membrane damage, or other direct physiological damage (Howarth, 2005). Long-term damage, on the other hand, may be a secondary result of increased respiration, compromised membrane integrity, enzyme inefficiency, or other impairment of metabolic activity, growth, or viability.

Recent estimates indicate that in India alone around 13.5 million ha of area is heat stressed (Joshi *et al.*, 2007a). Wheat productivity is the highest in fertile Indo-Gangetic plains of North India reaching 8 t ha^{-1} under irrigated conditions, while in Central India maximum yield ranged between 4 to 5 t ha^{-1} only. These differences in yield correspond to higher temperatures and longer day lengths prevailing in Central India during crop growth period as compared to North India (Ruwali and Prasad, 1991). Wheat is sown late on considerable area in north-western and north-eastern plains of India due to high cropping intensity and delayed harvest of previous crop. In spite of high productivity, the north-western plains are more prone to high temperature stress caused by dry winds from the Thar Desert. On the other hand, wheat grown in the north-eastern plains suffer high temperature coupled with high humidity (Joshi *et al.*, 2007b). Sinha and Swaminathan (1991) had studied the impact of high temperature on wheat and rice in North West India, and had reported that a rise in 1°C would reduce their growth duration by one week. This would in turn reduce yield by 4 to 5 quintals ha^{-1}. Myers (1999) has also put forward the finding that a mere 0.5°C increase in temperature could reduce wheat crop by 10% in India alone.

Effect on Photosynthetic Activity

Photosynthesis has been reported to be one of the most sensitive processes to heat stress (Conroy *et al.*, 1994; Demirevska-Kepova *et al.*, 2005), and both photosynthesis and respiration are known to be more sensitive in cool-season plants (Björkman *et al.*, 1980). High temperature influences the photosynthetic capacity of C_3 plants more strongly than in C_4 plants. In vegetative stage, damage to compensated leaf photosynthesis reduces CO_2 assimilation rates (Hall, 1992). Increased temperatures curtail photosynthesis and increase CO_2 transfer conductance between intercellular spaces and carboxylation sites. And generally the rate of photosynthesis decreases while dark- and photorespiration rates increase considerably under high temperatures. This affects the energy distribution and changes the activities of carbon metabolism enzymes, particularly Rubisco, thereby altering the rate of RuBP regeneration by the disruption of electron transport and inactivation of the oxygen evolving enzymes of PSII (Salvucci and Crafts-Brandner, 2004).

High temperature increases the radiant energy absorbed by the leaf affecting the process of photosynthesis (Krause and Santarius, 1975; Seeman *et al.*, 1984), and was also shown to accelerate the typical ultrastructural changes in chloroplasts of wheat flag leaves (Xu *et al.*, 1995). Thylakoid membranes and PSII are considered the most heat-labile cell structures (Santarius, 1974; Camejo *et al.,* 2005). The threshold temperature for denaturation of PSII was reported to be 35 to 41°C (Rekika *et al.*, 1997), which is generally encountered during maturation of wheat. Heat damages the nature of PSII through removal of the oxygen-evolving enhancer proteins from the thylakoids without damaging the PSI complex (Takeuchi and Thornber, 1994). Imbalance in the electron flow from the oxygen evolving complex (OEC) toward the acceptor side of PSII in the direction of PSI reaction centre is thus initiated (De Ronde *et al.*, 2004). Under high temperatures, PSI stromal enzymes and chloroplast envelopes are reported to be thermostable and in fact PSI-driven cyclic electron pathway, capable of contributing to thylakoid proton gradient, is activated (Bukhov *et al.*, 1999).

Strong light and elevated temperatures also causes the partitioning of the xanthophylls including violaxanthin, antheraxanthin and zeaxanthins between the light harvesting complexes and lipid phase of thylakoid membranes. This partitioning decreases thermostability of membranes (Havaux, 1998). Heat stress accelerates the kinetic energy and movement of molecules across membranes thereby loosening chemical bonds within molecules of biological membranes. This makes the lipid bilayer of biological membranes more fluid by either denaturation of proteins or an increase in unsaturated fatty acids (Savchenko *et al.*, 2002).

Effect on Source- Sink Relationship

Source activity is damaged by heat because both leaf area (Spiertz, 1974; Herzog, 1982) and photosynthesis is reduced (Kuroyanagi and Paulsen, 1985). This decrease of source capacity is a fundamental yield limiting factor, thus, lower photosynthetic capacity of canopy leads to decreased yield by shortening duration of grain filling period (Blum *et al.*, 1994). High temperature after anthesis may influence cell formation and development of the endosperm tissue. Endosperm cells that develop under these conditions are either smaller or fail to fill out due to reduced starch deposition, and temperatures, especially above 34°C, inhibit starch biosynthesis in the endosperm (Jenner, 1994). Besides, Wardlaw and Wrigley (1994) has contended that a sensitive stage of development with regard to yield appeared to be 10 to 15 DAA (i.e. the later stages of cell division and enlargement in the endosperm). Early investigations also report an increasing incidence of post-anthesis leaf senescence, which is coupled to a significant development of the respiration rates in the grain (Stoy, 1965; Spiertz, 1974; Berry and Björkman, 1980). These patterns, according to its extent, might trigger a decreasing carbohydrate availability for the grain (Thornley, 1971), justifying its weight decay. An enhanced rate of cell division in developing grains (Sharma, 1946) thus accelerating phaseic development (Frank and Bauer, 1997), and the subsequent reduction in grain filling duration (Al-Khatib and Paulsen, 1984) is well-reasoned out to affect the final grain yield. Reduction in grain filling period is often associated with a reduction in grain weight (Warrington *et al.*, 1977). Hall (1992) has opined the possibility of resources being diverted to tolerate heat stress and only limited photosynthates being available for reproductive development.

Effect on Grain Filling and Other Yield Attributing Characters

Heat stress during grain filling led to a significant reduction in the yield, biomass, grain number, harvest index and 1000-grain weight (Balla *et al.*, 2009). Al-Khatib and Paulsen (1984) investigated the effect of high temperature through comparisons at regimes of 25°C day/15°C night, 30°C day/20°C night, and 35°C day/25°C night from 5 days after anthesis to maturity. They concluded that a major effect of high temperature is acceleration of senescence, including cessation of vegetative and reproductive growth, deterioration of photosynthetic activities, and degradation of proteinaceous constituents. In later investigations they found that the grain yield of cultivars considered tolerant to high temperature was affected little by the same temperature that decreased grain yield of susceptible cultivars by >40%. They associated the better performance of tolerant cultivars with longer duration of photosynthesis, larger grains and higher harvest index (Al-Khatib and Paulsen, 1990). Ferris *et al.* (1998) examined spring wheat Chablis under field conditions, in which replicated crop areas were exposed to a range of temperatures for a 12-day period (70 to 82 DAS coinciding with anthesis), corresponding to an increase in mean temperature from 16 to 25°C at 82 DAS. The number of grains earhead· at maturity declined with increasing maximum temperature recorded over the mid-anthesis period (76 to 79 DAS) and, more significantly, with maximum temperature 1 d after 50% anthesis (78 DAS). Grain yield and harvest index also declined sharply with maximum temperature at 78 DAS, and corresponded to a 40% reduction in the number of grains per ear. The study emphasized that grain fertilization and grain set was most sensitive to the maximum temperature at mid-anthesis.

Heat stress also increased the duration of grain filling with reduction in grain growth leading to losses in grain density and weight (Reynolds *et al.*, 1994; Guilioni *et al.*, 2003). Gibson and Paulsen (1999) had reported yield reductions by 3 to 5% per 1°C increase above 15°C in wheat under controlled condition during the reproductive period. They cited that high temperature stress applied 15 DAA until maturity reduced grain yield by 18%. Similar treatment from 20 DAA decreased grain weight by 18%. Leithold *et al.* (1997) surveyed eleven spring wheat from diverse origin for responses to high temperatures at ear emergence and beginning of anthesis. Reductions in grain yield and grain number of the main earhead for all genotypes was observed. Stone and Nicolas (1998) exposed two wheat varieties, heat-tolerant Egret and heat-sensitive Oxley, to very high temperature (40°C day/19°C night) for a ten day duration. As little as one day of heat treatment reduced grain mass by 14% in Oxley, but only 5% in Egret. A linear decrease in individual grain mass was observed with increasing heat treatment duration, such that after the first day of heat stress, every additional day of treatment reduced mature individual grain mass by a further 1.6%. They also observed an increase in grain protein percentage despite the fact that protein content per grain was reduced by heat. High temperatures during grain filling can significantly elevate protein concentration while lowering the functionality of protein (Corbellini *et al.*, 1997).

Effect on Chlorophyll Content

Higher chlorophyll content or leaf greenness indicates a low degree of photoinhibition of the photosynthetic apparatus, therefore reducing carbohydrate losses for grain growth (Farquhar *et al.,* 1989). Leaves are reported to rapidly turn yellow at high temperature, with a gradual decrease in the chlorophyll content, suggesting a drastic reduction in photosynthetic activity (Bencze *et al.,* 2005). Rahman *et al.* (2009) found chlorophyll content of the flag leaf lower (10%) in magnitude in the case of late-sown crop, and had obtained a significant positive correlation (r = 0.4) between grain yield and chlorophyll content of the flag leaf.

Loss of chlorophyll is usually attributed to membrane damage and leaf senescence (Liu and Huang, 2000). When plants are exposed to high-temperature stress, chlorophyll biosynthesis is inhibited (Tewari and Tripathy, 1998). Fokar *et al.* (1998) found a significant and positive correlation between the rate of chlorophyll loss and photosynthetic stem reserves, indicating higher potential for utilization of stem reserves for grain filling associated with accelerated leaf senescence .

ACCLIMATION TO HIGH TEMPERATURE STRESS

Acclimation requires expression of diverse stress-responsive genes to maintain metabolic homeostasis during stress or to be able to re-establish, subsequent to the stress period (Hahn and Li, 1990). Heat tolerance is generally defined as the ability of the plant to grow and produce economic yield under high temperatures. A cell-membrane system that remains functional during heat stress appears central to adaptation of plants to high temperature (Raison *et al.*, 1980). According to Larcher (1987), stress first leads to the destabilisation of plant functions, followed by normalisation and an increase in resistance. Several studies show that an acclimated plant survives when exposed to a temperature, otherwise lethal to a non-acclimated plant (Key *et al.*, 1981; Hong *et al.*, 2003). This phenomenon is the major aspect of acclimation response termed as acquired thermotolerance.

Relevance of a physiological or biochemical trait for thermo-tolerance is established by the major principle of induction of an acclimation response through means of initially exposing seedlings/plants to a less severe temperature before they are challenged with severe temperature and subsequently recovery growth is measured. The seedling survival and recovery growth is considered as criteria to arrive at optimum acclimation stress levels. In Lima beans, acclimation of leaf at 37°C prior to challenging with temperature of 45°C resulted in significantly less injury as compared to leaf directly exposed to 45°C (Keeler *et al.*, 2000). Many recent studies show that genetic variability is seen only upon acclimation treatment prior to severe stress (Kumar *et al.*, 1999; Burke, 2001) and the variability observed when seedlings were exposed directly to severe stress was marginal (Uma *et al.*, 1995). Hahn and Li (1990) have also shown that thermo-protection was phenomenally high in seedlings exposed to prior acclimation stress.

STEM RESERVE REMOBILIZATION

Stem reserves potentially offer a powerful resource for grain filling under any type of stress that impairs current assimilation, including heat stress (Schnyder 1993, Blum *et al.* 1994, Blum 1998, Fokar *et al.* 1998) and water deficit (Davidson and Chevalier 1992, Palta *et al.* 1994, Yang *et al.* 2000). Recently, attempts have been made to explore the potential of remobilized stem reserves in supporting grain filling under heat-stress condition (Blum *et al.* 1994, Fokar *et al.* 1998, Yang *et al.* 2002), but more work is needed to aid wheat breeders to assess the value of stem reserve remobilization as a selection criterion under heat-stress environments.

Under optimum temperature conditions, photosynthesis provides the greatest part of grain yield, accounting for up to 90% of the carbohydrates in durum wheat while remobilization accounts for only 10 to 34% (Arduini *et al.,* 2006; Alvaro *et al.,* 2008). Grain dry matter accumulation and the canopy respiration are approximately identical sinks for the photosynthetic assimilates, and together their total demand for assimilates are greater in comparison with canopy photosynthesis during the grain filling stage (Blum, 1998). Therefore, stem reserves are eventually necessary for complete grain filling. Grain yield and grain N are influenced by both pre- and post-anthesis growth and uptake processes, and translocation processes during the post-anthesis period (Palta and Fillery, 1995; Flood and Martin, 2001). Primary data on the accumulation of both dry matter and N at various growth stages are necessary to understand the processes of assimilation and partitioning of C and N in the context of plant growth and development. The contribution of pre-anthesis assimilates to grain filling and grain weight depends on the amount of matter that is mobilized between anthesis and maturity and the efficiency of conversion of the mobilized matter in the grains (Gebbing *et al.*, 1999). This makes anthesis stage particularly important since it represents the transition from the vegetative to reproductive phase of growth.

In wheat it has been reported that pre-anthesis reserves contribute 8 to 27% of the carbon in carbohydrates and 30 to 47% of the carbon in proteins of the grain (Gebbing and Schnyder, 1999). Reserve accumulation and storage capacity in the stem strongly depend on the growing conditions before anthesis. Post-anthesis, the rate of grain growth depends on intercepted radiation during grain-filling, and the recycling of carbohydrates that are stored in the stem or produced by the breakdown of leaf tissue during senescence.

The phenomenon of mobilization and transfer of reserves to the grains was first described in detail and interpreted by Pierre (1869). Early studies by Austin *et al.* (1977a,b) clearly demonstrated the role of photosynthates, stored long-term in stems, in grain filling using $^{14}CO_2$

pulse-labelling. Post-anthesis, 42% of the initially fixed label was ultimately recovered in the grain, but only 22% reached the grain within the first 10 days. During the latter part of grain filling when stem sections can lose up to 50% of their dry weight (Bonnett and Incoll, 1992), the disappearance of WSC accounts for 90 to 100% of this weight loss in internodes of barley and wheat (Austin, 1980; Blacklow *et al.*, 1984; Gent, 1994).

The accumulation of reserves in the different internodes starts near the end of extension growth, i.e. when the internodes above it is elongating rapidly (Bonnett and Incoll, 1992). Stored reserves and their contribution to grain can be estimated by measuring post-anthesis changes in internode water-soluble carbohydrate content during the grain-filling period (Blum *et al.*, 1994; Shakiba *et al.*, 1996), or estimated by difference in canopy dry weight at anthesis and at maturity excluding the grains (Flood *et al.*, 1995).

a. Water Soluble Carbohydrates (WSC)

Water soluble carbohydrates (WSC), the dominant component of non-structural carbohydrates (NSC), and nitrogen in the form of protein and amino acids are the main components of the pre-anthesis biomass which is potentially available for grain filling (Spiertz and Ellen, 1978; Simpson *et al.*, 1983; Schnyder, 1993). Grain growth and development in wheat depend on carbon from three sources: (i) carbohydrate produced after anthesis and translocated directly to the grains, (ii) carbohydrate produced after anthesis but stored temporarily in the stem before being remobilized to the grains, and (iii) carbohydrate produced before anthesis stored mainly in the stem and remobilized to grains during grain filling (Daniels *et al.*, 1982; Kobata *et al.*, 1992). Stem biomass is a function of partitioning between stem structural dry matter (cell-wall polysaccharides) and stem carbohydrates.

WSC are excess carbohydrates stored in vegetative organs such as stem, sheaths, and tiller base of cool season cereals during the period from stem elongation to the early phase of grain filling (until 10 to 20 DAA) and serve as temporary carbohydrate reserves (Gebbing, 2003). The reserve of WSCs is likely to be determined by the activity of enzymes associated with both fructan and sucrose metabolism in the wheat stem, and can account for more than 40% of total wheat stem dry weight (McCaig and Clarke, 1982; Housley, 2000), and potentially 20% of grain yield under normal conditions (Blum, 1998; Wardlaw and Willenbrink, 2000).

WSC in wheat stems is mainly composed of fructans, sucrose, glucose and fructose, with fructans being the major component at the late stage of WSC accumulation phase (Ruuska *et al.*, 2006; Xue *et al.*, 2008). Starch is found in small amounts in wheat stems but is generally not mobilised (Kiniry, 1993). At the stage of maximum WSC content fructans and sucrose represented 85 and 10%, respectively, of the WSC in wheat stem internodes (Blacklow *et al.*, 1984), though variations may exist in the speed of translocation of assimilates and its concentration in the phloem sap (Fisher, 1990). Though these leaves diurnally accumulate mainly sucrose, and also to a lesser extent starch (Trevanion, 2000), when assimilate supply is in excess of demands for growth, fructan accumulation is initiated (Wang *et al.*, 2000). Fructan accumulation occurs mainly in the extended internodes and leaf sheaths and usually terminates within three weeks following anthesis, after which fructan is gradually lost until grain maturity. The predominant role for fructans is to bridge the temporal gaps between resource availability and demands.

Degree of polymerization (DP) of wheat stem fructan does not exceed 9 (Hendrix *et al.*, 1986). Low molecular weight fructans (i.e. fructans containing less than 10 hexose units per molecule) predominate in the stem (Blacklow *et al.*, 1984). Wardlaw and Willenbrink (1994) found that during the accumulation of WSC in wheat stems, the mass of fructan with a DP greater than five continued to increase, though the mass of fructans of DP 3 to 5 reached a maximum and then remained constant. The large concentration of fructan observed in the internode of wheat and barley suggests that fructans are believed to be synthesized from sucrose in the central vacuole of parenchyma cells (Frehner *et al.*, 1984), and most, if not all, is stored there itself (Wagner *et al.*, 1983). The mobilization of stem WSC generally starts during the period of near constant (i.e. maximum) rate of dry matter accumulation in grains and coincides with a strong decrease in the net assimilation rate (Spiertz and Ellen, 1978; Bell and Incoll, 1990). Differences in the accumulation of WSCs in stems among genotypes could potentially result from various factors such as photosynthetic capacity, carbon use efficiency and carbon partitioning between stem reserve deposition and other physiological processes like maintenance of respiration, growth and cell wall synthesis (Xue *et al.*, 2008).

b. Nitrogen

The role of N stored in the stem prior to anthesis has been properly investigated under heat stress. Over 80% of the total N content accumulates at anthesis and accounts for about 50–100% of the total N content of the wheat grains (Daigger *et al.* 1976, Simpson *et al.* 1983). Nitrogen in the form of protein and amino acids are components of the pre-anthesis reserves that is potentially available for grain filling (Van Sanford and MacKown, 1987) though some amount support structural functions (Lhuillier-Soundélé *et al.*, 1999). In cereals the main source of nitrogen for the grains is nitrogen remobilized from the vegetative parts (Simpson *et al.*, 1983), and this accounts for 60 to 92% of the nitrogen accumulated in the wheat grains at maturity (Austin *et al.*, 1977b; Papakosta and Garianas, 1991; Ehdaie and Waines, 2001).

The amount of nitrogen in the vegetative organs of the plant at flowering can be used to estimate the amount of nitrogen available for remobilization (Cox *et al.*, 1986). Bhatia and Rabson (1976) had reported that in wheat, a 1% increase in grain protein required approximately an additional 6% of N supplied to the grains on the basis of protein stoichiometry. The source of this additional N can be direct soil uptake during grain filling, or greater remobilization of N from leaves and other plant parts (Cox *et al.*, 1985).

The mass of nitrogen contained in the above-ground vegetative structures of wheat is usually near maximum at anthesis (as high as 80 to 100% of total plant nitrogen at maturity) decreasing thereafter (Heitholt *et al.*, 1990). Post-anthesis, plants continue the assimilation of N, and significant quantities of N can be assimilated during grain filling (Cox *et al.*, 1985). It has been estimated that two-thirds of the grain N in wheat is derived from N assimilated before anthesis and one-third from assimilation during grain development (Pavlov and Kolesnik, 1974). Nitrogen balance-sheets also corroborate the fact that 50 to 100% of the nitrogen accumulating in the grain does originate from mobilization in the vegetative tissues (Pearman *et al.*, 1977; Spiertz, 1977).

STEM RESERVE REMOBILIZATION UNDER HEAT STRESS

In wheat , stress that reduces plant water status and inhibits current photosynthesis during grain filling reduces assimilate storage and induces stem storage conversion into soluble sugars and the mobilization of sugars into the grains (Davidson and Chevalier, 1992). Wheat canopy respires rapidly during grain filling (McCullough and Hunt, 1989), and flag leaf photosynthesis alone cannot support both respiration and grain growth under terminal stresses (Rawson *et al.*, 1983). In cereals increased mobilization efficiency of reserves from leaves, stem or other plant parts has, therefore, been suggested as a potential strategy to improve grain filling and yield under heat stress, especially the period before anthesis and grain filling (Gaunt and Wright, 1992; Gent, 1994). Thus stem reserves becomes more important for grain yield in cereal crops under abiotic stress (Blum *et al.*, 1991).

To elucidate causal agents of reduced grain filling in wheat under high temperatures, Wardlaw (1974) had examined three main components of the plant system including source (flag leaf blade), sink (earhead), and transport pathway (peduncle). A photosynthetic decline was reported at temperatures above 30°C during which radio-labelled assimilate movement out of the flag leaf was also observed to be optimum, though the rate of movement through the stem was independent of temperature from 1 to 50°C. Tahir and Nakata (2005) increased maximum temperature from 30 to 38°C during the reproductive growth phase of eighteen wheat genotypes under controlled environment and observed significant reduction in N remobilization efficiency of most genotypes, and an increase in TNC remobilization efficiency and significant variation among genotypes was also noted. N remobilization efficiency significantly correlated with grain yield, grain weight and harvest index . TNC at 5 DAA negatively correlated with N at 5 DAA and harvest index, but the TNC remobilization efficiency under heat stress positively correlated with main stem grain yield, grain weight and harvest index. Chlorophyll loss rate from flag leaf also positively correlated with N and TNC remobilization efficiencies under heat stress suggesting a link between leaf senescence and remobilization efficiency. Studies on mobilized stem reserves through the years during grain filling in wheat has been shown to represent 22 to 65% of the dry matter that accumulates in the grain, and could account for increasing tolerance to high temperature (Bidinger *et al.*, 1977; Bell and Incoll, 1990; Davidson and Chevalier, 1992; Blum *et al.*, 1994). Ercoli *et al.* (2010) evaluated the effect of temperature during grain-filling in two Italian durum wheat varieties, Appio and Creso, grown with different nitrogen inputs. Dry matter remobilization during grain filling increased with the increase of N rate, but the increases were greater at 28/23°C compared to 20/15°C. However, the contribution of remobilized N to grain N content was higher at 20/15°C (78 to 95%) compared to 28/23°C (45 to 67%) and was not affected by supplied N rate. Higher temperature promoted remobilization of dry matter and restrained current photosynthesis but reduced grain yield, indicating that the loss of photosynthesis could not compensate for the gain from increased remobilization.

Recent studies have also attested to the significance of WSC in the stems as a physiological trait indicative of drought tolerance (Diab *et al.*, 2004; Ehdaie *et al.*, 2006). In barley pre-anthesis reserves contributed up to 74% to the grain yield (Gallagher *et al.*, 1975) and up to 57% in wheat (Gallagher *et al.*, 1976) when crops suffered from severe post-anthesis drought. Palta *et al.* (1994) found a reduction in post-anthesis assimilation by 57% and a 36% increase in remobilization of reserves. They estimated that stored C and N contributed for 64 and 81% to total grain C and N, respectively. High correlation between storage NSCs of wheat stems and yield among wheat cultivars under drought conditions has been reported (Takai *et al.*, 2006; Foulkes *et al.*, 2007)

Clarke *et al.* (1984) demonstrated very well that the impact of stem reserves should be evaluated only under stress conditions, which equally inhibit crop assimilation during grain filling in all materials tested. Evidence has pointed that the magnitude of the response can be modified by a change in the source-sink balance, with a greater response under source limiting conditions (Wardlaw *et al.*, 1980). Ugalde and Jenner (1990) and Willenbrink *et al.* (1998) demonstrated a decrease in fructan content in wheat peduncle during grain filling, which was more pronounced under source-limiting conditions, but increased under sink-limiting conditions. Augmenting the contribution of carbohydrates reserves accumulated during vegetative growth to grain filling may be worthwhile for improving harvest index (Loss and Siddique, 1994; Boras, 2004).

Conclusions and Future Perspectives

Heat tolerance is generally defined as the ability of the plant to grow and produce economic yield under high temperatures. A cell-membrane system that remains functional during heat stress appears central to adaptation of plants to high temperature (Raison *et al.*, 1980). According to Larcher (1987), stress first leads to the destabilisation of plant functions, followed by normalisation and an increase in resistance.

Current assimilation, as a source of carbon for grain filling, depends on the viable light intercepting green surfaces of the plant after anthesis. This source normally diminishes due to natural senescence and the effect of various stresses. At the same time the demand of the growing kernel increases, in addition to the demand due to maintenance and respiration of the live plant biomass. Hence, an important source of carbon for grain filling is stem reserve. Even under mild conditions, current assimilates may be limited for normal grain filling as observed in a three year study in Connecticut (Gent, 1994).

Breeding has resulted in considerable diversity of genotypes/cultivars that respond differently to the interaction of these environmental variables, and indeed, the genotype by environment interaction is important in selecting cultivars for specific environments. High-yielding genotypes do not perform on par with abiotic stress-adapted genotypes when yield is depressed below a crossover point (Blum, 1994). Therefore a genotype with stable and high yield across the environments would be more suitable as a cultivar and also as a donor parent for further breeding in hot environments that vary over the years and within a particular year across locations. Heritability of a trait expresses the reliability of the phenotypic value as a guide to the breeding value and the trait utility within the selection process. Today, crop production systems with their high yield outputs, improvement in grain filling has become more challenging than ever (Saini and Westgate, 2000; Zahedi and Jenner, 2003). Increasing sustainable food production, equitably, without compromising environmental integrity, therefore, remains a major challenge. High temperature stress is a major environmental stress limiting wheat (*Triticum aestivum* L.) productivity. Continuous high-temperature stress for wheat has been defined as when the mean average temperature of the coolest month is greater than 17.5°C (Fischer and Byerlee, 1991). Even short periods (5 to 6 days) of exposure of wheat crops to temperatures of 28 to 32°C result in significant decreases in yield of 20% or more (Stone and Nicolas, 1994). It is estimated that in India alone around 13.5 million ha of area under wheat is heat stressed (Joshi *et al.*, 2007a). According to the Recent Assessment Report of the Intergovernmental Panel on Climate Change, global earth temperatures have increased by 0.74°C in last century, and are likely to increase by 1.1 to 6.4°C by 2100 (IPCC, 2007) which will further aggravate the problem. Faced with the prospect of a rapidly changing climate, cultivars that are recommended for use at present might not be suitable if the climate changes. Breeding for a new cultivar usually takes 10 to12 years if the target traits are known and the environment in which to test new lines is available. Yield, which is a major selection criterion in drought and heat tolerance programmes, is complex as it is determined by several physiological, biochemical and metabolic plant processes whose genetics and associations are largely unclear (Liley and Ludlow, 1996). Breeders are confronted with difficulties in their selection programmes owing to heterogeneity in field-based bioassays and the uncertainty of environmental pressures. A further problem is that the heritability of yield is reduced under the conditions where yield is reduced (Ludlow and Muchow, 1990). The genes securing heat tolerance may be lost in the breeding programs which rely mainly on only empirical selection (Reynolds *et al.,* 1994). The ability to determine the thermo-sensitivity of a given genotype would provide a means to the genetic improvement of this trait. Screening techniques are required to characterize this variation, to identify appropriate breeding material, and ultimately to select for improved adaptation.

The gradual increase and then decrease in the temperature of the phytotrons mimics the actual situation encountered by heat-stressed plants under natural conditions. This has been proved more efficient than the sudden increase in studying the effect of heat stress on wheat (Stone and Nicolas 1998) and led to clear variations among genotypes.

Relative humidity was always in the range of 30–60% during the period of the heat-stress treatment although it was not controlled. Genotypic variation was observed in reaction to heat stress in terms of grain yield and its components. Significant genotypic variations were also found in the ability to accumulate N and TNC at 5DAA and later to remobilize them during grain filling under heat-stress conditions.

The use of gradual increase and then decrease in temperature might have helped in revealing these variabilities. Although harvest index of the main stem negatively correlated with the TNC concentration at 5DAA, it was positively and strongly correlated with TNC remobilization efficiency.

This suggests that TNC remobilization efficiency is more important than how much of TNC a genotype can accumulate at anthesis. The significant correlations of N and TNC remobilization efficiencies with the rate of chlorophyll loss under heat stress conditions suggested a link between N and TNC remobilization efficiencies and leaf senescence.

References

Al-Khatib, K. and Paulsen, G.M. (1984). Mode of high temperature injury to wheat during grain development, *Physiol. Plant*, 1: 363-368.

Al-Khatib, K. and Paulsen, G.M. (1990). Photosynthesis and productivity during high temperature stress of wheat genotypes from major world regions, *Crop Sci.*, 30: 1127-1132.

Alvaro, F., Isidro, J., Villegas, D., Garcia del Moral, L.F. and Royo, C. (2008). Breeding effects on grain filling, biomass partitioning and remobilization in Mediterranean durum wheat, *Agron. J.*, 100: 361-370.

Arduini, I., Masoni, A., Ercoli, L. and Mariotti, M. (2006). Grain yield, and dry matter and nitrogen accumulation and remobilization in durum wheat as affected by variety and seeding rate, *Eur. J. Agron.*, 25: 309-318.

Austin R.B., Edrich J.A., Ford M.A. and Blackwell R.D. (1977b). The nitrogen economy of winter wheat, *J. Agril. Sci.*, 88: 159-167.

Austin, R.B., Bringham, J., Blackwell, R.D., Evans, L.T., Ford, M.A., Morgan, C.L. and Taylor, M. (1980a). Genetic improvements in winter wheat yields since 1900 and associated physiological changes, *J. Agric. Sci.*, 94: 675-689.

Austin, R.B., Edrich, J.A., Ford, M.A. and Blackwell, R.D.(1977a). The fate of the dry matter, carbohydrates and ^{14}C lost from the leaves and stems of wheat during grain filling, *Ann. Bot.*, 41: 1309-1321.

Badaruddin, M., Reynolds, M.P. and Ageeb, O.A.A. (1999). Wheat management in warm environments: Effect of organic and inorganic fertilizer, irrigation frequency, and mulching, *Agron. J.*, 91: 975-983.

Balla, K., Bencze, S., Janda, T. and Veisz, O. (2009). Analysis of heat stress tolerance in winter wheat, *Acta Agronomica Hungarica*, 57(4): 437-444.

Bencze, S., Veisz, O. and Bedő, Z. (2005). Effects of elevated CO_2 and high temperature on the photosynthesis and yield of wheat, *Cereal Res. Commun.*, 33: 385-388.

Berry, J. and Björkman, O. (1980). Photosynthetic response and adaptations to temperature in higher plants, *Ann. Rev. Plant Physiol.*, 31: 491-532.

Bhatia, C.R. and Rabson, R. (1976). Bioenergetic considerations in cereal breeding for protein improvement, *Science*, 194: 1418-1421.

Bidinger, F., Musgrave, R.B. and Fischer, R.A. (1977). Contribution of stored pre-anthesis assimilate to grain yield in wheat and barley, *Nature*, 270: 431-433.

Björkman, O., Badger, M.R. and Armond, P.A. (1980). Response and adaptation of photosynthesis to high temperature, In: *Adaptation of plants to water and high temperature stress*, Turner, N.C. and Kramer, P.J. (Eds.), John Wiley & Sons, NY, pp. 233-249.

Blacklow, W.M., Darbyshire, B. and Pheloung, P. (1984). Fructans polymerised and depoylmerised in the internodes of winter wheat as grain-filling progressed, *Plant Sci. Lett.*, 36: 213-218.

Blum, A. (1998). Improving wheat grain filling under stress by stem reserve mobilization, *Euphytica*, 100: 77-83.

Blum, A., Klueva, N. and Nguyen, H.T. (2001). Wheat cellular thermotolerance is related to yield under heat stress, *Euphytica*, 117: 117-123.

Blum, A., Shpiler, L., Golan, G., Mayer, J. and Sinmena, B. (1991). Mass selection of wheat for grain filling without transient photosynthesis, *Euphytica*, 54: 111-116.

Blum, A., Sinmena, B., Mayer, J., Golan, G. and Shpiler, L. (1994). Stem reserve mobilisation supports wheat-grain filling under heat stress, *Aust. J. Plant Physiol.*, 21: 771-781.

Bonnett, G.D. and Incoll, L.D. (1992). The potential pre-anthesis and post-anthesis contributions of stem internodes to grain yield in crops of winter barley, *Ann. Bot.*, 69: 219-225.

Boras, L., Slafer, G.A. and Otegui, M.E. (2004). Seed dry weight response to source-link manipulations in wheat, maize and soybean, *Field Crops Res.*, 86: 131-146.

Bukhov, N.G., Wiese, C., Neimanis, S. and Heber, U. (1999). Heat sensitivity of chloroplasts and leaves: leakage of protons from thylakoids and reversible activation of cyclic electron transport, *Photosyn. Res.*, 59: 81-93.

Burke, J.J. (2001). Identification of genetic diversity and mutations in higher plant acquired thermotolerance, *Physiol. Plant*, 112: 167-170.

Camejo, D., Rodr´ıguez, P., Morales, M.A., Dell'amico, J.M., Torrecillas, A. and Alarc´on, J.J. (2005). High temperature effects on photosynthetic activity of two tomato cultivars with different heat susceptibility, *J. Plant Physiol.*, 162: 281-289.

Chatrath, R., Mishra, B. and Shoran, J., Yield potential survey-India. In: *Extended* abstracts of the international symposium on wheat yield potential, "Challenges to International Wheat Breeding".

Clarke, J.M., Townley-Smith, T.F., McCaig, T.N. and Green, D.G. (1984). Growth analysis of spring wheat cultivars of varying drought resistance, *Crop Sci.*, 24: 537-541.

Conroy, J.P., Seneweera, S., Basra, A.S., Rogers, G. and Nissen-Wooller, B. (1994). Influence of rising atmospheric CO_2 concentrations and temperature on growth, yield and grain quality of cereal crops, *Aust. J. Plant Physiol.*, 21: 741-758.

Corbellini, M., Carnevar, M.G., Mazza, L., Ciaffi, M., Lafiandra, E. and Borghi, B. (1997). Effect of the duration and intensity of heat shock during grain filling on dry matter and protein accumulation, technological quality, and protein composition in bread wheat and durum wheat, *Aust. J. Plant Physiol.*, 24: 245-260.

Cox, M.C., Qualset, C.O. and Rains, D.W. (1985). Genetic variation for nitrogen assimilation and translocation in wheat. II. Nitrogen assimilation in relation to grain yield andprotein, *Crop Sci.*, 25: 435-440.

Cox, M.C., Qualset, C.O. and Rains, D.W. (1986). Genetic variation or nitrogen assimilation and translocation in wheat. III. Nitrogen translocation in relation to grain yield and protein, *Crop Sci.*, 26: 737-740.

Daniels, R.W., Alcock, W.B. and Scarisbrick, D.H. (1982). A reappraisal of stem reserve contribution to grain yield in spring barley (*Hordeum vulgare* L.), *J. Agric. Sci.*, 98: 47-355.

Davidson, D.J. and Chevalier. P.M. (1992). Storage and remobilization of water-soluble carbohydrates in stem of spring wheat, *Crop Sci.*, 32: 186-190.

De Ronde, J.A.D., Cress, W.A., Kruger, G.H.J., Strasser, R.J. and Staden, J.V. (2004). Photosynthetic response of transgenic soybean plants containing an *Arabidopsis P5CR* gene, during heat and drought stress, *J. Plant Physiol.*, 61: 1211-1244.

Demirevska-Kepova, K., Hölzer,R., Simova-Stoilova, L. and Feller, U. (2005). Heat stress effects on ribulose-1,5-bisphosphate carboxylase/oxygenase, rubisco binding protein and rubisco activase in wheat leaves, *Biologia Plantarium*, 49: 521-525.

Diab, A.A., Teulat-Merah, B., Ozturk, N. and Benscher, Z.D. (2004). Identification of drought-inducible genes and differentially expressed sequence tags in barley, *Theor. Appl. Genet.*, 109: 1417-1425.

Ehdaie, B. and Waines, J.G. (2001). Sowing date and nitrogen rate effects on dry matter and nitrogen partitioning in bread and durum wheat, *Field Crops Res.*, 73: 47-61.

Ehdaie, B., Alloush, G.A.M., Madore, A. and Waines, J.G. (2006). Genotypic variation for stem reserves and mobilization in wheat, *Crop Sci.*, 46: 2093-2103.

Ercoli, L., Arduini, I., Mariotti, M. and Masoni, A. (2010). Post-anthesis dry matter and nitrogen dynamics in durum wheat as affected by nitrogen and temperature during grain filling, *Cereal Res. Comm.*, 38(2): 294-303.

Farquhar, G.D., Wong, S.C., Evans, J.R. and Hubick, K.T. (1989). Photosynthesis and gas exchange, In: *Plants under Stress,* Jones, H.G., Flowers, T.J. and Jones, M.B. (Eds), Cambridge University Press, Cambridge, pp. 47-69.

Ferris, R., Ellis, R.H., Wheeler, T.R. and Hadley, P. (1998). Effect of high temperature stress at anthesis on grain yield and biomass of field grown crops of wheat, *Plant Cell Environ.*, 34: 67-78.

Fischer, R.A. and Byerlee, D.B. (1991). Trends of wheat production in the warmer areas: Major issues and economic considerations. In*: Wheat for the non-traditional warm areas.* Proc. Conf., Iguazu, Brazil, CIMMYT, Mexico, DF, pp. 3-27.

Fisher, D.B. (1990). Measurement of phloem transport rates by an indicator-dilution technique, *Plant Physiol.*, 94: 455-462.

Flood, R.G. and Martin, P.J. (2001). Nitrogen accumulation and distribution at anthesis and maturity in ten wheats grown at three sites in north-western Victoria, *Aust. J. Exp. Agric.*, 44: 533-540.

Flood, R.G., Martin, P.J. and Gardener, W.K. (1995). Dry matter accumulation and partitioning and its relationships to grain yield in wheat, *Aust. J. Exp. Agric.*, 35: 495-502.

Fokar, M., Henry, T.N. and Blum, A. (1998b). Heat tolerance in spring wheat. II. Grain filling, *Euphytica*, 104: 9-15.

Foulkes, M.J., Sylvester-Bradley, R., Weightman, R. and Snape, J.W. (2007). Identifying physiological traits associated with improved drought resistance in winter wheat, *Field Crops Res.*, 103: 11-24.

Frank, A.B. and Bauer, A. (1997). Temperature effects prior to double ridge on apex development and phyllochron in spring barley, *Crop Sci.*, 37: 1527-1531.

Gallagher, J.N., Biscoe, P.V. and Hunter, B. (1976). Effects of drought on grain growth, *Nature*, 264: 541-542.

Gallagher, J.N., Biscoe, P.V. and Scott, R.K. (1975). Barley and its environment. V. Stability and grain weight, *J. App. Ecol.*, 12: 319-336.

Gebbing, T. (2003). The enclosed and exposed part of the peduncle of wheat (*Triticum aestivum*)- spatial separation of fructan storage, *New Phytol.*, 159: 245-252.

Gent, M.P.N. (1994). Photosynthate reserves during grain filling in winter wheat. *Agron. J.*, 86: 159-167.

Gibson, L.R. and Paulsen, G.M. (1999). Yield components of wheat grown under high temperature stress during reproductive growth, *Crop Sci.*, 39: 1841-1846.

Guilioni, L., W′ery, J. and Lecoeur, J. (2003). High temperature and water deficit may reduce seed number in field pea purely by decreasing plant growth rate, *Funct. Plant Biol.*, 30: 1151-1164.

Gupta, P.K., Mir, R.R., Mohan, A. and Kumar, J. (2008). Wheat genomics: present status and future prospects, *Int. J. Plant Genomics*, 2008: 896451.

Gupta, R.K. (2003). Quality of Indian wheat and infrastructure for analysis. In: A compendium of the training program on wheat improvement in eastern and warmer regions of India: Conventional and non-conventional approaches

Hahn, G.M. and Li, G.C. (1990). Thermo-tolerance, thermo-resistance and thermo-sensitization. In: *Stress proteins in biology and medicine*, Morimoto, R.I., Tissieres, A. and Georgopoulos, C. (Eds.), Cold Spring Harbor Press, NY, pp. 79-100.

Hall, A.E. (1992). Breeding for heat tolerance. *Plant Breed. Rev.*, 10: 129-168.

Havaux, M. (1998). Carotenoids as membrane stabilizers in chloroplasts *Trends Plant Sci.*, 3: 147-151.

Heitholt, J.J., Croy, L.I., Maness, N.O. and Nguyen, H.T. (1990). Nitrogen partitioning in genotypes of winter wheat differing in grain N concentration, *Field Crops Res.*, 23: 13-144.

Hendrix, J.E., Linden, J.C., Smith, D.H., Ross, C.W. and Park, I.K. (1986). Relationship of pre-anthesis fructan metabolism to grain number in winter wheat (*Triticum aestivum* L.). *Aus.J. Plant Physiol.*, 13: 391-398.

Herzog, H. (1982). Relation of source and sink during the grain filling period in wheat and some aspects of its regulation, *Physiol. Plant*, 56: 155-166.

Hong, S.W., Lee, U. and Vierling, E. (2003). Arabidopsis hot mutants define multiple functions required for acclimation to high temperatures, *Plant Physiol.*, 132: 757-767.

Housley, T.L. (2000). Role of fructans redistributed from vegetative tissues in grain filling of wheat and barley, *Develop. Crop Sci.*, 26: 207-222.

Howarth, C.J. (2005). Genetic improvements of tolerance to high temperature. In: *Abiotic Stresses: Plant Resistance Through Breeding and Molecular Approaches*, Ashraf, M. and Harris, P.J.C. (Eds.), Howarth Press Inc., NY.

IPCC, Climate Change. (2007). The Physical Science Basis, Summary for Policymakers, Contribution of Working Group I to the Fourth Assessment Report of the Intergovernmental Panel on Climate Change.

Joshi, A.K., Chand, R., Arun, B. and Singh, G. (Eds.) (2004). NATP project, (ICAR), BHU, Varanasi, India, 2004.

Joshi, A.K., Mishra, B., Chatrath, R., Ortiz Ferrara, G. and Singh, R.P. (2007a). Wheat improvement in India: Present status, emerging challenges, and future prospects, *Euphytica*, 157: 431-446.

Joshi, A.K., Ortiz-Ferrara, G., Crossa, J., Singh, G., Alvarado, G., Bhatta, M.R., Duveiller, E., Sharma, R.C., Pandit, D.B., Siddique, A.B., Das, S.Y., Sharma, R.N. and Chand, R. (2007b). Associations of environments in South Asia based on spot blotch disease of wheat caused by *Cochliobolus sativus*, *Crop Sci.*, 47: 1071-1081.

Keeler, S.J., Boettger, C.M. *et al.* (2000) Acquired thermotolerance and expression of the HSP100/ClpB genes of lima bean, *Plant Physiol.*, 123: 1121-1132.

Key, J.L, Lin, C.Y. and Chen, Y.M. (1981). Heat shock proteins of higher plants, *Proc. Nat. Acad. Sci.,* 70: 3526-3530.

Kiniry, J.R. (1993). Non-structural carbohydrate utilization by wheat shaded during grain growth, *Agron. J.*, 85: 844-849.

Krause, G.H. and Santarius, K.A. (1975). Relative thermostability of chloroplast envelope, *Planta*, 127: 285-299.

Kumar, G., Krishnaprasad, B.T. *et al.* (1999). Enhanced expression of heat shock proteins in thermotolerant lines of sunflower and their progenies selected on the basis of temperature induction response (TIR), *Theor. Appl. Gen.*, 99: 359-367.

Kuroyanagi, T. and Paulsen. G.M. (1985). Mode of high temperature injury to wheat II. Comparisons of wheat and rice with and without inflorescence, *Physiol. Plant*, 65: 203-208.

Larcher W. (1987). Stress bei Pflanzen (Stress in plants), *Naturwissenschaften*, 74: 158-167.

Larcher, W. (1995). Physiological Plant Ecology, 3rd ed. Springer, NY.

Leithold, B., Muller, G., Weber, W.E. and Westermann, T. (1997). Investigations on heat tolerance of spring wheat varieties of different origin under growth chamber conditions, *J. Agron. Crop Sci.*, 179: 115-122.

Lhuillier-Soundélé, A., Munier-Jolain, N.G. and Ney, B. (1999). Dependence of seed nitrogen concentration on plant nitrogen availability during the seed filling in pea, *Eur. J. Agron.*, 11: 157-166.

Liley, J.M. and Ludlow, M.M. (1996). Expression of osmotic adjustment and dehydration tolerance in disease rice lines, *Field Crop Res.,* 48: 185-197.

Lillemo, M., van Ginkel, M., Trethowan, R.M., Hernandez, E. and Crossa, J. (2005). Differential adaptation of CIMMYT bread wheat to global high temperature environments, *Crop Sci.*, 45: 2443-2453.

Liu, X. and Huang, B. (2000). Heat stress injury of creeping bentgrass in relation to membrane lipid peroxidation, *Crop Sci.*, 40: 503-510.

Loss, S.P. and Siddique, K.M.H. (1994). Morphological and physiological traits associated with wheat yield increase in Mediterranean environments, *Adv. Agron.*, 52: 229-276.

Ludlow, M.M. and Muchow, R.C. (1990). A critical evaluation of trait for improving crop yields in water-limited environments, *Adv. Agron.*, 42: 107-153.

McCaig, T.N. and Clarke, J.M. (1982). Seasonal changes in non-structural carbohydrate levels of wheat and oats grown in a semiarid environment, *Crop Sci.*, 22: 963-970.

McCullough, D.E. and Hunt, L.A. (1989). Respiration and dry matter accumulation around the time of anthesis in field stands of winter wheat (*Triticum aestivum*), *Ann. Bot.* 63: 321-329.

McMaster, G.S. (1997). Phenology, development, and growth of the wheat (*Triticum aestivum* L.) shoot apex: a review, *Adv Agron.*, 59: 63-118.

Morison J.I.L. (1996). Global environmental change impacts on crop growth and production in Europe. Implications of global environmental change for crops in Europe, *Aspects of Applied Biol.,* 45, 62-74.

Murchie, E.H., Pinto, M. and Horton, P. (2009). Agriculture and the new challenges for photosynthesis research, *New Phytol.*, 181: 532-552.

Myers, N. (1999). The next green revolution: Its environmental underpinnings, *Curr. Sci.*, 76: 507-513.

Palta, J.A. and Fillery, I.R.P. (1995). N application enhances remobilization and reduces losses of pre-anthesis N in wheat grown on a duplex soil, *Aust. J. Agric. Res.*, 46: 519-531.

Palta, J.A., Kobalta, T., Turner, N.C. and Fillery, I.L. (1994). Remobilisation of carbon and nitrogen in wheat as influenced by post-anthesis water deficits, *Crop Sci.*, 34: 118-124.

Papakosta, D.K. and Gagianas, A.A. (1991). Nitrogen and dry matter accumulation, remobilization, and losses for Mediterranean wheat during grain filling, *Agron. J.*, 83: 864-870.

Pavlov, A.N. and Kolesnik, T.I. (1974). Factors determining different levels of protein accumulation in the grain of high- and low-protein varieties of wheat, *Sov. Plant Physiol.*, 22: 267-272.

Pearman, I., Thomas, S.M. and Thorne, G.N. (1977). Effects of nitrogen fertilizer on growth and yield of spring wheat, *Ann. Bot.*, 41: 93-108.

Pierre, J.I. (1869). Recherches experimentales sur le development dublé et sur la repartition, dans ses differentes partes, des elements qui le constituent a diverses époques de son development, Mémoires de la Société Linneenne de Normandie, pp. 1-220.

Rahman, M.M., Hossain, A., Hakim, M.A., Kabir, M.R. and Shah, M.M.R. (2009). Performance of wheat genotypes under optimum and late sowing condition, *Int. J. Sustain. Crop Prod.*, 4(6): 34-39.

Raison, J.K., Berry, J.A., Armond, P.A. and Pike, C.S. (1989). Membrane properties in relation to the adaptation of plants to temperature stress, In: *Adaptation of plants to water and high temperature stress,* Turner, N.C. and Kramer, P.J. (Eds.), John Wiley & Sons, NY, pp. 261-273.

Rawson, H.M., Hindmarsh, J.H., Fischer, R.A. and Stockman, Y.M. (1983). Changes in leaf photosynthesis with plant ontogeny and relationships with yield per ear in wheat cultivars & 120 progeny, *Aust. J. Plant Physiol.*, 10:503-514.

Rekika, D., Monneveux, P. and Havaux, M. (1997). The *in vivo* tolerance of photosynthetic membranes to high and low temperatures in cultivated and wild wheats of the *Triticum* and *Aegilops* genera, *J. Plant Physiol.*, 150: 734-738.

Reynolds, M.P. and Godinez, D. (Eds), (2006). CIMMYT, Mexico, DF, p 53, 2006.

Reynolds, M.P., Balota, M., Delgado, M.I.B. Amani, I. and Fischer, R.A. (1994). Physiological and morphological traits associated with spring wheat yield under hot irrigated conditions, *Aus. J. Plant Physiol.*, 21: 717-730.

Ruuska, S.A., Rebetzke, G.J., van Herwaarden, A.F., Richards, R.A., Fettell, N.A., Tabe, L. and Jenkins, C.L.D. (2006). Genotypic variation in water-soluble carbohydrate accumulation in wheat, *Func. Plant Biol.*, 33: 799-809.

Ruwali, K.N. and Prasad, L. (1991). Effect of temperature stress on yield production of wheat, In: *Proc.* Int. Conf. Plant Physiol., BHU, Varanasi, pp. 271-277.

Saini, H.S. and Westgate, M.E. (2000). Reproductive development in grain crops during drought. *Adv. Agron.*, 68: 59-96.

Salvucci, M.E. and Crafts-Brandner, S.J. (2004). Inhibition of photosynthesis by heat stress: the activation state of Rubisco as a limiting factor in photosynthesis, *Physiol. Plant.*, 120: 179-186.

Savchenko, G.E., Klyuchareva, E.A., Abrabchik, L.M. and Serdyuchenko, E.V. (2002). Effect of periodic heat shock on the membrane system of etioplasts, *Russ. J. Plant Physiol.*, 49: 349-359.

Schnyder, H. (1993). The role of carbohydrate storage and redistribution in the source-sink relations of wheat and barley during grain filling - a review, *New Phytol.*, 123: 233-245.

Seeman, J.R., Berry, J.A. and Downton, W.J.S. (1984). Photosynthetic response and adaptation of high temperature in desert plants: A comparison of gas exchange and fluorescence methods for study of thermal tolerance, *Plant Physiol.,* 75: 364-368.

Shakiba, M.R., Ehdaie, B. Madore, M.A. and Waines, J.G. (1996). Contribution of internode reserves to grain yield in a tall and semi-dwarf spring wheat, *J. Gent. Breed.*, 50: 91-100.

Sharma, S.N. (1946). Thesis Assoc. Imp. Agril. Res. Inst., 1946.

Shpiler, L. and Blum, A. (1991). Heat tolerance for yield and its components in different wheat cultivars, *Euphytica*, 51: 257-263.

Siddique, M.R.B., Hamid, A. and Islam M.S. (2000). Drought stress effects on water relations of wheat, *Botanical Bulletin Academia Sinica*, 41: 35-39.

Simpson, R.D., Lambers, H. and Dalling, M.J. (1983). Nitrogen redistribution during grain growth in wheat (*Triticum aestivum* L.) IV. Development of a quantitative model of the translocation of nitrogen to the grain, *Plant Physiol.*, 71: 7-14.

Sinha, S.K. and Swaminathan, M.S. (1991). Deforestation, Climate Change and Sustainable Nutrition Security: A Case Study of India, *Climate Change*, 19: 201-209.

Spiertz, J.H.J. (1974). Grain growth and distribution of dry matter in the wheat plant as influenced by temperature, light energy and ear size, *Neth. J. Agric. Sci.*, 22: 207-220.

Spiertz, J.H.J. (1977). The influence of temperature and light intensity on grain growth in relation to the carbohydrate and nitrogen economy of the wheat plant, *Neth. J. Agric. Sci.*, 25: 182-197.

Spiertz, J.H.J. and Ellen, J. (1978). Effects of nitrogen on crop development and grain growth of winter wheat in relation to assimilation and utilization of assimilates and nutrients, *Neth. J. Agric. Sci.*, 26: 210-231.

Stone, P.J. and Nicholas, M.E. (1998). The effect of duration of heat stress during grain filling on two wheat varieties differing in heat tolerance: Grain growth and fractional protein accumulation, *Aust. J. Plant Physiol.*, 25: 13-20.

Stone, P.J. and Nicolas, M.E. (1994). Wheat cultivars vary widely in their responses of grain yield and quality to short periods of post-anthesis heat stress, *Aust. J. Plant Physiol.*, 21: 887-900.

Stoy, V.(1965). Photosynthesis, respiration, and carbohydrate accumulation in spring wheat in relation to yield, *Physiol. Plant. Suppl.*, 4: 1-125.

Tahir, I.S.A. and Nakata, N. (2005). Remobilization of nitrogen and carbohydrate from stems of bread wheat in response to heat stress during grain filling, *J. Agron. Crop Sci.*, 191: 106-115.

Takai, T., Shoji, M., Nishio, T., Ohsumi, A., Shiraiwa, T. and Horie, T. (2006). Rice yield potential is closely related to crop growth rate during late reproductive period, *Field Crops Res.*, 96: 328-335.

Takeuchi, T.S. and Thornber, J.P. (1994). Heat-induced alternations in thylakoid membrane protein composition in barley, *Aust. J. Plant Physiol.,* 21: 759-770.

Tewari, A.K. and Tripathy, B.C. (1998). Temperature-stress-induced impairment of chlorophyll biosynthetic reactions in cucumber and wheat, *Plant Physiol.*, 117: 851-858.

Thornley, J.H.M. (1971). Energy, respiration, and growth in plants, *Ann. Bot.*, 35: 721-728.

Trevanion, S.J. (2000). Photosynthetic carbohydrate metabolism in wheat (*Triticum aestivum* L.) leaves: optimization of methods for determination of fructose 2,6-bisphosphate, *J. Exp. Bot.*, 51: 1037-1045.

Ugalde, T.D. and Jenner, C.F. (1990). Substrate gradients and regional patterns of dry matter deposition within developing wheat endosperm. I. Carbohydrates, *Aust. J. Plant Physiol.*, 17: 377-394.

Uma, S., Prasad, T.G. and Udayakumar, M. (1995). Genetic variability in recovery growth and synthesis of stress proteins in response to polyethylene glycol and salt stress in Finger millet, *Ann. Bot.*, 76: 43-49.

Van Sanford, D.A. and MacKown, C.T. (1987). Cultivar differences in nitrogen remobilization during grain fill in soft red winter wheat, *Crop Sci.*, 27: 295-300.

Wagner, W., Keller, F. and Wiemken, A. (1983). Fructan metabolism in cereals: Induction in leaves and compartmentation in protoplasts and vacuoles, *Zeitschrift fur Pflanzenphysiologie*, 112: 359-372.

Wahid, A., Gelani, S., Ashraf, M. and Foolad, M.R. (2007). Heat tolerance in plants: An overview, *Environ. Expt. Bot.*, 61: 199-223.

Wang, C., Van den Ende, W. and Tillberg, J.E. (2000). Fructan accumulation induced by nitrogen deficiency in barley leaves correlates with the 6-SFT mRNA level, *Planta*, 211: 701-707.

Wardlaw, I.F. and Willenbrink, J. (1994). Carbohydrate storage and mobilisation by the culm of wheat between heading and grain maturity: the relation to sucrose synthase and sucrose phosphate synthase, *Aust. J. Plant Physiol.*, 21: 255-271.

Wardlaw, I.F. and Willenbrink, J. (2000). Mobilization of fructan reserves and changes in enzyme activities in wheat stems correlate with water stress during kernel filling, *New Phytol.*, 148: 413-422.

Wardlaw, I.F. and Wrigley, C.W. (1994). Heat tolerance in temperate cereals: an overview, *Aust. J. Plant Physiol.,* 21: 695-703.

Wardlaw, I.F., Scofield, I. and Cartwright, P.M. (1980). Factors limiting the rate of dry matter accumulation in grain of wheat grown at high temperature, *Aust. J. Plant Physiol.*, 7: 387-400.

Warrington, I.J., Dunstone, R.L. and Green, L.M. (1977). Temperature effects at three development stages on yield of the wheat ear, *Aust. J. Agric. Res.*, 28: 11-27.

Willenbrink, J., Bonnett, G.D., Willenbrink, S. and Wardlaw, I.F. (1998). Changes of enzyme activities associated with the mobilization of carbohydrate reserves (fructans) from the stem of wheat during kernel filling, *New Phytol.*, 139: 471-478.

Xu, Q., Paulsen, A.Q., Guikema, J.A. and Paulsen, G.M. (1995). Functional and ultrastructural injury to photosynthesis in wheat by high temperature during maturation, *Envir. Expt. Bot.*, 35: 43-54.

Xue, G.P., McIntyre, C.L., Jenkins, C.L.D., Glassop, D., van Herwaarden, A.F. and Shorter, R. (2008). Molecular dissection of variation in carbohydrates metabolism *New Phytol.*, 189: 941-479.

Yang, D.L., Jing, R., Chang, X.P. and Li, W. (2000). Identification of quantitative trait loci and environmental interactions for accumulation and remobilization of water-soluble carbohydrates in wheat (*Triticum aestivum* L.) stems, *Genetics*, 176: 571-584.

Zahedi, M. and Jenner, C.F. (2003). Analysis of effects in wheat of high temperature on grain-filling attributes estimated from mathematical models of grain-filling, *J. Agril. Sci.*, 141: 203-212.

7

SHORT TERM RESPONSES OF CROPS UNDER MERCURY CONTAMINATION AT HAZARDOUS WASTE SITES

Prasann Kumar, Padmanabh Dwivedi and A. Hemantaranjan

Department of Plant Physiology, Institute of Agricultural Sciences, Banaras Hindu University, Varanasi, 221005, U.P., India
Email address: prasann0659@gmail.com

INTRODUCTION

Accumulation of excess metal ions is toxic to diverse of plants. The impact of heavy metals on cell biochemistry is depends on their chemical nature. Data indicated that, approximately 90 elements are present in the crust of earth and among them maximum percentage are occupied by metals and 60% are heavy metals with density higher than 5 g cm^{-3}. Metals are defined as any element that has a silvery luster and is a good conductor of heat and electricity. There are many terms used to describe and categorize metals, including trace metals, transition metals, micronutrients, toxic metals, heavy metals. Bjerrum's (1959) definition of heavy metals is based upon the density of the elemental form of the metal, and he classifies heavy metals as those metals with elemental densities above 7 g cm^{-3}. In 1964, the editor of Van Nostrand's International Encyclopedia of Chemical Science, and in 1987, the editors of Grant and Hackh's Chemical Dictionary included metals with a density greater than 4 g $cm^{-3.}$ A little later in 1989, 1991, and 1992 Parker, Lozet, and Morris chose a defining density "greater than 5 g cm^{-3}". However, Streit (1994) used a density of 4.5 g cm^{-3} as his reference point, and Thornton (1995) chose 5 g/cm^3. The Rome Chemical Dictionary (Falbe and Regitz, 1996) gives 3.5 g cm^{-3} as a possible defining density. Accordingly, any idea of defining heavy metals on the basis of density must be abandoned as it creates confusion. Bennet (1986) and Lewis (1993) opted for atomic weight greater than of sodium (i.e., greater than 23), thus starting with magnesium, while Rand *et al.* (1995) preferred metals of atomic weight greater than 40, thus starting with scandium. Lewis (1993) suggested that forming soap with fatty acids is an important criterion of "heaviness". Thus, again we have no consistent basis for defining the term. Another group of definitions is based on atomic number. A fourth group of definitions is based on other chemical properties, with little in common, density for radiation screening, density of crystals, and reaction with dithizone. This brings us to definitions based vaguely on toxicity. One of these definitions (Hodson *et al.*, 1979) even refers to heavy metals as an "outdated term".

Based on the coordinate nature of the metals with biological system, Nieoboer and Richardson (1980), divided a biological relevant classification of metals into three categories (Table 1).

Table 1: Biological classification of metals

Sl. No.	Name of Categories	Affinities
1.	Class A elements	Affinity with O-containing ligands
2.	Class B elements	Affinity with N- or S-containing ligands
3.	Borderline elements	Intermediate between the two with affinity for all three groups of ligands i.e affinity with O, N, or S ligands with definite preference

(Source: Nieoboer and Richardson, 1980)

From the Table 1, it is clear that, a general manner in which different metal ions interact with biological system. Scientist reported that, Phytotoxicity of heavy metals in most parts can be attributed to symplastic accumulation of heavy metals particularly, in the plasmatic compartments of the cells, such as the cytosol and chloroplast stroma. Due to direct or indirect impairment of this heavy metal into the metabolism of cell results metal induced changes of the development of the plants (Van Assche and Clisters, 1990). Environmental contamination caused by mercury is a serious problem worldwide. Coal combustion, mercury and gold mining activities and industrial activities have led to an increase in the mercury concentration in soil. Mercury is regarded as one of the "priority hazardous substances" by the agency for toxic substance and disease registry because of its toxicity, mobility, and long residence time in the atmosphere (Agency for Toxic Substance and Disease Registry). Once mercury is released to the atmosphere it can be retained for between 6 to 24 months and be transported over tens of thousands of kilometers before eventual re-deposition on the earth's surface. The physical properties of mercury and its compounds are presented in Table 2.

Table 2: The physical properties of mercury and its compound

Properties	Hg°	$HgCl_2$	HgO	HgS	CH_3HgOH
Melting point (°C)	-38.8	277	500 (Decomposition)	584 (Sublimation)	137
Water-solubility (gL^{-1})	49.6×10^{-6} (20°C)	66(20°C)	0.053(25°C)	2×10^{-24} (25°C)	-
Boiling Point (°C)	356.7	303	-	-	-
Vapor Tension (pa)	0.18	0.009	9.2×10^{-12}	Not detectable	0.9

(Source: Schroeder and Munthe, 1998)

The contamination of soil with mercury has led to environmental concerns. Mercury can readily be taken up by the plants and be accumulated in human body through the food chain. It has been demonstrated that crops grown in mercury contaminated soil have total mercury concentration in their tissues. Qian *et al.*, 2009, found that the total mercury concentration in vegetables grown mercury-contaminated soil (0.09-0.54 mg kg^{-1}) ranged from 0.05 to 0.13 mg kg^{-1}. The contamination of food crops with mercury poses a direct health risk to local people. Available data clearly shows that mercury contamination of soil will result in an elevated exposure of mercury to local residents and will consequently pose a direct threat to the health of these people. Due to such environmental and public health concerns, there is great interest in the development of methods to remediate mercury contaminated soils. A recent study of mercury (Hg) poisoning in the Amazon rainforest has once again brought the metal under spotlight. A study has reported Minamata disease, a neurological disorder caused by mercury poisoning, in the Amazonian village. It is not a stray incidence since studied have shown high levels of mercury in several places of India and other countries. Mercury is released from rocks by geothermal and volcanic activities and 10 -50 per cent of the available mercury is contributed by human activities. Chlor-alkali plants and electrical industries are two major sources of the metal in the environment. Mercury toxicity can be characterized by several symptoms appears at the same time such as loss of coordination, slurred speech, loss of sensation at extremities of limbs and mouth, construction of visual field and loss of hearing. Severe poisoning can cause blindness, coma and death. Microbial bioremediation and phytoremediation are two approaches that can be used for removal of mercury from industrial wastes. Relying on the genes of *mer* operon, bacteria convert the mercury compounds to metallic mercury. The latter, being a liquid, is eliminated by the cell in vapour state. Genetically engineered plants with bacterial *mer* operon also seem to be promising for mercury removal. Hg exists in four major chemical forms:

[A] Elemental Hg (Hg°) in vapour state which is the least toxic form,

[B] Mercuric ion (Hg^{+2})

[C] Alkylated Hg, mainly monomethylmercury and dimethylmercury – the most toxic of Hg compounds and

[D] As legend with sulfide which is extremely insoluble in water and non-toxic.

Table 3: Toxicity of different form of mercury

Forms	Source	Transport and Fate	Toxicity Level
Elemental mercury (Hg°)	95% of atmospheric Hg is Hg° vapour	Tends to remain airborne, not easily deposited, may travel long distance before conversion to other forms	Minimum, almost non toxic
Inorganic mercury, mostly in divalent form (Hg^{+2})	Comprises only 5% of atmospheric Hg, mostly found in soil and water as complex ions	Easily deposited to earth's surface in dry form or in precipitation. Once in water may volatilize or partition into particles and may be transported to sediments.	High
Organo-mercurials (e.g. Methyl mercury)	Lipid soluble ion produced by bacteria in water or sediment, nearly all mercury in fish is methylated	Enters food chain through aquatic biota and uptake into fish, accumulates as it travels up the food chain, reaching highest concentration in organisms at the highest tropic level	Maximum

Source: Bhattacharyya *et al.*, 2008.

Most of the world's Hg is stored in the bedrock and deep-sea sediments. Mercury is found at several places on earth surface as cinnabar (HgS). Mercury gets volatilized from rocks, soil and water due to geothermal and volcanic activities. The total global input of Hg has been estimated to be 10^{10}g. About 10-50 percent of this flux originates from anthropogenic sources. Indiscriminate use of Hg compounds in industries and agriculture leads to an increase in Hg level. The highest amount of Hg comes from the chloralkali plants which use Hg electrodes in manufacturing chlorine and NaOH. The second largest users of Hg are electrical industries making Hg vapour lamps, electric switches, batteries, etc. The third largest consumers are the agricultural industries with mercury based pesticides and fungicides (Mitra 1986; Lenka *et al.*, 1992). The increasing prevalence of electrical and electronic manufacturing industry as a user and contributor of Hg in Asia is also worrying. Specifically, disposal of obsolete electrical and electronic wastes increase Hg load in ecosystem and biota. A single PC with recent technologies contains 500mg of Hg. After disposal, the entire amount comes to environment since the recycling efficiency is zero. All these increase the load of Hg waste in environment and pose a severe threat to living systems (Table 3).

Elemental Hg has been found to have a very long life span in environment. It has a residence time of one year in atmosphere. Other Hg compounds are washed by rainwater and are ultimately put in oceans where they survive for a long period. Mercury in the environment is an increasing health concern. Mercury has been reported to cause various neurodegenerative diseases (Mutter *et al.*, 2004). Scientist observed that elemental mercury and inorganic mercury compounds have been reported to damage the immune system and kidney. The greatest human health concerns are related to MeHg which has the ability to biomagnify in the food chain. MeHg can be taken up the aquatic organisms and further accumulate in fish. Therefore, populations living in areas of mercury contamination that consume a high amount of fish are greatest risk of MeHg poisioning (Harada 1995).

The toxicity of mercury to plants can be summarized as following:

[1] Affect on antioxidative system

[2] Affect on the photosynthetic system

[3] Inhibition of plant growth and yield production and an effect on nutrient uptake and homeostasis

[4] The inducement of genotoxicity

In terms of genotoxicity, a number of potentially reactive sites for mercury binding are present in DNA, depending on external conditions such as ionic strength, presence of different competing ions, and base composition. Mercury affects all groups of organisms and ecosystem processes, including microbial mediated processes and macro/mezzo fauna-mediated processes. The genetic structure and functional diversity of bacterial communities are sensitive to mercury expose (Tables 4 and 5).

Table 4: The concentration of mercury ions which affect the antioxidant system of the plants

Plant species	Substrate types	Mercury content
Artiplex condonocarpa	Hydroponics	0.05-1 mg l^{-1}
Medicago sativa	Semi hydroponics system	0-30μM
Pistacia lentiscus & *Tamarix gallica*	Hydroponics	0-100 μM
Lycopersicon esculentum Mill	Perlite, vermiculite	0-50 μM
Sesbania drummodii	Murashige-Skoog (MS) medium	0-50 μM
J. curcas	Murashige-Skoog (MS) medium	0-400 μM
Cucumis sativus	Hydroponics	0-500μM
Medicago sativa	Hydroponics	0-40μM

Source: Wang *et al.*, 2012

In the present context, it is summarized that mercury (Hg) is reported as one of the most toxic pollutant threatening our surrounding environment which has been introduced from gynogenic and anthropogenic source into our living system. Plants which are growing on contaminated sites will have the exposure of wide spectra of stresses and accordingly plant shows the responses. Phyto-toxicity of mercury in different parts can be attributed to symplastic accumulation of heavy metals particularly, in the plasmatic compartments of the cells, such as the cytosol and chloroplast stroma. Mercury is usually unconfined in the metal or ionic form, accumulating in the sediments where it get methylated by bacteria having anaerobic sulfate-reducing in nature to produce methyl mercury, a highly toxic organomercurial compounds. Chloroplast degradation have been seen frequently in plants growing in the mercury contaminated site leads to abnormalities in the vital metabolic processes like chlorophyll synthesis, Hill reaction, loosening of enzymes, membrane damage, lack of coordination with other metabolic processes and finally turn the plant towards death. Know a days, the most studies on the phytoremedation processes utilize the *mer A* and *mer B* genes to modify the plants via the chloroplast genome and nuclear genome for transformation of the plants. Therefore, in this review we are giving the glimpse of short responses of plants growing at contaminates site along with important remedial techniques and vital issues related in this connection.

IMPORTANT REMEDIAL TECHNIQUES BASED ON PHYSICAL AND CHEMICAL METHODS

At this movement, we have diverse approaches for the remediation of the mercury from contaminated sites. Generally, these techniques are based on chemical and physical methods. We feel, these are not eco-friendly but due to higher load of the contamination these are used frequently for remediation. So, in short we are giving few methodologies for the remediation of the mercurial elements.

Vitrification

Vitrification is an immobilization technique that is mainly used to remediate soils contaminated with heavy metals mixed with radioactive elements. Vitrification involves heating the wastes form a vitrified end product, into which the contaminants are incorporated and subsequently immobilized (USEPA 2008). The process may also cause contaminants to volatilize, thereby reducing their concentration in the soil and waste. Cicero and Bickford reported the use of vitrification technology to treat mercury-contaminated soil collected from a TNX pilot-plant facility at the Savannah River site. The results from this study showed that the soil was converted to a durable and leach resistant glass waste and that the optimum waste loading was demonstrated to be 60 wt % wastes with 24 wt % Na_2CO_3 and 16 wt% $CaCO_3$ used as the glass forming additives. Moreover, mercury in the off gas could be successfully captured by conventional condensers. The advantage of this method is that; [1] the treated waste is durable, resistible, and with good long- term stability; [2] the volume of the waste can be reduced; and [3] the treated waste product can be reused. The major disadvantages are high coat and the necessary capture and subsequent treatment of off gases.

Table 5: The affect of mercury on the photosynthetic system of plants

Plant species	Mercury content	Damage of photosynthesis system
Spirulina platensis	0-20 μM	The increase of Hg concentration led to a decrease in the maximal efficiency of PS II photochemistry, the efficiency of excitation energy capture by the open PSII reaction centers, and the quantum yield of PSII electron transport
Cucumis sativus	0-500μM	The presence of Hg in the substrate caused a linear decrease of chlorophyll content in the cotyledons; at 500 μM Hg, chlorophyll content was reduced by 59% and 94%, respectively, in 10- and 15-day-old seedlings.
Lycopersicon esculentum Mill	0-50 μM	Chlorophyll content decreased after 10-days Hg exposure in the first and the second leaves.
Pistacia lentiscus and *Tamarix gallica*	0-100 μM	Chlorophyll a decreased both in *P. lentiscus* and *T. gallica*; chlorophyll b was less affected leaves.

Source: Wang *et al.*, 2010

Thermal Desorption

Thermal desorption is a treatment technology that utilizes heat to increase the volatility of contaminants which are subsequently removed (separated) from the solid matrix (typically soil, sludge), but without combustion of media or contaminants. The desorbed contaminates are then treated in the off gas treatment system to control air emission (ITRC 1998). Traditional retorting as practiced by small-scale miners is an example of a thermal treatment method that involves distillation or dry distillation. In general, requirement for the retort system are very similar to thermal desorption, except that following desorption there should be some type of condensation system to collect volatilized mercury. Inorganic mercury is usually present in soil in the elemental state or as mercury (II) compounds such as HgS, HgO and $HgCO_3$. When the temperature reaches 600-800°C, these mercury compounds will be converted into gaseous elemental mercury, which can be recovered. There are many advantages of this method over other remediation methods, including the effective extraction and recovery of mercury, and safety. The major disadvantage is high energy costs and effectiveness only at rather high total soil mercury concentrations.

Nanotechnology

Nanotechnology involves the use of particles with at least on dimension in the range of 1-100 nm, to affect the mobility, toxicity and/or bioavailability of contaminants in their natural environment. Nano-sized particles are characterized by enhanced reactivity and a large surface area to volume ratio, which speeds up sorption kinetics, creating applications in the remediation of groundwater, surface water and the subsurface. For mercury, stabilized from sulfide (FeS) nanoparticles had been used to immobilize mercury in sediment. It must be noticed, however, that knowledge of the fate and transport of nanoparticles in the environment and the potential toxicological effects induced by the nanoparticles themselves is limited. Nanoparticle may indirectly influence human health when they inhaled, absorbed through skin, or ingested. Once nanoparticle enters the human body, they can migrate to or accumulate in places that larger particles cannot, such as the alveoli in the lungs. This is due to the minute size of nanoparticles. Furthermore an increase in the bioavailability of certain hydrophobic contaminants affected by some nanoparticles has been reported.

The major advantage of these methods for remediation is low cost, low energy demand and applicability to an in city treatment. However, the potential effects of nanoparticles on the environment have not been fully investigated and this method requires testing under field conditions before it can be considered a viable remediation techniques.

Electro-remediation

Electro-kinetic (ER) remediation consists of the controlled application of low intensity direct current through the soil between appropriately distributed electrodes. The system consists of three compartments; two electrode compartments and one soil compartments placed between these electrodes. The catholyte is separated from the soil by an anion exchange membrane. During electro kinetic soil treatment, ions can be removed from the soil by an anion exchange membrane. During electro-kinetic soil treatment, ions can be moved from the soil into the electrode compartment through ion exchange membranes, while no current carrying ions can transport from the electrode compartment into soil. Hydrogen ions (H^+) are generated at the anode due to water electrolysis, and migrate into the bulk of the soil. A low pH develops through the soil (except at the cathode where OH^- is generated), causing desorption of metallic contaminants from soil solid phases. The dissolved metallic ions moved to the electrode by ionic migration. Once metallic ions accumulate at electrode, they can pass through the ion exchange membranes and be removed by precipitation.

The efficiency of electro-remediation is greatly affected by soil properties, such as pH, the content of carbohydrates and organic matter which affect the degree of adsorption of mercury to soil constituents, and the geochemical fractionation of mercury particles in the soil. It is well known that mercury has a high affinity for organic matter; high organic matter content will decrease the efficiency of mercury removal from soil. One of the most important advantages of the electro-kinetic technique is its efficacy for the treatment of soil of low hydraulic permeability. The major disadvantages of these methods are: (1) the acidic conditions are required for remediation; (2) the amount of time required for remediation; and (3) the interfering effect of non target ions on remedial progress. Electro kinetic remediation is frequently used to remove metal ions from saturated soils or clay soils. However, very few applications have been reported under field conditions. The results showed that soluble mercury was almost completely removed from areas close to the cathode, but the residual concentration increased with distance in the direction of the anode.

PERMISSIBLE LIMITS

The Prevention of Food Adulteration Act and Rules 1955 has declared Hg as a poisonous metal. The Ministry of Commerce has limited its concentration in fish to 0.5 ppm (mg/kg) and in other food items to 1.0 ppm (mg/kg). The concentration of methylmercury, one of the most abundant organomercurials, has been restricted upto 0.25 ppm (mg/kg). The Bureau of Indian Standards has laid down its safety limits in drinking water at 0.001 ppm (mg/L). The permissible limit of Hg set by the Central Pollution Control Board during discharge of industrial waste in river is 0.01 ppm (mg/l)(Table 6, Figure 1).

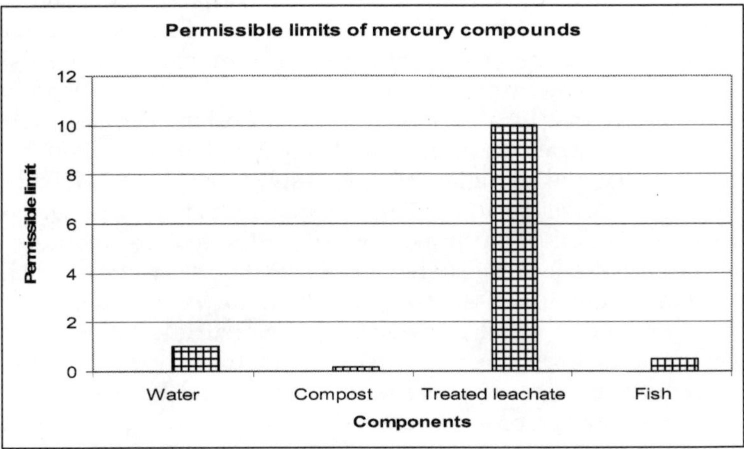

Fig. 1. Graph of permissible limits of mercury compounds
(Source: Ministry of Commerce and Ministry of Forest and Environment, Govt. of India)

Table 6: Permissible limits of mercury compounds

Component	Permissible Limit
Water	1.0 μg/L
Compost	0.15 mg/kg
Treated leachate	10 μg/L
Fish	0.5 mg/kg

Source: Ministry of Commerce and Ministry of Forest and Environment, Govt. of India

BIOTECHNOLOGICAL APPROACHES FOR REMEDIAL MEASURES

Removal of Hg from polluted sites includes incineration at high mercury concentrations, and safe storage in landfills or in underground storage at low concentrations. Bioremediation is one of the potent methods that can be used for Hg reduction. The method relies on the potential of the mer-operon- based resistant mechanism in bacteria, which functions by enzymatic breakdown of organomercurial and reduction of mercury ions (Hg^{+}, Hg^{+2}) to water-insoluble metallic Hg. The enzymes involved in this process are organomercurial lyase and mercuric reductase. The former convert an organomercurials compound to mercuric ions (Hg^{+2}) and the latter reduces Hg^{+2} to metallic mercury (Hg) with the help of NADPH. Hg thus produced diffuses out of cells and accumulates in pure form in the medium. The high vapour pressure of Hg ultimately results in its volatilization from media. Scientist isolated two actinomycete strains, CHR3 and CHR28, from metal-contaminated sediments from Baltimore Inner Harbour which were found to be resistant to $HgCl_2$ and phenylmercuric acetate (PMA). Hybridization experiments indicated that genes homologous to mercuric reductase and organomercurials lyase of *Streptomyces lividans* were present in the strains. *Lemna minor* (duckweed) is another prospective tool for absorbing Hg from polluted water. High concentrations of Hg has been found in the plant samples collected from Tiljala wetlands of the eastern fringes of Kolkata that serves as a sink for deposition of urban and industrial wastes. Natural unexposed population of duckweed, when cultivated in laboratory, was also found to accumulate Hg (Pahan *et al.*, 1991). Bizily *et al.*, 1999, have engineered a model plant *Arabidopsis thaliana* with a bacterial gene *merB*, the product of which convert organomercurials to inorganic Hg. Transgenic plants expressing the gene could tolerate a wide range of concentrations of mono-methyl mercuric chloride and PMA (Mukherjee *et al.*, 2004). The usual process for removal of mercury is the safe storage in landfills or in underground sites at low concentrations. In European countries, the process effluent from Chloralkali factories is currently cleaned by hydrogen sulfide precipitation, hydrazine precipitation or ion exchange columns (Bizily *et al.*, 1999). The processes ultimately lead to confined dumping of mercury instead of its removal (Table 7 and Figure 2).

Table 7: Annual Release of Hg by Chloralkali Plants

Year	Caustic soda production (MT)	Average annual Hg loss by a plant (MT)
1997-1998	676474	96.1
1998-1999	514852	73.1
1999-2000	475603	67.5
Average values (per year)	555643	78.9

Source: Bhattacharyya *et al.*, 2008

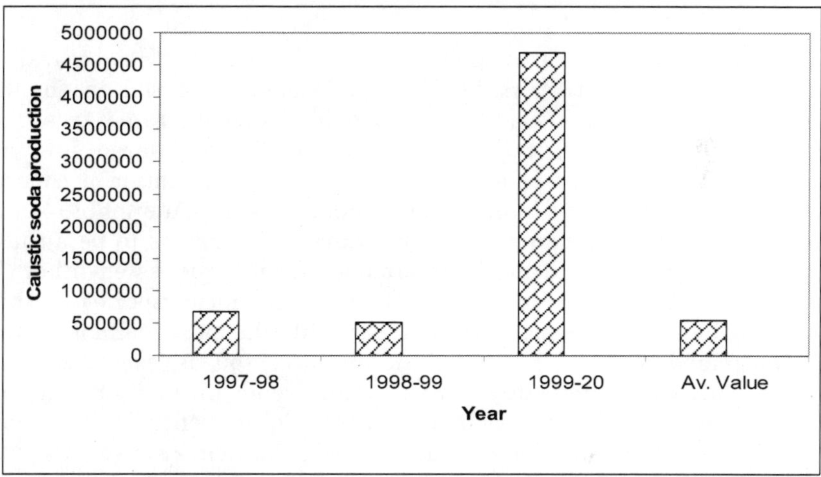

Fig. 2. Graphical representation of annual release of Hg by Chloralkali Plants (Source: Bhattacharyya *et al.*, 2008)

Table 8: Concentration of Total Hg in Biota Collected from Different Coastal Parts of Peninsular India

Location	Hg Concentration	Organism
Sagar Island, East coast	0.06-2.24µg/g dry wt	Bivalves
Mumbai, West coast	1.42-4.94µg/g dry wt	Crabs
Mumbai, West coast	0.03-0.82 µg/g dry wt	Fishes
Karwar, West coast	0.03-0.003 µg/g wet wt	Fishes
Madras, Southeast coast	0.08-0.14 mg/kg wet wt	Fishes

Source: Bhattacharyya *et al.*, 2008

Table 9: Mercury pollution outside India

Site	Sample	Mercury concentration
United State, Southern Alaska	[a] Stream sediment	[a] 5500 μg/g
	[b] Unfiltered stream water	[b] 500-2500 μg/L
	[c] Filtered stream water	[c] < 50 μg/L
New jersey	[a] Fish	[a] 100 μg/g
Guanabara Bay, Rio de Janeiro	[a] Estuarine water	[a] 0.7-5.23 ng/L
	[b] Mussel	[b] 11.6-53.5 μg/kg of wet weight
	[c] Carnivorous fish	[c] 199.5 μg/kg wet weight

Source: Bhattacharyya *et al.*, 2008

The technologies related to chemical engineering for Hg removal are expensive and disruptive. Excavation and roasting of soil in the Hg-contaminated sites is considered as the best demonstrated treatment technology (Bizily 1999), but it is impractical for very large sites and required a high amount of energy (Table 8, 9). Vitrification a concrete capping, which aims to stabilize Hg, renders the site uninhabitable for plants, insect and other organisms. Bioremediation seems to be an ideal treatment that would degrade organomercurials, sequester other Hg forms and convert Hg to the least toxic chemical species without disturbing biological productivity. Traditionally, the most common method for the remediation of mercury-contaminated soil is excavation and disposal. But these methods are costly and crude. In addition, they are only useful if the mercury is tightly localized. Consequently, recent efforts have focused on developing better remediation routes for soil contaminated with mercury. Phytoextraction is a cleanup technology that is still under development. For phytoextraction to be worthwhile, efforts should be conducted: (1) more studies to select mercury accumulator/ hyperaccumulator plant species; (2) more studies to understand fundamental mechanism involving plant uptake mercury; (3) more efforts should be conducted to investigate the emission of mercury to the air during the phytoextraction process; (4) more tests should be conducted to translate controlled experimental data to field conditions.

MERCURY SPECIATION IN SOIL

It is currently well known that the determination of a total mercury concentration is insufficient for understanding the biogeochemical cycle of the metal and for establishing an appropriate remediation method. This is mainly due to the lack of information concerning the reactivity, bioavailability and toxicity of mercury in soil afforded by such a single determination. The study of mercury speciation can be used to predict and explain the behavior of mercury in soil. Several approaches such as sequential extraction procedure (SEP), pyrolysis, X-ray absorption fine spectroscopy (XAFS), as well as nuclear magnetic resonance (NMR) and electron paramagnetic resonance (EPR) have been used to study mercury speciation in soil. Among these methods, SEP, Pyrolysis and XAPS are frequently used in mercury speciation analysis. Results on mercury speciation measured by pyrolysis and XAFS derivatives from different researchers are given in table 10. In general, soils from mercury mining areas have a low content of bioavailable mercury due to the majority of mercury being present as HgS. However, soil from gold mining areas and industry factories presents a relatively high proportion of bioavailable mercury because of the release of mercury into the soil as soluble form. Information on the mercury speciation in soil is crucial for establishing any remediation method. In general, a high proportion of bioavailability mercury in soil would demand the use of soil washing, phytoextraction and electro-remediation techniques through which mercury would be removed. For soil with a low bioavailable mercury concentration, metal available must first be increased to facilitate the use of these methods. For soil with high elemental mercury content, methods such as stabilization/ solidification, vitrification and immobilization are suitable remediation options. For soil with high non-available Hg content, thermal desorption technology may be suitable where mercury can be volatilized at high temperature (Table 10).

Table 10: The speciation of mercury in soil as determined by pyrolysis extraction methods

Samples	Operating conditions	Compounds released	Percentage of total mercury released
Soil	<180°C	Hg°, Hg_2Cl_2	10-30%
Soil (Hg mine)	<100°C	Metallic Hg	-
	150-250°C	Matrix-bound Hg	11%
	350°C	Cinnabar	89%
Tailing	<100°C	Hg°	Lack of data
	150-250°C	Matrix bound Hg	
	250-350°C	$HgSO_4$, HgO	
	>400°C	Fe-	
Sediment and soils	<100°C	Hg°	5%
	150-250°C	Non cinnabar Hg	40%
	250-350°C	Red cinnabar	5-30%
Soils (Chlor-alkali plant)	<100°C	Metallic Hg	18%
	150-250°C	Matrix-bound Hg	20%
	350°C	Cinnabar	80%

Soils	200-350°C	Non cinnabar Hg	79%
	250-350°C	Cinnabar	18%
Soils	25-350°C	Fe-oxyhydroxides/clay	20-50%
	453°C	Minerals bound Hg Cinnabar	50-80%

Source: Wang *et al.*, 2010

ENHANCEMENT OF MERCURY PHYTOREMEDIATION THROUGH GENETIC ENGINEERING

mer A or *mer B* are the two important target genes in plants which are utilized by the genetic engineers for modification purpose via nuclear or chloroplast genome. These genes expressing the mercuric ion reductase and (or) organomercurials lyase in the cytoplasm, endoplasmic reticulum or within the plastids (Ruiz and Daniell 2009). With the application of these genes the several plants species has been transform into transgenic plants. For instance these are tobacco, poplar, rice, peanut, salt marsh, chlorella or *Arabidopsis*. It is very interesting to say that Rice was the first plant in which transgenic has been produced. Heaton *et al.*, 2005, first develop *merA* transgenic rice resistant to 250 μM $HgCl_2$. It is reported that transgenic plants against Hg contamination will grew successfully in the area of contamination having range of 400μM PMA to 500μM $HgCl_2$ respectively (Ruiz and Daniell 2009). We interested to raise the one question that is, why we are concern with the study of phytoremediation of mercury. The answer is it is essential for study because Hg is one of the most toxic pollutants threatening our health and surrounding environment or ecosystem. Anthropogenic and natural disasters are the two major sources for the creation of higher concentration of heavy metals in the soil strata above than alarm critical limit in the soil. EPA 2009 estimates the annual global emissions between 4,800-8,300 tons per year. As we all know the fact that mercury is usually released in the metal or ionic form, accumulating sediments where it becomes methylated by anaerobic sulfate-reducing bacteria to produce methyl mercury a highly toxic organomercurial compounds. Their toxicity greatly depends on the form present in the environment. It is reported that inorganic form are usually less harmful than organic forms because they bind strongly to soil components that binds and reduce their availability and absorption. Due to hydrophobicity nature of organomercurial, it is observed that they are highly toxic which facilitates their transport along the cell membranes and accumulation in membrane bound organelles, inhibiting essential oxidative and photosynthetic pathways. It is reported that magnification of mercury become higher if we go from lower to higher tropic level subsequently (Harada 1995; Patra and Sharma 2000). Plants suffers from various abnormalities if they grow in mercury contaminated soil such as damage of membrane transporter, damage of Aquaporins structure, destruction of chloroplast structure and their normal metabolic functions like electron transport chain, oxygen evolution, photophosphorylation, Hill reaction, chlorophyll content and chlorophyll florescence (Kupper *et al.*, 1996; Sinha *et al.*, 1996).

MerAB Engineered through the Nuclear Genome

Bizily *et al.* (2003) reported that a well characterized phytoremediation system is the use of the bacterial *merA* and *merB* genes to genetically engineer plants for the remediation of mercury (Figure 3).

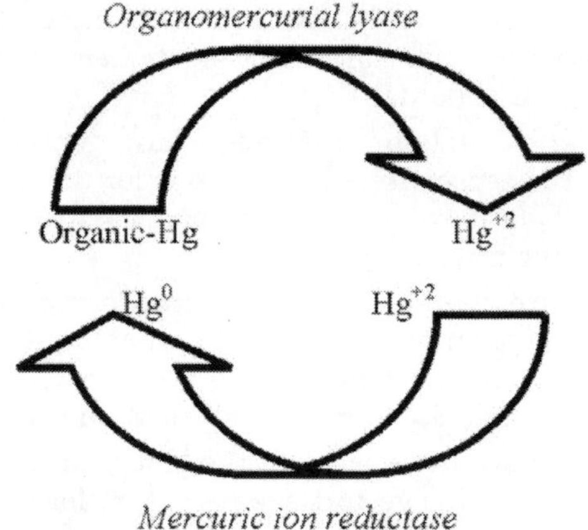

Fig. 3. Protonolysis of Organic-Hg and Hg respectively (Source: Drawn by author, unpublished)

- Rugh *et al.* (1998), first report the utilization of a forest tree for phytoremediation. He reported that if popular tree will transformed with *mer A* gene than plant become able to withstand the mercury concentration up to 50μM $HgCl_2$.
- Kim *et al.* (2005), demonstrated that plant roots is the foremost organ affected by Hg^{+2} and that root specific expression of the *mer A* gene can provide the resistant for mercury up to 80μM $HgCl_2$.
- Huang *et al.* (2006), demonstrated that transgenic *Chlorella* which expressing the *mer A* gene withstand the Hg concentration up to 40μM $HgCl_2$.
- Bizily *et al.* (2000), demonstrated the first report of nuclear genetic engineering for phytoremediation using *mer A* and *mer B* genes in *Arabidopsis thaliana.*

MerAB Engineered through the Chloroplast Autonomous Genomic System

It is reported that protected plastid will provide higher level of resistant against mercury toxicity (Ruiz and Daniell, 2009). Among the different type of plastids present in the plant cells, the chloroplast is reported more active metabolically and involves in the diverse activities like photosynthesis, evolution of oxygen, sulfur metabolism, nitrogen metabolism, synthesis of glucose, synthesis of starch, amino acids, fatty acids, photosynthetic pigments and vital amines (Kumar, 2014). We are giving the important discoveries related to *mer AB* engineered through the chloroplast Genome,

- Ruiz *et al.* (2003), integrated the mer A and mer B into the chloroplast genome and shows that this integration provides the resistant to very high concentration of phenylmercuric acetate up to the range of 400μM.
- Quesada *et al.* (2005), reported that, genetic engineering of chloroplast is especially advantageous for the development of the transgenic plant when multiple genes are required for the purpose of the phytoremediation.
- If we compare the approaches for transgenic using nuclear and chloroplast genome then we found that chloroplast genomic approaches is the more advantageous because multigenic pathways can be developed in a single transformation events.
- One of the major drawbacks of the chloroplast transformation is that, species specific vectors are necessary for transformation.
- Another limitation of the chloroplast transformation is achieving homoplasmy i.e., integration of transgenes into each chloroplast genome and elimination of untransformed chloroplast genome in respective crops.
- Hussein *et al.* (2007), worked on the phytoremediation of the mercury and organomercurial in chloroplast transgenic plants and find that chloroplast transgenic plants expressing *mer A* and *mer B* can accumulate Hg in roots to levels surpassing the concentration in soil, up to 2000μg g^{-1} without any detrimental effect and could accumulate 100 fold more Hg in leaves than untransformed plants.

So in general, transgenic plants are able to accumulate 100 folds and 4 fold more Hg in the shoot in the presence of organic mercury source or inorganic mercury source with respect to normal plants. Besides this one, it is interesting to say that it shown to volatilize Hg^0 efficiently, independent of other ionic form in the contaminated soil. Figure 4 represents an overview for the genetically modified plant cells for the optimal mercury phytoremediation. It utilizes the multiple genes for the purpose of the different processes like uptake, translocation and detoxification of the metals. Metals allocation in the different compartments like vacuoles and plastids also included the role of multiple genes for the same (Figure 4).

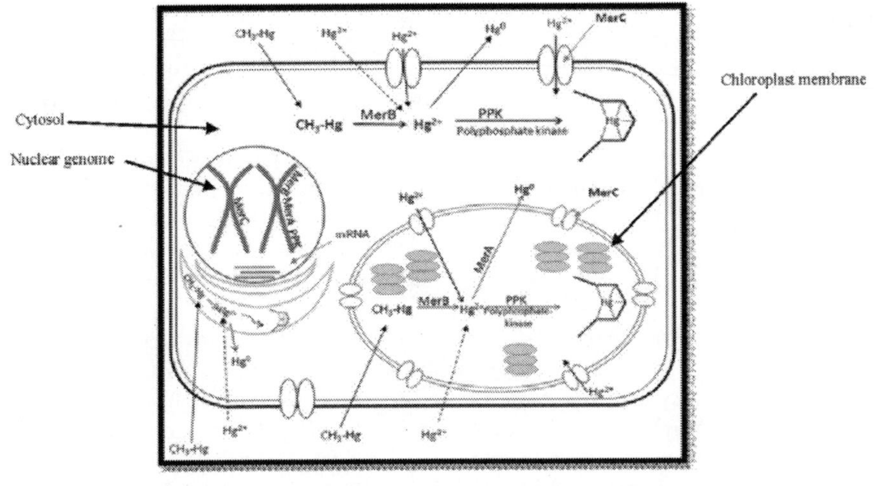

Fig. 4. Cell of genetically modified plants [Source: Ruiz and Daniell, 2009]

In the Figure 4, the solid line arrows indicate the rapid movement across the plasma membrane while dashed line across the membrane indicates the slow movement of Hg. The fast movement is possible only through the specialized Hg transporters while slow movement is possible only through the ion transporters. In the Table 11, we are giving the transgenic systems for phytoremediation of mercury with different levels of resistance and their special features.

Table 11: Examples of transgenic plants

Sl. No.	Species	Level of resistance (μM $HgCl_2$)	Special features	References
1.	Eastern cottonwood	10	*mer A* and *mer B* transformed into a forest tree	Lyyra *et al.*, 2007
2.	*Chlorella*	40	First transgenic algae for Hg bioremediation	Huang *et al.*, 2006
3.	*A. thaliana*	100	First report of *mer A* in plants	Rugh *et al.*, 1996
4.	Peanut	100	*merA* transformed into peanut	Yang H., 2003
5.	Salt marsh cordgrass	500	First wetland grass for Hg phytoremediation	Czako *et al.*, 2006
6.	Rice	250	First monocot for Hg phytoremedation	Heaton *et al.*, 2003
7.	Yellow poplar	50	*merA* transformed into a tree species	Rugh *et al.*, 1998

Conclusion

It is obvious that plants can be genetically modified to enhance their tolerance, uptake, translocation, accumulation and volatilization capabilities for Hg phytoremediation. Plants tend to accumulate most mercury in roots, and translocation to shoot tends to be major limitation dependent upon concentration and cellular integrity. Toxicity of methyl mercury is magnified due to its high accumulation in tissue. In general, we want to say that heavy metal contamination is of great concern in India and abroad. During the progress of civilization of human, degradation of natural resources like soil, air and water is one of the lethal rapidity approaching toward the development of society. So at the cost of nature; we can't make the rhythmic oscillation with development. Plants have its inherited capacity to minimize the adverse effects of exposed hazardous like Mercury. Various up-regulated and down-regulated metabolisms will operate for defending heavy metal stress. So, in general, plant for its short term responses, shows enzymatic cross talk especially oxidative enzymes like CAT, APX, SOD, Ascorbic acid. As we are being the professionals of plants physiology, looking into the whole physiology of stressed plants we find drastic changes. Heavy metals bound in the soil are leachable; they can be also spread via ground water. In view of avoiding the danger of contaminant spread, it is possible to use phytoremediation techniques, which can immobilize pollutant and resulting into decreased soil or water pollution. Some plant species are also able to accumulate heavy metals in their tissues, thus the contaminant can be removed from locality with plant harvest.

Future Insight

Scientific community should make an effort to address the most important questions that the application of phytoremediation and produce plants with the best capability of remediation along with the ability of large biomass, rapid growth rate, adaptation to wider climate conditions and expression of multiple genes in different cellular compartments.

Acknowledgement

Authors are thankful to University Grant Commission, New Delhi for providing Senior Research Fellowship to the first author.

References

Bennet H. 1986. (ed.) *Concise Chemical and Technical Dictionary*, 4th enlarged ed., Edward Arnold, London.

Bhattacharyya S., Chatterjee S., Basu S. 2008. Mercury pollution: The problem still persists. *Everyman's Sci.*, **XLII (5):** 279-286.

Bizily S., Rugh C.C., Meagher R.B. 2003. Phytoremediation of hazardous organomercurial by genetically engineered plants. *Nat. Biotechnol.*, **18:** 213-217.

Bizily S.P., Rugh C.L., Summers A.O., Meagher, R.B., 1999. Proc. Natl. Acad., Sci., USA, **96:** 6808.

Bjerrum J. 1959. Metal amine formation in aqueous solution. In: *Treatise on analytical chemistry*. Vol. I.J. Elving (eds.). New York: The Interscience Encyclopedia.

Clisters H., Van Assche, F. 1985. Inhibition of photosynthesis by heavy metals. *Photosynthesis Res.*, **7:** 31-40.

Czako M., Feng X., He Y., Liang D., Marton L. 2006. Transgenic *Spartina alterniflora* for phytoremediation. *Environ. Geochem. Hea.*, **28:** 103-110.

E. P. A. 2009. Mercury Human Exposure. US Environmental Protection Agency 2008. Retrieved February 8, 2009 from www.epa.gov/mercury/exposure.htm.

Falbe J., Regitz M. 1996. (eds.) *Roempp Chemie Lexikon*, Georg Thieme, Weinheim.

Grant R., Grant, C. 1987. (eds.) *Grant and Hackh's Chemical Dictionary*, McGraw-Hill, New York.

Harada M. 1995. Minamata disease research group Miniamata disease: methylmercury poisoning in Japan caused by environmental pollution. *Crit. Rev. Toxicol.*, **25:** 1-24.

Heaton A.C.P., Rugh C.L., Wang N.J., Meagher R.B. 2005. Physiological response of transgenic *merA*-TOBACCO (*Nicotiana tabaccum*) to foliar and root mercury exposure. *Water Air Soil Poll.*, **161:** 137-155.

Hodson P.V., Borgman U., Shear H. 1979. Toxicity of copper to aquatic biota. *In* "Copper in Environment Part 2: Health Effects" (ed) By Nrigau J.O., John Wiley, NY. pp 307-372.

Huang C.C., Chen M.W., Hsieh J.L., Lin W.H., Chen P.C., Chien L.F., 2006. Expression of mercuric reductase from Bacillus megaterium MB1 in eukaryotic microalga *Chlorella* sp., *Appl. Microbiol. Biotechnol.*, **72:** 197-205.

Hussein H., Ruiz O.N., Terry N., Daniell H. 2007. Phytoremediation of mercury and organomercurial in chloroplast transgenic plants: Enhanced root uptake, translocation to shoot and volatilization. *Environ. Sci. Technol.*, **41:** 8439-8446.

Kim T., Balish R.S., Heaston A.C., McKinney E.C., Dhankher O.P., Meagher R.B. 2005. Engineering a root-specific, repressor-operator gene complex. *Plant Biotechnol. J.*, **3:** 571-582.

Kumar P., Dwivedi P., Mandal B. 2014. Phytoremediation for Defending Heavy Metal Stress in Weed Flora. *Intl. J. Agri., Environ.* and *Biotechnol.*, **6(4):** 587-595.

Kupper H., Kupper F., Spiller M. 1996. Environmental relevance of heavy metal substituted chlorophylls using the example of water plants. *J. Exp. Bot.*, **47:** 259-266.

Lenka M., Panda K.K., Panda B.B. 1992. Monitoring and Assessment of Mercury Pollution in the Vicinity of a Chloralkali Plant. *Arch. Environ. Cont. Toxicol.*, **22:** 195-202.

Lewis R.J. Sr. 1993. (ed.) *Hawley's Condensed Chemical Dictionary*, 12th ed., Van Nostrand Reinhold, New York.

Lozet J., Mathieu C. 1991. *Dictionary of Soil Science*, 2nd ed., A.A. Balkema, Rotterdam.

Morris C. 1992. (ed.) Academic Press Dictionary of Science and Technology, Academic Press, San Diego.

Lyyra S., Meagher R.B., Kim T., Heaton A., Montello P., Balish R.S., Merkle S.A. 2007. Coupling two mercury resistance genes in Eastern cottonwood enhances the processing the organomercurial. *Plant Biotechnol.*, **5:** 254-62.

Ministry of Commerce and Ministry of Forest and Environment, Govt. of India

Mitra, S. 1986. Mercury pollution: Poisoning and Therapy, in Mercury in the ecosystem, Transtech Publications, Switzerland, P. 195.

Mukherjee S., Bhattacharyya P., Dattagupta A.K. 2004. Distribution of Heavy Metals in Valuable Coastal Fishes. *Environ. Intl.*, **30:** 811.

Mutter J., Naumann J., Sadaghiani C., Schneider R., Walach H. 2004. Alzheimer Disease: Mercury as pathogenetic factor and apolipoprotein E as a moderator. *Neuroendocrin. Lett.* **25**.

Nieober E., Richardson D.H.S. 1980. The replacement of the nondescript term heavy metals by a biologically and chemically significant classification of metal ions. *Envion. Pollut. Ser.* B: **1:** 3-26.

Pahan K., Gachhui R., Ray S., Chaudhuri J., Mandal A.1991. A PMA degrading constitutive organomercurial lyase in broad spectrum mercury. *Ind. J. Exp. Biol.*, **29:** 1147.

Parker, S.P., 1989. (ed.) *McGraw-Hill Dictionary of Scientific and Technical Terms*, 4th ed., McGraw- Hill, New York.

Patra M., Sharma A. 2000. Mercury Toxicity in Plants. *Bot. Rev.*, **66:** 379-422.

Prevention of Food and Adulteration Act and Rules 1955.

Qian J.H., Zayed A., Zhu Y.L., Yu M., Terry N. 1999. Phytoaccumulation of trace elements by wetland plants III. Uptake and accumulation often trace elements by twelve plant species. *J. Environ. Qual.*, **28 (5):** 1448-1455.

Quesada T., Ruiz O.N., Daniell H. 2005. Characterization of heterologous multigene operons in transgenic chloroplasts: transcription, processing and translation. *Plant Physiol.*, **138:** 1-17.

Rand G.M., Wells P.G., McCarty L.S. 1995. "Introduction to aquatic toxicology", in *Fundamentals of Aquatic Toxicology*, G. M. Rand (ed.), Taylor and Francis, Washington, DC.

Rugh C.L., Senecoff J.F., Meagher R.B., Merkle S.A. 1998. Development of transgenic yellow poplar for mercury phytoremediation. *Nat. Biotechnol.*, **16:** 925-928.

Ruiz O.N., Daniell H. 2009. Genetic Engineering to enhance mercury phytoremediation. *Curr. Opin Biotechnol.*, **20(2):** 213-219.

Ruiz O.N., Hussein H.S., Terry N., Daniell H. 2003. Phytoremediation of organomercurial via the chloroplast genetic engineering. *Plant Physiol.*, **132:** 1344-1352.

Schroeder W.H., Munthe, J. 1998. Atmospheric Mercury – An Overview. *Atmos. Environ.*, **29:** 809-822.

Sinha S., Gupta M., Chandra P. 1996. Bioaccumulation and biochemical effect of mercury in the plant *Bacopa monnieria* (L.) *Environ. Toxicol. Water Qua.*, **11:** 105-112.

Streit B. 1994. *Lexikon der Okotoxikologie*, VCH, Weinheim.

Thornton I. 1995. Metals in the Global Environment—Facts and Misconceptions, ICME, Ottawa.

United States Environmental Protection Agency, 2007. Cleaning up the Nation's Waste Sites: Markets and Technology Trends. EPA/542/R-96/005. Office of Solid Waste and Emergency Response, Washington, DC.

U.S.E.P.A. 2008. Mercury in medical waste. Mercury fact sheet n. 1, Tech. rep., Environmental Protection Agency, Region 5, Air and radiation Division, www.epa.gov/nscep/Van Nostrand. 1964. International Encyclopedia of Chemical Science. Van Nostrand, New Jersey.

Wang S.X., Zhang L., Li G.H., Wu Y., Hao J.M., Pirrone N., Sprovieri F., Ancora, M.P. 2010. Mercury emission and speciation of coal-fired power plants in China, *Atmos. Chem. Physiol.*, **10:** 1183-2010.

Wyszkowska J., Kucharski J., Lajszner W. 2005. Enzymatic activities in different soils contaminated with copper. *Polish J. Environ. Std.*, **14:** 659-664.

Yang H., Nairn J., Ozias-Akins P. 2003. Transformations of peanut using a modified bacterial mercuric ion reductase gene driven by an actin promoter from *Arabidopsis thaliana*. *J. Plant Physiol.*, **160:** 945-52.

8

MOLECULAR GENETICS AND PHYSIOLOGICAL MECHANISMS OF SUBMERGENCE TOLERANCE IN RICE

D. Vijayalakshmi[1] and M. Raveendran[2]

[1]Department of Crop Physiology, Tamil Nadu Agricultural University, Coimbatore.
[2]Department of Plant Biotechnology, Tamil Nadu Agricultural University, Coimbatore.
**Corresponding author email: vijiphysiology@gmail.com*

INTRODUCTION

Rice (*O. sativa* and *O. glaberrima*) is one of the world's most important crops, particularly in Asia, Africa and Latin America. It is being grown under diverse ecosystems ranging from upland, rainfed lowland, deep water and tidal wetlands. Present and anticipated global food demands necessitate a significant increase in rice productivity in marginal farmlands. Climate projections suggest that temperatures, precipitation and flooding, and sea level rise are likely to increase, with adverse impacts on crop yield and farm income in Southeast Asia (Unnikrishnan *et al.*, 2006; Wassmann *et al.*, 2009; INCCA 2010). Extremes in water availability (droughts and floods) have increased in frequency and intensity over the past 50 years in farming regions throughout the world (Bailey-Serres *et al.*, 2012). The popular "mega-varieties" grown in large areas of Asia are sensitive to flooding stress and even under normal rainfall conditions, about 3 mha of rice area is subjected to submergence or flooding stress wherein rice plants are completely submerged for 1–2 weeks, resulting in partial or even complete crop failure.

There are two different flood types, flash flood and deepwater flood. Flash-floods are highly unpredictable and may occur at any growth stage of the rice crop and the yield loss may be anywhere between less than 10 and 100% depending on factors such as water depth, duration of submergence, temperature, turbidity of water, rate of nitrogen fertilization, light intensity and age of the crop. In Cauvery delta zones of Tamil Nadu submergence is a major problem during the monsoon season, where three lakh hectares of paddy are affected every year.

A major challenge for plant physiologists is to incorporate knowledge of the biochemistry, molecular biology and biophysics of cells and organs to understand the abilities of the whole plant to withstand flooding stresses and associated oxygen shortages in the soil. The review article focuses on various morphological, physiological traits and genetic and molecular mechanisms that are necessary for submergence tolerance and the role of plant hormones in submergence tolerance and methods for improving submergence tolerance.

ENVIRONMENTAL CHARACTERIZATION OF FLOOD TYPES

Submergence tolerance is defined as 'the ability of a rice plant to survive 10–14 d of complete submergence and renew its growth when the water subsides. Flooding levels vary, depending on the amount and duration of rain, underlying geological formations and distance from the water. Among these factors, there are two typical kinds of flood. One is short duration over a few weeks and not very deep, termed a '**flash flood**' and the other is deep flooding that lasts for a long time, called as '**deepwater flood**' (Fig. 1). Flash floods are unexpected and uncontrollable, and its flooding water level can reach 50 cm in the rainfed lowlands of the humid and semihumid tropics of South and Southeast Asia. In these areas, flash floods at the seedling stage of rice cause severely reduced yields of rice grain (Hattori *et al.*, 2011)

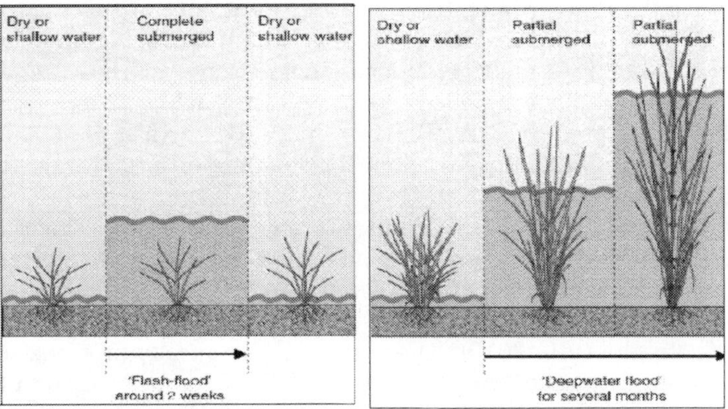

Fig. 1. Types of flooding (Source: Hattori *et al.,* 2011).

India has the largest area (i.e. 17.2 million hectares) under rainfed lowland amongst the South-East Asian countries (Khush and Brar, 1998). In these areas flash floods or short-term submergence is a major abiotic stress and flash floods at the seedling stage of rice cause severely reduced yields of rice grain (Hattori *et al.,* 2011). Flash flooding can cover the entire plant for prolonged periods, and most rice cultivars die within 7 days of complete submergence (Xu *et al.*, 2006; Bailey- Serres *et al.,* 2010).

Molecular Genetics of Submergence Tolerance

Submergence tolerance is rare, genetically determined trait with relatively high heritability that is controlled by one or few genes with major effects and minor modifiers (Suprihatno and Coffman, 1981; Mohanty and Khush, 1985; Sinha and Saran, 1988; Haque *et al.,* 1989). Several factors such as duration of submergence, water level, temperature, water turbidity and soil conditions affect the tolerance of rice plant to complete submergence (Adkin *et al.,* 1990; Jackson and Ram, 2003). Varietal differences in degree of tolerance have been noted (Supapaoj *et al.,* 1982) and these genetic resources have been used in several breeding programmes (Mohanty and Khush, 1985).

The characterization of quantitative trait loci (QTLs) in rice demonstrated the importance of ethylene in flooding responses. They are called SUBMERGENCE1 (SUB1) locus isolated from a submergence-tolerant landrace (FR13A) of eastern India and the SNORKEL (SK) locus from a deepwater landrace (C9285) from Bangladesh (Xu *et al.,* 2006; Hattori *et al.,* 2009). Whereas SUB1 controls the tolerance of complete submergence by dampening underwater growth, SK controls the avoidance of submergence by promoting underwater elongation growth.

Quiescence Strategy of Submergence-Tolerant Rice

The SUB1 locus on chromosome 9 confers up to 69% of phenotypic variation in the tolerance of complete submergence of vegetative plants. Plants with the SUB1 region from FR13A are capable of surviving 2 weeks or longer of complete inundation (Xu and Mackill 1996; Nandi *et al.,* 1997; Xu *et al.,* 2000). This multigenic locus includes two or three genes of the group VII subgroup of ethylene-response factor (ERF) transcription factors that were designated SUB1A, SUB1B, and SUB1C (Xu *et al.,* 2006). It was determined that SUB1A is sufficient for submergence tolerance. Although SUB1B and SUB1C appear to be invariably present at the SUB1 locus, they are evidently not determinants of submergence tolerance by quiescence. Among indica and aus accessions of rice with SUB1A, those with the SUB1A-1 allele are typically submergence tolerant, whereas those with the SUB1A-2 allele are typically submergence intolerant (Xu *et al.,* 2006; Singh *et al.,* 2010). These two alleles encode proteins that only differ at a single amino acid, Ser-186 in SUB1A-1 and Pro-186 in SUB1A-2, a putative mitogen-activated protein kinase (MPK) phosphorylation site in SUB1A-1 (Xu *et al.,* 2006).

Submergence tolerant (Sub1)	*aus indica*	Ser$_{186}$ *SUB1A-1*	*SUB1B-1*	*SUB1C-1*
Submergence intolerant	*indica*	Pro$_{186}$ *SUB1A-2*	*SUB1B-1,3,6,7*	*SUB1C-3,5,8*
	indica japonica O. nivara O. rufipogon		*SUB1B-2,3,4,5,8*	*SUB1C-2,6*

Fig. 2. Ethylene responsive factor (ERF) gene composition (haplotype) at the SUB1 locus of submergence tolerant and intolerant rice. Gene allele forms are indicated with a number following the gene name. Alleles are defined based on the nucleotide sequence encoding the ERF. The two alleles of SUB1A have a single amino acid difference and are distinct in the level of transcript induction during submergence. Arrows above genes indicate transcript levels in leaves following submergence. (Bailey-Serres *et al.*, 2012).

Physiological Traits for Submergence Tolerance

The adverse effects of flooding or submergence are multiple and complex. Rice cultivars differ in their tolerance to submergence, and cultivars with adequate submergence tolerance do exist. FR 13 A, a cultivar from Eastern India, can survive for more than two weeks completely submerged (Mazaredo and Vergara, 1982). However, submergence tolerant cultivars are poor in yields, with low phenotypic acceptability by farmers due to inferior quality of grain. Understanding the physiological mechanism (Fig. 2) for cultivar differences in submergence tolerance may facilitate future breeding programmes.

Some of the important physiological traits for submergence tolerance are as follows:

1. Maintaining higher gaseous diffusion and photosynthetic rates
2. Sustained sugar supply and energy metabolism
3. Limited underwater elongation
4. Reduced GA formation
5. Elevated alcholoic fermentation pathways
6. Elevated active scavenging systems
7. Aerenchyma formation

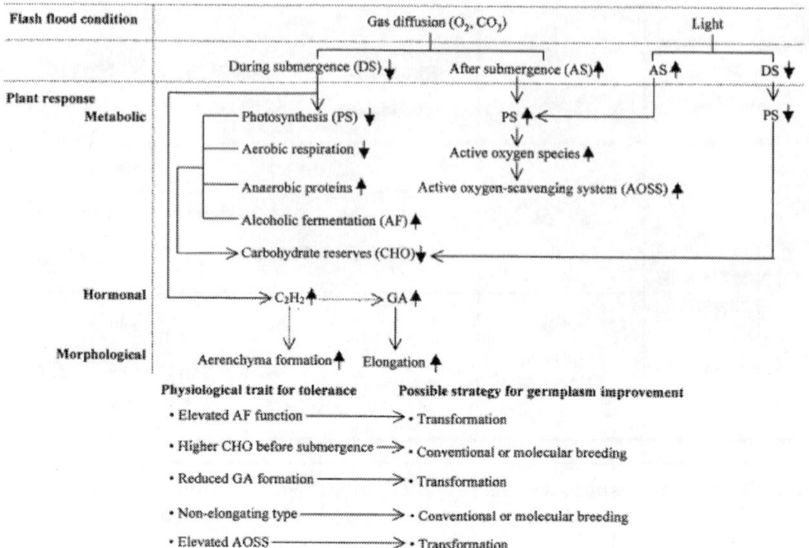

Fig. 3. Physiological responses of rice plants to flash flood and physiological mechanisms for improving performance under submergence conditions (Source: Australian Weeds Committee, 2012)

Gas Diffusion and Photosynthetic Rates

Limited gas diffusion is the most important factor during flooding. Since gas diffusion is 104-fold slower in solution than in air , the depletion of O_2 is a major feature of the flooded field which creates a condition of low O_2 (hypoxia) or no O_2 at all (anoxia) around the plant tissues such as seeds or root apices and stele, even if the latter contain aerenchyma. The maximum photosynthetic rates of six rice cultivars decreased during seven days of submergence with FR13A, a submergence-tolerant cultivar, maintaining 32% of that of the non-submerged control while IR42, a submergence-intolerant cultivar, had 0% (Smith *et al.,* 1987). Generally, photosynthetic rate and stomatal conductance decreased in Swarna and Swarna Sub1 cultivars, during the progression of submergence as compared to control plant. Submergence also alters the photo-system (PS) II activity, as reflected in a decrease in the values of Fo, Fm and the Fv/Fm ratio and degradation of chlorophyll contents. However, Swarna Sub1 improves photosynthetic activity, lesser degradation of chlorophyll, higher stomatal conductance and efficient PS II activity compared to Swarna (Panda and Sarkar, 2011).

Sustained sugar supply and energy metabolism

High amounts of water soluble carbohydrates and starch prior to submergence and slower rates of depletion during submergence are adaptive traits. High initial stem carbohydrates and ability to sustain CO_2

assimilation during flooding are probably essential for providing energy for rapid growth to keep pace with the rising water. Initiation of new leaves and their subsequent growth requires availability of non-structural carbohydrates (Sarkar *et al.*, 2006). The maintenance of greater quantities of NSC at the end of submergence depended on the level of NSC contents before submergence and their low consumptive used during submergence. The cultivars that maintained a higher carbohydrate content at the end of submergence were found to develop new leaves very quickly and accumulated greater biomass during re- emergence. The carbohydrate content of plants was found to be significantly and positively associated with re-generation growth (Panda and Sarkar, 2011). Cultivars that maintained more than 6% of their initial non-structural carbohydrate at the time of re-aeration were found to be capable of developing new leaves. High carbohydrate status after submergence is the consequence of its level before submergence. Seedlings of tolerant species have 30–50% more non-structural carbohydrates compared to the susceptible cultivars. These non-structural carbohydrates are utilized during submergence to supply the required energy for growth and maintenance metabolism.

Shoot elongation

Minimal underwater elongation during submergence is an important tolerance trait because elongating plants would tend to lodge as soon as the water level recedes. Rapid leaf elongation during submergence competes with maintenance respiration for carbon sources. Hence, minimum underwater elongation is associated with increased survival. The possible mechanism of submergence induced shoot elongation was mediated by an interaction between ethylene and GA. This negative relationship between submergence tolerance and elongation during submergence was confirmed using IRRI gene bank database of 903 cultivars (Setter and Laureles, 1996). Quiescence requires strong induction of SUB1A by ethylene, which is increased as a consequence of submergence and biosynthesis. SUB1A is either poorly induced or absent in lowland rice varieties that attempt to escape submergence. The presence or absence of SUB1A does not affect the decline in ABA observed in response to submergence. In tolerant varieties, ethylene-promoted induction of SUB1A results in limited responsiveness to bioactive gibberellin (GA) because of increases in the GA response repressors SLR1 and SLRL1. The SUB1C gene, always present at the SUB1 locus, is triggered by de-repression of GA responsiveness. The underwater escape promoted by GA includes increased starch and soluble sugar catabolism and increased cell wall expansin proteins that promote expansibility (Fukao and Bailey-Serres, 2008). But, in quiescence strategy, ethylene blocks GA responses which in turn deactivates the Sub1C gene and limits elongation growth.

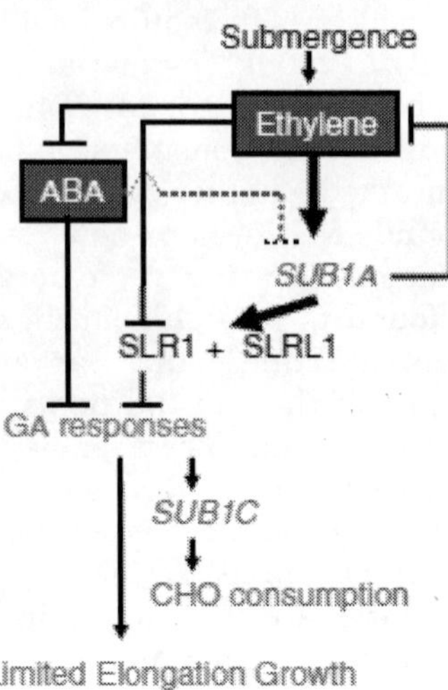

Fig. 4. Quiescence strategy showing limited underwater elongation (Source: Fukao *et al.,* 2006).

Hormonal regulation

Plant hormones are known to play a role in survival under flooded conditions by regulating shoot elongation and leaf senescence. The fast growth of cultivars adapted to deep water conditions seems to be mediated by the action of three different plant hormones, ethylene, GA and ABA. Under anaerobic conditions, ethylene concentration increases in plant tissue because of both increased synthesis and entrapment. This causes a reduction in ABA concentration with a concomitant increase in GA level as well as responsiveness, resulting in enhanced shoot elongation. Conversely, tolerance of flash flooding is associated with a increased level or responsiveness to ethylene. Since enhanced stem elongation is not desired under flash flood conditions, blocking GA synthesis could enhance survival by reducing elongation and conserving energy for maintenance and regeneration growth during and after submergence, respectively (Setter and Laureles, 1996).

The submergence tolerance of FR13A is linked to a major quantitative trait locus (QTL), known as Submergence1 (*Sub1*), on chromosome 9. This locus (200 kb) explains a large proportion of the variation (35– 69%) in flooding tolerance between indica (tolerant) and japonica (intolerant) rice cultivars (Xu and Mackill, 1996). *Sub1* region encodes three transcription factors (*Sub1A*, *Sub1B* and *Sub1C*) belonging to the B-2 subgroup of the ethylene response factors (ERFs)/ethylene-responsive element binding proteins (EREBPs)/apetala 2-like proteins

(AP2). The transcription of both *Sub1A* and *Sub1C* is strongly up regulated by submergence and down regulated by de-submergence.

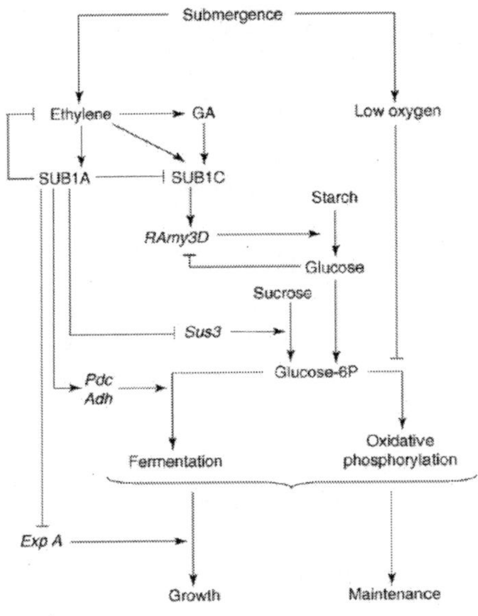

Fig. 5. Schematic model linking *Sub 1A* to submergence tolerance (Source: Perata and Voesnek, 2007)

However, *Sub1C* is less upregulated in the presence of *Sub1A* than in its absence, suggesting that *Sub1A* might downregulate *Sub1C*. The third ERF, *Sub1B* is only slightly regulated upon submergence. Submergence tolerance is strongly correlated with the presence of the *Sub1A-1* allele, whereas intolerance to submergence is associated with the *Sub1A-2* allele or with the complete absence of the *Sub1A* gene. Increased ethylene levels inside submerged plant tissues result in an accumulation of *Sub1A* transcript in submergence- tolerant rice. The *Sub1A* protein increases transcription of genes associated with ethanolic fermentation and represses transcription of genes associated with cell elongation (*Exp A*) and carbohydrate catabolism (*Sus & Amy*) (Fig. 5). Furthermore, *Sub1A* is responsible for a feedback restriction in ethylene production *Sub1A* expression inhibits the increased GA responsiveness during submergence through elevation of the GA signaling repressors, SLR1 and SLRL1 (Jung *et al.*, 2010). Negative regulation of GA responses aids in avoidance of carbohydrate starvation, allowing plants to endure submergence. Upon re-aeration, leaf dehydration occurs. *Sub 1A*-mediated enhancement of ROS amelioration limits oxidative damage and chlorophyll degradation during reoxygenation. Increased ABA responsiveness by *Sub1A* induces expression of LEA mRNAs and suppresses leaf water loss. *Sub 1A* also up regulates transcripts encoding ERFs associated with acclimation to drought (e.g., DREB1s). The *Sub1A*-mediated responses to water deficit can independently enhance recovery from submergence or provide protection during drought, both of which induce cellular dehydration (Fig. 6)

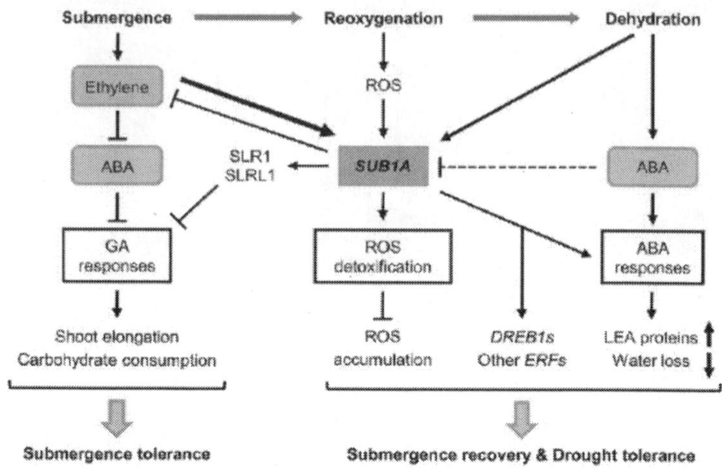

Fig. 6. Model of *Sub 1A*-mediated response during flooding in rice (Source: Jing *et al.,* 2007)

Use of GA Inhibitors on Submergence Tolerance: GA application to plants 48 h before submergence resulted in more elongation during submergence, with consequent reduction in plant survival. In contrast, when GA biosynthetic inhibitor, paclobutrazol was applied, shoot elongation under submergence was reduced resulting in higher survival percentage. Furthermore, dwarf mutants exhibiting little or no capacity for GA biosynthesis showed submergence tolerance on par with the highly tolerant landrace FR 13A, when plants of the same mass or carbohydrate content were used in the experiment. (Sarkar *et al.,* 2006)

Alcoholic fermentation

Submergence can shift aerobic respiration to the less efficient anaerobic fermentation pathway. This pathway depends on continued supply of substrate (glucose) and the two key enzymes, alcohol dehydrogenase (ADH) and pyruvate decarboxy lase (PDC). Alcoholic fermentation is the key catalytic pathway for recycling NAD to maintain glycolysis and substrate level phosphorylation in the absence of oxygen. Increased alcoholic fermentation is thus one way to alleviate the adverse effect of anoxia on reduced production of energy for growth and maintenance purposes.

Aerenchyma

The presence of gas filled spaces, known as aerenchyma, in roots of numerous plant species is considered to be an important anatomical adaptation for survival under flooded conditions (Fig. 7). Aerenchyma provides a diffusion path of low resistance for the transport of oxygen from aerial plant parts to roots or rhizomes in a waterlogged, Oxygen deficient environment. Aerenchyma also provide a path for diffusion of volatile compounds such as ethylene, methane, CO, ethanol and acetaldehyde that are produced in anaerobic soil and plant tissues during flooding. Aerenchyma in some plants is constitutive while in others it is induced in response to flooding. Oxygen is supplied through aerenchyma for root respiration and oxygenation of rhizosphere facilitating detoxification of chemically reduced iron, manganese and hydrogen.

In general, aerenchyma can be classified into two types: (i) schizogenous aerenchyma, which develops by cell separation and differential cell expansion that creates spaces between cells, in e.g. Rumex palustris; and (ii) lysigenous aerenchyma, formed by the death and subsequent lysis of some cells, e.g., in rice, maize and wheat. In the roots, lysigenous aerenchyma forms in the cortex, whereas in the stems it can form in the cortex and pith cavity (Armstrong, 1979). In some wetland plant species such as rice, root lysigenous aerenchyma is constitutively formed under drained soil conditions, and its formation can be further enhanced during soil waterlogging.

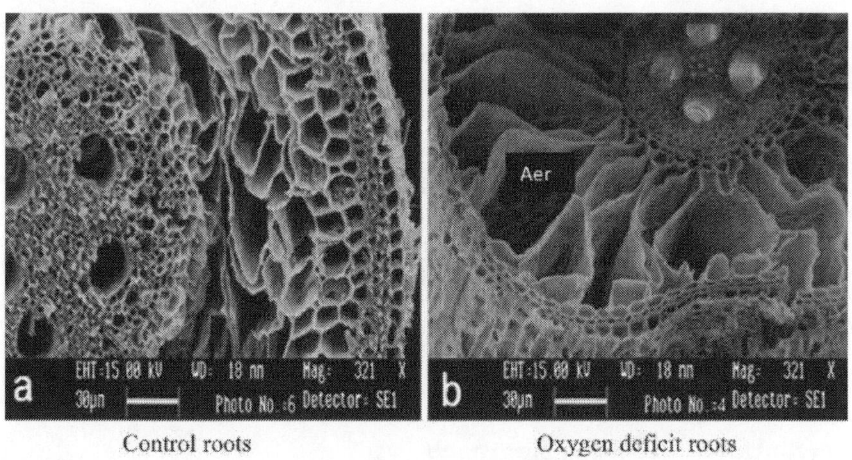

Fig. 7. SEM image of Rice roots showing aerenchyma cells under flooding stress (Source: Joshi and Kumar, 2012).

In general, aerenchyma can be classified into two types: (i) schizogenous aerenchyma, which develops by cell separation and differential cell expansion that creates spaces between cells, in e.g. Rumex palustris; and (ii) lysigenous aerenchyma, formed by the death and subsequent lysis of some cells, e.g., in rice, maize and wheat. In the roots, lysigenous aerenchyma forms in the cortex, whereas in the stems it can form in the cortex and pith cavity (Armstrong, 1979). In some wetland plant species such as rice, root lysigenous aerenchyma is constitutively formed under drained soil conditions, and its formation can be further enhanced during soil waterlogging.

Hence, selection for submergence tolerance may include selecting plants with high initial carbohydrate and reduced shoot elongation during submergence. However, these traits may need further testing by using larger rice populations from the field so that they could finally be utilized by breeders for large-scale screening.

METHODS FOR IMPROVING SUBMERGENCE TOLERANCE

Germplasm Improvement

Water-control measures in submergence-prone areas can help reduce the damage caused by flooding, but this normally entails huge investment beyond the reach of resource poor farmers normally living in these areas. The effect of flash flooding in rice interacts with growth stage with germination being highly sensitive, and with tolerance increasing progressively with seedling age. Most of the existing rice cultivars are seriously damaged if they are completely submerged for more than three days. However, a few tolerant landraces were identified that can withstand complete submergence for 10–14 days. Examples of these are 'FR 13A', 'FR 43B', 'Goda Heenati', 'Kurkaruppan' and 'Thavalu'. However, these tolerant landraces have poor combining ability and agronomic characters and possess the same locus that controls most of the submergence-tolerance phenotype (*Sub1*).

Recently, some new rice landraces have been identified with reasonably higher levels of tolerance to submergence but with better agronomic traits than the landraces identified before. These lines are 'Atiranga', 'Khoda', 'Khadara', 'Kusuma' and 'Kalaputia'. However, these lines are yet to be tested for their combining ability as well as the dominant mechanisms associated with submergence tolerance.

IMPROVEMENT OF SUBMERGENCE TOLERANCE

Conventional Breeding Approach

Promising breeding lines with good plant type and submergence tolerance such as 'IR 26702-25-3' ('FR13A' source) and 'IR 31406-333-1' ('Kurkaruppan' source) have been used as donors in breeding programmes. Subsequent breeding efforts at IRRI led to evolution of tolerant lines with good agronomic traits. The line 'IR 49830-7-2-2' combines high tolerance levels with higher yield potential and resistance to diseases and insect pests and it has been extensively used as a donor parent in the breeding programme. 'Sudhir' is another variety which has been developed from the 'FR13A' 'Biraj' crosses. This variety has been released by the 'Central Variety Release Committee, Indian Council of Agricultural Research, Government of India' in 1999.

Major attempts have also been made in breeding for improved submergence tolerance through the use of double haploid lines (DHLs), developed using crosses between submergence tolerant and sensitive rice cultivars. The role of DHLs is to screen up several hundred lines within each population for submergence tolerance in representative areas of flash-flooding. Two DHL populations for submergence tolerance have been developed at IRRI using the cross combination of 'IR 49830 ′ CT 6241'and 'FR13A ′ IR 42'.

Plant Genetic Engineering

Identification/isolation/cloning of genes which are associated with improved flooding stress tolerance appear to be the limiting factors. A great deal of research in this context is focused on carbohydrate metabolism for the obvious reason that a reduced O_2 supply directly hampers normal respiration resulting in decreased levels of ATP synthesis. The pathway of respiration switches over from oxidation to fermentation mode during anaerobiosis. The induction of ethanolic fermentation pathway is considered to be an important component of responses against flooding stress.

The availability of cloned *pdc* and *adh* genes which encode two enzymes (i.e. pyruvate decarboxylase or PDC and alcohol dehydrogenase or ADH) involved in ethanolic fermentation pathways have been used in transgenic experiments. Apart from PDC and ADH, several other enzymes play a role in mobilization of carbon from complex carbohydrate forms (i.e.starch, sucrose etc.) to simpler six-carbon forms which readily enter glycolysis (i.e. glucose) and metabolizing the six-carbon simpler sugars to pyruvate. These include glucose phosphate isomerase, phosphofructo-kinase, aldolase, triose phosphate isomerase, glyceraldehyde phosphate dehydrogenase, enolase, pyruvate kinase and phosphoglycerate kinase. These genes need to be cloned and transformed.

Marker Assisted Backcross Breeding (MAB Breeding)

Research shows that the mega varieties could be efficiently converted to a submergence tolerant variety in three backcross generations, involving a time of two to three years. Polymorphic markers for foreground and recombinant selection were identified to develop a wider range of submergence tolerant varieties to meet the needs of farmers in the flood-prone regions. This approach demonstrates the effective use of marker assisted selection for a major QTL in a molecular breeding programme.

The basis of a marker-assisted backcrossing (MABB) strategy is to transfer a specific allele at the target locus from a donor line to a recipient line while selecting against donor introgressions across the rest of the genome. The use of molecular markers, which permit the genetic dissection of the progeny at each generation, increases the speed of the selection process, thus increasing genetic gain per unit time. The main advantages of MAB are:

(1) Efficient foreground selection for the target locus,
(2) Efficient background selection for the recurrent parent genome,
(3) Minimization of linkage drag surrounding the locus being introgressed, and
(4) Rapid breeding of new genotypes with favorable traits.

The effectiveness of MAB depends on the availability of closely linked markers and/or flanking markers for the target locus, the size of the population, the number of backcrosses and the position and number of markers for background selection. MAB has been successful towards transferring the *sub1* loci conferring tolerance against submergence into the susceptible genotypes. Submergence tolerant versions of mega varieties like Swarna, Samba Mahsuri, Savithri, CR 1009 and IR 64 have been developed through MAB breeding.

Research Highlights at Tamil Nadu Agricultural University:

The authors acknowledge the Department of Biotechnology (DBT), New Delhi for funding the research project on "Marker assisted introgression of Sub1 locus conferring tolerance to flooding into elite rice varieties of Tamil Nadu". The salient findings of the project are as follows:

- CO 43 is a medium duration variety popularly grown in the Cauvery delta region of Tamil Nadu. It is known for its superior level of salinity tolerance. Improving the submergence tolerance of this genotype will be having an added advantage. Hence Co 43 was selected as a target genotype for marker assisted introgression of *Sub1* locus.
- Crosses were made between Co 43 x FR 13A and true hybrids were identified by SSR markers. F1s were maintained by ratooning and back crossed with Co 43. On the other hand, seeds collected from F1s were used for rising F2 generation.
- Parents (Co 43 and FR13A) were surveyed for polymorphism using SSR markers linked to *Sub1* locus. SSR marker namely RM219 and CAPS marker namely ART5 have been identified for Foreground Selection.
- To identify SSR markers for back ground selection, a total of 224 markers were surveyed. Out of these, 83 SSRs showed polymorphism between Co 43 and FR13A. These markers are being used for back ground selection.
- Repeated backcross breeding and whole genome selection was done and superior BC_3F1 plants possessing >85% of CO43 genome were back crossed with CO 43 to develop BC_4F_1. BC_4F_1 progenies are in germination stage.

- Backcross progenies were evaluated for their tolerance against flooding/submergence. BC_2F_3 families were grown in tanks for 21 days and submerged for 13 days. After 13 days of submergence, tanks were desubmerged and recovery was scored after 2 weeks. Survived plants were genotyped again and it was found that most of the plants were either heterozygous or homozygyous for FR13A allele.

Future outlook

The characterization of the *Sub1* locus has resulted in a big step forwards in our understanding of the regulation of flooding tolerance but the difference in submergence tolerance between and within species is only partly resolved by the identification of the *Sub1A* gene. More basic research on regulation of flooding response in terms of transcriptional activation of stress-responsive genes and signal transduction mechanisms is the need of the hour. Screening methods for flooding tolerance has been created which will take a great lead in developing varieties which are flood tolerant.

Breeding improved submergence tolerant cultivars has been going on for more than three decades. However, improved cultivars have generally not been adopted by the farmers in submergence prone areas. The development of submergence-tolerant versions of popular varieties with high grain quality and wide adaptability offers a model for an alternative to the time-consuming, labor-intensive task of developing new varieties in a conventional crossing program.

With the availability of large numbers of microsatellite markers, high throughput genotyping and fine mapping of the major QTLs, virtually any major QTL can now be introgressed into a variety without changing the desirable agronomic characteristics. Thus, there is a need to use MAB strategy for introgression of *Sub1,* a major QTL into a set of popular varieties of India.

References

Adkins, SW; Shiraishi, T and McComb, JA. 1990. Submergence tolerance of rice – A new glasshouse method for the experimental submergence of plants. *Physiol Plant.,* 80: 642-646.

Armstrong, W. 1979. Aeration in higher plants. *Adv Bot Res.,*7: 236–332

Bailey, SJ; Fukao, T; Ronald, PA; Ismail, S; Heuer and Mackill, D. 2010. Submergence tolerant rice: *SUB1's* journey from landrace to modern cultivar. *Rice,* 3: 138–147.

Bailey, SJ; Fukao, T; Gibbs, DJ; Holdsworth, MJ; Lee, SC; Licausi, F; Perata, P; Voesenek, LACJ and .Dongen, JTV. 2012. Making sense of low oxygen sensing. *Trends Plant Sci.,* 17:129-138

Fukao, T; Bailey-Serres, J. 2008. Submergence tolerance conferred by Sub1A is mediated by SLR1 and SLRL1 restriction of gibberellin responses in rice. *Proceedings of the National Academy of Sciences USA.*, 105:16814–16819

Haque, QA; Hille Ris Lambers, D; Tepora, NM and Dela Cruz, QD.1989. Inheritance of submergence tolerance in rice. *Euphytica.*, 41: 247-251.

Hattori, Y; Nagai, K and Ashikari, M. 2011. Rice growth adapting to deepwater. *Curr. Opin. Plant Biol.,* 14: 100-105.

Hattori, Y; Nagai, K; Furukawa, S; Song, XJ; Kawano, R; Sakakibara, H; Wu, J; Matsumoto, T; Yoshimura, A; Kitano, H; Matsuoka, M; Mori, H and Ashikari. 2009. The ethylene response factors *SNORKEL1* and *SNORKEL2* allow rice to adapt to deep water. *Nature.,* 460: 1026-1030.

Hattori, Y; Nagai, K and Ashikari, M. 2011. Rice growth adapting to deepwater. *Current Opinion in Plant Biol.*, 14:100–105

INCCA (Indian Network for Climate Change Assessment) 2010. A 4x4 assessment, a sectoral and regional analysis for 2030s: *Ministry of Environment and Forests, Government of India.*

Jackson, MB and Ram, PC. 2003. Physiological and Molecular Basis of Susceptibility and Tolerance of Rice Plants to Complete Submergence. *Ann. Bot.,* 91: 227-241.

Jung, KH; Young-Su Seo; Cao; Takeshi Fukao; Canlas, PE; Fawn Amonpant; Julia Bailey-Serres and Ronald PC. 2010. The Submergence Tolerance Regulator *Sub1A* Mediates Stress-Responsive Expression of AP2/ERF Transcription Factors. *Plant Physiol.,* 152:1674–1692

Khush, GS and Brar, DS. 1998. The application of biotechnology to rice. In: Ives CL, Bedford BM, editors. Agricultural biotechnology in international development. Wallingford (UK): *CAB Internatl.*, 97-121.

Mazaredo, AM and Vergara, SBS. 1982. Physiological differences in rice varieties tolerant of and susceptible to submergence. In: *Proceedings of the 1981 International Deepwater Rice Workshop.* IRRI, Los Banos, 327-341.

Mohanty, HK and Khush, GS. 1985. Diallele analysis of submergence tolerance in rice. *Theor Appl Genet.,* 70: 467- 473.

Nandi, SP; Subudhi, K; Senadhira, D; Manigbas, NL; Sen-Mand, S and Huang, N. 1997. Mapping QTLs for submergence tolerance in rice by AFLP analysis and selective genotyping. *Mol. Gen. Genet.,* 255(1): 1-8.

Panda Debabrata and Sarkar Ramani Kumar. 2011. Improvement of Photosynthesis by *Sub1* QTL in Rice Under Submergence: Probed by Chlorophyll Fluorescence OJIP Transients. *J. of Stress Physiol & Biochem.,* 7(3): 250-259.

Sarkar, RK; Reddy, JN; Sharma, SG and Ismail, AM. 2006. Physiological basis of submergence tolerance in rice and implications for crop improvement. *Curr. Sci.,* 91: 899-906.

Setter, TL and Laureles, EV. 1996. The beneficial effect of reduced elongation growth on submergence tolerance of rice. *J. Exp. Bot.*, 47: 1551–1559.

Singh, N; Dang, T; Vergara, G; Pandey, D; Sanchez, D; Neeraja, C; Septiningsih, E; Mendioro, M; Tecson-Mendoza, R; Ismal, A; Mackill, D and Heuer, S. 2010. Molecular marker survey and expression analyses of the rice submergence-tolerance genes SUB1A and SUB1C. *Theor Appl Genet.* doi:10.1007/s00122-010-1400-z.

Sinha, MM and Saran, S. 1988. Inheritance of submergence tolerance in lowland rice. *Oryza* 25: 351-354

Smith, PA; Kupkanchanakul, K; Emes, MJ and Cutter, EG. 1987. Changes in fluorescence and photosynthesis during submergence of deepwater rice. In: *Proceedings of the 1987 International deepwater rice workshop.* IRRI, Los Bonos, 327 -335.

Supapoj, N; Boonwrite, C; O'Toole, JC; De Datta, SK and Jackson, BR. 1982. Using check varieties to adjust visual scores for non-uniform soil drying in drought resistance field screening. *International Rice Research Newsletter.,* 7: 12.

Suprihatno, B and Coffman WR. 1981. Inheritance of submergence tolerance in rice (*Oryza sativa L.*). *SABRAO J.*, 13: 98–102.

Unnikrishnan, AS; Kumar, RK; Fernandes, SE; Michael, GS and Patwardhan, SK. 2006. Sea level changes along the Indian coast: Observations and projections. *Curr Sci.,* 90:362–368.

Wassmann, R; Jagadish, SVK; Heuer, S; Ismail, AM; Redona, E; Serraj, R; Singh, RK; Howell, G; Pathak, H and Sumfleth, K. 2009. Climate Change Affecting Rice Production: The Physiological and Agronomic basis for Possible Adaptation Strategies. *Adv. Agron.*, 101:60–122.

Xu, K; Xu, X; Fukao, T; Canlas, P; Maghirang-Rodriguez, R; Heuer, S; Ismail, AM; Bailey- Serres, J; Ronald, PC and Mackill, DJ. 2006. *Sub1A* is an ethylene-response-factor-like gene that confers submergence tolerance to rice. *Nature.*, 442: 705-708.

Xu, K; Xu, X; Ronald, PC and Mackill, DJ. 2000. A high-resolution linkage map in the vicinity of the rice submergence tolerance locus *Sub1. Mol. Gen. Genet.,* 263: 681-689.

Xu, KN and Mackill, DJ. 1996. A major locus for submergence tolerance mapped on rice chromosome 9. *Mol. Breed.*, 2: 219–224.

9

PACLOBUTRAZOL: A POTENTIAL GROWTH REGULATOR UNDER ABIOTIC STRESSES

Preeti Singh*, Dinesh K. Yadav, Vijai P. and A. Hemantaranjan

Department of Plant Physiology, Institute of Agricultural Sciences, Banaras Hindu University, Varanasi-221005
**Corresponding author's email: physiologist.15aug@gmail.com*

INTRODUCTION

Our global environment is going to change day by day which is harmful to the crop plants by reducing their productivity. In India there was a waste damage of crops is take place due to the abiotic stresses. Stress is an altered physiological condition caused by factors that tend to disrupt the equilibrium. Strain is any physical and chemical change produced by a stress (Gaspar *et al.*, 2002). Stress being a constraint or highly unpredictable fluctuations imposed on regular metabolic patterns cause injury, disease or aberrant physiology. Plants are frequently exposed to many stresses such as drought, low temperature, salt, flooding, heat, oxidative stress and heavy metal toxicity, while growing in nature. The growth and development of plants is adversely affected by these environmental threats via alterations in the normal metabolic functions which leads to the death of the plant under severe conditions. Under these adverse environmental conditions the ROS (reactive oxygen species) are generated which caused oxidative damage to the cells by hampering membrane properties. Plant growth regulators has major vital role in the growth and development. There are several plant growth regulator/hormones are well established and having growth regulatory effects in the plants under normal as well as under stress environmental conditions. They involved in the signal transduction pathway in a very minor quantity and also with the cross talk of other metabolic functions. Paclobutrazol (PBZ) is one of the important growth regulators of the trizoles group, which has considerable influence on aboitic stress tolerance in various crops along with tree plants by accelerating or by inhibiting metabolic processes. PBZ-treated plants have a more efficient free-radical scavenging system that enables them to detoxify active oxygen (Kopyra and Gwozdz, 2003). Even though PBZ induced stress tolerance and stress protection has been attributed to an increase in some antioxidant enzymes (Lurie *et al.,* 1994; Paliyath and Fletcher, 1995; Pinhero *et al.,* 1997). PBZ has been reported to confer protection to plants by reducing oxidative damage via the elevation of antioxidants or the reduction of oxidative enzyme activity. This article reviews crucial aspects of paclobutrazole and its mitigating effect on the harmful effects caused by a number of abiotic stresses including UV-radiation on plant species carried out in the last two decades.

PACLOBUTRAZOL (PBZ)

Paclobutrazol (PBZ) ((2RS, 3RS)-1- 4-(chlorophenyl)-4, 4-dimethyl-2-1, 2, 4-triazol-1-yl-penten-3-ol) is a member of the triazole family. Triazoles have both fungi toxic and plant-growth regulatory effects. In addition, they can also protect plants against various stresses, including drought, low and high temperatures, waterlogging, air pollutants, and fungal pathogens. The structure of paclobutrazol contains two asymmetric carbon atoms and the groups at these chiral centres at carbon atoms 2 and 3 can exist in two enantiomeric forms that are mirror images of each other (Fig. 1). The presence of two chiral centres makes possible two pairs of enantiomers (2RS, 3SR and 2RS, 3RS) but it is the 2RS3RSdiastereoisomer which is formed predominantly during the stereo-selective synthesis of paclobutrazol. This diastereoisomer can be resolved, with difficulty, into its constituent 2R, 3R- and 2S, 3S-enantiomers.

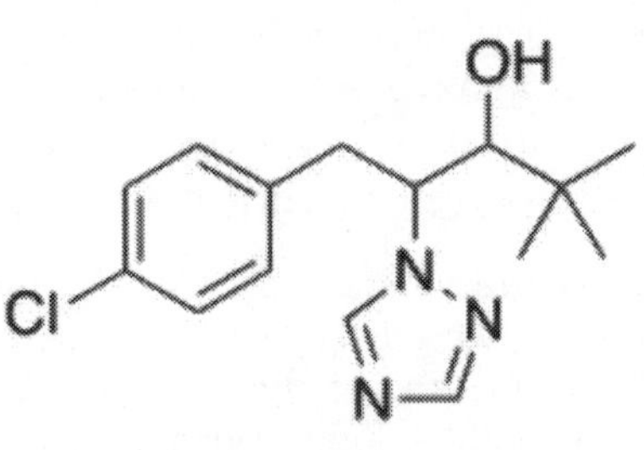

OH

Cl N N N

Fig. 1. Chemical structure of paclobutrazol (Source: http://en.wikipedia.org/wiki/paclobutrazol)

The growth properties of paclobutrazol are mediated by changes in the balance of important plant hormones including the gibberellins, ABA and cytokinins (Fletcher *et al.*, 1990). Many ways of its application to plants have been done, for example, spraying on leaves or stems and on soil drench.

Mode of Action

PBZ is a cell elongation and internode extension inhibitor that retards plant growth by inhibition of gibberellins biosynthesis. Gibberellins stimulate cell elongation. When gibberellins production is inhibited, cell division still occurs, but the new cells do not elongate. The result is shoots with the same numbers of leaves and internodes compressed into a shorter length. Another response of plants to treatment with PBZ is increased production of the hormone abscisic acid and the chlorophyll component phytol, both beneficial to plant growth and health. PBZ may also induce morphological modifications of leaves, such as smaller stomatal pores, thicker leaves, and increased number and size of surface appendages, and increased root density that may provide improved environmental stress tolerance and disease resistance (Chaney, 2005). The activity of antioxidant enzymes such as SOD, POX, PPO and APX are increased under the abiotic stress situations which provides antioxidant defense mechanisms to the plants.

Mobility

PBZ has been characterized as a compound with a moderate potential for mobility in soil and water environments (U.S. EPA, 2007B).

Persistence

PBZ has been characterized as an environmentally stable compound in soil and water environments (U.S. EPA, 2007B). Laboratory studies with US loam and silt-loam soils indicated that PBZ degraded with a half-life of more than 1 year under both aerobic and anaerobic conditions. PBZ interferes with gibberellin biosynthesis by inhibiting oxidation of ent-kaurene to ent-kauronoic acid in endoplasmic reticulum through disabling cytochrome P 450-dependent oxygenase (Srivastav *et al.,* 2010).

EFFECT OF PACLOBUTRAZOL UNDER WATERLOGGING STRESS

Oxygen limitation is the primary stress to plants growing in flooded soils. Plants are severely affected by flooding, because under such stress the production and the quenching of the reactive oxygen species (ROS) in plants will fail to be kept in a balanced state (Crawford and Brandle, 1996). One of the major consequences of soil flooding in plants is oxygen deficiency; consequently, the roots suffer from periodic or prolonged deprivation of oxygen and this interferes with respiration. The lack of suitable electron acceptors leads to the saturation of redox chains, the accumulation of NAD(P)H, and a decline in ATP generation (Kennedy *et al.*, 1992; Perata and Alpi, 1993). In plant cells, oxidative stress reactions are associated with toxic free radicals yielded by the reduction of molecular oxygen leading to the production of superoxide radicals (O2 -), singlet oxygen ($^1O2$), hydroxyl radicals (OH) and hydrogen peroxide (H2O2). These radicals can deactivate various Calvin-cycle enzymes and are involved in oxidative systems, in which they mark proteins for degradation (Chaudiere and Ferrari-Ilious, 1999). The toxic radicals can be removed through the mobilization of antioxidant reserves, which react both enzymatically and chemically with the toxic molecular species and their products. Chemical constituents have been identified to scavenge free radicals and thus protect active plant cells against oxygen toxicity (Sairam *et al.*, 1998; Ahmed *et al.*, 2002). In order to counter the hazardous effects of ROS under stress, plants have evolved a complex antioxidative defense system composed of both antioxidant enzymes and metabolites, such as ascorbate peroxidase (APX), catalase (CAT), superoxide dismutase (SOD), glutathione reductase (GR), ascorbic acid or ascorbate (ASA), reduced glutathione (GSH), oxidized glutathione (GSSG), malondialdehyde (MDA) and vitamin E (Sairam *et al.*, 2002).

The interaction between the production and the scavenging of activated oxygen species maintains the plants in a relatively stable state. Plants are adapted to minimize radical damage using their natural defense mechanisms. Thus, maintaining the balance between the formation and detoxification of activated oxygen species is critical to cell survival during flooding stress. When roots are submerged, the anoxic conditions inhibit aerobic respiration and less energy is yielded. The roots transport less water to the leaves. The antioxidative enzymes, hormones, and other solutes entering the leaves via the transpiration stream may also undergo changes (Grichko and Glick, 2001; Voesenek *et al.*, 2004). These changes may constitute physiologically active messages that modify leaf physiology and development; such modifications may include chlorophyll breakdown, protein degradation, decreased stem elongation coupled with increased stem thickness, a decrease in membrane permeability, peroxidation, slower leaf expansion, petiole epinasty and stomatal closure. Stomatal closure causes a decrease in internal CO2 concentration. Subsequently, a concomitant decline in photosynthesis results from diminished availability of CO_2 for carbon fixation (Carvalho and Amancio, 2002). Reduction of CO2 concentration increases the amount of harmful ROS within the leaf due to ongoing light reactions that can lead to senescence and even the death of the plant (Else *et al.*, 1995; Kato *et al.*, 2001). Flooding tolerance can be achieved using several adaptive mechanisms, and the modulation of antioxidative system levels may be part of the whole mechanism. High levels of some antioxidant enzymes and metabolites, such as SOD, CAT, APX, GR, GSH, ASA, MDA, AOX or GPX, are important in tobacco (Hurng and Kao, 1994a, b), corn (Yan and Dai, 1996), wheat (Biemelt *et al.*, 1998), soybean (VanToai and Bolles, 1991), and rice (Ushimaro *et al.*, 1992)

Paclobutrazol affect the isoprenoid pathway, and alter the levels of certain plant hormones by inhibiting gibberellins synthesis, reducing ethylene evolution, and increasing cytokinin levels (Kamountsis and Chronopoulou-Sereli, 1999). Some morphological changes observed in triazole-treated plants include the inhibition of plant growth, decreased intermodal elongation, increased chlorophyll levels, enlarged chloroplasts, thicker leaf tissue, increased root to shoot ratio and elevated levels of epicuticular wax formation (Watson and Himelick, 2004). The chloroplast is a major site of free-radical production and paclobutrazol in plants, protecting them by boosting antioxidant defense systems. PBZ-treated plants have a more efficient free-radical scavenging system that enables them to detoxify active oxygen (Kopyra and Gwozdz, 2003). Even though PBZ induced stress tolerance and stress protection has been attributed to an increase in some antioxidant enzymes (Lurie *et al.*, 1994; Paliyath and Fletcher, 1995; Pinhero *et al.*, 1997)

The effects of PBZ on the sweet potato subjected to flooding stress were monitored by tracking changes in both antioxidative enzyme activities (CAT, GR, APX and SOD) and antioxidant content (ASA, total ASA, total glutathione and MDA) on various days of treatment. Plants receiving PBZ that are under environmental stress have significantly higher antioxidative system levels compared to plants under non stressed conditions (Lurie *et al.*, 1994; Paliyath and Fletcher, 1995). Results support the findings presented in Fletcher *et al.* (2000), namely that PBZ protected wheat seedlings from injury due to flooding, and that PBZ treatment reduced all symptoms of damage at various stages of growth. According to these findings, paclobutrazol would be a beneficial pre-treatment for plants that are grown in wetlands, lowlands or areas that are prone to flooding, such as where there is natural glorying of the soil, or in areas subject to short and intense rainfall.

EFFECT OF PACLOBUTRAZOL UNDER DROUGHT STRESS

Environmental stresses represent the most limiting factors for agricultural productivity. Nowadays, a big concern is the water deficit, extremes in temperature and low atmospheric humidity leading to drought stress which is one of the most limiting factors for plant performance (Hit and Shinozaki, 2004). Water stress or drought is a major problem in agriculture and the ability to withstand such stress is of immense economic importance. Plants are subjected to several environmental stresses that adversely affect growth, metabolism and yield (Lawlor, 2002). Drought, one of the environmental stresses, is the most significant factor restricting plant growth and crop productivity in the majority of agricultural fields of the world (Tas and Tas, 2007). Drought is a meteorological term and is commonly defined as a period without significant rainfall. Generally drought stress occurs when the available water in the soil is reduced and atmospheric conditions cause continuous loss of water by transpiration or evaporation. Drought stress tolerance is seen in almost all plants but its extent varies from species to species and even within species (Jaleel *et al.,* 2007b-e) Tolerance to abiotic stresses is very complex, due to the intricate of interactions between stress factors and various molecular, biochemical and physiological phenomena affecting plant growth and development (Razmjoo *et al.*, 2008). In many cases, high yield potential can contribute to yield in moderate stress environment (Blum, 1996). Drought stress is considered to be a moderate loss of water, which leads to stomatal closure and limitation of gas exchange. Desiccation is much more extensive loss of water, which can potentially lead to gross disruption of metabolism and cell structure and eventually to the cessation of enzyme catalyzed reactions (Smirnoff, 1993; Jaleel *et al.,* 2007d). Water deficit affects crop growth, depending on the stage of growth and the degree or intensity of water stress (Clavel *et al.*, 2005). Drought stress inhibits cell enlargement more than cell division. It reduces plant growth by affecting various physiological and biochemical processes, such as photosynthesis, respiration, translocation, ion uptake, carbohydrates, nutrient metabolism and growth promoters (Jaleel *et al.,* 2008a-e; Farooq *et al.*, 2008).

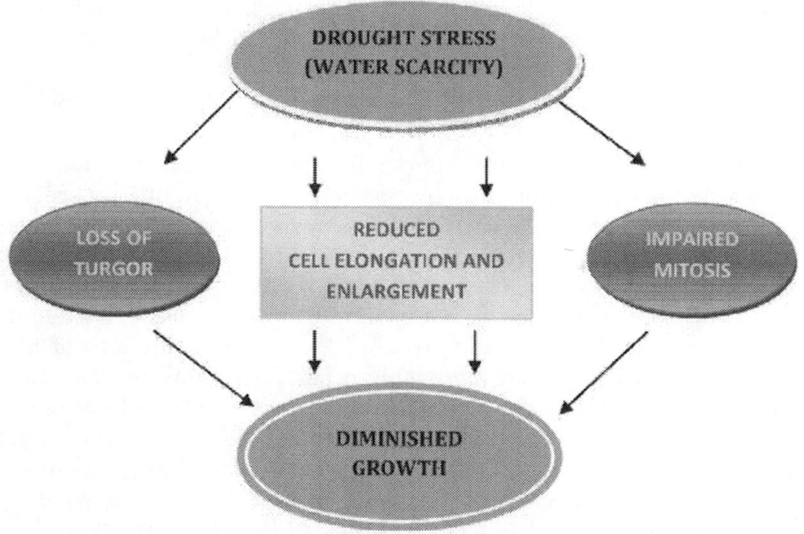

Fig. 2: Some important causes of growth reduction in plants under drought stress (Source: Jaleel *et al.,* 2009; *Int. J. Agric. Biol.* 11: 100-105)

Water stress tolerance is seen in almost all plant species but its extent varies from species to species. Water deficit stress is a global issue to ensure survival of agricultural crops and sustainable food production (Shao *et al.*, 2006). Currently, protection of plants from abiotic stresses through application of plant growth regulators (PGR) attracts more attention (Manivannan *et al.*, 2007a; Jaleel *et al.*, 2007a). Tolerance to abiotic stresses is very complex, due to the complexity of interactions between stress factor and various molecular, biochemical and physiological phenomena affecting plant growth and development (Ashraf and Harris, 2004). The plant growth regulating properties of triazoles are mediated by their ability to alter the balance of important plant hormones including gibberellic acid, ABA and cytokinins (Kamountsis and Chronopoulon-Sereli, 1999).

Triazoles induce a variety of morphological and biochemical responses in plants, including inhibited shoot elongation, stimulated root growth, increased cytokinin synthesis and a transient rise in ABA, as well as conferring protection from various environmental stresses (Jaleel 2007a; Fletcher *et al.*, 2002; Jaleel 2007b). Protections of plants from apparently unrelated stresses by triazole is also mediated by a reduction in free radical damage and increase in the antioxidant potential and have an efficient free-radical scavenging system that enables them to detoxify active oxygen (Kopyra and Gwozdz, 2003). The growth regulating properties of paclobutrazol are mediated by changes in the balance of important plant hormones including the Gibberellins, ABA and cytokinins (Fletcher and Hofstra, 1990). Paclobutrazol has been proved as an agent in stress amelioration in medicinal plants (Jaleel 2007a). Paclobutrazol increased the diameter and length of fibrous roots and enhanced the lateral root formation in apple seedlings (Steffens *et al.*, 1983) and tomato plants (Berova and Stoeva, 2002).

Protection of plants from apparently unrelated stress by triazoles is mediated by a reduction in free-radical damage and increase in antioxidant potential (Jaleel *et al.*, 2007). Paclobutrazol (PBZ) is one of the triazole compounds that is a widely used as growth retardant, inducing mild stress tolerance in seedlings and adult plants. More specifically, PBZ has been reported to protect plants against drought stress (Fletcher *et al.*, 2000). In ornamental crops, PBZ is used for reducing the size of plants, improving compactness, and increasing other functional aspects, such as the ability to resist both abiotic and biotic stresses. Several studies have demonstrated the efficiency of this compound to reduce the negative effects of water stress (Banőń *et al.*, 2001; Berova and Zlatev, 2003).

The effects of triazoles on hormonal changes and photosynthetic rate have been reported by different researchers (Ye *et al.*, 1995; Zhou and Ye, 1996). On the other hand, differences in sensitivity to drought between different species and/or cultivars (Zollinger *et al.*, 2006) and even between growth stages have been demonstrated for many plants (Mingeau *et al.*, 2001). Paclobutrazol decreased electrolyte leakage in wheat seedlings (Kraus *et al.*, 1995) during drought stress from loss of membrane integrity and ion leakage on wheat seedlings (Orabi and Mekki, 2008). PBZ induces large elastic module in cell walls in PBZ treated plants in white spruce, which in turn maintains turgor and thus increases the plant tolerance to drought stress (Fletcher *et al.*, 2000). The obvious increases in antioxidant enzymes activities which in turn enhanced scavenging of harmful free radicals and also, the inhibition of electrolyte leakage and lipid peroxidation is correlated with the proved ability of triazoles in maintaining the membrane integrity (Fletcher *et al.*, 2000).

Water stress and PBZ treatment enhanced free proline content. Keyvan (2010) found that under water stress there was an increase in proline content in wheat cultivars. In addition, the primary response of drought stressed wheat plantlets was osmotic adjustment through proline accumulation, which is well established in many plant species (Ahmad *et al.*, 2007). The higher proline content could be due to enhance activity of ornithine aminotransferase (OAT) and pyrroline 5-carboxylate reductase (P5CR), the enzyme involved in proline biosynthesis as well as due to the inhibition of proline oxidase, proline catabolising enzymes (Debnath, 2008). Also, triazole induces a transient raise in ABA content and this increased ABA content due to triazole treatment could be the reason for increased proline and amino acid content of triazole treated carrot plants (Gopi *et al.*, 2007). Sankar *et al.* (2007) who found that, combination of drought and PBZ treatments caused higher level of reduced glutathione content in both varieties of *Arachis hypogaea* L. plants when compared to control.

Changes in intracellular glutathione status may therefore be expected to have important consequence for the cell, through modification of the cell metabolic functions associated with glutathione regulated genes (Szalai *et al.*, 2009). Lipid peroxidation levels in wheat seedlings, determined as the content of MDA. Tatar and Gevrek (2008); Zhang (2007) who observed an increase in MDA concentration in both wheat and soybean plants exposed to water stress. On the other hand, uniconazole reduced MDA accumulation of stressed soybean. According to Fletcher *et al.* (2000), inhibition of lipid peroxidation may be one of the mechanisms responsible for the anti-senescence effects of the triazoles.

Zhang *et al.* (2007), who found that under water stressed conditions, PBZ increased the soluble sugars content in leaves of soybean. Accumulations of soluble carbohydrates increase the resistance to drought stress in plants (Keyvan 2010). Earlier reports mentioned that sugars protect the cells during drought by the following mechanism; the hydroxyl groups of sugars may substitute for water to maintain hydrophilic interactions in membranes and proteins during dehydration. Thus, sugars interact with proteins and membranes through hydrogen bonding, thereby preventing protein denaturation (Al- Rumaih, and Al- Rumaih, 2007). Improvement of photosynthesis by paclobutrazol is the main contribution for better yield regardless of water conditions, as was indicated by other studies (Qiu *et al.*, 2005). In this study, the increase in photosynthetic pigments due to application of PBZ might be attributed to more stimulation of stomatal regulation (Navarro *et al.*, 2007).

EFFECT OF PACLOBUTRAZOL UNDER SALINITY STRESS

A major challenge towards world agriculture involves production of 70% more food crop for an additional 2.3 billion people by 2050 worldwide (FAO, 2009). Salinity is a major stress limiting the increase in the demand for food crops. More than 20% of cultivated land worldwide (about 45 hectares) is affected by salt stress and the amount is increasing day by day. Plants on the basis of adaptive evolution can be classified roughly into two major types: the halophytes (that can withstand salinity) and the glycophytes (that cannot withstand salinity and eventually die). Majority of major crop species belong to this second category. Thus salinity is one of the most brutal environmental stresses that hamper crop productivity worldwide (Flowers, 2004; Munns and Tester, 2008.

Salinity stress involves changes in various physiological and metabolic processes, depending on severity and duration of the stress, and ultimately inhibits crop production Jamesetal, 2011; Rozema and Flowers, 2008. Initially soil salinity is known to represses plant growth in the form of osmotic stress which is then followed by ion toxicity (Jamesetal, 2011; Rahnama *et al.*, 2010). During the initial phases of salinity stress, water absorption capacity of root systems decreases and water loss from leaves is accelerated due to osmotic stress of high salt accumulation in soil and plants, and therefore salinity stress is also considered as hyper osmotic stress (Munns, 2005). Osmotic stress in the initial stage of salinity stress causes various physiological changes, such as interruption of membranes, nutrient imbalance, impairs the ability to detoxify reactive oxygen species (ROS), differences in the antioxidant enzymes and decreased photosynthetic activity, and decrease in stomatal aperture (Munns and Tester, 2008; Rahnama *et al.*, 2010). One of the most detrimental effects of salinity stress is the accumulation of Na+ and Cl^- ions in tissues of plants exposed to soils with high NaCl concentrations. Plants develop various physiological and biochemical mechanisms in order to survive in soils with high salt concentration. Principle mechanisms include, but are not limited to, (1) ion homeostasis and compartmentalization, (2) ion transport and uptake, (3) biosynthesis of osmoprotectants and compatible solutes, (4) activation of antioxidant enzyme and synthesis of antioxidant compounds, (5) synthesis of polyamines, (6) generation of nitric oxide (NO), and (7) hormone modulation.

Increasing facts demonstrates the roles of a Salt Overly Sensitive (SOS) stress signaling pathway in ion homeostasis and salt tolerance (Hasegawa *et al.*, 2000; Sanders, 2000) The SOS signaling pathway (Fig. 3) consists of three major proteins, SOS1, SOS2, and SOS3. *SOS1,* which encodes a plasma membrane Na+/H+ antiporter, is essential in regulating Na^+ efflux at cellular level. Over expression of this protein confers salt tolerance in plants (Shi *et al.*, 2000; Shi *et al.*, 2002). *SOS2* gene, which encodes a serine/threonine kinase, is activated by salt stress elicited Ca+ signals. This protein consists of a well-developed N-terminal catalytic domain and a C-terminal regulatory domain (Liu *et al.*, 2000). The third type of protein involved in the SOS stress signaling pathway is the SOS3 protein which is a myristoylated Ca^{2+} binding protein and contains a myristoylation site at its N-terminus. This site plays an essential role in conferring salt tolerance (Ishitani *et al.*, 2000). C-terminal regulatory domain of SOS2 protein contains a FISL motif also known as NAF domain), which is about 21 amino acid long sequence, and serves as a site of interaction for Ca2+ binding SOS3 protein (Fig. 3). This interaction between SOS2 and SOS3 protein results in the activation of the kinase (Guo *et al.*, 2004). The activated kinase then phosphorylates SOS1 protein thereby increasing its transport activity which was initially identified in yeast (Quintero *et al.*, 2002). SOS1 protein is characterized by a long cytosolic C-terminal tail, about 700 amino acids long, comprising a putative nucleotide binding motif and an autoinhibitory domain. This autoinhibitory domain is the target site for SOS2 phosphorylation (Fig. 3). Besides conferring salt tolerance it also regulates pH homeostasis, membrane vesicle trafficking, and vacuole functions (Quintero *et al.*, 2011; Oh *et al.*, 2010). Thus with the increase in the concentration of Na+ there is a sharp increase in the intracellular Ca^{2+} level which in turn facilitates its binding with SOS3 protein. Ca^{2+} modulates intracellular Na+ homeostasis along with SOS proteins. The SOS3 protein then interacts and activates SOS2 protein by releasing its self-inhibition. The SOS3-SOS2 complex is then loaded onto plasma membrane where it phosphorylates SOS1 (Fig. 3). The phosphorylated SOS1 results in the increased Na+ efflux, reducing Na+ toxicity (Martinez-Atienza *et al.*, 2007).

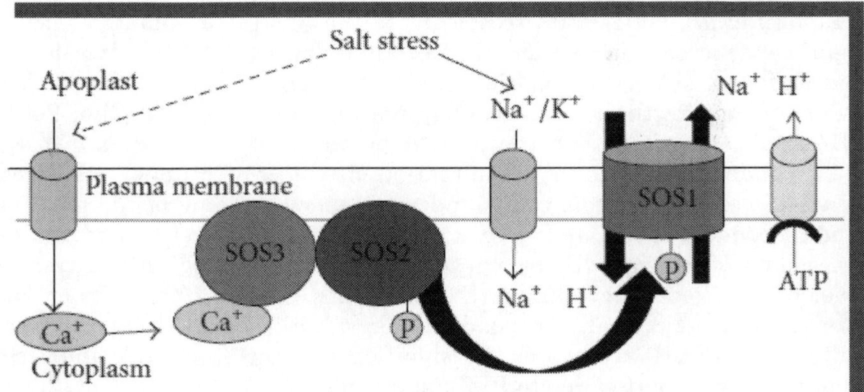

Fig. 3. Model of SOS pathway for salinity stress responses (Source: Bhaskar and Bingru, 2014; http://dx.doi.org/10.1155/2014/701596)

Plant adaptations to salt stress are complex and affected by internal constitutive salt tolerance mechanisms and external environmental factors. One of the plant responses to environment stresses is the rapid and increased generation of reactive oxygen species (ROS), including hydrogen peroxide (H_2O_2), superoxide anion ($O2^-$) and hydroxyl radicals (OH^-), which can damage membrane lipids, proteins, and nucleic acids (Liang *et al.* 2003; Patade *et al.* 2011). Different plants develop several protection mechanisms to eliminate ROS and prevent oxidative damage. Among of them, enzymatic antioxidants such as superoxide dismutase (SOD), catalase (CAT), peroxidase (POX) and ascorbate peroxidase (APX) provide the first line of defense against ROS (Demiral and Turkan 2004; Jebara *et al.* 2005) and non enzymatic antioxidants ascorbic acid (AA) and reduced glutathione (GSH).

Some researchers have used PGRs for reducing or eradicating the negative effects of salinity (Kabar, 1987; Mutlu and Bozcuk, 2000). Phytohormones suggested playing important roles in stress responses and adaptation (Sharma *et al.*, 2005; Shaterian *et al.*, 2005). It is thought that the repressive effect of salinity on seed germination and plant growth could be related to a decline in endogenous levels of phytohormones (Zholkevich and Pustovoytova, 1993; Jackson, 1997; Debez *et al.*, 2001). Manipulation of crop production with chemicals is one of the most important advancement achieved in agriculture. Triazole compounds induce a variety of morphological and biochemical responses in plants including stimulation of root growth, reduction in free-radical damage, and increase in antioxidant potential (Jaleel, Gopi, Manivannan and Panneerselvam 2007; Manivannan *et al.*, 2007).

Paclobutrazol was also studied in the medicinal plant *Catharanthus roseus* belonging to the family Apocyanaceae. *Catharanthus roseus* contain many medicinally important alkaloids which have anticancer activities (Filippini *et al.*, 2003). PBZ treatment to the unstressed plants caused a significant increase in root length, fresh and dry weight. The plant height and leaf area showed a significant reduction under PBZ treatment. The reduction in growth may be an adaptive response to stress (Zhu, 2001). PBZ application improved the growth potential of *Catharanthus roseus* plants under salinity injury. Stimulation of root growth under salt stress by triadimefon, a triazole compound was reported in one of the previous studies conducted in our lab by Panneerselvam *et al.* (1997) in radish. Increasing of root growth by triazoles is also associated with increased level of endogenous cytokinin (Fletcher and Arnold, 1986). Under salt stress, the non-enzymatic antioxidants like ascorbic acid (AA) and reduced glutathione (GSH) were affected significantly. Increased AA and GSH contents were reported due to PBZ application in lemon (Jain *et al.*, 2002). An increase in AA and GSH content due to PBZ treatment is the basis of the stress protection ability of PBZ (Fletcher *et al.*, 2000). PBZ and NaCl treatments led to changes in the activities of APX and CAT in roots and leaves of *Catharanthus roseus*. The extractable specific activity of APX increased in salt-stressed root and leaves of *Catharanthus roseus* plants when compared to control. PBZ treatments to the unstressed plants caused an increase in APX activity in both roots and leaves. The activity of

CAT was decreased under NaCl stress in roots. But in leaves, there was no significant change in the activity when compared to control plants. PBZ treatments to the salt-stressed and unstressed plants caused an increase in CAT activity to a significant level when compared to NaCl stressed and control plants. Increased APX activity was reported in wheat under long-term salinity stress (Sairam and Srivastava, 2002). Increased activities of antioxidant enzymes with triadimefon treatment were reported in radish under salinity (Muthukumarasamy *et al.*, 2000). Triazole compounds like PBZ can inhibit GA biosynthesis and in turn make substrate available for the synthesis of other isoprenoid pathway compounds like ABA and/or cytokinin (Kabi and Pujari, 1980). Cytokinin and ABA have been used to reverse the salinity induced deleterious effects (Katz *et al.*, 1978, Singh *et al.*, 1987). Protection of plants from apparently unrelated stress by triazoles is mediated by a reduction in free radical damage and increase in antioxidant potentials and enzyme activities (Fletcher and Hofstra, 1990). From these results, it can be concluded that, the PBZ at 15 mg L^{-1} plant^{-1}application could well be used for partial amelioration of salinity stress in *Catharanthus roseus* plant. It has been shown that triazole compounds may counteract the effect of salinity e.g., in sunflower and mungbean seedlings it has been observed that the pretreatment of seeds with LAB 150978 (a triazole compound) counteracted the inhibitory effect of salinity on root growth (Saha and Gupta, 1993).

EFFECT OF PACLOBUTRAZOL UNDER LOW TEMPERATURE STRESS

Chilling sensitivity has been looked upon as an important issue in plant physiological and biophysical studies, particularly in tropical and sub-tropical species, which show characteristic damage symptoms when subjected to low, above zero temperatures (Prasad *et al.*, 1994). Plants native to warm regions such as maize, tomato, cucumber, rice, bean, and sweet potato are generally injured at temperatures below 10 °C (Saltveit, 2000). Symptoms of chilling injury include wilting, reduced growth and photosynthetic capacity, chlorosis, necrosis, discoloration, abnormal ripening, increased disease susceptibility, leakage of ions from cell membranes, and changes in respiration and ethylene production (Rab and Saltveit, 1996; Sato *et al.*, 2001).

There is increasing evidence that chilling causes elevated levels of reactive oxygen species (ROS), which contribute significantly to chilling damage (Prasad *et al.*, 1996; Wise, 1995). The development of chilling injury symptoms is, in part, due to the failure to keep in balance the production and quenching of the ROS in plants (Crawford and Brandle, 1996). In plant cells, oxidative stress reactions are associated with toxic free radicals from the reduction of molecular oxygen to superoxide radicals ($O2^{-}$), singlet oxygen ($^{1}O2$), hydroxyl radicals (.OH), and hydrogen peroxide (H_2O_2). The toxic radicals can be removed through the mobilization of antioxidant reserves, which react both enzymatically and chemically with the toxic molecular species and their products. Chemical constituents have been identified that scavenge free radicals and thus protect active plant cells against oxygen toxicity. To counter the hazardous effects of ROS under chilling stress, plants have evolved a complex antioxidative defense system composed of antioxidative enzymes (i.e., APX, CAT, SOD, GR) and metabolites (i.e., ASA, GSH, GSSG, MDA and vitamin-E) (Perata and Alpi, 1993; Sairam *et al.*, 2002). High levels of some antioxidant enzymes and metabolites are important to the survival of corn (Anderson *et al.*, 1995; Hodges *et al.*, 1997; Pinhero *et al.*, 1997), wheat (Mishra *et al.*, 1993; Scebba *et al.*, 1999), rice (Saruyama and Tanida, 1995; Kuk *et al.*, 2003), tomato (Venema *et al.*, 2000), pea (Edwards *et al.*, 1994), spinach (Schoner and Krause, 1990), cucumber (Kang *et al.*, 2002), squash (Wang, 1996), banana (Kang *et al.*, 2003) and Arabidopsis (O'Kane *et al.*, 1996) under the oxidative stress resulting from being subjected to different levels of chilling. Past studies (Sala, 1998; Shen *et al.*, 1999; Walker and Mckersie, 1993) have demonstrated that chilling sensitive species and cultivars have a lower antioxidant capacity than do tolerant species and cultivars. Since periods of chilling temperature seem to be accompanied by the appearance of an alternative electron sink, we hypothesized that, under chilling conditions, the ROS were more enhanced in chilling-sensitive than in chilling-tolerant varieties. To prove this, low molecular weight antioxidants and scavenging enzymes were measured in both chilling-sensitive and chilling-tolerant genotypes.

Triazoles have been characterized as plant protectant under low temperature stress (Senaratna *et al.*, 1988; Upadhyaya *et al.*, 1989; Kraus and Fletcher, 1994; Pinhero and Fletcher, 1994; Kraus *et al.*, 1995; Voesenek *et al.*, 2003). In PBZ treated tomatoes, apart from the increase in the antioxidants α-tocopherol and ascorbate, free fatty acids were higher and there was a reduction in the loss of membrane phospholipids, as compared to the untreated controls. Triazole-induced tolerance to low temperature stress has been associated with increased levels of endogenous ABA (Fletcher *et al.*, 2000), which has been reported to trigger the genetic processes for hardening (Zeevaart and Creelman, 1988). In field studies, winter survival of peas and cereal crops (Davis *et al.*, 1988) and resistance to frost damage in corn and tomatoes (Fletcher and Kraus, 1995) were enhanced by triazoles.

It is known that low temperature stress alone can inhibit plant growth. Increase in growth rate such as shoot length, fresh and dry weights after the temperature stress was higher in PBZ-treated seedlings compared to the controls in tomato plant. It was established that paclobutrazol protect corn plants from chilling injury and a similar increase in plant growth parameters was observed (Pinhero and Fletcher, 1994). An increase of MDA accumulation following low temperature stress was observed in both the control and the treated variants, i.e., a state of oxidative stress is induced related to membrane damage. According to Fletcher *et al.* (2000), inhibition of lipid peroxidation may be one of the mechanisms responsible for the anti-senescence effects of the triazoles. Lipid peroxidation mediated by activated toxic oxygen species should be accompanied by changes in activities of enzymes involved in oxygen metabolism (Cakmak and Horst, 1991). Peroxidase is one of the major systems for the enzymatic removal of H_2O_2 in plants. The higher activity of peroxidase in the PBZ-treated seedlings immediately prior to the stress treatment suggests, that paclobutrazol probably prepares the cell to meet and overcome stress by stabilizing membranes and forming a potential of higher antioxidant capacity under low temperature stress. It was established that paclobutrazol prevent the decline in total chlorophyll content in corn plants after exposure to chilling temperatures (Pinhero and Fletcher, 1994). Some of the free radical scavenging enzymes is reported to increase in wheat (Kraus and Fletcher, 1994), corn (Pinhero and Fletcher, 1994) and cucumber (Upadhyaya *et al.*, 1989) plants after triazole treatments and their activities are conserved even after exposure to extreme temperature. The triazole compounds enhance the free radical scavenging capacity of treated plants including the levels of carotenoids, ascorbate, superoxide dismutase and ascorbate peroxidase (Senaratna *et al.*, 1988; Kraus *et al.*, 1995). In this study we suggest that the protection caused by paclobutrazol was due to a similar mechanism of enhanced free-radical scavenging systems. In plants the chloroplast is a major site of free radical production, and PBZ protects plants by increasing antioxidant defense systems. PBZ treated plants have a more efficient free radical scavenging system to detoxify ROS (Kopyra and Gwozda, 2003). Some attempts to increase chilling tolerance and avoid chilling injury have been tried such as breeding for increased tolerance, modifying crop management practices, and applying chemicals. PBZ has been reported to confer protection to plants experiencing stress by reducing oxidative damage via elevation of antioxidants or reducing the activity of oxidative enzymes. Physiological effects of PBZ on plants including changes in growth (Arzani and Roosta, 2004; Arzani *et al.,* 2009) and morphology (Sugavanam, 1984), antioxidant activity (Kraus and Fletcher, 1994), increasing levels of proline (Baninasab, 2009) and carbohydrates (Coleman and Estabrooks, 1992) play key roles in withstanding the low temperature stress.

EFFECT OF PACLOBUTRAZOL UNDER HIGH TEMPERATURE STRESS

Plants are exposed to a wide range of environmental conditions and one of the major forces that shape the structure and function of plants are temperature stresses, which include high temperature stress and considered as major abiotic stress for crop plants. Due to global climate change, high temperature stress is becoming the major area of concern for the researchers worldwide. The reactions of plants to these stresses are complex and have devastating effects on plant metabolism, disrupting cellular homeostasis and uncoupling major physiological and biochemical processes. High temperature stress disrupts photosynthesis and increase photorespiration thereby altering the normal homeostasis of plant cells. The constancy of temperature, among different metabolic equilibria present in plant cells, depends to a certain extent on a homeostatically regulated ratio of redox components, which are present virtually in all plant cells. Several pathways, which are present in plant cells, enable correct equilibrium of the plant cellular redox state and balance fluctuations in plant cells caused by changes in environment due to stressful conditions. The responses of plants to high temperature stress vary with extent of temperature increase, its duration and the type of plant.

A direct result of stress-induced cellular changes is overproduction of reactive oxygen species (ROS) in plants which are produced in such a way that they are confined to a small area and also in specific pattern in biological responses. Reactive oxygen species are highly reactive and toxic and cause damage to proteins, lipids, carbohydrates which ultimately results in oxidative stress. ROS may also serve as signaling molecules in mediating important signal transduction pathways that coordinate an astonishing range of diverse plant processes under temperature stress. To counter temperature induced oxidative stress, plants up regulate a variety of enzymatic and non-enzymatic antioxidants which are also information-rich redox buffers and important components of redox signaling that interact with bio-membrane related compartments. They provide essential information on cellular redox state, and regulate gene expression associated with stress responses to optimize defense and survival, stress acclimation and tolerance. The work done by various researchers has explored a direct link between ROS scavenging and plant tolerance under high temperature extremes in various crops which include legumes, cereals, oil crops and vegetables. There is ample need to develop high temperature tolerance in crop plants by exploring suitable strategies to manage oxidative stress and maintain cellular redox state. Paclobutrazol protects wheat and corn against extreme temperatures, where these effects appears and are associated with enhanced activity of free radical scavenging systems (Kraus and Fletcher, 1994).

EFFECT OF PACLOBUTRAZOL UNDER UV-B RADIATIONS

Reduction in global stratospheric ozone has led to increased solar UV-B radiation (290 - 315 nm) reaching the earth's surface. High levels of UV-B may constitute a significant environmental stress for terrestrial plants limiting plant yield and distribution in many regions of the world. UV-B radiation induces a multitude of rapid and long term morphogenic responses which are mediated by an unknown number of unidentified photosensory systems (Jansen, 2002). Some UV-B driven morphogenetic responses may not involve a dedicated photosensory system, but rather are a consequence of UV-B induced change in secondary metabolism. Studies have demonstrated several detrimental effects of UV-B on plants, including DNA damage, inhibition of photosynthetic activity and growth (Mark and Tevini, 1997). In order to prevent these harmful effects of UV-B radiation, plant have developed several defense mechanisms, including repair of inflicted damage and screening of the internal tissues against the radiation. Numerous studies have shown that UV-B radiation induces reductions in plant growth and crop yield in several species. Several mechanisms for tolerance to UV-B have been proposed, including enhanced synthesis of UV-B absorbing pigments, leaf thickening, increased synthesis of cuticular wax and reduced leaf expansion (Caldwell *et al.*, 1995).

Enhanced free radical scavenging capacity to control oxidative damage incurred from UV-B radiation is also thought to contribute to UV-B tolerance. These are, in part, due to the direct effects of UV-B radiation on photosynthesis, nucleic acids, enzymes, pigments and growth regulators or to its indirect effects administered through the formation of reactive oxygen species (ROS) like superoxide and hydroxyl radicals and hydrogen peroxide. High doses of UV-B light induce the production of ROS including peroxidase and free radicals, causing damage to proteins, lipids and DNA, and affecting the cell integrity, morphology, and physiology of plants (Frohnmeyer and Staiger, 2003). The level of damage caused by these factors depends on the effectiveness of the cellular defense mechanism, which includes scavenging enzymes such as superoxide dismutase, peroxidase and catalase, and non-enzymatic components such as ascorbic acid and reduced glutathione.

Triazole compounds such as paclobutrazole, triadimefon, hexaconazole etc. are widely used as growth regulator that has a profound influence on hormonal balance, photosynthetic rate, enzyme activities, lipid peroxidation and yield component in various plants (Kishorekumar *et al.*, 2007). They affect the isoprenoid pathway and alter the levels of certain plant hormones by inhibiting gibberellin synthesis, reducing ethylene evolution and increasing cytokinin levels (Manivannan *et al.*, 2008. Triazoles enhance activities of antioxidant systems to efficiently scavenge free radicals and thus enable plants to better cope with suboptimal environmental constraints. Therefore, protection of plants from apparently unrelated stress by triazoles is mediated by a reduction in free-radical damage and increase in antioxidant potential.

Hexaconazole (HEX), a broad spectrum fungicide of the triazole group, is reported to induce morphological and physiological changes such as reduction in shoot elongation, stimulation of rooting, inhibition of gibberellin synthesis, altered biochemical constituents including defense-related enzyme and carbohydrate status, and a transient raise in abscisic acid content (Gopi *et al.*, 2007; Johnson *et al.*, 2008). Some triazoles have commercially important applications. They have been useful in controlling growth, lodging resistance, cold hardiness, and yield of some important crop plants. However, there is little information on the effects of triazole treatment on the ability of plants to tolerate elevated levels of UV-B radiation in crop plants in order to predict further effects on crop yield and productivity.

Fig. 4. Effects of Hexaconazole (HEX) applications on leaf morphology of cucumber (*Cucumis sativus* L.) under UV-B radiation. The image was taken at 21 day after the end of treatment. 10 days old seedlings were sprayed with 100 ppm HEX and exposed to 6 kJ m^{-2} day^{-1} UV-B radiation alone or in combination. (Source: Tee and Jung, 2012; http://dx.doi.org/10.5322/JES.2012.21.12.1435)

The effect of hexaconazole (HEX), a triazole compound, on the growth, yield, photosynthetic response and antioxidant potential in cucumber (*Cucumis sativus* L.) plants subjected to UV-B stress. UV-B radiation and HEX were applied separately or in combination to cucumber seedlings. The growth parameters were significantly reduced under UV-B treatment; however, this growth inhibition was less in HEX treated plants. HEX caused noticeable changes in plant morphology such as reduced shoot length and leaf area, and increased leaf thickness. HEX was quite persistent in inhibiting shoot growth by causing a reduction in shoot fresh and dry weight. HEX noticeably recovered the UV-B induced inhibition of biomass production. Significant accumulation in anthocyanin and flavonoid pigments in the leaves occurred as a result of HEX or UV-B treatments. HEX permitted the survival of more green leaf tissue preventing chlorophyll content reduction and higher quantum yield for photosynthesis under UV-B exposure. HEX treatment induced a transient rise in ABA levels in the leaves, and combined application of HEX and UV-B showed a significant enhancement of ABA content which activates

H_2O_2 generation. UV-B exposure induced accumulation of H_2O_2 in the leaves, while HEX prevented UV-B induced increase in H_2O_2, indicating that HEX serves as an antioxidant agent able to scavenge H_2O_2 to protect cells from oxidative damage. An increase in the ascorbic acid was observed in the HEX treated cucumber leaves affecting many enzyme activities by removing H_2O_2 during photosynthetic processes. The activities of antioxidant enzymes including catalase (CAT), ascorbate peroxidase (APX), superoxide dismutase (SOD) and peroxidase (POD) in the leaves in the presence of HEX under UV-B stress were higher than those under UV-B stress alone. These findings suggest that HEX may participate in the enhanced tolerance to oxidative stress. From these results it can be concluded that HEX moderately ameliorate the effect of UV-B stress in cucumber by improving the components of antioxidant defense system.

The foliage of triazole-treated plants exhibited intense dark green, mostly due to enhanced chlorophyll content. High rates of triazoles may result in some leaf curling or cupping but usually does not cause chlorosis or necrosis. Similar results were observed in UV-B stressed soybean treated with paclobutrazol (Kraus *et al.*, 1995).

Coclusion and Future Insights

The triazoles are the largest and most important group of systemic compounds, developed in 1960s for the control of fungal diseases in plants and animals. Commercial triazole derivatives (such as paclobutrazol [(2RS, 3RS)-1-(4-chlorophenyl)-4,4- dimethyl-2-(1,2,4-triazolyl)-pentan-3-ol]) have been recommended for use as either fungicides or plant growth regulators. Characteristic of morphological and anatomical effects of the triazole include reduced shoot elongation and trichome length, increased epicuticular wax, larger chloroplasts and increased root growth. Biochemical effects of the triazole include detoxification of active oxygen, increased levels of proline, antioxidants and chlorophyll content. More recently, it was found that triazole compounds are able to protect plants from the environmental stress conditions, e.g. drought, extreme temperature, gaseous sulphur dioxide and fungal infections. The triazole mediated stress protection is often explained in terms of hormonal changes such as an increase in cytokinins, a transient rise in ABA and a decrease in ethylene. With the growing knowledge, the new avenues open the door to gather insight for judicious utilization of PBZ in view of its immense potential for mitigating prevalent abiotic stresses. It is suggestible that genotypic screening of crop plants in response to varied concentrations of different sources and different modes of applications of PBZ under abiotic stresses could be of considerable value in crop protection for yield, in general, and understanding the underlying mechanisms, as modified by PBZ, at physiological, biochemical and molecular levels.

References

Ahmad, M.S.A., Javed, F., Ashraf, M. 2007. Iso-osmotic effect of NaCl and PEG on growth, cations and free proline accumulation in callus tissue of two indica rice (*Oryza sativa* L.) genotypes. *Plant Growth Regul.,* **53**, 53-63.

Ahmed, S., Nawata, E., Hosokawa, M., Domae, Y. and Sakuratani, T. 2000. Alterations in photosynthesis and some antioxidant enzymatic activities of mungbean subjected to waterlogging. *Plant Sci.*, **163**, 117–23.

Al- Rumaih, M.M. and Al- Rumaih, M.M. 2007. Physiological response of two species of datura to uniconazole and salt stress. *J. Food Agric. Environ.*, **5**, 450-453.

Anderson, M.D., Prasa, T.K. and Stewart, C.R. 1995. Changes in isozyme profiles of catalase peroxidase and glutathione reductase during acclimation to chilling in mesocotyls of maize seedlings. *Plant Physiol* ., **109**, 1247–1257.

Arzani, K. and Roosta, H.R. 2004. Effects of paclobutrazol on vegetative and reproductive growth and leaf mineral content of mature apricot (*Prunus armeniaca* L.) trees. *J. Agr. Sci. Technol.,* **6**, 43-56.

Arzani, K., Bahadori, F. and Piri, S. 2009. Paclobutrazol reduces vegetative growth and enhances flowering and fruiting of mature 'J.H. Hale' and 'Red Skin' peach trees. *Hort. Environ. Biotechnol.*, **50**, 84-93.

Asare-Boamah, N. and Fletcher, R. 1986.Protection of bean seedlings against heat and chilling injury by triadimefon. *Physiol Plant.*, **67**, 353–358.

Ashraf, M. and Harris, P.J. 2004. Potential water biochemical indicators of salinity tolerance in plants. *Plant Sci.*, **166**, 3-16.

Baninasab, B. 2009. amelioration of chilling stress by paclobutrazol in watermelon seedlings. *Sci. Hort.* **121**, 144-148.

Banon, S., Ochoa, J., fernandez, J.A., Franco, J.A., 2001. *Paclobutrazol as aids for Phillyrea angustifolia. Hort Sci.,* **36**, 449-500.

Berova, M. and Zlatev, Z. 2003. Physiological response of paclobutrazol-treated triticale plants to water stress. *Biol. Plant.,* **46**, 133-136.

Berova, M., Zlatev, Z. and Stoeva, N. 2002. Effect of paclobutrazol on wheat seedlings under low temperature stress. *Bulg. J. Plant Physiol.*, **28**, 75-84.

Biemelt, S., Keetman, U. and Albrecht, G. 1998. Re-aeration following hypoxia or anoxia leads to activation of the antioxidative defense system I roots of wheat seedlings. *Plant Physiol.*, 116, 651–8.

Blum, A. 1996. Constitutive traitsaffecting plant performance under stress. *In*: Edmeades, G.O., M. Banziger, H.R. Mickelson and C.B. Pena-Valdivia (eds.), *Developing Drought and Low N Tolerant Maize*, pp. 131–35. Proceedings of the Symposium, Mexico D.F.: Cimmyt.

Cakmak, W. and Horst, J. 1991. Effects of aluminium on lipid peroxidation, superoxide dismutase, catalase, and peroxidase activities in root of soybean (*Glycine max*). *Physiol. Plant*, **83**, 463–468.

Caldwell, M.M., Teramura, A.H., Tevini, M., Bornman, J.F., Bjorn, I.O. and Kulandaivelu, G. 1995. Effects of increased solar ultraviolet radiation on terrestrial plants. *Ambio.*, **24**, 166-173.

Carvalho, L.C. and Amancio, S. 2002. Antioxidant defence system in plantlets transferred from in vitro to ex vitro: effects of increasing light intensity and CO2 concentration. *Plant Sci.,* **162**, 33–40.

Chaney, W.R. 2005. Growth retardants: A promising tool for managing urban trees. *Purdue Extension document*, FNR-252.

Chaudiere, J. and Ferrari-Ilious, R. 1999. Intracellular antioxidants from chemical to biochemical mechanisms. *Food Chem. Toxicol.,* **3**, 949–62.

Clavel, D., Drame, N.K., Roy-Macaule, H., Braconnier, S. and Laffray, D. 2005. Analysis of early responses to drought associated with field drought adaptation in four Sahelian groundnut (*Arachis hypogaea* L.) cultivars. *Environ. Exp. Bot.*, 54, 219–230.

Coleman, W.K. and Estabrooks, E.N. 1992. Enhancement of cold hardiness in apple tree by pbz, thidiazuron and flurprimidol. *Can. J. Plant Sci.*, **72**, 1267-1274.

Crawford, R.M.M. and Brandle, R. 1996. Oxygen deprivative stress in a changing environment. *J. Exp. Bot.*, **47**, 145–59.

Davis, T. and Curry, E. 1991. Chemical regulation of vegetative growth. *Critic Rev. Plant Sci.*, **10**, 151–188.

Davis, T., Steffens, G. and Sankhla, N. 1988. Triazole plant growth regulators. *Hort. Rev.*, **10**, 63–105.

Debez, A., Chaibi, W. and Bouzid, S. 2001. Effect du NaCl et de regulatoeurs de croissance sur la germination d' *Atriplex halimus* L. *Cah Agric* **10**,135-138.

Debnath, M. 2008. Responses of Bacopa monnieri to salinity and drought stress in vitro. *J. Medicinal Plants Res.,* **11**, 347-351.

Demiral, T., Turkan, I. 2004. Does exogenous glycinebetaine affect antioxidative system of rice seedlings under NaCl treatment. *J. Plant Physiol.*, **161**, 1089–1100.

Edwards, E.A., Enord, C., Creissen, G.P. and Mullineaux, P.M. 1994. Synthesis and properties of glutathione reductase in stressed peas. *Planta*, **192**, 137–143.

Else, M.A., Hall, K.C., Arnold, G.M. and Jackson, M.B. 1995. Export of abscisic acid, 1-aminocyclopropane-1-carboxylic acid, phosphate, and nitrate from roots to shoots of flooded tomato plants. *Plant Physiol.,* **107**, 377–84.

FAO, 2009. High Level Expert Forum—How to Feed theWorld in 2050, Economic and Social Development, Food and Agricultural Organization of the United Nations, Rome, Italy.

Farooq, M., Basra, S.M.A., Wahid, A., Cheema, Z.A,. Cheema, M.A. and Khaliq, A. 2008. Physiological role of exogenously applied glycine betaine in improving drought tolerance of fine grain aromatic rice (*Oryza sativa* L.). *J. Agron. Crop Sci.*, **194**, 325–333.

Filippini, R., Caniato, R., Piovan, A., Cappelletti, E.M. 2003. Production of anthocyanins by Catharanthus roseus. *Fitoterapia,* **74**, 62–67.

Fletcher, F.A., Gilley, A., Sankhla, N. and Davis, T.D. 2000. Triazoles as plant growth regulator and stress protectants. *Hort. Review*, **24**, 55–138.

Fletcher, R.A. and Arnold, V. 1986. Stimulation of cytokinins and chlorophyll triadimefon stimulates rooting in bean hypocotyls. *Physiol Plant*, **73**, 401–405.

Fletcher, R.A. and Hofstra, G. 1990. Improvement of uniconazole induced protection in wheat seedlings. *J. Plant Growth Regul.*, **9**, 207-217.

Fletcher, R.A. and Hofstra, G. 1988. Triazoles as potential plant protectants. In: Berg K, Plempel M (ed) Sterol biosynthesis inhibitors: pharmaceutical and agrochemical aspects. *Ellis Harwood, Ltd.* London.

Fletcher, R. and Kraus,, T. 1995. Triazoles: Protecting plants from environmental stress. In: Agrifoot Research, Eds. R. Meerveld, Ontario, *Queens Printer*, **18**, 15–19.

Flowers, T.J. 2004. Improving crop salt tolerance, *J. Exp. Botany*, **55** (396), 307–319.

Frohnmeyer, H. and Staiger, D. 2003. Ultraviolet-B radiation responses in plants : Balancing damage and protection. *Plant Physiol.*, **133**, 1420-1428.

Gao, J., Hofstra, G. and Fletcher, R. 1988. Anatomical changes induced by triazoles in wheat seedlings. *Can. J. Bot.*, **66,** 1178–1185.

Gaspar, T., Franck, T., Bisbis, B., Kevers, C., Jouve, L., Hausman, J.F. and Dommes, J. 2002. Concepts in plant stress physiology. Application to plant tissue cultures. *Plant Growth Regul.*, **37**, 263–285.

Gopi, R., Jaleel, C.A., Sairam, R., Lakshmanan, G.M.A., Gomathinayagam, M. and Panneerselvam, R., 2007. Differential effects of hexaconazole and paclobutrazol on biomass, electrolyte leakage, lipid peroxidation and antioxidant potential of *Daucus carota* L., *Colloids Surf. B: Biointerf.*, **60**, 180-186.

Grichko, V.P. and Glick, B.R. 2001. Ethylene and flooding stress in plants. *Plant Physiol Biochem*, **39**, 1–9.

Grossmann, K. 1990. Plant growth retardants as tools in physiological research. *Physiol. Plant.*, **78**, 640–648.

Guo, Y., Qiu, Q.S., Quintero, F.J. *et al.*, 2004. Transgenic evaluation of activated mutant alleles of SOS2 reveals a critical requirement for its kinase activity and C-terminal regulatory domain for salt tolerance in *Arabidopsis thaliana. Plant Cell*, **16** (2), 435–449.

Hasegawa, P.M., Bressan, R.A., Zhu, J.K. and Bohnert, H.J. 2000. Plant cellular and molecular responses to high salinity, *Annual Review of Plant Biology*, **51**, 463–499.

Hit, H. and Shinozaki, K. 2004. Plant Responses to Abiotic Stress. *Springer,* Berlin, Heidelberg New York, Hong Kong, London, Milan, Paris, Tokyo.

Hodges, D.K., Andrew, C.J., Johnson, D.A. and Hamilton, R.I. 1997. Antioxidant enzyme responses to chilling stress in differentially sensitive inbred maize lines. *I. Exper. Bot.*, **48**, 1105–1113.

Hurng, W.P. and Kao, C.H. 1994a. Effect of flooding on the activities of some enzymes of activated oxygen metabolism, the levels of antioxidants, and lipid peroxidation in senesencing tobacco leaves. *Plant Growth Regul.*, **14**, 37–44.

Hurng, W.P. and Kao, C.H. 1994b. Lipid peroxidation and antioxidative enzymes in senesencing tobacco leaves following flooding. *Plant Sci.,* **96**, 41–4.

Ishitani, M., Liu, J., Halfter, U., Kim, C.S., Shi, W. and Zhu, J.K., 2000. SOS3 function in plant salt tolerance requires Nmyristoylation and calcium binding, *Plant Cell*, **12** (9), 1667–1677.

Jackson, M. 1997. Hormones from roots as signals for the shoots of stressed plants. *Elsevier Trends J.,* **2**: 22–28.

Jain, S.K., Singh, R. and Misra, K.K. 2002. Effect of paclobutrazol on growth yield and fruit quality of lemon (*Citrus limon*), *Indian J. Agric. Sci.*, **72**, 488-489.

Jaleel, C.A., Gopi, P. and Panneerselvam, R. 2007a. Responses of antioxidant defense system of *Catharanthus roseus* (L.) G. Don. to paclobutrazol treatment under salinity. *Acta Physiol. Plantarum*, **29**, 205-209.

Jaleel, C.A., Gopi, R. and Panneerselvam, R. 2008b. Growth and photosynthetic pigments responses of two varieties of *Catharanthus roseus* to triadimefon treatment. *Comp. Rend. Biol.*, **331**, 272–277.

Jaleel, C.A., Gopi, R., Manivannan, P. and Panneerselvam, R. 2007. Responses of antioxidant defense system of *Catharanthus roseus* (L.) to paclobutrazol treatment under salinity. *Act. Physiol. Planta.*, **29**, 205-209.

Jaleel, C.A., Gopi, R., Manivannan, P., Kishorekumar, T.A. 2007. Gomathinayagam, M. and Panneerselvam, R., Changes in biochemical constituents and induction of early sprouting by triadimefon treatment in white yam (*Dioscorea* m*rotundata* Poir.) tubers during storage. *J. Zhejiang Univ. Sci. B.*, **8(4),** 283-288.

Jaleel, C.A., Gopi, R., Sankar, B., Gomathinayagam, M. and Panneerselvam, R. 2008b. Differential responses in water use efficiency in two varieties of *Catharanthus roseus* under drought stress. *Comp. Rend. Biol.*, **331**, 42–47.

Jaleel, C.A., Manivannan, P., Lakshmanan, G.M.A., Gomathinayagam, M. and Panneerselvam, R. 2008c. Alterations in morphological parameters and photosynthetic pigment responses of *Catharanthus roseus* under soil water deficits. *Colloids Surf. B: Biointerfaces,* **61**, 298–303.

Jaleel, C.A., Manivannan, P., Murali, P.V., Gomathinayagam, M. and Panneerselvam, R. 2008d. Antioxidant potential and indole alkaloid profile variations with water deficits along different parts of two varieties of *Catharanthus roseus. Colloids Surf. B: Biointerfaces*, **62**, 312–318.

Jaleel, C.A., Manivannan, P., Sankar, B., Kishorekumar, A., Gopi, R., Somasundaram, R. and Panneerselvam, R. 2007b. Water deficit stress mitigation by calcium chloride in *Catharanthus roseus*; effects on oxidative stress, proline metabolism and indole alkaloid accumulation. *Colloids Surf. B: Biointerfaces*, **60**, 110–116.

Jaleel, C.A., Manivannan, P., Sankar, B., Kishorekumar, A., Gopi, R., Somasundaram, R. and Panneerselvam, R. 2007d. Induction of drought stress tolerance by ketoconazole in *Catharanthus roseus* is mediated by enhanced antioxidant potentials and secondary metabolite accumulation. *Colloids Surf. B: Biointerfaces,* **60**, 201–206.

Jaleel, C.A., Manivannan, P., Sankar, B., Kishorekumar, A., Sankari, S. and Panneerselvam, R. 2007e. Paclobutrazol enhances photosynthesis and ajmalicine production in *Catharanthus roseus. Process Biochem.*, **42**, 1566–1570.

Jaleel, C., Manivannan, P., Sankar, B. and Kishorekumar, A. 2007. Responses of antioxidant defense system of Catharanthus roseus (L.) G. Don. to paclobutrazol treatment under salinity. *Acta Physiol. Plant.,* **29**, 205-209.

James, R.A., Blake, C., Byrt, C.S. and Munns, R. 2011. Major genes for Na+ exclusion, Nax1 and Nax2 (wheat HKT1;4 and HKT1;5), decrease Na+ accumulation in bread wheat leaves under saline and waterlogged conditions, *J. Exp. Botany*, **62**, (8), 2939–2947.

Jansen, M.A.K. 2002. Ultraviot-B radiation effects on plants: Induction of morphogenic responses, *Physiol. Plant*, **116**, 423-429.

Jebara, S., Jebara, M., Limam, F., Elarbi Aouani, M. 2005. Changes in ascorbate peroxidase, catalase, guaiacol peroxidase and superoxide dismutase activities in common bean (*Phaseolus vulgaris*) nodules under salt stress. *J. Plant Physiol.*, **162**, 929–936.

Johnson, I., Marimuthu, T., Samiyappan, R. and Cha, B. 2008. Influence of hexaconazole on biochemieal constituents of groundnut, *Kor. J. Pesticide Sci.*, **12**, 335-341.

Kabar, K. 1987. Alleviation of salinity stress by plant growth regulators on seed germination. *J. Plant Physiol.,* **128**, 179-183.

Kabi, T. and Pujari, R. 1980. Effect of IAA, GA3 and CCC at different levels of inorganic phosphate on rooting of Ipomoea batatus L. leaves. *Geobios*, **7**, 170–171.

Kamountsis, A.P. and Chronopoulon-Sereli, A.G. 1999. Paclobutrazol affects growth and flower bud the production in gardenia under different light regimes. *Hort. Sci.*, **34**, 674-675.

Kang, G., Wang, C., Sun, G. and Wang, Z. 2003. Salicylic acid changes activities of H_2O_2-metabolizing enzymes and increases chilling tolerance of banana seedlings. *Envir. Exper. Bot.*, **50**, 9–15.

Kang, H.M., Saltveit, M.E. 2002. Reduced chilling tolerance in elongation cucumber seedling radical is related to their reduced antioxidant enzyme and DPPH-radical scavenging activity. *Physio Planta*, **115**, 244–250.

Kato, C., Ohshima, N., Kamada, H. and Satoh, S. 2001. Enhancement of the inhibitory activity for greening in xylem sap of squash root with waterlogging. *Plant Physiol Biochem.*, **39**, 513–9.

Katz, A., Dehan, K. and Itai, C. 1978. Kinetin reversal of NaCl effects. *Plant Physiol.*, 62, 836–837.

Kennedy, R.A., Rumpho, M.E., Fox, T.C. 1992. Anaerobic metabolism in plants. *Plant Physiol.*, **100**, 1–6.

Keyvan, S. 2010. The effects of drought stress on yield, relative water content, proline, soluble carbohydrates and chlorophyll of bread wheat cultivars. *J. Anim. Plant Sci.*, **8**, 1051- 1060.

Kishorekumar, A., Jaleel, C.A., Manivannan, P., Sankar, B., Sridharan, R. and Panneerselvam, R. 2007. Comparative effects of different triazole compounds on growth, photosynthetic pigments and carbohydrate metabolism of *Solenostemon rotundifolius, Colloids Surf. B: Biointerf.*, **60**, 207-212.

Kopyra, M. and Gwozdz, E.A. 2003. Antioxidant enzymes in paraquat and cadmium resistant cell lines of horseradish. *Biol. Lett.*, **40**, 61-69.

Kraus, T. and Fletcher, R. 1994. Paclobutrazol protects wheat seedlings from heat and paraquat injury. Is detoxification of active oxygen involved. *Plant Cell Physiol.*, **35**, 45–52.

Kraus, T.E., Evans, R.C., Fletcher, R.A. and Pauls, K.P. 1995. Paclobutrazol enhances tolerance to increased levels of UV-B radiation in soybean (*Glycine max*) seedlings, *Can. J. Bot.*, **73**, 797-806.

Kraus, T.E., Mckersie, B.D., Fletcher, R.A., 1995. Paclobutrazol induced tolerance of wheat leaves to paraquat may involve increased antioxidant enzyme activity. *J. Plant Physiol.,* **145**, 570-576.

Kuk, Y.I., Shin, J.S., Burgos, N.R., Hwang, T.E., Han, O. and Cho, B.H. 2003. Antioxidative enzymes offer protection from chilling damage in rice plants. *Crop Sci.*, **43**, 2109–2117.

Lawlor, D.W. and Cormic, G. 2002. Photosynthetic carbon assimilation and associated metabolism in relation to water deficit in higher plants. *Plant Cell Env.*, **25**, 275-294.

Liang, Y, Chen, Q., Liu, Q., Zhang, W. and Ding, R. 2003. Exogenous silicon (Si) increases antioxidant enzyme activity and reduces lipid peroxidation in roots of salt-stressed barley (*Hordeum vulgare* L.). *J. Plant Physiol.*, **160**, 1157–1164.

Liu, J., Ishitani, M., Halfter, U., Kim, C.S. and Zhu, J.K. 2000. The *Arabidopsis thaliana* SOS2 gene encodes a protein kinase that is required for salt tolerance, *Proc. Nat.Acad. Sci. USA*, **97** (7), 3730– 3734.

Lurie, S., Lipsker R.Z., Aloni, B. 1994. Effects of paclobutrazol and chilling temperatures on lipids, antioxidants and ATPase activity of plasma membrane isolated from green bell pepper fruits. *Physiol. Plant.,* **91**, 593–8.

Mackay, C., Hall, J., Hofstra, G. and Fletcher, R. 1990. Uniconazole-induced changes in abscisic acid, total amino acids and proline in *Phaseolus vulgaris. Pesti. Biochem. Physiol.*, **37**, 74–82.

Manivannan, P., Jaleel, C.A., Kishorekumar, A. and Sankar, B. 2008. Protection of *Vigna unguicula* (L.) Walp. plants from salt stress by paclobutrazol, *Colloids Surf. B: Biointerf.*, **61**, 315-318.

Manivannan, P., Jaleel, C.A., Kishorekumar, A., Sankar, B., Somasundaram, R., Sridharan, R. and Panneerselvam, R. 2007. Changes in antioxidant metabolism of *Vigna unguiculata* (L.) Walp. By propiconazole under water deficit stress. *Colloids Surf. Biointerfaces*, **57**, 69–74.

Manivannan, P., Jaleel, C.A., Kishorekumar, A., Sankar, B., Somasundaram, R., Sridharan, R. and Panneerselvam, R. 2007a. Propiconazole induced changes in antioxidant metabolism and drought stress amelioration in *Vigna unguiculata* (L.) Walp. *Colloids and Surf. B: Biointerfaces*, **57**, 69-74.

Mark, U. and Tevini, M. 1997. Effects of solar ultraviolet-B radiation, temperature and CO_2 on growth and physiology of sunflower and maize seedlings. *Plant Ecology*, **128**, 225-234.

Martinez-Atienza, J., Jiang, X. Garciadeblas, B. *et al.,* 2007. Conservation of the salt overly sensitive pathway in rice. *Plant Physiol.,* **143** (2), 1001–1012.

Mingeau, M., Perrier, C. and Ameglio, T. 2001. Evidence of drought-sensitive periods from flowering to maturity on high bush blueberry. *Sci. Hort.*, **89**, 23-40.

Mishra, N.P., Mishra, R.K. and Singhal, G.S. 1993. Changes in the activities of anti-oxidant enzymes during exposure of intact wheat leaves to strong visible light at different temperatures in the presence of protein synthesis inhibitions. *Plant Physiol.,* **102**, 903–910.

Munns, R. and Tester, M. 2008. Mechanisms of salinity tolerance, *Annu. Rev. of Plant Biol.,* **59**, 651–681.

Munns, R. 2005. Genes and salt tolerance: bringing them together, *New Phytologist*, **167** (30), 645–663.

Muthukumarasamy, M., Dutta Gupta, S., Panneerselvam, R. 2000. Enhancement of peroxidase, polyphenol oxidase and superoxide dismutase activities by triadimefon in NaCl stressed *Raphanus sativus* L. *Biol Plant* , **43**, 317–320.

Mutlu, F., Bozcuk, S. 2000. Tuzlu kos¸ullarda aycicegi tohumların c¸imlenmesi ve erken buyume uzerine dıssal spermin'in etkileri. *Turkish J. Biol* ., **24**, 635-643.

Navarro, A., Sanchez-blanco, J.M., Banon, S. 2007. Influence of paclobutrazol on water consumption and plant performance of Arbutus unedo seedlings. *Scientia Hortic.*, **111**, 133-139.

O'Kane, D., Gill, V., Boyd, P. and Burdon, R. 1996. Chilling, oxidative stress and antioxidant responses in *Arabidopsis thaliana* callus. *Planta*, **198**, 371–377.

Oh, D. H., Lee, S.Y., Bressan, R.A., Yun, D.J. and Bohnert, H.J. 2010. Intracellular consequences of SOS1 deficiency during salt stress, *J. Exp. Bot.*, **61** (4), 1205–1213.

Orabi, S.A., Mekki, B.B. 2008. Root yield and quality of sugar beet (*Beta vulgaris* L.) in response to ascorbic acid and saline irrigation water. *American-Eurasian J. Agric. Environ.,* **4**, 504-513.

Paliyath, G. and Fletcher, R.A., 1995. Paclobutrazol treatment alters peroxidase and catalase activities in heat stressed maize coleoptiles. *Physiol. Mol. Biol. Plant*, **1**, 171–8.

Panneerselvam, R., Muthukumarasamy, M. and Karikalan, L. 1997. Triadimefon enhances growth and net photosynthetic rate in NaCl stressed plants of *Raphanus sativus* (L.). *Photosynthetica,* **34**, 605–609.

Patade, V.Y., Bhargava, S., Suprasanna, P. 2011. Salt and drought tolerance of sugarcane under iso-osmotic salt and water stress: growth, osmolytes accumulation, and antioxidant defense. *J Plant Interact.* **6**, 275–282.

Perata, P. and Alpi, A. 1993. Plant responses to anaerobiosis. *Plant Sci.*, **93**, 1–17.

Pinhero, R. and Fletcher, R. 1994. Paclobutrazol and ancymidol protect corn seedlings from high and low temperature stresses. *Plant Growth Regul.*, **15**, 47–53.

Pinhero, R.G., Rao, M.V., Paliyath, G., Murr, D.P. and Fletcher, R.A. 1997. Changes in activities of antioxidant enzymes and their relationship to genetic and paclobutrazol-induced chilling tolerance of maize seedlings. *Plant Physiol.,* **114**, 695–704.

Prasad, T.K., Anderson, M.D., Martin, B.A. and Stewart, C.W. 1994. Evidence for chilling induced oxidative stress in maize seedlings and a regulatory role for hydrogen peroxide. *Plant Cell,* **6**, 65–74.

Prasad, T.K. 1996. Mechanisms of chilling-induced oxidative stress injury and tolerance: changes in antioxidant system, oxidation of proteins and lipids and protease activities. *Planta,* **10**, 1017–1026.

Qiu, J., Wang, R.M., Yan, J.Z. and Hu, J. 2005. Seed film coating with uniconazole improves rape seedling growth in relation to physiological changes under water logging stress. *Plant Growth Regul.,* **47**, 75– 81.

Quintero, F.J., Martinez-Atienza, J., Villalta, I. *et al.,* 2011. Activation of the plasma membrane Na/H antiporter salt overly sensitive 1 (SOS1) by phosphorylation of an auto-inhibitory C-terminal domain, *Proceedings of the National Academy of Sciences of the United States of America,* **108** (6), 2611–2616.

Quintero, F.J., Ohta, M., Shi, H., Zhu, J.K. and Pardo, J.M. 2002. Reconstitution in yeast of the *Arabidopsis* SOS signalling pathway for Na+ homeostasis, *Proc. Nat. Acad. Sci. USA,* **99** (13), 9061–9066.

Rab, A. and Saltveit, M.E., 1996. Sensitivity of seedling radicals to chilling and heat-shock induced chilling tolerance. *J. Amer. Soc. Hort. Sci.,* **121**, 711–715.

Rahnama, A., James, R.A., Poustini, K. and Munns, R. 2010. Stomatal conductance as a screen for osmotic stress tolerance in durum wheat growing in saline soil. *Functional Plant Biology,* **37** (3), 255–263.

Razmjoo, K., Heydarizadeh, P. and Sabzalian, M.R. 2008. Effect of salinity and drought stresses on growth parameters and essential oil content of *Matricaria chamomile. Int. J. Agric. Biol.,* **10**, 451–454.

Rozema, J. and Flowers, T., 2008. Ecology: crops for a salinized world, *Science,* **322**, (5907), 1478–1480.

Saha, K., Gupta, K., 1993. Effect of LAB-150978-a plant growth retardant on sunflower and mungbean seedlings under salinity stress. *Indian J Plant Physiol.,* **36**, 151-154.

Sairam, R.K. and Srivastava, G.C. 2002. Changes in antioxidant activity in sub-cellular fractions of tolerant and susceptible wheat genotypes in response to long term salt stress. *Plant Sci .,* **162**, 897–904.

Sairam, R.K., Deshmukh, P.S., Saxena, D.C. 1998. Role of antioxidant systems in wheat genotypes tolerance to water stress. *Biol Plantarum,* **41**, 387–94.

Sairam, R.K., Rao, K.V. and Srivastava, G.C. 2002. Differential response of wheat genotypes to long term salinity stress in relation to oxidative stress, antioxidant activity and osmolyte concentration. *Plant Sci.,* **163**, 1037–46.

Sala, J.M. 1998. Involvement of oxidative stress in chilling injury in cold-stored mandarin fruits. *Postharv. Bio. Tech.,* **13**, 255–261.

Salveit, M.E. 2000. Chilling injury is reduced in cucumber and rice seedlings and in tomato pericarp discs by heat-shocks applied after chilling. *Postharv. Bio. Tech.,* **21**, 169–177.

Sanders, D. 2000. Plant biology: the salty tale of *Arabidopsis. Curr. Biol.,* **10** (13), R486–R488.

Sankar, B., Abduljaleel, C.A., Manivannan, P., Kishorekumar, A., Somasundaram, R., Panneerselvam, R. 2007. Effect of paclobutrazol on water stress amelioration through antioxidants and free radical scavenging enzymes in *Arachis hypogaea* L. Colloids and Surfaces B*: Biointerfaces,* **60**, 229-235.

Saruyama, H. and Tanida, M. 1995. Effect of chilling on activated oxygen-scavenging enzymes in low temperature- sensitive and tolerant cultivars of rice. *Plant Sci.*, **109**, 105–113.

Sato, Y., Murajami, T., Funatsuki, H., Matsuta, S. and Tanida, M. 2001. Heat shock-mediated APX gene expression and protection against chilling injury in rice seedlings. *Exp. Bot.*, **52**, 145–151.

Scebba, F., Sebastiani, L. and Vitagliano, C. 1999. Protective enzymes against activated oxygen species in wheat seedling: responses to cold acclimation. *J. Plant Physiol.*, 155, 762–768.

Schoner, S. and Krause, G.H. 1990. Protective systems against active oxygen species in spinach: response to cold acclimation in excess light. *Planta,* **180**, 383–389.

Senaratna, T., Mackay, C., McKersie, B. and Fletcher, R. 1988. Uniconazole-induced chilling tolerance in tomato and its relationship to antioxidant content. *J. Plant Physiol.*, **133**, 56–61.

Shao, H.B., Shao, M.A. and Liang, Z.S. 2006. Osmotic adjustment comparison of 10 wheat (*Triticum aestivum* L) genotypes at soil water deficits. *Colloids Surf. B: Biointerfaces*, **47**, 132-139.

Sharma, N., Abrams, S.R. and Waterer, D.R. 2005. Uptake, movement, activity, and persistence of an abscisic acid analog (80 acetylene ABA methyl ester) in marigold and tomato. *J. Plant Growth Regul.,* **24**, 28-35.

Shaterian, J., Waterer, D., De Jong, H. and Tanino, K. K. 2005. Differential stress responses to NaCl salt application in early- and late maturing diploid potato (*Solanum sp.*) clones. *Environ. Exp. Bot.,* **54**, 202-212.

Shen, W., Nada, K. and Tachibana, S. 1999. Effect of cold treatment on enzyme and non-enzyme antioxidant activities in leaves of chilling-tolerant and chilling sensitive cucumber cultivars. *J. Jap. Soc. Hort. Sci.,* **68**, 967–973.

Shi, H., Ishitani, M., Kim, C. and Zhu, J.K., 2000. The *Arabidopsis thaliana* salt tolerance gene SOS1 encodes a putative Na+/H+ antiporter, *Proc. Nat. Acad. Sci. USA,* **97** (12), 6896–6901.

Shi, H., Quintero, F.J., Pardo, J.M. and. Zhu, J.K. 2002. The putative plasma membrane Na+/H+ antiporter SOS1 controls long distance Na+ transport in plants, *Plant Cell*, **14** (2), 465–477.

Singh, N.K., La Rosa, P.C., Handa, A.K., Hasegawa, P.M., Bressan, R.A. 1987. Hormonal regulation of protein synthesis associated with salt tolerance in plant cells. *Proc. Natl. Acad. Sci., USA*, **84**, 739–743.

Smirnoff, N. 1993. The role of active oxygen in the response of plants to water deficit and desiccation. *New Phytol.*, **125**, 27–58,

Srivastav, M., Kishor, A., Anil, D. and Sharma, R.R. 2010. Effect of paclobutrazol and salinity on ion leakage, proline content and activities of antioxidant enzymes in mango (*Mangifera indica* L.). *Scientia Hort.*, **125**, 785-788.

Steffens, G.L., Wang, S.Y., Steffens C.L. and Brennan, T. 1983. Influence of paclobutrazol (PP 333) on apple seedling growth and physiology. *Proc. Plant Growth Regul. Soc. Amer.*, **10**, 195-205.

Sugavanam, B. 1984. Diastereoisomers and Enantiomers of PBZ: Their Preparation and Biological Activity. *J. Pesticide Sci.*, **15**, 296-302.

Szalai, G., Kellos, T., Galiba, G., Kocsy, G. 2009. Glutathione as an antioxidant and regulatory molecule in plants under abiotic stress conditions. *J. Plant Growth Regul.*, **28**, 66-80.

Tas, S. and Tas, B. 2007. Some physiological responses of drought stress in wheat genotypes with different ploidity in Turkiye. *World J. Agri. Sci.*, **3**, 178–183.

Tatar, O. and Gevrek, M.N., 2008. Influence of water stress on proline accumulation, lipid peroxidation and water content of wheat. *Asian J. Plant Sci.,* **7**, 409-412.

Upadhyaya, A., Davis, T., Walser, R., Galbraith, A. and Sankhla, N. 1989. Uniconazole-induced alleviation of low temperature damage in relation to antioxidant activity. *Hort. Sci.*, **24**, 955–957.

Ushimaro, T., Shibasaka, M. and Tsuji, H. 1992. Development of O_2 detoxification system during adaptation to air of submerged rice seedlings. *Plant Cell Physiol.*, **33**, 1065–71.

VanToai, T.T. and Bolles, C.S. 1991. Postanoxia injury in soybean (*Glycine max*) seedlings. *Plant Physiol.*, **97**, 588–92,

Venema, J.H., Villerius, L. and Van Hasselt, P.R. 2000. Effect of acclimation to suboptimal temperature on chillinginduced photo-damage: comparison between a domestic and a high altitude wild *Lycopercison* species. *Plant Sci.*, **152**, 153–163.

Voesenek, L.A.C.I., Benschop, J.J., Bou, J., Cox, M.C.H., Vreeburg, R.A.M. and Peeters, A.J. M. 2003. Interactions between plant hormones regulate submergence-induced shoot elongation in the flooding-tolerant dicot Rumex palustris. *Ann. Bot.*, **91**, 205–11.

Voesenek, L.A., Rijnders, J.H., Peeters, A.J., Van, De Steeg, H.M., De Kroon, H. 2004. Plant hormones regulates fast shoot elongation under water: from genes to communities. *Ecol Soc. Am.*, **85**, 16–27.

Walker, M.A. and Mc Kersie, B.D. 1993. Role of the ascorbate glutathione antioxidant system in chilling resistance of tomato. *J. Plant Physiol.*, **141**, 234–239.

Wang, C.Y. 1996. Temperature precondition affects ascorbate antioxidant system in chilled zucchini squash. *Postharv. Bio. Tech.*, **8**, 29–36.

Wang, C. 1985. Modification of chilling susceptibility in seedlings of cucumber and zucchini squash by the bioregulator paclobutrazol. *Sci. Hort.*, **26**, 293–298.

Watson, G.W., Himelick, E.B. 2004. Effects of soil pH, root density, and tree growth regulator treatments on pin oak chlorosis. *J. Arboriculture*, **30**, 172–7.

Wise, R.R. 1995. Chilling-enhanced photooxidation: the production, action and study of reactive oxygen species produced during chilling in the light. *Photosyn. Res.*, **45**, 79–97.

Yan, B., Dai, Q., Flood-induced membrane damage, lipid oxidation and activated oxygen generation in corn leaves. *Plant Soil*, **179**, 261–8, 1996.

Ye, Q.F., Zhou, W.J., Xi, W.F. and Fang, J.Y. 1995. Effects of S-3307 on levels of endogenous (IAA, ABA and ZT) and physiological of rape seedlings. *Acta. Agri. Zhejiang,* **7**, 45-456.

Zeevaart, J. and Creelman, R. 1988. Metabolism and physiology of abscisic acid. *Annu. Rev. Plant Physiol.*, **39**, 439–475.

Zhang, M., Duan, L., Tian, X., He, Z., Li, J., Wang, B., Li, Z. 2007. Uniconazole-induced tolerance of soybean to water deficit stress in relation to changes in photosynthesis, hormones and antioxidant system. *J. Plant Physiol.,* **164**, 709-717.

Zholkevich, V.N. and Pustovoytova, T.N. 1993. The role of *Cucumis sativum* L leaves and content of phytohormones under soil drought. *Russ. J. Plant Physiol* ., **40**, 676-680.

Zhou, W. and Ye, Q. 1996. Physiological and yield effects of uniconazole on winter rape *(Brassica napus* L.). *J. Plant Growth Regul.*, **15**, 69-73.

Zhu, J. K. 2001. Plant salt tolerance. *Trends Plant Sci.*, **6**, 66–71.

Zollinger, N., Kjelgren, R., Cerny-koenig, T., Kopp, K. and Koenig, R. 2006. Drought responses of six ornamental herbaceous perennials. *Sci. Hort.,* **109**, 267-274.

Section III:

Microbial Diversity and Molecular Strategies in Plant Nutrition

10

SOIL MICROBIAL DIVERSITY AND NUTRIENT TRANSFORMATION AS INFLUENCED BY INTRODUCTION OF TRANSGENIC PLANTS

B. Basak[1], Sharmistha Pal[2] and Debarati Bhaduri[3*]

[1] *ICAR-Directorate of Medicinal and Aromatic Plants Research, Boriavi, Anand 387 310, Gujarat, India*
[2] *ICAR-Central Soil and Water Conservation Research and Training Institute, Research Centre, Madhya Marg, Chandigarh 160 019, India*
[3] *ICAR- Directorate of Groundnut Research, Junagadh-362 001, Gujarat*
**e-mail: debarati.ssiari@gmail.com*

INTRODUCTION

One of the potential environmental effects of the recent rapid increase in the global agricultural area cultivated with transgenic crops is a change in soil microbial-mediated processes and functions. Among the many essential functions of soil biota are soil organic matter decomposition, nutrient mineralization and immobilization, oxidation–reduction reactions, biological N fixation, and solubilization. However, relatively little research has examined the direct and indirect effects of transgenic crops and their management on microbial-mediated nutrient transformations in soils. There is an urgent need to review the available literature related to the environmental effects of transgenic crops and their management on soil microbial-mediated nutrient transformations, and to consider soil properties and climatic factors that may affect the impact of transgenic crops on these processes. Targeted genetic traits for improved plant nutrition include greater plant tolerance to low Fe availability in alkaline soils, enhanced acquisition of soil inorganic and organic P, and increased assimilation of soil N. Among the potential direct effects of transgenic and their management are changes in soil microbial activity due to differences in the amount and composition of root exudates, changes in microbial functions resulting from gene transfer from the transgenic crop, and alteration in microbial populations because of the effects of management practices for transgenic crops, such as pesticide applications, tillage, and application of inorganic and organic fertilizer sources. Possible indirect effects of transgenic crops, include changes in the fate of transgenic crop residues and alterations in land use and rates of soil erosion, deserve further study. Despite widespread public concern, no conclusive evidence has yet been presented that currently released transgenic crops, including both herbicide and pest resistant crops, are causing significant direct effects on stimulating or suppressing soil nutrient transformations in field environments. Further consideration of the effects of a wide range of soil properties, including the amount of clay and its mineralogy, pH, soil structure, and soil organic matter, and variations in climatic conditions, under which transgenic crops may be grown, is needed in evaluating the impact of transgenic crops on soil nutrient transformations. Future environmental evaluation of the impact of the diverse transgenic crops under development could lead to an improved understanding of soil biological functions and processes.

TRANSGENIC PLANTS: A BRIEF OVERVIEW

Transgenic plants or genetically modified (GM) plants are those plants possess a gene or genes that have been transferred from different species. GM plants generally developed for a variety purpose e.g., disease and pest resistance, longer shelf life, herbicide tolerance, nutrition improvement and resistance to abiotic stress. In the year 1994, first GM crop approved for commercial cultivation in USA was the FlavrSavr tomato having higher shelf life. Since the commercialization of GM crop, the planting of GM crops has consistently increased by 10% or more each year world-wide. Different kinds of gene are transferred in different plant as per the requirement which will be discussed here in brief. Till now, *cry* gene from *Bacillus thuringiensis*, commonly known as *Bt* have been inserted into many plant species such as maize, cotton, potato, tomato, rice, eggplant and canola to produce their own *Cry* protein which is responsible for their resistance against insect.

Bt maize is the most widely grown *Bt* crop in the world. The "first-generation" *Cry* proteins engineered into maize were *Cry1Ab* and *Cry9C*. Monsanto develop the variety called 'YieldGard' corn, transforming event MON810, expressing *Cry1Ab*, is the most widely grown *Bt* crop in the world today (Benedict and Ring, 2004). The transgenic maize expressing *Cry1Ab* was initially developed to control European Corn Borer (ECB), latter it was found to effective against several other lepidopteran pests (Gonzales-Nunez *et al.*, 2000; Dutton *et al.*, 2005). Recently Monsanto has developed MON863 *Bt* maize under the trade name 'YieldGard' effective against rootworm. The event MON863 expresses *Cry3Bb1* protein throughout the plant including leaf, grain, pollen and root at concentration ranges from 3.2 to 93 μg g^{-1} of plant tissue (USEPA 2007).

The first generation *Bt* cotton varieties were developed by Monsanto under the commercial name 'Ingard' in Australia and 'Bollgard' in USA. The Bollgard event MON531 and 'Ingard' event MON757 both express *Cry1Ac* protein. These transgenic cotton varieties are effective against bollworm mainly pink bollworm. A second generation Bt cotton variety was developed by inserting another gene *Cry2Ab* under the name of Bollgard II which is more effective against pest than Bollgard (Monsanto Company, 2002).

New Leaf Potato, the first commercial variety of transgenic potato was developed by Monsanto to control Colorado potato beetle, which is resistant to most classes of chemical pesticides. The variety expresses *Cry3A* protein and active against certain species of beetle (Coleoptera). Due to public pressure, many food producers have chosen not use *Bt* potatoes in their product. As a result, since 2001 *Bt* potatoes are no longer available (Shelton *et al.*, 2002).

FlavrSavr became the first commercially grown genetically engineered food to be granted for human consumption. A second copy of the tomato gene polygalacturonase was inserted into the tomato genome in the antisense direction. Polygalacturonase enzyme is responsible for degradation of pectin which is component of tomato cell wall, causing the fruit to soft. The FlavrSavr failed to achieve commercial success and was withdrawn from the market in 1997. DNA Plant Technology (DNAP), Agritope and Monsanto developed tomato that delayed ripening by preventing the production of ethylene, a hormone that triggers ripening of fruit. In these tomatoes ethylene production was inhibited by reducing the amount of 1-aminocyclopropane-1-carboxylic acid (ACC). Monsanto tomato was engineered with the *ACC deaminase* gene. Monsanto's tomato was engineered with the *ACC deaminase* gene from the soil bacterium *Pseudomonas chlororaphis* that lowered ethylene levels by breaking down ACC.

It is well known that Golden rice is a variety of rice produce through genetic engineering to biosynthesis beta-carotene, a precursor of pro-vitamin A in the edible part of rice.Golden rice was created by Ingo Potrykus of the Institute of Plant Sciences at the Swiss Federal Institute of Technology, working with Peter Beyer of the University of Freidburg. Golden rice designed to produce beta-carotene in the edible part of rice, the endosperm. The plant doses not normally produce the pigment in the endosperm, since photosynthesis does not occur in endosperm. Golden rice was created by transforming rice with two beta-carotene biosynthesis gene: psy (phytoene synthase) from daffodil (*Narcissus pseudonarcissus)* and crtl from the soil bacterium *Erwinia uredovora*. The *psy* and *crt1* genes were transformed into the rice nuclear genome and placed under the control of an endosperm-specific promoter so they are only expressed in the endosperm.

Similarly *Bt* rice was developed by transferring a *cry* gene from *Bacillus thuringiensis* expressing *Cry1Ac* toxin effective against yellow stem borer (YSB). The successful lines caused 100% mortality when consumed by yellow stem borer larvae. Production of toxin at high levels (approaching 1% of the cellular protein) is considered important in preventing resistance in the target insect, but that might have some effect on human being and environment.

The *Bt* brinjal is a suite of transgenic brinjal (also known as eggplant) created by inserting a crystal gene (*Cry1Ac*) from soil bacterium *Bacillus thuringiensis* into the genome of various brinjal cultivar. Apart from the insect resistant (*Bt*), so many transgenic crop was develop for different purpose like disease resistance, herbicide tolerance, stress tolerance etc (Table 1).

Table 1: Global scenario of transgenic plant under commercial cultivation.

Types of Crops	Name of crop	Growing Countries	Purpose
Field crops	Soybean	9 countries- Canada, USA, Argentina, South Africa, Brazil, East Europe, Uruguay, Paraguay and Chile	Herbicides resistance
	Cotton	8 countries- USA, Australia, Argentina, Mexico, China, South Africa, India and Colombia	Insect resistance and herbicide tolerance
	Maize	8 countries- Canada, USA, Argentina, South Africa, Western Europe, Uruguay, Philippines and Chile	Insect resistance
	Canola	2 countries-Canada and USA	Insect resistance
Vegetables	Tomato	China	Insect and disease resistance, edible vaccines and delayed ripening
	Squash	USA	Virus resistant
	Pepper	China	Virus tolerant
Fruits	Papaya	Hawaii (USA)	Disease resistance
Other crops	Tobacco	USA	Insect resistance

MECHANISMS BY WHICH TRANSGENIC PLANT AFFECT SOIL MICROBES

There are a variety of ways through which transgenic plants may affect the soil microorganisms both directly and indirectly. Direct effect will depend upon extent of activity of transgene proteins (Oger *et al.*, 1997) and quantities of protein that accumulate in soil environment. On the other hand, indirect effects may be due to change in plant proteins and root exudates composition as a result of modifying the metabolic pathways of plants.

The transgene proteins in GM crop for pest and disease resistance can involve production of chemical substances are potentially toxic to non target soil microbes and may affect the beneficial soil microbes like N-fixer, mycorrhizal fungi and microbes involves in nutrient recycling. Indirect effect may be due to change in root exudates composition and subsequently change in microbial community structure.

Accumulation of Transgenic Protein in Soil

Accumulation of transgenic protein in soil may potentially influenced by many factors including amount of protein present in plant, persistence of the protein in soil, soil physical and chemical environment that influence the bioavailability of protein in the soil. Sometimes constitutive promoters are used to express the toxin (protein) in parts of plant at all the growth stages of the transgenic plant (Odell *et al.*, 1985). In this way large quantity of transgenic protein accumulates in soil environment which potentially effect non-target soil microorganisms. Since the transgene protein express in different plant parts, plant litter is potentially one of the largest sources introductions of transgene protein. The crystal toxin protein produce by transgenic plants expressing the gene of *Bacillus thuringiensis* (*Bt*) is one of the most common transgene protein that has been released into environment in large quantities. *Bt* toxin can enter soil through decomposition of plant litter, turnover of root tissue and root exudates. Experimental studies have shown that *Bt* toxin as well as another common transgene protein T4 lysozyme, are not only present in root exudates but that they maintain biological activity after entering the soil (Ahrenholtz *et al.*, 2000; de Vries *et al.*, 1999; Saxena and Stotzky, 2000; Saxena *et al.*, 1999, 2002). If the expression of transgene in transgenic plants can be controlled temporally or specially by organ/tissue specific promoter, the potential environmental risk caused by transgenic plant can be reduced (Maizel and Weigel, 2004; Wilkinson *et al.*, 1997).

The stability of protein released in soil after decomposition depends upon their interaction with soil i.e. some protein can bind to the surface active particles and reduce their availability to microbes. The potential impact of protein on soil microbes will depend on the persistence of the transgene protein in soil after decomposition of transgenic plants. For example, when purified *Bt* toxin was added to non-sterile soil, activity against larvae was still detected after 234 days (Saxena and Stotzky, 2000). Soil with high clay content and low pH increase the persistence of *Bt* toxin in soil. During the first month no degradation of the *Cry1Ab* toxin in transgenic *Bt* corn leaves was observed, after that toxin concentration decrease 20% of initial value during the second month. No degradation in winter and toxin continued to degrade slowly when temperature rises spring. So it is clear from the results that the transgene protein can remain in soil for a long time and continuous cultivation of transgenic plants over many years lead to accumulation of transgenic protein in soil which might have harmful effect on soil microorganisms.

Changes in Chemical Composition in Transgenic Plants

In the process of genetic manipulation, particularly during cell and tissue culture (especially when antibiotics are used as selecting agents) some unintentional change may occur in plant characteristics. For example, expression of an antisense GIGANTEA gene fragment in transgenic radish causes delayed bolting and flowering (Curtis *et al.*, 2002). This unintentional change may occur in plant tissue as well as in root exudates of transgenic plants. It has been observed that significant difference in fructose and soluble carbohydrate content between transgenic plant and parental plants (Escher *et al.*, 2000). Transgenic alfalfa was engineered to over express the nodule-enhanced malate dehydrogenase (neMDH) gene but it also significantly increased the exudation of citrate, oxalate, malate, succinate and acetate as compared to non-transgenic plant (Tesfaye *et al.*, 2001). The compositions of soil microbial communities in rhizosphere are influenced by root exudates in a selective manner. The changes in root exudates composition between transgenic plants and non-transgenic plants may produce different effects on microbial communities. Though the understanding of the relationship between microbial communities in rhizosphere and plant growth is limited, but in future this will become an important task for soil microbiologist and plant breeder to focus on optimization of plant beneficial microbial population in rhizosphere.

EFFECT OF TRANSGENIC PLANTS ON SOIL MICROORGANISMS

Microorganisms are the dominant organisms in soil both in terms of biomass and activity, and involved numerous important processes including decomposition of organic matter, nutrient mineralization, regulation of plant pathogen, decomposition of agricultural chemicals and improvement of soil structure (Gupta and Yeates, 1997; Bruinsma *et al.*, 2003). The rhizosphere contains the majority of microbiota in soil (10 fold more than the microbiota in bulk soil) and plant-microbe interaction in the rhizosphere is the most important factor that influence the health and growth of plant. It is well known that root exudates determine the types of microorganisms reside in the rhizosphrere (Lynch, 1994; Bardgett *et al.*, 1999). Therefore, any change in root exudates in terms of quantity and quality could potentially modify the composition (biodiversity) and activity of soil microorganisms. Since the introduction of transgenic plant or GM crop in cultivation, it has been seen an increasing number of studies devoted to assessing the effect of GM crop on soil microbial diversity (Table 2) as summarized below. Due to limits of the review, it is not possible to discuss all these studies individually, and therefore, a number of illustrative examples have been chosen that typify the studies performed to date. In particular, focus will be on changes in soil microbial diversity, activity, processes mediated by microbes in soil and of course the technique to asses these.

Table 2: Soil microbial diversity as affected by cultivation of transgenic plant

Organisms	Transgenic plant	Protein	Effect on biodiversity	Reference
Culturable bacteria and fungi	*Bt* cotton	*Cry1Ac*	Significant, increase in number in soil with *Bt* cotton	Donegan *et al.*, 1995
Culturable aerobic bacteria and fungi	*Bt* potato	*Cry3A*	Minimal difference in number	Donegan *et al.*, 1995
Rhizosphere bacterial community	*Bt* maize	*Cry1Ab*	No difference in community structure between *Bt* and non- *Bt* maize by CLCP but difference in community structure by ARISA	Brusetti *et al.*, 2004
Arbuscular mycorrhizal (AM) fungi	*Bt* maize	*Cry1Ab*	Significantly lower presymbiotic hyphal growth of mycorrhizae in soil with *Bt* maize than in soil with non- *Bt* maize	Turrini *et al.*, 2004
Culturable bacteria	*Bt* maize	*Cry1Ab*	Lower ratio of gram-positive to gram-negative bacteria in soil with *Bt* maize than in soil with non-*Bt* maize	Xue *et al.*, 2005
Culturable functional bacteria (potassium dissolving, phosphate solubilizing and Nitrogen fixing)	*Bt* cotton	*Cry1Ac*	Increasing number of functional bacteria in soil *Bt* cotton than in soil with non-*Bt* cotton	Rui *et al.*, 2005

Microbial biomass	*Bt* maize	*Cry3Ab1*	No adverse effect of *Bt* maize on microbial biomass	Devare *et al.*, 2007
Soil Bacteria	Potato	T4 lysozyme	Sensitive to most gram in the soil with transgenic potato than in soil with non- transgenic potato	Ahrenholtz *et al.*, 2000
AM fungi	Tobacco	β- 1, 3-glucanases	Delayed or sometimes affect colonization in soil with transgenic tobacco than in soil with non-transgenic tobacco	Vierheilig *et al.*, 1993
Root endophyte	Rape	5-enolpyruvylshikimate-3-phosphate synthase and glyphosate oxidoreductase	Diversity lower in soil with transgenic rape than soil with non-transgenic rape	Siciliano and Germida, 1999
Microbial functional diversity	Alfalfa	Alpha-amylase and lignin peroxidase	Functional diversity higher in the rhizosphere of transgenic alfalfa than non-transgenic alfalfa	Tesfayea *et al.*, 2003

Soil Microbial Diversity and Transgenic Plants

There are numerous studies already been carried out, with different methods (e.g. functional and structural composition of soil microbial communities) and different *Bt* plants (e.g. maize, cotton, rice, canola, potato and tobacco), about the effect of *Bt* crops on soil microbial communities (Table 3) (Fig. 1). It was revealed from studies that results ranging from no effect to minor and significant effect, of these plants on soil microbial community (Donegan *et al.*, 1999; Turrini *et al.*, 2004; Xue *et al.*, 2005; Rui *et al.*, 2005). Significant reduction in hyphal growth of AM fungi *Glumas mosseae* was found in soil containing *Bt* maize root exudates as compared non-*Bt* maize (Turrini *et al.*, 2004). Sharp differences were also observed in the composition of microbiota in soils of Australia associated with residue of *Bt* cotton and non-*Bt* cotton. More extensive colonization by fungi (spores and hyphae) was observed in the residue of *Bt* cotton than non-*Bt* cotton. Higher ratios of fungi to bacteria were found in *Bt* cotton residues than non-*Bt* cotton by using SIR (substrate induced respiration) technique (Gupta *et al.*, 2002; Gupta and Watson, 2004).

Table 3: Summary of the effect of transgenic crop on soil enzyme activity.

Activity	Transgenic plant	Protein	Effect on biodiversity	Reference
Dehydrogenase and phosphatase	*Bt* rice straw	*Cry1Ab*	Increased enzyme activity after application of rice straw in flooded soil	Wu *et al.*, 2004
Arylsulfatase, phosphatase, dehydrogenase, and protease	*Bt* maize	*Cry1Ab*	No differences in the activities of enzymes between soil with *Bt* and non-*Bt* maize	Flores *et al.*, 2005
Urease, alkaline phosphatase, dehydrogenase, phenol oxidase and protease	*Bt* cotton	*Cry1Ac*	No differences in the activities of enzymes between soil with *Bt* and non-*Bt* cotton	Shen *et al.*, 2005
Urease, acid phosphomono-esterases, invert-ases, cellulases, and arylsulfatase	*Bt* cotton	*Cry1Ac*	Stimulate the activities of all the enzymes except that of arylsulfatases which was inhibited	Sun *et al.*, 2007
Dehydrogenase and alkaline phosphatase	Transgenic alfalfa	Lignin peroxidase	Enzyme activity lower in soil with transgenic alfalfa than soil with non- transgenic alfalfa	Giovanni *et al.*, 1999

A significant difference between *Bt* maize (expressing *Cry1Ab* gene) non-*Bt* maize in both total and metabolic-active 16S rRNA fraction of culturable rhizosphere heterotrophic bacteria by denaturing gradient gel electrophoresis (DGGE) and micorrhizal colonization (e.g. significant lower level of *G. mossaea* was detected in *Bt* root) was reported by Castaldini *et al.* (2005). The ratio of gram-positive to gram negative bacteria was lower in the soil with *Bt* maize than in the soil with non-*Bt* maize, ratio was higher in soil with *Bt* potato than with non-*Bt* potato (Xue *et al.*, 2005). It was observed by Rui *et al.* (2005) that more number of cultural functional groups of bacteria (potassium dissolving bacteria, phosphate solubilizing bacteria and nitrogen fixing bacteria) in the rhizosphere soil of non-*Bt* cotton than rhizosphere soil of *Bt* cotton in the early and middle stage of crop growth. Most of the studies indicated that *Bt* plants cause no or only minor change in microbial community structure in soil. Brusetti *et al.* (2004) compared the rhizosphere bacterial community associated with *Bt* maize expressing the *Cry1Ab* protein and near-isogenic counterpart (ICP) using several techniques, including viable counts, community- level catabolic profiling (CLCP) and PCR-based automated ribosomal inter-genic spacer analysis (ARISA), found that viable count and CLCP did not show any difference between *Bt* and non-*Bt* maize, but ARISA showed the community structure differed with age of plants between *Bt* and non-Bt maize, suggesting that root exudates could select different bacterial communities. Wei *et al.* (2006) reported that field release transgenic papaya affect microbial communities and enzymes activities in soil. Total colony forming unit (CFUs) of bacteria, fungi and actinomycetes in soil planted with transgenic papaya (mutant gene of papaya ringspot virus) were significantly higher than non-transgenic papaya. No significant effect of *Cry1Ab* protein release to soil from *Bt* maize root exudates on culturable bacteria, fungi and protozoa (Flores *et al.*, 2005). Most of the study suggested that no significant effect of transgenic plants on soil microbial diversity even if change in microbial diversity found but not consistent.

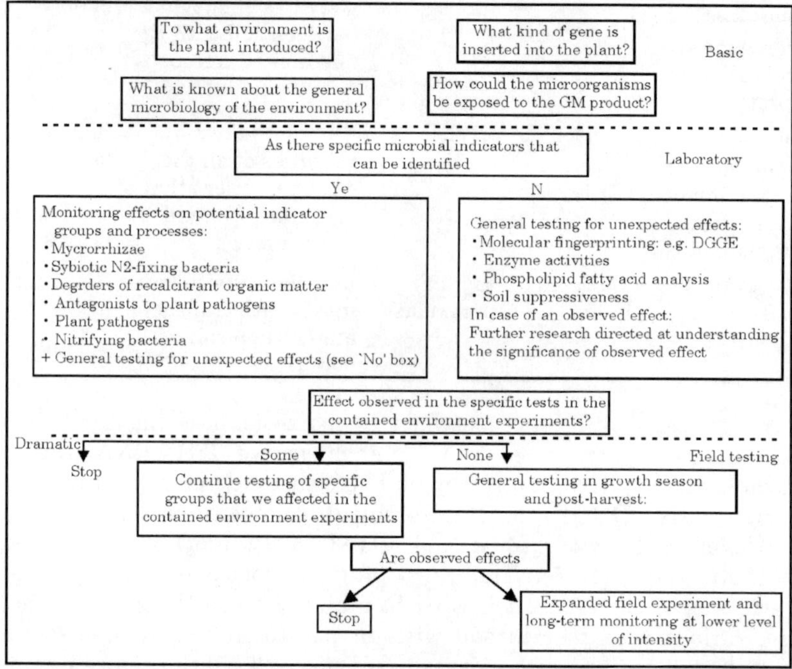

Fig. 1. Schematic approach outlining a new framework for measuring the potential effects of genetically modified plants on soil-borne microbial communities and functions (modified from Kowalchuk *et al.*, 2003)

Plants alter the composition and diversity of soil microbial communities in a selective manner (Nehl *et al.*, 1997), indicating that plant type influences which microorganisms colonize in their rhizosphere. It was found that the selective effect of plants on rhizosphere community can occur even at the cultivar level (Rengel *et al.*, 1998). In an experiment in which root growth solution, collected from *Bt* and non-*Bt* maize grown in hydroponic culture, were added to soil indicated that root exudates of *Bt* maize led to development of bacterial communities in soil that differed from those associated with exudates of non-*Bt* maize. Experiment with some culture independent method revealed that minor or no effect of *Bt* specific effects on soil microorganisms, where as plant types and age appeared to be the major factor affecting microbial diversity (Donegan *et al.*, 1995; Griffiths *et al.*, 2005, 2006; Naef *et al.*, 2006). For example, it was found that the endotoxin from *B. thuringiensis* subsp *kurstaki*, both purified and produced in the transgenic plants did not have a direct effect on soil microorganisms, and that the effects observed, were related to plant varieties use, may have been caused by unexpected change in the plant characteristic that resulted from genetic manipulation or tissue culturing of GM plants. Siciliano *et al.* (1998) and Siciliano and Germida (1999) reported that composition of root associated microbial community differed between transgenic canola and non transgenic canola, as determined by CLPP and fatty acid methyl ester (FAME) analyses and suggested that these effects could have been caused by altered chemical composition of the plant tissue as result of genetic manipulation. Changes in composition in crop residues as result of the introduction of transgenic trait have been observed in transgenic *Bt* crops (Masoero *et al.*, 1999; Escher *et al.*, 2000; Saxena and Stotzky, 2001; Folmer *et al.*, 2002; Mungai *et al.*, 2005). Other studies also have shown that the effects of GM plants on microbial communities are subject to more seasonal variation or other environmental factor such as soil type and agricultural practices than to expression of transgenic protein in plants (Lottmann *et al.*, 1999; Heuer *et al.*, 2002; Griffiths *et al.*, 2000, 2005; Fang *et al.*, 2005).

Soil Microbial Activity and Transgenic Plants

The term microbial activity used to indicate a wide range of activities carried out by microorganisms in soil. Various methods have been used to determine microbiological activity: e.g. soil respiration (CO_2 evolution), reflects the rate of catabolism; enzymatic activities reflect catalyzing the transfer of hydrogen and electron from one compound to another (Nannipieri *et al.*, 1990). Soil enzymatic activity mainly contributed by soil microbes and has been reported to be affected by the presence of transgenic crop (Table 4).

Soil respiration is measured in terms of CO_2 evolution rate and it did not show any change during the growth of *Bt* maize, where as there was a lower evolution of CO_2 from soil amended with residues of *Bt* maize (Castaldini *et al.*, 2005). By contrast Dinel *et al.* (2003) reported 30.5% lower evolution of CO_2 from soils planted with *Bt* maize than from soils planted with near- isogenic non-*Bt* maize. Significant lower evolution (20-39%) of CO_2 from the soils amended with biomass of *Bt* maize, canola, cotton, potato, rice and tobacco than from the soil with their near-isogenic non-*Bt* counterparts (Flores *et al.*, 2005), which they attributed to a higher lignin content in the biomass of *Bt* plants. Modification of lignin content could result in effects that may have ecological implication (Halpin *et al.*,

2004). Lignin content in plant may be beneficial, as it can provides greater resistant to attack by insect particularly the second generation of European corn borer, reduce susceptibility to molds (Masoero *et al.*, 1999), retard litter degradation and decomposition by microbes (Tovar-Gomez *et al.*, 1997), as lower evolution of CO_2 from soils amended with biomass of *Bt* corn than with biomass of non-Bt corn (Flores *et al.*, 2004). The longer persistence of the biomass of *Bt* plants may extent the time that toxin is present in soil, thereby enhance the hazard to nontarget organisms and selection of toxin-resistant target insects. However, Hopkins and Gregorich (2003) did not observed any differences in the decomposition of plant material from *Bt* and non-*Bt* maize as determined by CO_2 evolution.

Table 4: Summary of the effect of transgenic plant on microbial mediated process in soil.

Processes	Transgenic plant	Protein	Effect on biodiversity	Reference
Carbon mineralization (CO_2 evolution)	*Bt* maize	*Cry1Ab*	Lower mineralization (CO_2 evolution) in soil with *Bt* maize than soil with non-*Bt* maize	Dinel *et al.*, 2003
Carbon mineralization (CO_2 evolution)	*Bt* maize, potato, rice, cotton, canola, and tobacoo	*Cry1Ab* *Cry1Ac* *Cry3A*	Lower mineralization (CO_2 evolution) in soil with *Bt* crops than their non-*Bt* maize counterparts	Flores *et al.*, 2005
Decomposition rate	*Bt* maize	*Cry1Ab*	Very small differences in rate of decomposition of *Bt* and non-*Bt* maize	Gupta and Watson, 2004
N mineralization	*Bt* maize	*Cry3Bb1*	No differences in potential aerobic N mineralization	Devare *et al.*, 2004
N mineralization	*Bt* maize residue	*Cry1Ab*	No consistent differences in N mineralization	Mungai *et al.*, 2005
Potential N mineralization and nitrification	*Bt* maize	*Cry3Bb1*	No adverse effects of *Bt* maize on N mineralization potential and on nitrification	Devare *et al.*, 2007
Chemical composition	*Bt* maize	*Cry1Ab*	Higher content of starch and lignin and lower content of protein and soluble N than non-*Bt* maize	Masoero *et al.*, 1999
Lignin content	*Bt* maize	*Cry1Ab*	Higher lignin content in *Bt* maize than non-*Bt* maize	Flores *et al.*, 2005; Poerschmann *et al.*, 2005; Saxena and Stotzky, 2001

Phosphatases and dehydrogenases activities were found to increase after addition of *Bt* rice straw to flooded soil (Wu *et al.,* 2004). It was found that the addition of leaves and stem of two *Bt* cotton varieties to soil stimulated the activities of the soil ureases, acid phosphomonoesterases, invertases, and cellulases, whereas the activities of soil arylsulfatases was inhibited (Sun *et al.*, 2007). In contrast, Flores *et al.* (2005) reported that the activities of these enzymes were not consistently different in soil amended or non-amended with biomass of *Bt* or non-*Bt* maize. Similarly, Shen *et al.* (2006) found no consistent significant difference in the activities of some enzymes (urease, alkaline phosphatasess, dehydrogenase, phenol oxidase, and proteases) between soil of *Bt* and non-*Bt* cotton, no significant difference was found in the activity of some enzymes (catalase, dimethylesulphoxide, reductase and β-glycosides) at four trail sites of *Bt* and non-*Bt* maize over 4 consecutive years of maize cultivation (Lang *et al.*, 2006). These results were corroborated with the findings of Icoz *et al.* (2007), who also found no consistent significant differences in the activities of representative enzymes involved in the biodegradation of plant biomass between soils with *Bt* maize and their ICP during 4 consecutive years of maize cultivation. On the other hand field released papaya showed contrasting affects on different soil enzymes activities (Wei *et al.*, 2006). Soil planted with the transgenic papaya had significant higher activities of arylsulfatase, alkaline phosphatase, invertase but lower enzymes activities of urease and polyphenol oxidase than soil planted with non transgenic papaya.

Microbial-Mediated Processes as Influenced by Transgenic Plants

Soil microbes play important role in regulating many processes in soils that are essential for plant growth, soil health and sustain productivity. Among the many essential functions of soil microbes are decomposition of organic matter and mineralization and immobilization of nutrient (Gupta and Watson, 2004) (Fig. 2). However, there has been relatively little research on the influence of transgenic plant and *Cry* proteins on these microbial mediated processes in soil (O'Callaghan *et al.*, 2005). Here some of the promising examples are being summarized (Table 5) to provide an overall picture of microbial mediated process influenced by GM crops. Cultivation of GM crops may have several direct and indirect effects on microbial-mediated processes in soil ecosystem. The primary indirect effects of transgenic plants on such processes are mainly due to changes in amount and composition of crop residue from GM crops. Differences in chemical composition of crop residues have been observed in *Bt* crops. For example, it has been found that two maize hybrids had starch and lignin and lower protein and soluble N content than their ICP (Masoero *et al.*, 1999). Similarly, Saxena and Stotzky (2001) found a 33-97% higher lignin content in Bt maize (events: Bt 11, MON810 and 176) than their ICP. In contrast, Escher at al. (2000) found a lower lignin content, as well as a lower C:N ratio, and a higher content of soluble carbohydrates in the leaves of Bt maize than their ICP. Other studies on the chemical composition of Bt crop residues have shown no significant differences. In general, higher lignin content decreases the rate of decomposition of crop residue (Parton *et al.*, 1996; Hopkins *et al.*, 2001) and, thus, may affect the residue-associated microbial populations.

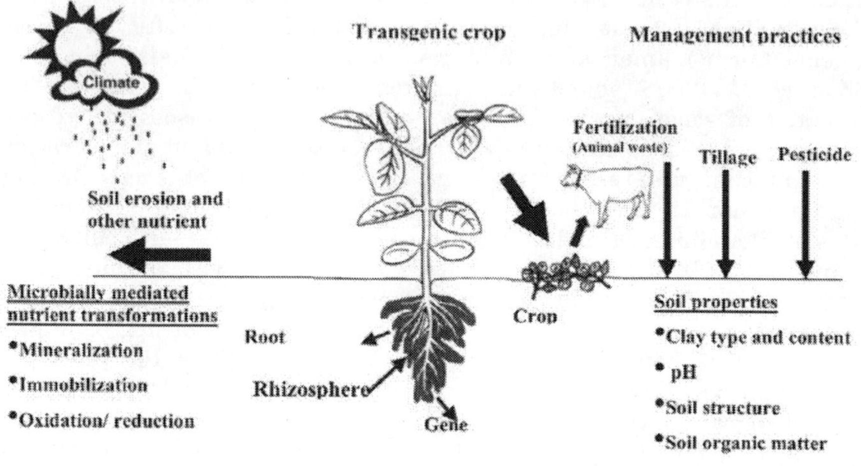

Fig. 2. Potential non-targeted direct and indirect effects of transgenic crops on microbial-mediated soil nutrient transformations (modified from Motavalli *et al.*, 2004)

Table 5: Examples of targeted transgenic traits for improved plant nutrient acquisition from soil.

Trait(s)	Mode of action	Crop	Reference
Tolerance to low Fe availability in alkaline soils	root exudation of mugineic acid phytosiderophores	rice (*Oryza sativa* L.), sorghum [*Sorghum bicolor* (L.) Moench], corn (*Zea mays* L.)	Takahashi *et al.*, 2001
Tolerance to low P availability in acidic and alkaline soils	root exudation of organic acids (e.g., citric and malic acids) to solubilize soil P	corn and other monocots	Lo´ pez-Bucio *et al.*, 2000
	root exudation of phytase extracellular enzyme to utilize organic P in soil	-	Richardson *et al.*, 2001
Improved assimilation of soil N	increased activity of NADH-glutamate synthase in plant	tobacco (*Nicotiana tabacum* L.)	Chichkova *et al.*, 2001

Net N mineralization occurs if the C:N ratio is less than 30 and net N immobilization occurs if the ratio is greater than 30 (Wanger and Wolf, 1999). There for lignin content in GM crop residue important determining factor for organic N mineralization in soil. However, incorporation of *Bt* residue did not show any significant difference in level of inorganic N in soil under either till or non-till condition, and N dynamics was not found change with incorporation of *Bt* and non-*Bt* maize residue in both laboratory and field condition (Mungai *et al.*, 2005). Difference in the potential for anaerobic N-mineralization was reported in the soils collected from fields planted with *Bt* maize or their ICP (Devare *et al.*, 2004).

Methods used for Assessing the Effect of Transgenic Plant on Soil Microorganisms

Our knowledge about soil microbial diversity is limited in the sense of our inability to study soil microorganisms. It has also been estimated that approximately 1% of soil bacterial population can be cultured by standard laboratory techniques. Four types of experimental methods have been employed to assess the effect of transgenic plants on different aspect of soil microorganisms. Plate count method is simple and useful to identify and characterized the changes of specific strain but not effective to detect the significant changes of microbial community due to introduction of transgenic plants (Donegan *et al.*, 1995; Wu *et al.*, 2004; Saxena and Stotzky, 2001). The reason might be that majority of soil microbes remain unidentified because they are not culturable with standard method (Ward *et al.*, 1990). Other method like most probable number method (MPN), direct microscopic count, biochemical assay (fatty acid) and molecular based techniques can be employed to study the soil microbial diversity. These above mentioned methods have become less popular not only because these method not able to provide the whole picture but also because the procedure are laborious. The plating method and most probable method (MPN) mainly used to detect effect of transgenic plants on specific soil microorganisms such as symbiotic nitrogen fixing bacteria, decomposer of organic matter and nitrifying bacteria (Bruinsma *et al.*, 2003).

The substrate utilization assay with Biolog GN microplates is one of the methods for metabolic fingerprinting. This method having the same limitation as in case of culture based methods. Sometimes it is very difficult to determine the microorganisms responsible for the observed metabolic fingerprint with this method (Haack *et al.*, 1995; Giovanni *et al.*, 1999). However, this method often used and found effective in studying the effects of transgenic plants on soil microbial communities (Donegan *et al.*, 1999; Giovanni *et al.*, 1999; Tesfayea *et al.*, 2003).

Due to limitation of culture based techniques, molecular based techniques are in increasing popularity to study soil microbial diversity in recent time. DNA fingerprint analysis based on PCR (Polymerase Chain Reaction) amplified rRNA genes from soil extracted DNA is cultivation-independent analysis microbial community through detection and characterization of microbial nucleic acid sequence within soil sample

(Amann *et al.*, 1995). It is more sophisticated and comprehensive method for determining the difference in microbial community structure than culture-dependent method. Now-a-days, several DNA fingerprint methods such as DGGE, RFLP, ARDRA and SSCP have been employed to evaluate the effect of transgenic plant on soil microorganisms (Donegan *et al.*, 1999; Giovanni *et al.*, 1999; Heuer *et al.*, 2002).

Similarly to DNA based method, microbial lipid analysis such as FAME and PLFA are cultivation-independent and able to provide a quantitative measure potential viable biomass. These methods often can detect the effect of transgenic plants on soil microbial diversity (Cowgill *et al.*, 2002; Dunfield and Germida, 2003). Many microorganisms share common fatty acids and may change their fatty acid composition depending on the availability of the nutrients and other environmental factor (Drijber *et al.*, 2000; Kozdroj and van Elsas, 2001).

Substrate induced respiration and soil enzyme activity are activity based assay which have been use to monitor and the change of microbial community caused by transgenic plants (Donegan *et al.*, 1999; Giovanni *et al.*, 1999; Wu *et al.*, 2004). These methods can provide information on general microbial activities, but not about the specific microbial group that actually contribute to different soil enzymatic activities. More over the activities of soil enzymes and respiration may come from not only microbial communities but also other organisms and influenced by the soil temperature and water content (Lu, 1999). So it is conceptually wrong to assume a simple relationship between soil enzymes activity and activity of microbes in soil (Nannipieri *et al.*, 2002).

Based on the advantage and disadvantage of these methods, it is appeared that the application of multiple methods will provide more comprehensive information about microbial diversity than application of any single method. However uses of multiple methods often make it difficult to interpret experimental results. For example, significant differences were detected among transgenic alfalfa line and parent line in metabolic fingerprint in Biolog GN plates, culturable bacteria and some soil enzymes activities, but no difference in DNA fingerprint (Donegan *et al.*, 1999). Though transgenic plants have some effect on soil ecosystem but it is difficult to measure the changes of microbial diversity in terms of their activity in soil ecosystem.

Conclusion and Future Insights

Although genetically modified plants (GMPs) offer many benefits, their engineering and cultivation has also fuelled considerable debate regarding possible undesirable environmental effects (Hails, 2000). Plants provide the primary source of energy to life, both above- and belowground; however, most environmental risk assessment studies for GMPs have focused on potential aboveground effects. Microorganisms dominate soil-borne communities, accounting for 80% of the total biomass (excluding roots) and largely determine ecosystem functions, such as nutrient cycling and decomposition. Their direct and indirect interactions with plants create strong feedback mechanisms, influencing primary production and vegetation dynamics (Putten and Peter, 1997). That the underground component is often neglected in risk assessments of GMPs stems mainly from the inherent difficulties of studying soil-borne microbes. Soil systems are complex and heterogeneous, and we are currently unable to culture most soil microorganisms. However, recent methodological advances, especially new molecular biological techniques (Stephen and Kowalchuk, 2002), are now helping us understanding these communities, sparking an increase in research on the underground effects of GMP cultivation.

References

Ahrenholtz, I; Harms, K; de Vries, J and Wackernagel, W 2000. Increased killing of *Bacillus subtilis* on the hair roots of transgenic T4 lysozyme-producing potatoes. *Appl. Environ. Microbiol.*, 66: 1862–1865.

Amann, RI; Ludwig, W and Schleifer, KH 1995. Phylogenetic identification and in situ detection of individual microbial cells without cultivation. *Microbiol. Rev.*, 59: 143–169.

Bardgett, RD; Dentin, CS and Cook, R 1999. Below-ground herbivory promotes soil nutrient transfer and root growth in grassland. *Ecol Let.*, 2: 357–360.

Benedict, JH and Ring DR 2004. Transgenic crops expressing Bt proteins: current status, challenges and outlook. In: Koul, O and Dhaliwal, GS (eds.), Transgenic Crop Protection: Concepts and Strategies. Science Publishers, Inc. Enfield, NH, USA. pp. 15–83.

Bruinsma, M; Kowalchuk, GA and van Veen, JA 2003. Effects of genetically modified plants on microbial communities and processes in soil. *Biol Fertil Soils*, 37: 329–337.

Brusetti, L *et al.* 2004. Bacterial communities associated with the rhizosphere of transgenic Bt 176 maize (*Zea mays*) and its non-transgenic counterpart. *Plant Soil*, 266: 11–21.

Castaldini, M *et al.* 2005. Impact of Bt corn on rhizospheric and soil eubacterial communities and on beneficial mycorrhizal symbiosis in experimental microcosms. *Appl. Environ. Microbiol.*, 71: 6719–6729.

Chichkova, S; Arellano J; Vance CP and Herna´ndez, G 2001. Transgenic tobacco plants that overexpress alfalfa NADH-gluta- mate synthase have higher carbon and nitrogen content. *J. Exp. Biol.*, 52: 2079–2087.

Cowgill, SE; Bardgett, RD and Kiezebrink, DT 2002. The effect of transgenic nematode resistance on non-target organisms in the potato rhizosphere. *J. Appl. Ecol.*, 39: 915–923.

Curtis, IS; Nam, HG; Yun, JY and Seo, KH 2002. Expression of an antisense GIGANTEA (GI) gene fragment in transgenic radish causes delayed bolting and flowering. *Transgenic Res.*, 11: 249– 256.

de Vries, J; Harms, K; Broer, I; Kriete, G; Mahn, A; Düring, K and Wackernagel, W 1999. The bacteriolytic activity in transgenic potatoes expressing a chimeric T4 lysozyme gene and the effect of T4 lysozyme on soil- and phytopathogenic bacteria. *Syst. Appl. Microbiol.*, 22: 280–286.

Devare, MH; Jones, CM and Thies, JE 2004. Effect of Cry3Bb transgenic corn and tefluthrin on the soil microbial community. *J. Environ Qual*, 33: 837–843.

Dinel, H; Schnitzer, M; Saharinen, M; Meloche, F; Pare, T; Dumontet, S; Lemee, L and Ambles, A 2003. Extractable soil lipids and microbial activity as affected by Bt and non-Bt maize grown on a silty clay loam soil. *J. Environ Sci Health*, 38: 211–219.

Donegan, KK; Seidler, RJ; Doyle, JD; Porteous, LA; Digiovanni, G; Widmer, F and Watrud, LS. 1999. A field study with genetically engineered alfalfa inoculated with recombinant *Sinorhizobium meliloti*: Effects on the soil ecosystem. *J. Appl Ecol.*, 36: 920–936.

Donegan, KK *et al.* 1995. Changes in levels, species, and DNA fingerprints of soil microorganisms associated with cotton expressing the *Bacillus thuringiensis* var. kurstaki endotoxin. *Appl. Soil Ecol.*, 2: 111–124.

Drijber, RA; Doran, JW; Parkhurst, A and Mand Lyon, DJ 2000. Changes in soil microbial community structure with tillage under longterm wheat-fallow management. *Soil Biol. Biochem.*, 32: 1419– 1450.

Dunfield, KE and Germida, JJ 2003. Seasonal changes in the rhizosphere microbial communities associated with field-grown genetically modified canola (*Brassica napus*). *Appl. Environ Microbiol.*, 69: 7310–7318.

Dutton, A; Romeis, J and Bigler, F 2005. Effects of Bt-maize expressing Cry1Ab and Bt-spray on *Spodoptera littoralis. Entomol Exp et Applicata*, 114: 161–169.

Escher, N; Kach, B and Nentwig, W 2000. Decomposition of transgenic *Bacillus thuringiensis* maize by microorganisms and woodlice *Porcello scaber* (Crustacea: Isopoda). *Bas. Appl. Ecol.*, 1: 161–169.

Escher, N; Kach, B and Nentwig, W 2000. Decomposition of transgenic Bacillus thuringiensis maize by microorganisms and woodlice Porcello scaber (Crustacea: Isopoda). *Basic Appl Entomol*, 1: 161–169.

Fang, M; Kremer, RJ; Motavalli, PP and Davis, G 2005. Bacterial diversity in rhizospheres of nontransgenic and transgenic corn. *Appl Environ Microbiol.*, 71: 4132–4136.

Flores, S; Saxena, D and Stotzky, G 2005. Transgenic Bt plants decompose less in soil than non-Bt plants. *Soil Biol Biochem*, 37: 1073–1082.

Folmer, JD; Grant RJ; Milton, CT and Beck, J 2002. Utilization of Bt corn residues by grazing beef steers and Bt corn silage and grain by growing beef cattle and lactating dairy cows. *J. Animal Sci.*, 80: 1352–1361.

Giovanni, GD; Watrud, LS; Seidler, RJ and Widmer, F 1999. Comparison of parental and transgenic alfalfa rhizosphere bacterial communities using biolog GN metabolic fingerprinting and enterobacterial repetitiveintergenic consensus sequence-PCR (ERICPCR). *Microb. Ecol.*, 37: 129–139.

Gonzales-Nunez, M; Ortego, F and Castanera, P 2000. Susceptibility of Spanish populations of the Corn Borers *Sesamia nonagrioides* (Lepidoptera: Noctuidae) and *Ostrinia nubilalis* (Lepidoptera: Crambidae) to a Bacillus thuringensis endotoxin. *J Econ. Entomol.* 93: 459–463.

Griffiths, BS *et al.* 2005. A comparison of soil microbial community structure, protozoa, and nematodes in field plots of conventional and genetically modified maize expressing the Bacillus thuringiensis Cry1Ab toxin. *Plant Soil*, 275: 135–146.

Griffiths, BS; Geoghegan IE and Robertson, WM 2000. Testing genetically engineered potato, producing the lectins GNA and Con A, on notarget soil organisms and processes. *J. Appl. Ecol.*, 37: 159–170.

Gupta, VVSR; Roberts, GN; Neate, SM; McClure, SG; Crisp, P and Watson, SK 2002. Impact of Bt-cotton on biological processes in Australian soils. In: Akhurst, RJ; Beard, CE and Hughes, PA (eds.), Proceedings of the Fourth Pacific Rim Conference on the Biotechnology of Bacillus thuringiensis and Its Environmental Impacts. CSIRO, Australia. pp. 191–194.

Gupta, VVSR and Watson, S 2004. Ecological impacts of GM cotton on soil biodiversity: Below-ground production of Bt by GM cotton and Bt cotton impacts on soil biological processes. Australian Government Department of the Environment and Heritage, CSIRO Land and Water. pp. 1–72.

Gupta, VVSR and Yeates, GW 1997. Soil microfauna as bioindicators of soil health. In: Pankurst, CE (ed.) Biological Indicators of Soil Health. CAB International. New York. pp. 201–233.

Haack, SK; Garchow, H; Klug, MJ and Forney, LJ 1995. Analysis of factors affecting the accuracy, reproducibility, and interpretation of microbial community carbon source utilization patterns. *Appl. Environ. Microbiol.*, 61: 1458–1468.

Hails, RS 2000. Genetically modified plants – the debate continues. *Trends Ecol. Evol.* 15: 14–18.

Halpin, C *et al.* 1994. Manipulation of lignin quality by down regulation of cinnamyl alcohol dehydrogenase. *Plant J.*, 6: 339–350.

Heuer, H; Kroppenstedt, RM; Lottmann, J; Berg, G and Smalla, K 2002.Effects of T4 lysozyme release from transgenic potato roots on bacterial rhizosphere communities are negligible relative to natural factors. *Appl. Environ. Microbiol.*, 68: 1325–1335.

Hopkins, DW and Gregorich, EG 2003. Detection and decay of the Bt endotoxin in soil from a field trial with genetically modified maize. *Europ J Soil Sci.*, 54: 793–800.

Hopkins, DW; Webster, E.A., Chudek, J.A., Halpin, C., 2001. Decomposition in soil of tobacco plants with genetic modifications to lignin biosynthesis. *Soil Biol Biochem*, 33: 1455–1462.

Icoz, I; Saxena, D; Andow, D; Zwahlen, C and Stotzky, G 2008. Microbial populations and enzyme activities in soil in situ under transgenic corn expressing Cry proteins from Bacillus thuringiensis, *J Environ Qual.*, 37(2): 647-662.

Kowalchuk, GA; Bruinsma, M and Veen, JA 2003. Assessing responses of soil microorganisms to GM plants. *Trends Ecol Evol.*, 18 (8): 403-410.

Kozdroj, J and van Elsas, J D 2001. Structural diversity of microbial communities in arable soils of a heavily industrialised area determined by PCR-DGGE fingerprinting and FAME profiling. *Appl. Soil Ecol.*, 17: 31–42.

Lang, A; Arndt, M; Beck, R; Bauchhenss, J and Pommer, G 2006. Monitoring of the environmental effects of the Bt gene. Bavarian State Research Center for Agriculture, No. 2006/10. Vo¨ ttinger Strasse 38, 85354 Freising-Weihenstephan. /http://www.LfL.bayern.deS.

Lo´ pez-Bucio, J; Nieto-Jacobo, MF; Ram´ırez-Rodr´ıguez, V and Herrera-Estrella, L 2000. Organic acid metabolism in plants: From adaptive physiology to transgenic varieties for cultivation in extreme soils. *Plant Sci.* 160:1–13.

Lottmann, J; Heuer, H; Smalla, K and Berg, G 1999. Influence of transgenic T4 lysozyme-producing potato plants on potentially beneficial plant-associated bacteria. *FEMS Microbiol. Ecol.*, 29: 365–377.

Lu, RK 1999. The Analytical Method of Agricultural Soil Chemistry. Agricultural Science and Technology Press of China. Beijing. pp. 228–264.

Lynch, J 1994. The rhizosphere – form and function. *Appl. Soil Ecol.*, 1: 193–198.

Maizel, A and Weigel, D 2004. Temporally and spatially controlled induction of gene expression in *Arabidopsis thaliana*. *Plant J.*, 38: 164–71.

Masoero, F; Moschinim, M; Rossi, F; Prandini, A and Pietri, A 1999. Nutritive value, mycotoxin contamination, and in vitro rumen fermentation of normal and genetically modified corn (Cry1Ab) grown in northern Italy. *Maydica*, 44: 205–209.

Monsanto Company 2002. Insect Efficacy Testing with Bollgards II Cotton. Public Interest Document, submitted to EPA. Monsanto Company, St. Louis, MO, USA.

Motavalli, PP; Kremer, RJ; Fang, M and Means, NE 2004. Impact of Genetically Modified Crops and Their Management on Soil Microbially Mediated Plant Nutrient Transformations. *J. Environ. Qual.* 33: 816–824.

Mungai, NW; Motavalli, PP; Nelson, KA and Kremer, RJ 2005. Differences in yields, residue composition and N mineralization dynamics of Bt- and non-Bt-maize. *Nutr. Cycl. Agroecosystems*, 73: 101–109.

Naef, A; Zesiger, T and Defago, G 2006. Impact of transgenic Bt maize residues on the mycotoxigenic plant pathogen *Fusarium graminearum* and the biocontrol agent *Trichoderma atroviride*. *J. Environ. Qual.* 35: 1001–1009.

Nannipieri, P; Grego, S and Ceccanti, B 1990. Ecological significance of the biological activity in soil. In: Bollag, JM and Stotzky, G (eds.) Soil Biochemistry, Vol. 6. Marcel Dekker, NY, pp. 293–355.

Nannipieri, P; Kandeler, E and Ruggiero, P 2002. Enzyme activities and microbiological and biochemical processes in soil. In: Burns, RG and Dick, R (eds.) Enzymes in the Environment. Marcel Dekker, New York. pp. 1–33.

Nehl, DB; Allen SJ and Brown, JF 1997. Deleterious rhizosphere bacteria: an integrating perspective. *Appl. Soil Ecol.*, 5: 1–20.

O'Callaghan, M; Travis, RG; Burgess, EPJ and Malone, LA 2005. Effects of plants genetically modified for insect resistance on nontarget organisms. *Ann Rev Entomol.*, 50: 271–292.

Odell, JT; Nagy, F and Chua, NH 1985. Identification of DNA sequences required for activity of the cauliflower mosaic virus 35S promoter. *Nature*, 313: 810–812.

Oger P, Annik P and Yves D 1997 Genetically engineered plants producing opines alter their biological environment. *Nat. Biotechnol.* 15, 369–372.

Parton, WJ; Ojima, DS and Schimel, DS 1996. Models to evaluate soil organic matter storage and dynamics. In: Carter, MR and Stewart, BA (eds.) Structure and Organic Matter Storage in Agricultural Soils. CRC Press, Boca Raton, FL. pp. 421–448.

Putten, VWH and Peters BAM 1997. How soil-borne pathogens may affect plant communities. *Ecology*, 78: 1785–1795.

Rengel, Z; Ross, G and Hirsch, P 1998. Plant genotype and micronutrient status influence colonization of wheat roots by soil bacteria. *J Plant Nutr.*, 21: 99–113.

Richardson, AE; Hadobas PA and Hayes JE 2001. Extracellular secretion of *Aspergillus* phytase from *Arabidopsis* roots enables plants to obtain phosphorus from phytate. *Plant J.*, 25:641–649.

Rui, YK; Yi, GX; Zhao, J; Wang, BM; Li, ZH; Zhai, ZX; He, ZP and Li, Q 2005. Changes of Bt toxin in the rhizosphere of transgenic Btcotton and its influence on soil functional bacteria. *World J Microbiol Biotechnol.*, 21: 1279–1284.

Saxena, D and Stotzky, G 2000. Insecticidal toxin from *Bacillus thuringiensis* is released from roots of transgenic Bt corn *in vitro* and *in situ*. *FEMS Microbiol. Ecol.*, 33, 35–39.

Saxena, D; Flores, S and Stotzky, G 1999. Transgenic plants: Insecticidal toxin in root exudates from Bt corn. *Nature* 402: 480.

Saxena, D and Stotzky, G 2001. *Bt* corn has a higher lignin content than non-*Bt* corn. *Am. J. Bot.*, 88(9): 1704-1706.

Saxena, D; Flores, S and Stotzky, G 2002. Bt toxin is released in root exudates from 12 transgenic corn hybrids representing three transformation events. *Soil Biol. Biochem.*, 34: 133–137.

Shelton, AM; Zhao, JZ and Roush, RT 2002. Economic, ecological, food safety and social consequences of the deployment of Bt transgenic plants. *Ann Rev Entomol.*, 47: 845–881.

Shen, RF; Cai H and Gong, WH 2006. Transgenic Bt cotton has no apparent effect on enzymatic activities or functional diversity of microbial communities in rhizosphere soil. *Plant Soil*, 285, 149–159.

Siciliano, SD and Germida, JJ 1999. Taxonomic diversity of bacteria associated with the roots of field-grown transgenic Brassica napus cv. Quest, compared to the non-transgenic B. napus cv. Excel and B. rapa cv. Parkland. *FEMS Microbiol. Ecol.*, 29: 263–272.

Siciliano, SD; Theoret, CM; de Freitas, JR; Hucl, PJ and Germida, JJ 1998. Differences in the microbial communities associated with the roots of different cultivars of canola and wheat. *Cand. J. Microbiol.*, 44: 844–851.

Stephen, JR and Kowalchuk, GA 2002. Ribotyping methods for assessment of in situ microbial community structure. In: Bitton, G (ed.) Encyclopedia of Environmental Microbiology (Vol. 5). John Wiley & Sons. pp. 2728–2741.

Sun, CX; Chen, LJ; Wu, ZJ; Zhou, LK and Shimizu, H 2007. Soil persistence of Bacillus thuringiensis (Bt) toxin from transgenic Bt cotton tissues and its effect on soil enzyme activities. *Biol Fertil Soils*, 43: 617–620.

Takahashi, M; Nakanishi, H; Kawasaki, S; Nishizawa, NK and Mori S 2001. Enhanced tolerance of rice to low iron availability in alkaline soils using barley nicotianamine aminotransferase genes. *Nat. Biotechnol.*, 19: 466–469.

Tesfaye, M; Temple, SJ; Allan, DL; Vance, CP and Samac, DA 2001. Overexpression of malate dehydrogenase in transgenic alfalfa enhances organic acid synthesis and confers tolerance to aluminum. *Plant Physiol.*, 127, 1836–1844.

Tovar-Gomez, MR; Emile, JC; Michalet-Doreau, B and Barriere, Y 1997. In situ degradation kinetics of maize hybrid stalks. *Animal Feed Sci Technol.*, 68: 77–88.

Turrini, A; Sbrana, C; Nuti, MP; Pietrangeli, B and Giovannetti, M 2004. Development of a model system to assess the impact of genetically modified corn and aubergine plants on arbuscular mycorrhizal fungi. *Plant Soil* 266: 69–75.

US Environmental Protection Agency (EPA) 2007. Pesticides: Regulating Biopesticides, Plant Incorporated Protectants, Current & Previously Registered Section 3 PIP Registrations. /http://www.epa.gov/pesticides/biopesticides/pips/pip_list.htm.

Vierheilig, H; Alt, M; Neuhaus, J; Boller, T and Wiemken, A 1993. Colonization of transgenic Nicotiana sylvestris plants, expressing different forms of Nicotiana tabacum chitinase by the root pathogen Rhizoctonia solani and by the mycorrhizal symbiont Glomus mosseae. *Mol Plant Microbe Interact.*, 6: 261–264.

Wagner, GH and Wolf, DC 1999. Carbon transformations and soil organic matter formation. In: Sylvia, DM *et al.* (eds.) Principles and Applications of Soil Microbiology. Prentice-Hall, Englewood Cliffs. NJ. pp. 218–258.

Ward, DM; Weller R and Bateson, MM 1990. 16S rRNA sequences reveal numerous uncultured microorganisms in a natural community. *Nature*, 345: 63–65.

Wei, XD; Zou HL; Chu, LM; Liao, B; Ye, CM and Lan, CY 2006. Field release transgenic papaya affects microbial communities and enzymes activities in soil. *Plant Soil*, 285: 347–358.

Wilkinson, JE, Twell, D and Lindsey, K 1997. Activities of CaMV 35S and nos promoters in pollen: implications for field release of transgenic plants. *J. Exp. Bot.*, 48: 265–275.

Wu, WX; Ye QF; Min, H; Duan, XJ and Jin, WM 2004. Bt transgenic rice straw affects the culturable microbiota and dehydrogenase and phosphatase activities in a flooded paddy soil. *Soil Biol Biochem.*, 36: 289–295.

Xue, K; Luo, HF; Qi, HY and Zhang, HX 2005. Changes in soil microbial community structure associated with two types of genetically engineered plants analyzing by PLFA. *J. Environ Sci.* (China) 17: 130–134.

11

AN INSIGHT OF IRON CHLOROSIS IN HORTICULTURAL CROPS: PHYSIOLOGICAL AND MOLECULAR BASIS, AND POSSIBLE MANAGEMENT STRATEGIES

Binayak Chakraborty[1], Koushik Chakraborty[2*] and Debarati Bhaduri[2]

[1]*College of Agriculture, Navasri Agricultural University, Waghai – 394 730, The Dangs, Gujarat, INDIA*
[2]*ICAR-Directorate of Groundnut Research, Ivnagar Road, Junagadh – 362 001, Gujarat, INDIA*
**Corresponding author: koushik_iari@rediffmail.com*

INTRODUCTION

Among the essential micronutrients, iron (Fe) is one of the most important mineral ions required as an integral component during the entire life cycle of the plants. It is an important constituent of several enzymes in plant systems which are mainly involved in electron transport redox reactions accomplished in cytochromes and iron-sulfur (Fe-S) proteins. Iron is also a component of non-heme iron proteins which are required during photosynthesis, respiration and N_2 fixation (Taiz and Zeiger, 1991). Iron is also vital for the establishment and function of symbiotic root nodules of legumes involved in nitrogen fixation (Udvardi and Day, 1997). When plants are grown under iron-deficient conditions, the young leaves develop a characteristic greenish-yellow colour, often referred to as iron chlorosis (López-Millán *et al.*, 2001).

Chlorosis is a general term which denotes loss of chlorophyll, resulting yellowing of leaves. Under severe deficiency, the entire plant may be affected turning the leaves white, resulting in abscission from the tree and die-back of young shoots (Wallace and Lunt, 1960). Though iron is not a constituent of the chlorophyll molecule, but iron deficiency leads to a reduction in chlorophyll biosynthesis in the chloroplasts (Abadia, 1992). It has been suggested that iron is not directly involved in the chlorophyll biosynthesis but it is required indirectly in the biosynthetic pathway as it controls the formation of the precursor molecule i.e., δ-aminolevulinic acid (Marschner, 1995). Iron is believed to have low mobility within the plant system probably due to its precipitation in the older leaves as insoluble oxides or phosphates or formation of complex with phyto-ferritin (Bienfait and Van der Mark, 1983). Thus, the long distance transport of iron through phloem is hampered.

Iron deficiency is a widespread nutritional disorder in many crops. Despite the abundant occurrence of iron, the low solubility of iron compounds in many soils prevents sufficient uptake of iron by the plants leading to the deficiency of iron (Lindsay, 1984). Like other living organisms, plants exhibit different rhizospheric strategies to cope with iron insolubility like acidification of rhizosphere, excretion of reductants or chelators (Rogers and Guerinot, 2002), and maintaining a high root reductase activity at the plasma membrane and in the cell wall (apoplast) of apical root zones (Bienfait *et al.*, 1983). There are two mechanisms of iron uptake in plants as physiological response to iron deficiency *viz.* *Strategy I* (dicotyledonous and monocotyledonous plant except *Poaceae* family): a reduction-based strategy that comprises three principal steps: Fe solubilisation; reduction of Fe(III) to Fe(II); and subsequent Fe(II) import into the cell, and *Strategy II* (monocotyledonous plant of *Poaceae* family): chelation-based strategy to form soluble Fe(III) complexes, which are then taken up by the root (Römheld and Marschner 1986; Marschner and Römheld, 1994; Brumbarova *et al.*, 2015). However, the roots do not exclusively control the ion uptake. Chlorotic leaves of efficient genotypes might also send a signal to the roots to induce the reactions and interpret the “chlorosis signal” (Salazar-Garcia, 1999). Under iron deficiency conditions, various types of changes occur in the morphology of root, iron uptake capacity, physiological, biochemical and mineral composition of leaves and flowers in both the plant types. The effect of iron deficiency also results in some changes in the quality and yields of fruits. Several perennial, deciduous, as well as evergreen fruit crops of horticultural importance, come under non-*Graminaceae* families, and hence their response may be classified as *Strategy I* response.

Iron chlorosis is a common disorder occurring in fruit crops mainly in alkaline and calcareous soils throughout the world. In calcareous soils, iron chlorosis is not only due to deficiency of iron *per se*, but mostly due to its presence as unavailable form. In calcareous soils, high pH and abundant $CaCO_3$ often creates ionic imbalance and poor soil physical conditions such as soil compaction, poor aeration, and extreme soil temperature, which all contribute to the development of iron deficiency

(Köseoğlu, 1995; Başar, 2000; Lucena, 2000). In these soils, the concentration of bicarbonate may also be as high as up to 10 mM (Boxma, 1982), which might again be responsible for iron chlorosis (Mengel, 1995). Bicarbonate is absorbed through the roots for the synthesis of organic acids, which further accumulate in the roots (Lee and Woolhouse, 1969; Gout *et al.,* 1993). Hence, iron-chlorosis in calcareous soil is often termed as "Lime-induced chlorosis". Iron deficiency in the horticultural plants is a major concern of study at present for better yields and productivity.

EFFECT OF IRON CHLOROSIS IN PLANTS

Effect on Overall Physiology and Morphology

Changes in leaves

Leaves are the principle sites where iron deficiency can be visually recognized. Maldonado-Torres *et al.* (2006) found morphological changes in the leaf at cellular and ultracellular levels in iron deficient Mexican lime (*Citrus aurantifolia*). Chlorotic leaves are thicker than green ones, due to increase in palisade and spongy parenchyma cell length and thickness (Maldonado-Torres *et al.*, 2006). However, no significant effects were reported about leaf thickness in case of peach and pear under iron deficiency conditions (Morales *et al.*, 1998). Iron deficiency leads to a reduction in epicuticular waxes on both the adaxial and abaxial surfaces in the leaves of peach. However, in pear leaves, these differences were much less. Further, upto 35% loss in weight of cuticule from the lower surface of the leaves was reported in deficient pear leaves than control leaves. The soluble lipid content may also be affected by iron deficiency and peach leaves showed upto 41% of reduction of soluble lipids under iron deficiency (Fernández *et al.*, 2008).

Changes in root

Various types of characteristic changes are displayed by the *Strategy I* plant species under iron stress. Morphological changes of roots are very critical which mainly include: root tip thickening, inhibition of root elongation, production of profuse root hairs, release of reducing compounds and the differentiation of transfer cells (Bienfait, 1988; Brown and Ambler, 1974; Brown and Jolley, 1989; Bavaresco *et al.*, 1994, Chaney *et al.*, 1992; Römheld and Marschner, 1981). Differentiation of transfer cells occurs for excretion of protons and reduction of Fe(III) (Marschner, 1995).These responses are normally found in cases of iron chlorosis tolerant genotypes (Socias-i-Company *et al.*, 1995). Formation of cluster root has been reported in low iron containing soils in case of *Lupinus albus, L. consentinii* and *Ficus benjamina* (Gardner *et al.*, 1982; Rosenfield *et al.*, 1991).

Effect on Growth and Vigour

Iron deficiency is usually characterized by leaf chlorosis, but occasionally it might result in small green leaves due to reduced leaf growth and suppression of leaf formation (Kosegarten *et al.*, 1998). A decline in growth under iron deficiency sometimes may happen before the occurrence of chlorosis as meristematic growth is more sensitive to iron deficiency than chlorophyll synthesis and chloroplast development (Gruber and Kosegarten, 2002). A reduction in growth of younger leaves is one of the common symptoms of iron deficiency (Abadía, 1992; Donnini *et al.*, 2003). Both iron deficiency and toxicity can limit the growth of the plants in respect to biomass accumulation, shoot and root length, area and number of leaves, relative growth rate, etc. (Nenova and Stoyanov, 1993; Snowden and Wheeler, 1993; de la Guardia and Alcántara, 2002b; Batty and Younger, 2003). Up to 43 per cent reduction of dry weight was reported over control when iron deficiency was imposed in MS medium of peach rootstock Mr.S 2/5 after 20 days. Dry weight of plantlets grown in bicarbonate containing MS medium was also found to be reduced by up to 62 per cent over control (Lombardi *et al.*, 2003). It was found that under complete iron deficiency, pea plants exhibited a lower shoot/root ratio and a higher dry biomass per unit of fresh biomass in roots (Nenova, 2009). Researchers suggested that application of 0.4% ferrous sulphate significantly increases the terminal shoot length, number of leaves and also the leaf area in guava (Arora and Singh, 1970). Application of Fe-humate also increases the flush growth and tree size of Hamlin orange and Flame grapefruit in high pH soils (Alava and Obreza, 1997).

Effect on Fruit Yield and Quality

Iron deficiency pattern in woody fruit species is more complex rather than in annual crops (Tagliavini, and Marangoni, 2002). There are very few reports on effect of iron chlorosis on fruit yield and quality since most of the studies have been carried out in pot culture. Most studies on the effects of iron chlorosis on fruit and quality have focused mainly on fruit size, which is increased by iron fertilization. Application of iron sources on iron deficient fruit plants increases the fruit yield in several fruits like peach (Sanz *et al.*, 1997), olive (Fernández-Escobar *et al.*, 1993), kiwifruit (Loupassaki *et al.*, 1997) and citrus (Pérez-Sanz *et al.*, 2002; Sites *et al.*, 1953; Pestana *et al.*, 2001). Iron deficiency was also reduces the yield of strawberry (Nelson and Jolly, 1984) and improved economic yield can be obtained with application of iron sources like Fe-EDDHA to the iron insufficient cultivars of strawberry (Kafkas *et al.*, 2007). Álvarez-Fernández *et al.* (2003) found smaller fruit size in two cultivars of peach under iron deficient conditions without deleterious effect on firmness of the fruits. Iron deficiency leads to decrease in chlorophyll concentration which affects directly on the photosynthesis, thereby heavily reducing the fruit yield. Alava and Obreza (1997) found significant increase in fruit yield with the application Fe-amendments in high pH soils than un-amended trees of both Hamlin orange and Flame grapefruit.

Iron deficiency can delay fruit ripening in citrus (Iglesias *et al.*, 2007; Pestana *et al.*, 2001) and peach (Álvarez-Fernández *et al.*, 2003). Application of iron is found to increase the common quality attributes of fruits like increase in soluble solids, juice content, improvement in the skin colour development and low titratable acid content in orange trees (Sites *et al.*, 1953). However, application of Fe-amendment to Hamlin orange and Flame grapefruit did not significantly influence the fruit quality (Alava and Obreza, 1997). Iron deficiency generally leads to increase in vitamin C, phenolic compounds and low total sugar/total organic acid ratios in peach (Álvarez-Fernández *et al.*, 2003). Thus, organoleptic scores of fruit get improved with application of iron on deficient trees.

Effect on Pigment Concentration

Iron deficiency leads to a progressive decrease in leaf chlorophyll concentrations in kiwi regardless of its genotypes (Rombolà *et al.*, 2002). Iron is directly or indirectly involved in the production of chlorophyll pigments and a deficiency of iron irreversibly damages the chlorophyll biosynthesis (Jacobson and Oertli, 1956). Good correlations have been reported between iron and chlorophyll contents by several workers (Bennett, 1945 and Jacobson and Oertli, 1955). The appearance of chlorosis in the leaves is due to an imperfect development of chloroplast under iron deficiency (Gris, 1857). Leaf chlorophyll concentration can be a good tool to monitor the iron status in plants (Rashid *et al.*, 1990; Chakraborty *et al.*, 2012b; Chakraborty *et al.*, 2014b). Álvarez-Fernández *et al.* (2003) concluded that chlorosis could only be due to iron deficiency because

a. Other essential elements were in adequate supply
b. Leaves showed characteristic changes like:
    i. Increase in concentration of K and organic acids
    ii. Displacement of the xanthophyll cycle pigments towards zeaxanthin
    iii. Increase in non-photochemical quenching and energy dissipation
c. With the application of $FeSO_4$ the leaves become green (Abadia, 1992; Morales *et al.*, 1998b and Igartua, 2000).

Under field conditions, it is very difficult to find a significant positive correlation between leaf iron concentration and the degree of iron chlorosis expressed as chlorophyll content (Pastana *et al.*, 2001). Often the iron concentration is found to be more in chlorotic leaves when expressed as dry weight (Morales *et al.*, 1998b and Abadía, 1992). This phenomenon is named as "Iron Paradox" (Römheld, 2000), which indicates the existence of iron pools in the chlorotic leaves which are some how inactive (Morales *et al.*, 1998b; Morales *et al.*, 2000b). However, an increased Chl a/Chl b ratio has also been found in iron deficient leaves with no significant difference between morning and midday sampling (Morales *et al.*, 2000). Thus, it can be concluded that the measurement of chlorophyll content for determination of iron deficiency is a very useful technique only under experimental conditions but not under field conditions. Iron content in the leaves varies with respect to genetic make up of the variety. A good correlation model drawn between chlorophyll content and leaf iron concentration could be an appropriate way of monitoring leaf and plant health with respect to iron fertilization.

Effects on Photosynthesis

Many iron-containing compounds in plants are involved in photosynthesis, such as the cytochrome oxidase complex, heme, ferredoxin, and iron-sulfur protein. In iron deficient conditions, the amounts of those compounds are less, and iron-dependent enzyme activities are lower (Agarwal *et al.*, 1965). This undoubtedly affects normal photosynthesis. In addition, ferrous iron deficiency results in disruption of the electron transport chain in photosynthesis, which affects the photolysis of water and blocks photosynthesis. Ferrous iron deficiency therefore, reduces the efficiency of plant photosynthesis and affects the synthesis of assimilation products (Li *et al.*, 2011). Besides, iron deficiency destroys the structure and functioning of the chloroplast. About 80 per cent of the leaf iron is present in chloroplast while 60 per cent of leaf iron is used during photosynthetic electron transport chain (Terry and Abadía, 1986). A decrease in photosynthetic rate of iron deficient leaf can be observed (Terry, 1980) due to reduction in number of photosynthetic units per unit area (Spiller and Terry, 1980) and also the remaining photosynthetic unit showed a lower PS II efficiency (Morales *et al.*, 1998) under iron deficient condition.

Under iron deficiency, the photosynthesis activity of peach cv. BigTop gets decreased (Tognetti and Sebastiani, 2004). Larbi *et al.* (2006) also reported that iron deficiency decreases the photosynthetic and transpiration rates, instantaneous transpiration efficiencies and stomatal conductance. However, it also increase the substomatal CO_2 concentrations in pear (*Pyrus communis* L.), sugar beet (*Beta vulgaris* L.) and peach (*Prunus persica* L.). Molassiotis *et al.* (2006) exposed two different rootstocks of peach (GF-677; tolerant to Iron deficiency and Cadaman; sensitive to Iron deficiency) for four months to iron deficiency imposed by omission of Fe (−Fe) or by the inclusion of bicarbonate (supplied as 20 mM $NaHCO_3$) in the nutrient solution, which showed a decline in photosynthetic rate (P_n) and stomatal conductance (g_s). However, there was no significant change in intercellular CO_2 concentration (C_i). The above changes were noted for the bicarbonate-added nutrient solution pots of Cadaman. Iron application to the iron-deficient sugar beet leaves exhibited increased photosynthetic rates and actual PS II efficiency (Larbi *et al.*, 2004; Chakraborty *et al.*, 2012a). This evidence suggests that iron deficiency reduces the photosynthetic rates to a greater extent mainly due to damage in the chlorophyll structure.

Effects on Photosystem II Efficiency and Chlorophyll Fluorescence

Change in chlorophyll fluorescence can be used as an alternative tool for identifying iron deficiency in fruit trees (Val *et al.*, 1995). It is preferred by the researchers because of its non-destructive, non-intrusive approach and also ensures stress conditions before the development of visual symptoms. The actual photosystem II efficiency at the steady-state photosynthesis is decreased by iron deficiency both in the early morning and at midday, due to closure of photosystem II reaction centers and decrease in the intrinsic photosystem II efficiency (Morales *et al.*, 2000a; Morales *et al.*, 1999). The Kautsky effect i.e., the shape of leaf chlorophyll fluorescence induction kinetics during a dark light transition is also markedly decreased by iron deficiency (Morales *et al.*, 2001). This kind of effect is found only in the case of extremely iron deficient leaves but in moderate and in severe iron deficient leaves a very little effect is seen (Abadia *et al.*, 1999). Iron deficiency decreases not only absorbance, but also increases the relative amount of light dissipated thermally by the photosystem II antenna (Morales *et al.*, 1999). Iron deficiency induces larger changes in the chlorophyll fluorescence lifetimes in an increased manner in sugar beet (Morales *et al.*, 1998a).

Pestana *et al.* (2005) found that the basal florescence (F_0) was not affected with the supply of iron in the tolerant Troyer citrange rootstock while in iron-susceptible rootstock Taiwanica orange rootstock, an intermediate behavior was noted. Basal fluorescence is generally not affected in the tolerant genotypes A reduction in the maximum quantum yield of PSII (F_v/F_m) and the ratio between variable to initial fluorescence F_v/F_0 has been reported in Cadaman rootstock of peach (susceptible to Iron deficiency), when Iron deficiency was induced by the inclusion of bicarbonate (supplied as 20 mM $NaHCO_3$) in the nutrient solution (Molassiotis *et al.*, 2006).

Effect on Enzymatic Activity

Iron deficiency has been reported to change the relative activities of many enzymes in plants. There are many enzymes whose activities are altered under iron scarcity. A brief note of these enzymes which can be used as a diagnostic tool of iron deficiency is presented here.

Ferric chelate reductase (FCR, EC 1.16.1.7)

Plants can be classified as iron-efficient or iron-inefficient based on their ability to induce root ferric chelate reductase (FCR) activity in response to iron deficiency. Increased root FCR activity is one of the adaptive mechanisms exhibited by plants to improve iron uptake under iron deficiency (Rombolà *et al.*, 2002). The reduction process of chelated Fe(III) to Fe(II) at the root surface, prior to iron uptake in dicotyledonous plant is carried out by a plasma membrane-bound ferric chelate reductase (FCR) that transports electrons from cytosolic NADH to apoplastic Fe(III) (Brüggemann *et al.* 1990; Curie and Briat, 2003). The second reduction step occurs in mesophyll cells, for which another FCR (Brüggemann *et al.* 1993) is required that is similar to the enzyme present at the root surface (Rombolà *et al.*, 2002). It has been assumed that alkalization of the apoplast may decrease the FCR activity in the plasma membranes of mesophyll cells. In sugar beet leaves, it was observed that beyond pH 6.0, the activity of FCR in mesophyll protoplasts is decreased markedly in both iron-sufficient and iron-deficient protoplasts (González-Vallejo *et al.*, 2000).

Leaf mesophyll FCR activity was found to be light dependent and increased when epidermis was removed (Larbi *et al.*, 2001). This is one of the most accepted analogies between the root FCR activity and leaf mesophyll FCR activity (Mengel, 1995). However whether, root FCR or shoot FCR plays an important role in iron-sensing mechanism is still controversial. In potato, iron-stress responses were regulated in root (Bienfait *et al.*, 1987). In contrast, based on whole plant and split-root experiments, it has been suggested that FCR activity is controlled by a putative shoot-born signal, transmitting shoot iron status to the roots (Grusak and Pezeshgi 1996; Schikora and Schmidt 2001). In *Arabidopsis*, it was found that reduced iron is transported by IRT1 (Iron Regulated Transporter) (Eide *et al.*, 1996; Vert *et al.*, 2001) while the reduction of iron occured with the help of H^+ from NADH which is catalyzed by FRO2 (Scimidt, 2003). Besides this, three transporter proteins designated as AtNramp1, 3, and 4 were also found to be induced by iron deficiency in root epidermal cells (Curie *et al.*, 2000; Thomine *et al.*, 2000). These genes were thought to be controlled both by the local induction from the root iron pool and through a systemic regulation involving a shoot-born signal (Vert *et al.*, 2003). In red clover roots, FCR activity was induced in the different phases of FCR induction by both roots and shoots (Zheng *et al.*, 2003). It has been proposed that stress signal-transduction for Iron-reduction in sugar beet is under control of ascorbate (Asc) and reduced glutathione (GSH) status of the roots (Zaharieva *et al.*, 2004). Therefore, the signal for induction of FCR activities originates in different parts and also varies among the plant species (Molassiotis *et al.*, 2005). Recent studies have shown that ferric chelate reductases encoded a multigene family (FRO_s). In *Arabidopsis*, it was suggested that FROs may function at organellar membranes. FRO2 reduces iron at the root surface and encoded Fe(III)-chelate reductase (Robinson *et al.*, 1999) so that it may be transported across the plasma membrane. But in the case of pea (*Pisum sativum*), FRO1encodes a Fe(III)-chelate reductase (Waters *et al.*, 2002). More recent studies in *Arabidopsis* suggest that in particular, FRO7 plays an essential role in iron delivery to chloroplasts, while two other FROs (FRO3 and FRO8) localized at the mitochondria and might therefore, contribute to mitochondrial iron homeostasis (Jeong and Connolly, 2009).

Iron-efficient dicotyledonous plants respond to iron deficiency by enhancing root FCR activity (Moog and Brüggemann, 1994). Resupply of iron to the iron deficient plants reduces the FCR activity. Reduction of FCR activity to the extent of about 19 and 88 per cent was observed after 24 and 96 h, respectively with the supplementation of iron in nutrient solution in sugar beet (López-Millán *et al.*, 2001). However, iron tolerant rootstock of peach Titan × Nemaguard and Montclar did not show any significantly higher iron stress induced FCR activity as compared to the less tolerant peach rootstock Nemaguard (Egilla *et al.*, 1994). Rombolà *et al.* (2002) found that iron deficiency increased the FCR activity in tolerant kiwi rootstock D1 than in susceptible Hayward kiwi during the progression of deficiency condition. Therefore, the study of root FCR activity in iron deficient plants offers a lot of research opportunities. Leaf mesophyll protoplast FCR activity has also been reported to decrease under iron deficiency condition in sugar beet (Elena *et al.*, 2000). FCR activity at the root tips can be effectively used as physiological index to screen iron chlorosis tolerant rootstocks of pear (Tagliavini *et al.*, 1995) and grapes (Dell'Orto *et al.*, 2000). Through the implementation of this technique, Ma *et al.* (2006) suggested that *P. xerophila* is tolerant to iron chlorosis of pear than *P. betulaefolia* seedlings, since, the former showed a two fold higher FCR activity at the root tip than the later. Conversely, de la Guardia and Alćantara (2002a) suggested that determining root FCR capacity is not the best method to screen out genotype tolerance to iron deficiency stress. Chouliaras *et al.* (2004) found an induced FCR activity in (-Fe)- treated roots of *Citrus sinensis* L. cultivars Valencia and New Hall grafted on *C. aurantium* and *Swingle citrumelo* rootstocks whereas a depressed FCR activity was noted in bicarbonate-treated roots. Rombolà *et al.* (2002) found iron deficiency tolerant kiwi genotype D1 exhibits a higher FCR activity than iron chlorosis susceptible genotype Hayward.

Catalase (CAT, EC 1.11.1.6)

Catalase is known to be one of the most potent catalysts. Till now, more than 300 catalyses are known and this group of enzymes is divided into different classes viz. monofunctional catalases (>225), bifunctional catalase-peroxidases (>50) and manganese-containing catalases (>25) (Chelikani *et al.*, 2004). Monofunctional, heme-containing catalases are widespread in the nature and extensively characterized. This catalases have been further subdivided into large (>75 kDa) and small (<60 kDa) on the basis of size of subunits (Klotz *et al.*, 1997). Four porphyrin heme (iron) groups in catalases allow reacting with the H_2O_2. Catalase catalyses conversion of hydrogen peroxide, a powerful and potentially harmful oxidizing agent, to water and molecular oxygen (Loew, 1900). Catalase also oxidizes potential cellular toxic compounds like phenol, formaldehyde, formic acid and alcohols with the help of H_2O_2. Some heme-containing catalases are bifunctional, acting as a catalase and also as a peroxidase. In these bifunctional catalase-peroxidases, a variety of organic substances can be used as a hydrogen donor. These bifunctional catalases are closely related to plant peroxidases (Chelikani *et al.*, 2004). Plant peroxidases exhibit a very low level of catalase activity, attributable to the presence of heme, which alone exhibits catalatic activity (Nicholls *et al.*, 2001).

Iron chlorosis leads to a depressed catalase activity in several plant species like *Averrhoa carambola* Linn., *Pithecobium dulce* Benth., *Tabernaemontana coronaria* Willd., *Rosa* sp. (*creeper*) *Lagerstroemia indica* Linn., *Agle marmelos* Correa., *Ixora arborea* Roxb., *Murraya paniculata* (Linn.) Jack., *Gardenia jasminoides* Ellis. and *Bougainvillaea spectabilis* Willd. However, the activity of this enzyme can be restored by external foliar application of iron solution to (Agarwala and Mehrotra, 1977). Based on their solution culture experiment in the absence of iron (-Fe) and presence of bicarbonate decreased catalase activities in iron deprived leaves was reported in *Citrus sinensis* L. cultivars Valencia and New Hall grafted on *C. aurantium* and *Swingle citrumelo* rootstocks (Chouliaras *et al.*, 2004).

Peroxidase (POD, EC 1.11.1.7)

Peroxidase has numerous biochemical and physiological roles in higher plants like plant growth, differentiation and development processes, including auxin catabolism, ethylene biosynthesis, plasma membrane redox systems, generation of H_2O_2, cell wall acidification, lignification and suberization, as well as response to pathogen infections. Peroxidase oxidizes a variety of hydrogen donors at the expense of peroxide or molecular oxygen. Here, iron acts as a pro-oxidant factor (Ranieri *et al.*, 2001) and this reaction is popularly known as Fenton reaction (Minotti and Aust, 1987). Recently, the role and activity of peroxidase in H_2O_2-mediated signaling processes under iron deficiency has become a prime area of research in this direction.

Peroxidase enzyme activity, a Fe-hemoprotien, was suggested as a reliable tool for diagnosing iron deficiency in citrus (Bar-Akiva, 1961) since the decrease in activity was found to be specific for iron deficiency (Bar-Akiva and Lavon, 1967) and application of iron sources increased the peroxidase activity in deficient bael (*Agle marmelos*) and low-chill peaches (Agarwala and Mehrotra, 1977; Chakraborty *et al.*, 2012a).

Superoxide dismutase (SOD, EC 1.15.1.1)

Four classes of SODs are known which are distinguished by the respective metal prosthetic groups: Cu/Zn, Fe, Mn and Ni. Iron and Mn-SODs constitute a structural family (Parker *et al.*, 1987) that is unequally distributed throughout the kingdoms of living organisms and is located in different cellular compartments (Grace, 1990). Superoxide dismutases

(SODs) scavenge the reactive oxygen species (ROS) and covert O_2^- to H_2O_2. These are key components of the cellular defense against O_2^- stress (Fridovich, 1995). Iron is present in the active site of superoxide dismutase (Elstner and Osswald, 1994) and plays a dual role in cell metabolism either as an antioxidant or a pro-oxidant factor. Further, the oxidative stress is negatively correlated with iron availability (Mohamed and Aly, 2004). Molassiotis *et al.* (2005) found an enhanced SOD activity in five different peach rootstocks under iron deficiency. Iron resupply mediated enhanced SOD activity was reported in iron deficient low-chill peach cultivars (Chakraborty *et al.*, 2012a).

3. MOLECULAR STRATEGY FOR IRON UPTAKE AND POSSIBLE INTERVENTIONS FOR EFFICIENT Fe-USE

Most of dicotyledonous plants, which utilize the Fe-efficient *Strategy I* for iron uptake by their root systems via an increase in activity of an ATPase-driven proton pump results in the release of protons into the soil which increases the solubility of ferric iron; membrane-bound ferric chelate reductase reduces ferric iron to the more soluble ferrous iron; ferrous iron is transported across the plasma membrane into the root cells by a specific Fe(II)-transporter protein (Hell and Stephan, 2003; Grotz and Guerinot, 2006)). Subsequently, ferrous iron is oxidized and transported as an Fe(III)-citrate complex for long-distance transport in the xylem from the roots to the shoots. For assimilation by the leaves and other tissues, ferric iron is reduced again by FROs (Hell and Stephan, 2003). Expression of the genes encoding the major strategy I Fe-acquisition proteins is up-regulated under conditions of Fe deficiency, making transcription a key regulatory factor for the performance of the whole system. The central transcriptional regulator of Fe uptake is the basic helix–loop–helix (bHLH) protein Fer-like Iron Deficiency Induced Transcription Factor (FIT). FIT was identified based on its homology to the previously identified tomato (*Solanum lycopersicum*) protein FER (Ling *et al.*, 2002; Bauer *et al.* 2004; Yuan *et al.*, 2005).

Being the main Fe importer and owing to its affinity for other divalent metals, Iron Regulated Transporter 1 (IRT1) is subject to tight regulation at the transcriptional and post-transcriptional levels (Potocki *et al.*, 2013). Under conditions of Fe limitation the activity of the FIT regulatory system leads to increased IRT1 gene expression. Interestingly, the presence of low numbers of IRT1 transcripts in fit mutants suggests that the basic IRT1 expression is also under the control of other transcriptional regulators (Brumbarova *et al.*, 2015). Analysis of the *pye* mutant in Arabidopsis revealed yet another transcriptional regulatory network, apart from the FIT network, defined by the bHLH transcription factor POPEYE (PYE). Combinations of transcriptome- and genome-scale DNA-binding studies have demonstrated that the PYE regulatory network has a role in assuring the redistribution of already imported Fe (Long *et al.*, 2010).

It has been shown recently that different genes in different tissues are involved in the Fe(III) reduction process. In the *Arabidopsis thaliana* genome, a FRO gene family with eight putative members (AtFRO1–AtFRO8) was identified (Wu *et al.*, 2005; Mukherjee *et al.*, 2006). The AtFRO2 gene encodes the inducible low-iron ferric chelate reductase responsible for ferric iron reduction at the root surface (Robinson *et al.*, 1999). AtFRO2 expression is regulated by iron nutrient level and overexpression of AtFRO2 confers tolerance to growth under low iron conditions compared to wild-type plants. In transgenic soybean plants, overexpression of AtFRO2 enhances ferric reductase activity in roots and leaves, and reduces iron-deficiency chlorosis (Vasconcelos *et al.* 2006). FRO5 is primarily expressed in roots and located on the plasma membrane, FRO3 is expressed at high levels in roots and shoots, FRO8 is mainly expressed in shoots; both FRO3 and FRO8 are located in the mitochondria (Jeong and Connolly, 2009). FRO7 is localized to chloroplasts, and plays a role in iron acquisition in chloroplasts and is required for efficient photosynthesis in young seedlings under iron-depleted conditions (Jeong *et al.*, 2008). FRO6 is specifically expressed in green tissues (Mukherjee *et al.*, 2006) and has been shown to reside on the plasmamembrane (Jeong *et al.*, 2008). Light-responsive elements within the FRO6 promoter region mediate gene activation on exposure to light and protoplasts prepared from fro6 plants show reduced ferric chelate reductase activity (Jeong and Connolly, 2009). FRO6 may therefore be the best candidate for a leaf plasma membrane ferric reductase. The fro6 plants however, do not display any obvious phenotypes, indicating that FRO6 may function redundantly in FRO family, or that ferric chelate reductase activity is not necessary for iron uptake by leaf cells (Jeong and Connolly, 2009).

MANAGEMENT OF IRON CHLOROSIS IN HORTICULTURAL CROPS

Iron is the fourth most abundant elements in the earth's crust (Beard and Johnson, 2004). But low bioavailability of iron in the soil surface leads to the disorder due to Fe deficiency in the perennial crop plants. Sometimes the leaves may possess comparatively fair amount of inactive Fe ions which may also be responsible for the iron deficiency (Römheld, 2000). These types of phenomena build this nutritional disorder more complex than the deficiency of other essential elements in plants. Hence, iron deficiency is also termed as "Iron paradox". From the last two decades extensive research works have been undergone to find out the most effective, economical and eco-friendly way to correct iron deficiency in horticultural crops. It is well- established fact that supplementation of iron to the plants improves the deficiency status of this nutritional disorder. A variety of iron compounds have been tested and so many methodologies have been designed to apply iron sources in horticultural crops to counteract this problem. The different useful strategies to manage iron chlorosis in horticultural crops are discussed below.

Soil Application of Iron Sources

Soil application of iron sources to alleviate iron chlorosis is a common practice in fruit orchards. The agronomic performance of iron deficiency improvement status with soil application of iron sources largely depends upon the soil types and the chemical properties of the used iron sources. Application of Fe(II)Sulphate in calcareous or alkaline soil is uneconomical. Since HCO_3^- acts upon the Fe^{2+} ion and make it unavailable for the plants (Başar, 2000). Only synthetic iron chelates are suitable to correct iron chlorosis through soil application in these types of soil. The chelated iron compounds are too expensive to be afforded by the growers. However, the inexpensive chelates like FeEDTA or FeDTPA have resulted meager improvement in iron deficiency status in perennial horticultural trees (Cantera *et al.*, 2002; Lucena *et al.*, 1987). Studies are still continued on chemical characterization, soil reaction and agronomic performance of these chelated iron sources. The chealted molecules consisted of polyamine–carboxylic acids with phenolic groups like ethylendiamine di (*o*-hydroxyphenylacetic) acid (EDDHA) and ethylendiamine di(2-hydroxy-4-methylphenylacetic) acid (EDDHMA), are found to be the most effective and commonly used in correction of iron deficiency. The ortho-ortho isomer of these chelated compounds is the most active form than other isomers (Hernández-Apaolaza *et al.*, 1997). The effectiveness of EDDHMA/Fe^{3+} was reported to be almost similar as EDDHA/Fe^{3+} when tested in peaches and grapes (Reed *et al.*, 1888). Now a days, ethylendiamine di(2-hydroxy-5-sulfophenylacetic) acid derivates (EDDHSA/Fe^{3+}) are being used to correct iron deficiency and found most promising than EDDHA/Fe^{3+} and EDDHMA/Fe3+ in peach (cv. Sudanell) and pear (cv. Conference) (Álvarez-Fernández *et al.*, 2005).

Foliar Application of Iron Sources

Foliar fertilization of iron sources is another approach to provide iron in the plants. This method is more efficient than soil application. Since the response is faster than soil application. Secondly, the required quantity of a particular nutrient which has to be supplied through foliar sprays will be less than the soil application. Therefore, it is an economical practice. However, not all forms of iron are suitable for foliar fertilization. It is quite obvious that the sources/salts which contain ferrous (Fe^{2+}) form of iron are more suitable for foliar fertilization while, ferric (Fe^{3+}) forms are less effective (Chakraborty *et al.*, 2014b). Usually, the higher valence of a cation lowers its penetration capacity into the cell (Mengel, 2002). As described earlier, a reduction step is required during leaf penetration of ferric form for conversion into ferrous by the leaf FCR (EC 1.16.1.7).

Additionally, under iron deficiency the reducing capacity of leaf FCR is also decreases (González-Vallejo *et al.*, 2000). Further, the molecular radius of chelated iron sources is much higher than other divalent iron sources and therefore, entry through ectodesmata pores with a diameter of less than 1 nm is difficult (Schőnherr, 2001). The molecular radius of chelated iron sources are reported to be more than 0.7 nm (Furia, 1972).

The concentration of iron sources in spray solution also plays an important role in foliar fertilization. Because, higher concentration of salts may burn the leaf surfaces or defoliation and too lower concentration may increase the frequency of spray. For example, one spray of 0.5% Fe(II) Sulphate is found to be effective to manage iron chlorosis in low-chill peaches like Saharnpur Prabhat Shan-e-Punjab and Pratap at young nursery stage (Chakraborty *et al.*, 2014b). Use of adjuvants and surfactants are becoming popular in foliar fertilization. Adjuvants are added with spray solution to accelerate cuticular and foliar penetration (Perkins *et al.*, 2005) while surfactants improve the retention capacity of spray droplet. Some surfactants also improve the plasma membrane permeability either by solubilizing cuticular waxes or may hydrate the cuticle (Tamura *et al.*, 2001; Wang and Liu, 2007). Researchers also suggested that slightly acidic nature of spray solution enhances the iron diffusion capacity through the cuticular surfaces. Pear leaves exhibited more pronounced recovery status when Fe(II)-Sulphate applied as sprays in addition with weak acids like citric acid than only Fe(II)-Sulphate (Álvarez-Fernández *et al.*, 2004). Therefore, so many independent factors are involved in foliar iron application to make this process too complex to understand. Hence, standardized technologies are required to establish for foliar fertilization of iron sources in each horticultural crops mentioning the concentration, sources, adjuvants, surfactants and number of sprays required to overcome the economic threshold level against this nutritional disorder.

Trunk Injection of Iron Sources

It is another possible remediation strategy to curb iron chlorosis in horticultural crops. Among the various Fe-containing fertilizers, ferrous sulfate gave the best results at economic rate. Phytotoxicity is the main problem associated with this technique, results in foliar burning at higher Fe concentration. Fernández-Escobar *et al.* (1993) recommended injecting ferrous sulfate at concentrations ≤ 1% to reduce the risk of foliar burning while using the method for olive and peach. This treatment is most effective if applied in the early spring during bud break. The trunk injection system is of two types- 1. high-pressure (Reil and Beutel, 1976) and 2. low-pressure (Navarro *et al.*, 1992). The commercial use of high-pressure trunk-injection system is limited due to requirement of special injecting equipment, high labour cost, and the high pressure may injure some trees. Of late the low-pressure trunk-injection method came into existence and became popular as it is easy to use, economical, and does not require special equipment. This injecting method has successfully been applied to alleviate iron chlorosis in many established deciduous fruit trees (Barney *et al.,* 1985). Effects can be expected to last for two or more years, after that retreatment may be necessary. Iron compounds in dry or liquid form can be placed directly into holes drilled into the lower trunk. Systems also are available that use plastic tubing and tees, capsules of various types, or a hypodermic-like tool to place iron materials into the tree. Though this technique can be quite effective, they injure the trunk; hence best possible care should be taken to minimize the trunk injury by using modified methods. Using the method requires some precautions *viz.* injecting materials on hot, dry, windy days since leaves may blacken or burn symptoms should be avoided and the tree is well-watered for several days before and several weeks after injection treatments.

Use of Resistant Genotypes and Rootstocks

Apart from the use of iron sources, another way to prevent iron chlorosis is the use of tolerant genotypes or rootstocks. Plant breeding methods have made it possible to develop some outstanding varieties as well as rootstocks which are capable to resist this nutritional disorder. Earlier works were mainly focused on screening of resistant rootstocks since this part of the plant actively involved in iron acquisition from soil. However, genotypes of scion variety may also influence on the uptake of iron from soil by the iron tolerant rootstock (Ksouri *et al.*, 2005). Therefore, it is necessary to categorize the different varieties of a species against iron deficiency. Regarding these aspects, some preliminary works have been reported in different horticultural crops (Tables 1 & 2). Further attention is required to develop more and more resistant sources either through conventional breeding approaches or application of biotechnology (somaclonal variants) in horticultural crops (Piagnani *et al.*, 2003; Piagnani and Zocchi, 1997).

Table 1: List of iron containing fertilizers recommended for management of iron chlorosis

Type of source	Empirical formula	Per cent iron	Suitability on method of application		
			Soil	Foliar	Trunk injection
A. Inorganic iron salts					
1. Ferrous sulphate	$FeSO_4.7H_2O$	20.00	×	√	√
2. Ferrous ammonium sulphate	$(NH_4)_2Fe(SO_4)_2\ 6H_2O$	14.00	×	√	√
3. Ferrous ammonium phosphate	$Fe(NH_4)PO_4$	30.00	×	√	√
B. Chelated iron sources					
1. Fe-Citrate	$C_6H_5FeO_7$	22.70	√	×	√

2. Hampene Iron (NaFe(III)EDTA)	$C_{10}H_{12}FeN_2NaO_8 \cdot 3H_2O$	9.00	√	×	√
3. Seqestrenee 330 (NaFe(III)DTPA)	$C_{14}H_{19}FeN_3NaO_{10}$	10.00	√	×	√
4. Ferriplex 138/ Seqestrenee 138 (NaFe(III)EDDHA)	$C_{18}H_{16}O_6N_2NaFe$	6.00	√	×	Not Tested
5.Ferriplus (NaFe EDDHMA)	$C_{20}H_{20}O_6N_2FeNa_3$	13.50	√	×	Not Tested
6. Ferrilene (NaFe EDDHSA)	$C_{18}H_{16}O_{12}N_2S_2FeNa_3$	8.70	√	×	Not Tested
7. Versonol Ag Fe (NaFeHEDTA)	$C_{10}H_{15}N_2O_7FeNa_3$	5.00	√	×	Not Tested

Table 2: List of some genotypes in horticultural crops tested against iron deficiency

Crop	Genotype	Reaction to iron deficiency	References
Grape (*Vitis vinifera*)	Cabernet Sauvigon, Khamri and Mahdaoui when grafted on 140Ru	Moderately tolerant	Piagnani *et al.* (2003) Ksouri *et al.* (2005)
	Saouadi, Arich Dressé, Blanc 3 when grafted on S.O. 4	Moderately tolerant	Ksouri *et al.* (2005)
	Balta4, Beldi, and Cardinal	Susceptible	Ksouri *et al.* (2005)
Low-chill peach	Saharnpur Prabhat	Tolerant	Chakraborty *et al.* (2014a); Chakraborty *et al.* (2012a)
	Shan-e-Punjab	Moderately tolerant	
	Pratap	Susceptible	
Olive	Nevadillo Negro Pajarero and Manzanilla de Sevilla	Tolerant	Alcántara *et al.* (2003)
	Cornezuelo de Jaén	Moderately tolerant	
	Leccino, Arbequina, Lechín de Sevilla and Galega	Susceptible	
Strawberry	Tango, Spadeka	Tolerant	Kafkas *et al.* (2007)
	Osmanli	Moderately tolerant	
	Camarosa, Dorit	Susceptible	

Table 3: Examples of some iron deficiency tolerant rootstocks in horticultural crops

Crop	Sources of resistance	References
Apple	1. *Malus xiaojinensis*	Han *et al.* (1994) Han *et al.* (1998)
Apricot	1. Myrabolan B (*P.cerasifera*) 2. Mariana: G.F. 8-1 (*P. cerasifera* × *P. munsoniana*) 3. Damas 1869 (*Prunus domestica* × *P. spinosa*)	Loreti, (1994)
Cheery	1. Mahaleb SL 64 (*Prunus mahaleb*) 2. Adara (*Prunus cerasifera*)	Moreno *et al.* (1995b) Moreno *et al.* (1996)
Citrus	1. Rangpur lime (*Citrus limonia*) 2. Rough lemon (*C. jambhiri*) 3. Alemow (*C. macrophylla*) 4. Citrange "Troyer"(*C. sinensis* × *Poncirus trifoliata*)	Manthey *et al.* (1994) Pestana *et al.* (2005)
Grape	1. *V. vinifera* 2. *V. rupestris* 3. *Vitis berlandieri* 4. 420 A (*V. berlandieri* × *V. riparia*) 5. 140 Ruggeri (*V. berlandieri* × *V. rupestris*) 6. Cabernet Sauvignon (*V. vinifera*)	Tagliavini and Rombolà (2001); Brancadoro *et al.* (1995) Chauvet and Reynier (1979) Jiménez *et al.* (2007)
Oilve	Hojiblanca	Alcántara *et al.* (2003)
Peach	1. Myrobalan (*P. cerasifera*): Mr.S. 2/5, Mr.S 1/3, Mr.S 1/6, Mr.S 2/8 2. St. Julien: Adesoto 101 (*P. insititia*) 3. Damas 1869 (*Prunus domestica* × *P. spinosa*) 4. I.S. 5/18, I.S. 5/29, I.S. 5/22 (Sirio) 5. GF 677 6. Titan × Nemaguard (Almond × Peach) 7. Adesoto, Felinem, Krymsk 86 and PAC 9921-07	Loreti and Massai, (1998) Byrne *et al.* (1990) Moreno *et al.* (1995a) Cinelli *et al.* (1995) Syrgiannidis (1985) Molassiotis *et al.* (2006); Byrne (1988) Jiménez *et al.* (2008)
Pear	1. Fox 16 (B 21), Fox 11 (A 28), OH×F 333, OH×F 51 OH×F 87, OH×F 69 (*Pyrus communis*) 2. Other wild spices (*P. xerophila, P. syriaca P. amygdaliformis, P. elaegrifolia, P. cordata*,. *P. gharbiana, P. caucasica, P. longipes, P. regeli* and *P. salicifolia*)	Bassi *et al.* (1994) Westwood and Lombard 1983; Procopiou and Wallace (2000) Lombard and Westwood (1987) Ma *et al.* (2005)
Plum	1. Myrobalan (*P. cerasifera*): Mr.S. 2/5, Ademir (Myrabolan 599 AD) 2. Mariana: G.F. 8-1 (*P. cerasifera* × *P. munsoniana*)	Moreno *et al.* (1995b) Loreti (1994)

Intercropping or Sod Culture

Intercropping or sod culture (covering of orchard floor with grass species) with graminaceous species could be suggested as an integrated management approach to correct iron chlorosis in fruit orchards. It clearly indicated that phytosiderophores released by graminaceous grass species help in iron uptake by the dicot plant species (Kamal *et al.*, 2000; Cesco *et al.*, 2006). Monocot graminaceous grass species are more efficient in iron acquisition from soil than dicot plant species. Phytosiderophores (PS) released by graminaceous plant species are helpful in mobilizing iron slowly from the soil iron sources. The dicot plants are able to reduce the PS-Fe(III) complexes and iron is taken up by the roots. Intercropping of guava (*Psidium guajava* L.) with either sorghum or corn was found suitable for management of iron chlorosis in guava plants (Kamal *et al.*, 2000). The iron chlorosis symptoms of citrumelo "Swingle" (*Poncirus trifoliate* × *Citrus paradisi*) can be alleviated with graminaceous plant species like *Poa pratensis* L. and *Festuca rubra* L (Cesco *et al.*, 2006).

Adjustment of Rhizosphere Soil pH

The bio-availability iron is reported to be more at lower pH ranges of the soil. Therefore, acidification of saline or calcareous soil may improve the bio-available iron status at the tree rhizosphere. There are many methods of lowering the pH of soil. Acidified peat or acidifiers may be placed in small holes near the feeder root zone of the trees otherwise; a layer of soil may be excavated and replaced with acidified peat supplemented with iron or ammonium sulphate (Sommer, 1993). The later process is quite practical during establishment of a new orchard at calcareous or saline soil site. Localized placement of strong acids along with drip water directly at the root zone of the trees may be suggested under chronic deficiency symptom. The most commonly used acidifiers are gypsum ($CaSO_4.2H_2O$), elemental sulphur (S) and sulphuric acid (H_2SO_4). The pH of the soil may be lowered with application of gypsum as a result of reduction in Na^+ ions. Elemental sulphur forms sulphuric acid in soil through microbial actions. This sulphuric acid reacted with lime ($CaCO_3$) and produces gypsum and lowers the soil pH. In addition, ammonium-containing fertilizers, calcium chloride and phosphogypsum can also be used to lower the rhizosphere soil pH (Gharaibeh *et al.*, 2014). However, application of phosphogypsum may raise the environmental issues. Because, this material contains relatively high levels of natural uranium-series radionuclides and other impurities such as fluoride, sulfate and trace elements (Rutherford *et al.*, 1994; Burnetta and Elzermanb 2001). However, this operation seems totally impractical in perennial fruit orchards (Wallace, 1991). Since, long lasting decrease in pH of the calcareous or saline soil is difficult to achieve.

Conclusions and Future Perspective

Presently, fighting with the negative impacts of iron deficiency in horticultural plants has been a prime concern of research throughout the world. Different methods like soil and foliar iron fertilization are applied to correct iron chlorosis in horticultural plants. However, a deeper insight is required for application of different iron sources combined with surfactants and adjuvants to rectify iron chlorosis. None the less, the influence of environmental, physiological and morphological factors of plants and also interaction of these processes at both cellular and molecular level need to be studied in details to give a more methodical approach for management of iron chlorosis. Apart from chemical management iron chlorosis, recent studies suggested adoption of transgenic approach either by overexpressing particular iron uptake related genes or by altering the whole regulatory network could be a more powerful strategy in order to combat the problem. Thus rectification of iron chlorosis needs multidisciplinary approach for a complete and stable solution in future.

References

Abadía, J 1992. Leaf responses to Fe deficiency: A review. *J. Plant Nutr.*, 15: 1699-1713.

Abadía, J; Morales, F and Abadía, A 1999. Photosystem II efficiency in low chlorophyll, iron-deficient leaves. *Plant Soil*, 215:183-192.

Agarwal, SC, Sharma, CP, and Farooq, S (1965) Effect of iron supply on growth, chlorophyll, tissue iron and activity of certain enzymes in maize and radish. *Plant Physiol.,* 40: 493–499.

Agarwala, SC and Mehrotra, NK 1977. Catalase and peroxidase in the leaves of iron deficient plants. *Proc. Ind. Acad. Sci.*, 86b(1):55-60.

Alava, AK and Obreza, TA 1997. Correction of iron deficiency in orange and grape fruit trees in high pH soils. *Proc. Flo. Sta. Hort. Soc.*, 110:32-36.

Alcántara, E; Cordeiro, AM and Barranco, D 2003. Selection of olive varieties for tolerance to iron chlorosis. *J. Plant Physiol.*, 160:1467-1472.

Álvarez-Fernández, A; García-Laviña, P; Fidalgo, C; Abadía J and Abadía, A 2004. Foliar fertilization to control iron chlorosis in pear (*Pyrus communis* L.) trees. *Plant Soil*, 263:5-15.

Álvarez-Fernández, A; García-Marco, S and Lucena, JJ 2005. Evaluation of synthetic iron(III)-chelates (EDDHA/Fe3+, EDDHMA/Fe3+ and the novel EDDHSA/Fe3+) to correct iron chlorosis. *Europ. J. Agron.*, 22:119-130.

Álvarez-Fernández, A; Paniagua, P; Abadía, A and Abadía, J 2003. Effects of Fe deficiency chlorosis on yield and fruit quality in peach (*Prunus persica* L. Batsch). *J. Agri. Food Chem.*, 51:5738-5744.

Arora, JS and Singh, JR 1970. Some effect of iron spray on growth, yield and quality of guava fruit (*Psidium guajava* L.). *J. Jap. Hort. Sci.*, 39:41-45.

Bar-Akiva, A 1961. Biochemical indications as a means of distinguishing between iron and manganese deficiencies in citrus plant. *Nature*, 190: 647-648.

Bar-Akiva, A and Lavon, R 1967. Visible symptoms and some metabolic patterns in micronutrient deficient Eureka lemon leaves. *Israel J. Agri. Res.*, 17: 7-16.

Barney, DL; Walser, RH; Davis, TD and Williams, CF 1985. Trunk injection of iron compounds as a treatment for overcoming iron chlorosis in apple tree. *Hortsci.,* 20:236-238.

Başar, H 2000. Factors affecting iron chlorosis observed in peach trees in the Bursa region. *Turk. J. Agri. For.*, 24: 237-245.

Bassi, D; Tagliavini, M and Marangoni, B 1994. Selection of clonal rootstocks of *Pyrus communis* (L.). *Acta Hortic.*, 367:364–371.

Batty, LC and Younger, PL 2003. Effects of external iron concentration upon seedling growth and uptake of Fe and phosphate by the common reed *Phragmites australis* (Cav.) Trin ex. Steudel. *Ann. Bot.* (Lond.), 92: 801–806.

Bauer, P. *et al.* 2004. Analysis of sequence, map position, and gene expression reveals conserved essential genes for iron uptake in Arabidopsis and tomato. *Plant Physiol.,* 136: 4169–4183.

Bavaresco, L; Fregoni, M and Perino, A 1994. Physiological aspects of lime-induced chlorosis in some *Vitis* species. I. Pot trial on calcareous soil. *Vitis*, 33:123-126.

Beard, BL and Johnson, CM 2004. Fe isotope variations in the modern and ancient earth and other planetary bodies. *Rev. Mineral. Geochem.*, 55:319-357.

Bennett, JP 1945. Iron in leaves. *Soil Sci.*, 60:91-105.

Bienfait, H.F., R.J. Bino, V.D.A.M. Bliek, J.F. Duivenvoorden, and J.M. Fontain, 1983. Characterization of ferric reducing activity in roots of Fe deficient *Phaseolus vulgaris. Physiol. Plant.*, 59: 196-202.

Bienfait, HF 1988. Mechanisms in Fe-efficiency reactions of higher plants. *J. Plant Nutr.*, 11: 605-629.

Bienfait, HF and Mark, VD 1983. Phytoferritin and its role in iron metabolism, In: Robb, DA and Pierpoint WS (eds.), Metals and Micronutrients: Uptake and Utilization by Plants. Academic Press, New York, USA. pp.111-123.

Boxma, R 1982. Bicarbonate as the most important soil factor in lime-induced chlorosis in the Netherlands. *Plant Soil*, 37:233-243.

Brancadoro, L; Rabotti, G; Scienza, A and Zocchi, G 1995. Mechanisms of Fe-efficiency in roots of *Vitis* spp. In response to iron deficiency. *Plant Soil*, 171: 229-234.

Brown, JC and Ambler, JE 1974. Iron stress response in tomato (*Lycopersicon esculentum*). I. sites of Fe reduction, absorption and transport. *Plant Physiol.*, 31: 221-224.

Brown, JC and Jolley, VD 1989. Plant metabolic responses to iron deficiency stress. *BioScience.*, 39:546-551.

Brüggemann, W; Maas-Kantel, K and Moog, PR 1993. Iron uptake by leaf mesophyll cells: the role of the plasma membrane-bound ferric-chelate reductase. *Planta*, 190:151-155.

Brüggemann, W; Moog, PR; Nakagawa, H; Janiesch, P and Kuiper, PJC 1990. Plasma membrane-bound Fe^{3+}-EDTA reductase and iron deficiency in tomato

(*Lycopersicon esculentum*). Is there a Turbo reductase? *Physiol. Plant.*, 79:339-346.

Brumbarova, T., Bauer, P., and Ivanov, R. 2015. Molecular mechanisms governing *Arabidopsis* iron uptake. *Trend. Plant Sci.,* 20: 124–133.

Burnetta, WC and Elzermanb, AW 2001. Nuclide migration and the environmental radiochemistry of Florida phosphogypsum. *J. Environ. Radioact.*, 54:27-51.

Byrne, DH 1988. Comparative growth of two peach seedling rootstocks under alkaline soil conditions. *J. Plant Nutri.*,11:1663-1669.

Byrne, DH; Bacon, TA and Egilla, JNA 1990. Patrones de frutales de hueso tolerantes a suelos calizos. In: Socias I Company R. (Ed.), Estado actual de los patrones frutales. Inf. *Tecn. Econ. Agrar. Extra* 9:117-133.

Cantera, RG; Zamarreño, AM and García-Mina, JM 2002. Characterization of commercial iron chelates and their behavior in an alkaline and calcareous soil. *J. Agric. Food Chem.*, 50:7609-7615.

Cesco, S; Rombolà, AD; Tagliavini, M; Varanini, Z and Pinton, R 2006. Phytosiderophores released by graminaceous species promote ^{59}Fe-uptake in citrus. *Plant Soil*, 287: 223-233.

Chakraborty, B; Singh, PN; Kumar, S and Srivastava, PC 2014a. Uptake and distribution of iron from different iron sources applied as foliar sprays to chlorotic leaves of low-chill peach cultivars. *Agric Res.*, 3:293-301.

Chakraborty, B; Singh, PN; Shukla, A and Mishra DS 2012a. Physiological and biochemical adjustment of iron chlorosis affected low-chill peach cultivars supplied with different iron sources. *Physiol. Mol. Biol. Plants*, 18:141–148.

Chakraborty, B; Singh, PN; Singh, AK and Srivastava, PC 2014b. Evaluation of different iron sources for iron chlorosis recovery in low-chill peach cultivars. *J. Plant Nutri.*, 37: 224-231.

Chakraborty, B; Singh, PN; Singh, AK and Yadav, V 2012b. Assessment of the recovery of physiological status in iron deficient low-chill peach plants with SPAD meter. *Res. J. Agric. Sci.*, 3:156-158.

Chaney, RL; Chen, Y; Green, CE; Holden, MJ; Bell, PF; Luster, DG and Angle, JS 1992. Root hairs on chlorotic tomatoes are an effect of chlorosis rather than part of the adaptive Fe-stress response. *J. Plant Nutri.*, 15:1857-1875.

Chauvet, M and Reynier, A 1979. Manuel de Viticulture. Bailliere, Paris, France.

Chelikani, P; Fitab, I and Loewen, PC 2004. Diversity of structures and properties among catalases. *Cell Mol. Life Sci.*, 61:192-208.

Chouliaras, V; Therios, I; Molassiotis, A and Diamantidis, G 2004. Iron chlorosis in grafted sweet orange (*Citrus sinensis* L.) plants: physiological and biochemical responses. *Biol. Plant.*, 48:141-144.

Cinelli, F; Viti R; Byrne, DH and Reed, DW 1995. Physiological characterization of two peach seedlings rootstocks in bicarbonate nutrient solution. Root iron reduction and iron uptake. In: Abadia, J. (Ed.), Iron Nutrition in Soils and Plants. Kluwer Academic Publishers, Dordrecht, The Netherlands, pp. 323–328.

Curie, C and Briat, JF 2003. Iron transport and signaling in plants. *Ann. Rev. Plant Biol.*, 54: 183-206.

Curie, C; Alonso, JM; LeJean, M; Echker, JR and Briat, JF 2000. Involvement of NRAMP1 from *Arabidopsis thaliana* in iron transport. *Biochem. J.*, 347:749-755.

de la Guardia, MD and Alćantara, E 2002a. A comparison of ferric-chelate reductase and chlorophyll and growth ratios as indices of selection of quince, pear and olive genotypes under iron deficiency stress. *Plant Soil.*, 241: 49-56.

de la Guardia, MD and Alćantara, E 2002b. Bicarbonate and low iron level increase root to total plant weight ratio in olive and peach rootstock. *J. Plant Nutri.*, 25: 1021-1032.

Dell'Orto, M; Brancadoro, L; Scienza, A and Zocchi, G 2000. Use of biochemical parameters to select grapevine genotypes resistant to iron-chlorosis. *J. Plant Nutri.*, 23:1767-1775.

Donnini, S; Castagna, A; Guidi, L; Zocchi, G and Ranieri, A 2003. Leaf responses to reduced iron availability in two tomato genotypes: T3238FER (iron efficient) and T3238fer (iron inefficient). *J. Plant Nutri.*, 26: 2137-2148.

Egilla, JN; Byrne, DH and Reed, DW 1994. Iron stress response of three peach rootstock cultivars: ferric iron reduction capacity. *J. Plant Nutri.*, 17: 2079-2103.

Eide, D; Broderius, M; Fett, J and Guerinot, ML 1996. A novel iron-regulated metal transporter from plants identified by functional expression in yeast. *Proc. Nat. Acad. Sci.*, USA, 93: 5624-5628.

Elena, B; Fermín, GV; Morales, LC; Abadía, A and Abadía, J 2000. Iron deficiency decreases the Fe(III)-chelate reducing activity of leaf protoplasts. *Plant Physiol.*, 122: 337-344.

Elstner, F and Osswald, W 1994. Mechanism of oxygen activation during plant stress. *Proc. Res. Sci. Edinb.*, 102B: 31-154.

Fernández, V; Eichert, T; Río, VD; López-Casado, G; Heredia-Guerrero, JA; Abadía, A; Heredia, A and Abadía, J 2008. Leaf structural changes associated with iron deficiency chlorosis in field-grown pear and peach: physiological implications. *Plant Soil,* 311:161-172.

Fernández-Escobar, R; Barranco, D and Benlloch, M 1993. Overcoming iron chlorosis in olive and peach trees using a low-pressure trunk injection method. *HortScience.*, 28:192-194.

Fridovich, I 1995. Superoxide radical and superoxide dismutases. *Ann. Rev. Biochem.*, 64:97-112.

Furia, TE 1972. Sequestrans in foods. CRC handbook of food additives. CRC Press Inc., Boca Raton

Gardner, WK; Parberry GD and Barber, DA 1982. The acquisition of phosphorus by *Lupinus albus* L. II. The effect of varying phosphorus supply and soil type on some characteristics of the soil/root surface. *Plant Soil,* 68: 33-41.

Gharaibeh, MA; Rusan, MJ; Eltaif, NI and Shunnar, OF 2014. Reclamation of highly calcareous saline-sodic soil using low quality water and phosphogypsum. *Appl. Water Sci.*, 4:223-230.

González-Vallejo, EB; Morales, F; Cistué, I; Abadía, A and Abadía, J 2000. Iron deficiency decreases the Fe(III)-chelate reducing activity of leaf protoplasts. *Plant Physiol.*, 122:337-344.

Gout, E; Bligny, R; Pascal, N and Douce, R 1993. ^{13}C Nuclear magnetic resonance studies of malate and citrate synthesis and compartimentation in higher plant cells. *J. Biol. Chem.*, 268: 3986-3992.

Grace, SC 1990. Phylogenetic distribution of superoxide dismutase supports an endosymbiotic origin for chloroplasts and mitochondria. *Life Sci.*, 47: 1875-1886.

Gris, A 1857. Recherches microscopiques sur la chlorophylle. *Annels Sci. Nat. Bot. Ser.,* 4,7: 179-219.

Grotz, N., and Guerinot, M.L. 2006. Molecular aspects of Cu, Fe and Zn homeostasis in plants. *Biochim. Biophys. Acta,* 1763: 595–608.

Gruber, B and Kosegarten, H 2002. Depressed growth of non-chlorotic vine grown in calcareous soil is an iron deficiency symptom prior to leaf chlorosis. *J. Plant Nutri. Soil Sci.*, 65:111-117.

Grusak, MA and Pezeshgi, S 1996. Shoot-to root signal transmission regulates root Fe(III) reductase activity in dgl mutant of pea. *Plant Physiol.*, 110: 329 -334.

Han, ZH; Shen, T; Korcak, RF and Baligar, VC 1994. Screening for iron-efficienct species in the genus *Malus. J. Plant Nutri.*, 17: 579-592.

Han, ZH; Shen, T; Korcak, RF and Baligar, VC 1998. Iron Absorption by iron-efficient and -inefficient species of apples. *J. Plant Nutri.*, 21:181-190.

Hell, R., Stephan, U.W. 2003. Iron uptake, trafficking and homeostasis in plants. *Planta,* 216: 541–551

Hernández-Apaolaza, L; Barak, P and Lucena, J. 1997. Chromatographic determination of commercial Fe(III) chelates of ethylenediaminetetraacetic acid, ethylenediamine(*o*-hydroxyphenylacetic) acid and ethylenediaminedi(*o*-hydroxy-*p*-methilphenylacetic) acid. *J. Chromatogr., A* 789, 453-460.

Igartua, E; Grasa, R; Sanz, M; Abadía A and Abadía, J 2000. Prognosis of iron chlorosis from the mineral composition of flowers in peach. *J. Hort. Sci. Biotech.*, 75: 111-118.

Iglesias, DJ; Cercós, M; Colmenero-Flores, JM; Naranjo, MA; Ríos, G; Carrera, E; Ruiz-Rivero, O; Lliso, I; Morillon, R; Tadeo, FR and Talon, M 2007. Physiology of citrus fruiting: Review. *Braz. J. Plant Physiol.*, 19: 333-362.

Jacobson, L and Oertli, JJ 1956. The relation between iron and chlorophyll contents in chlorotic sunflower leaves. *Plant Physiol.*, 31:199-204.

Jeong, J and Connolly, EL 2009. Iron uptake mechanisms in plants: Functions of the FRO family of ferric reductases. *Plant Sci.*, 176: 709-714.

Jeong, J., Cohu, C., Kerkeb, L., Pilon, M., Connolly, E.L., and Guerinot, M.L. 2008. Chloroplast Fe(III) chelate reductase activity is essential for seedling viability under iron limiting conditions. *Proc. Natl. Acad. Sci. USA,* 105: 10619–10624.

Jiménez, S; Gogorcena, Y; Hévin, C; Rombolà, AD and Ollat, N 2007. Nitrogen nutrition influences some biochemical responses to iron deficiency in tolerant and sensitive genotypes of *Vitis.*, *Plant Soil*, 290: 343-355.

Jiménez, S; Pinochet, J; Abadía, A; Moreno, MÁ and Gogorcena, Y 2008. Tolerance response to iron chlorosis of *Prunus* selections as rootstocks. *HortScience*, 43: 304-309.

Kafkas, E; Silberbush, M and Paydas, S 2007. Physiological characterization of strawberry cultivars with differential susceptibility to iron deficiency. *World J. Agri. Sci.*, 3:196-203.

Kamal K, Hagagg, L and Awad, F 2000. Improved Fe and Zn acquisition by guava seedlings grown in calcareous soils intercropped with graminaceous species. *J. Plant Nutri.,* 23:2071-2080.

Klotz, MG, Klassen, GR and Loewen, PC 1997. Phylogenetic relationships among prokaryotic and eukaryotic catalases. *Mol. Biol. Evol.*, 14:951-958.

Kosegarten, H; Schwed, U; Wilson, GH and Mengel, K 1998. Comparative investigation on the susceptibility of faba bean (*Vicia faba* L.) and sunflower (*Helianthus annuus* L.) to Fe-chlorosis. *J. Plant Nutri.*, 21:1511-1528.

Köseoğlu, AT 1995. Effects of iron chlorosis on mineral composition of peach leaves *J. Plant Nutr.*, 18:765-776.

Ksouri, R; Gharsalli, M and Lachaal, M 2005. Physiological responses of Tunisian grapevine varieties to bicarbonate-induced iron deficiency. *J. Plant Physiol.*, 162:335-341.

Larbi, A; Abadía, A; Abadía, J and Morales, F 2006. Down coregulation of light absorption, photochemistry, and carboxylation in Fe-deficient plants growing in different environments. *Photosynth. Res.*, 89:113-126.

Larbi, A; Abadía, A; Morales, F and Abadía, J 2004. Fe resupply to Fe-deficient sugar beet plants leads to rapid changes in the violaxanthin cycle and other photosynthetic characteristics without significant *de novo* chlorophyll synthesis. *Photosynth. Res.*, 79:59-69.

Larbi, A; Morales, F; López-Millán, AF; Gogorcena, Y; Abadía, A; Moog, PR and Abadía, J 2001. Technical advance: reduction of Fe(III)-chelates by mesophyll leaf disks of sugar beet. Multi-component origin and effects of Fe deficiency. *Plant Cell Physiol.*, 42:94-105.

Lee, JA and Woolhouse, HW 1969. Root growth and dark fixation of carbon dioxide in calcicoles and calcifuges. *New Phytol.*, 68:247-255.

Li, Y.L., Cai Q., Yu, D., and Guo, C. 2011. Overexpression of AtFRO6 in transgenic tobacco enhances ferric chelate reductase activity in leaves and increases tolerance to iron-deficiency chlorosis. *Mol. Biol. Rep.,* 38: 3605–3613.

Lindsay, WL 1984. Soil and plant relationships associated with iron deficiency with emphasis on nutrient interactions. *J. Plant Nutri.*, 7:489-500.

Ling, H.Q. *et al.* 2002. The tomato fer gene encoding a bHLH protein controls iron-uptake responses in roots. *Proc. Natl. Acad. Sci. U.S.A.,* 99: 13938–13943.

Loew, O 1900. Physiological studies of Connecticut leaf tobacco. U.S.D.A. Rep. 56:5-57.

Lombard, PB and Westwood, MN 1987. Pear rootstocks. In: Rom, RC, Carlson, RF (Eds.), Rootstocks for Fruit Crops. Wiley, New York USA.

Lombardi, L; Sebastiani, L and Vitagliano, C 2003. Physiological, biochemical, and molecular effects of in vitro induced iron deficiency in peach rootstock Mr.S 2/5. *J. Plant Nutr.,* 26:2149-2163.

Long, T.A. *et al.* 2010. The bHLH transcription factor POPEYE regulates response to iron deficiency in Arabidopsis roots. *Plant Cell,* 22: 2219–2236.

López-Millán, AF, Morales, F; Gogorcena, Y; Abadía, A and Abadía, J 2001. Iron resupply-mediated deactivation of Fe-deficiency stress responses in roots of sugar beet. *Aus. J. Plant Physiol.,* 28:171-180.

Loreti, F 1994. Attuali conoscenze sui principali portinnesti degli alberi da frutto. *Frutticoltura,* 9:9-62.

Loupassaki, MH; Lionakis, SM and Androulakis, II 1997. Iron deficiency in kiwi and its correction by different methods. *Acta Hort.*, 444:267-271.

Lucena, JJ 2000. Effects of bicarbonate, nitrate and other environmental factors on iron deficiency chlorosis: A review. *J. Plant Nutri.,* 23:1591-1606.

Lucena, JJ; Gárate, A and Carpena, O 1987. Iron-chelates evaluation in a calcareous soil. *Plant Soil*, 103:134-138.

Ma, C; Tanabe, K; Itai, A; Tamura, F; Chun JP and Teng, Y 2006. Tolerance to lime-induced chlorosis of Asian pear rootstocks (*Pyrus spp.*). *J. Japan Soc. Hort. Sci.*, 74:419-423.

Ma, C; Tanabe, K; Itai, A; Tamura, F; Teng, Y and Chun JP 2006. Responses of two Asian pear rootstocks (Pyrus spp.) to Fe-deficiency chlorosis induced by addition of bicarbonate to nutrient solution. *J. Jap. Soc. Hort. Sci.,* 75:219-223.

Maldonado-Torres, R; Etchevers-Barra, JD; Alcántar-González, G; Rodriguez-Alcazar, J and Colinas-León, MT 2006. Morphological changes in leaves of Mexican lime affected by iron chlorosis. *J. Plant Nutri.,* 29:615-628.

Manthey, JA; McCoy, DL and Crowley, DE 1994. Stimulation of rhizosphere iron reduction and uptake in response to iron deficiency in Citrus rootstocks. *Plant Physiol. Biochem.*, 32: 211-215.

Marschner, H 1995. Functions of mineral nutrients: micronutrients. Iron. In Mineral Nutrition of Higher Plants. Academic Press: Cambridge, U.K. pp313-324.

Marschner, H. and V. Römheld, 1994. Strategies of plants for acquisition of iron. *Plant Soil.*, 165, 261–274.

Mengel, K 1995. Iron availability in plant tissues-iron chlorosis on calcareous soils. *Plant Soil.,* 165:275-283.

Mengel, K 2002. Alternative or complementary role of foliar supply in mineral nutrition. *Acta Hortic.*, 594:33-47.

Minotti, G and Aust, SD 1987. The requirement for iron (III) in the initiation of lipid peroxidation by iron (II) and hydrogen Peroxide. *J. Biol. Chem.,* 262:1098-1104.

Mohamed, AA and Aly, AA 2004. Iron deficiency stimulated some enzymes activity, lipid peroxidation and free radicals production in *Borage officinalis* induced *in vitro*. *Int. J. Agri. Biol.*, 6:179-184.

Molassiotis, AN; Tanou, G; Diamantidis, G; Patakas, A and Therios, I 2006. Effects of 4-month Fe deficiency exposure on Fe reduction mechanism, photosynthetic gas exchange, chlorophyll fluorescence and antioxidant defense in two peach rootstocks differing in Fe deficiency tolerance. *J. Plant Physiol.*, 163:176-185.

Molassiotis, AN; Grigorios, C; Diamantidis, IN; Therios, VT and Kortessa, ND 2005. Oxidative stress, antioxidant activity and Fe(III)-chelate reductase activity of five *Prunus* rootstocks explants in response to Fe deficiency. *J. Plant Growth Regul.*, 46:69-78.

Moog, PR and Brüggemann, W 1994. Iron reductase systems on the plant plasma membrane: A review. *Plant Soil,* 165: 241-260.

Morales, F; Abadía, A and Abadía, J 1998a. Photosynthesis, quenching of chlorophyll fluorescence and thermal energy dissipation in iron-deficient sugar beet leaves. *Aus. J. Plant Physiol.*, 25:403-412.

Morales, F; Belkhodja, R; Abadía, A and Abadía, J 2000a. Photosystem II efficiency and mechanisms of energy dissipation in iron-deficient, field grown pear trees (*Pyrus communis* L.). *Photosynth. Res.*, 63:9-21.

Morales, F; Belkhodja, R; Goulas, Y; Abadía, J and Moya, I 1999. Remote and near-contact chlorophyll fluorescence during photosynthetic induction in iron-deficient sugar beet leaves. *Remote Sens. Environ.*, 69:170-178.

Morales, F; Grasa, R; Abadía, A and Abadía, J 1998b. Iron chlorosis paradox in fruit trees. *J. Plant Nutr.*,21:815-825.

Morales, F; Grasa, R; Gogorcena, Y; Abadía, A and Abadía, J 2000b. Where is Fe located in iron-chlorotic peach leaves? 10th International Symposium on Iron Nutrition and Interactions in Plants Houston, Texas, USA. p.99.

Morales, F; Moise, N; Quilez, R; Abadía, A; Abadía, J and Moya, I 2001. Iron deficiency interrupts energy transfer from a disconnected part of the antenna to the rest of Photosystem II. *Photosynth. Res.,* 70:207-220.

Moreno, MA; Montañés, L; Tabuenca, MC and Cambra, R 1996. The performance of Adara as a cherry rootstock. *Sci. Hortic.*,65:85-91.

Moreno, MA; Tabuenca, MC and Cambra, R 1995a. Adesoto 101, a plum roootstock for peaches and other stone fruit. *HortScence*, 30:1314-1315.

Moreno, MA; Tabuenca, MC and Cambra, R 1995b. Ademir, a Myrobalan rootstock for plums. *HortScience*, 30:1475-147.

Mukherjee, I., Campbell, N.H., Ash, J.S., and Connolly, E.L. 2006. Expression profiling of the Arabidopsis ferric chelate reductase (FRO) gene family reveals differential regulation by iron and copper. *Planta,* 223: 1178–1190.

Navarro, C; Fernandez-Escobar, R and Benlloch, M 1992. A low-pressure, trunk injection method for introducing chemical formulations into olive trees. *J. Amer. Soc. Hort. Sci.,* 117:357-360.

Nelson, SD and Jolly, VD 1984. Effect on acidified iron-sulfur-rich mining residue (Iron-Sol) on soil and plant nutrient relationships in drip irrigated raspberry and strawberry. *J. Plant Nutr.*, 7:251-257.

Nenova, V. and I. Stoyanov, 1993. Physiological and biochemical changes in maize plants under iron deficiency I. Growth and photosynthesis. *J. Plant Nutr.,* 16, 835–849.

Nenova, V.R., 2009. Growth and photosynthesis of pea plants under different iron supply. *Acta Physiol. Plant.*, 31, 385–391.

Nicholls, P; Fita I and Loewen, PC 2001. Enzymology and structure of catalases. *Adv. Inorganic Chem.*, 51:51-106.

Ortiza, PR; Meza, BIC; Requena, FRG; Flores, GM and Barra, JDE 2007. Evaluation of different iron compounds in chlorotic Italian lemon trees (*Citrus lemon*). *Plant Physiol. Biochem.*, 45:330-334.

Parker, MW; Blake, CC; Barra, D; Bossa, F; Schinina, ME; Bannister, WH and Bannister, JV 1987. Structural identity between the iron and manganese containing superoxide dismutases. *Protein Eng.*, 1:393-397.

Pastana, M., P.J. Correia, A. de Varennes, J. Abadía and E.A. Faria, 2001. The use of floral analysis to diagnosis the nutritional status of orange trees. *J. Plant Nutri.*, 24, 1913-1923.

Pérez-Sanz, A; Álvarez-Fernández, A; Casero, T; Legaz, F and Lucena, JJ 2002. Fe enriched biosolids as fertilizers for orange and peach trees grown in field conditions. *Plant Soil*, 241:145-153.

Perkins, MC; Roberts, CJ; Briggs, D; Davies, MC; Friedmann, A; Hart, C and Bell, G 2005. Macro and microthermal analysis of plant wax/surfactant interactions: plasticizing effects of two alcohol ethoxylated surfactants on an isolated cuticular wax and leaf model. *Appl. Surf. Sci.,* 243:158-165.

Pestana, M; David, M; Varennes, AD; Abadía, J and Faria, EA 2001. Responses of "Newhall" orange trees to iron deficiency in hydroponics: effects on leaf chlorophyll, photosynthetic efficiency, and root ferric chelate reductase activity. *J. Plant Nutri.*, 24: 1609-1620.

Pestana, M; de-Varennes A; Abadía, J and Faria, EA 2005. Differential tolerance to iron deficiency of citrus rootstocks grown in nutrient solution. *Sci. Hortic.*, 104:25-36.

Pestana, M; Domingos, I; Gama, F; Dandlen, S; Miguel, MG; Pinto, JC; Varennes, AD and Correia, PJ 2011. Strawberry recovers from iron chlorosis after foliar application of a grass-clipping extract. *J. Plant Soil Sci.*, 174:473-479.

Piagnani, C and Zocchi, G 1997. Physiological responses of grapevine callus cultures to iron deficiency. *J. Plant Nutri*, 20:1539-1549.

Piagnani, C; Nisi, PD; Espen, L and Zocchi, G 2003. Adaptive responses to iron-deficiency in callus cultures of two cultivars of *Vitis* spp. *J. Plant Physiol.*, 160: 865-870.

Potocki, S. *et al.* 2013. The extracellular loop of IRT1 ZIP protein – the chosen one for zinc? *J. Inorg. Biochem.* 127: 246–252.

Ranieri, A; Castagna, A; Baldan, B and Soldatini, GF 2001. Iron deficiency differently affects peroxides isoforms in sunflower. *J. Exp. Bot.,* 52(354):25-35.

Rasid, A; Couvillon, GA and Jones, JB 1990. Assessment of Fe status of peach rootstocks by techniques used to distinguish chlorotic and non-chlorotic leaves. *J. Plant Nutri.*,13:285-307.

Reed, DW; Lyons Jr, CG and Ray, G 1988. Field evaluation of inorganic and chelated iron fertilizers as foliar sprays and soil application. *J. Plant Nutri*, 11:1369-1378.

Reil, WO and Beutel, JA 1976. A pressure machine for injecting trees. *Calif. Agric.* 30:4-5.

Robinson, NJ; Procter, CM; Connolly, EL and Guerinot, ML 1999. A ferric-chelate reductase for iron uptake from soils. *Nature*, 397:694-697.

Rogers, EE and Guerinot, ML 2002. Iron acquisition in plants. In: Templeton D. (ed.) Molecular and cellular Iron Transport, New York: Marcel Dekker. pp.359-373.

Rombolá, AD; Brüggemann, W; López-Millán, AF; Tagliavini, M; Abadía, J; Marangoni, B and Moog, PR 2002. Biochemical responses of tolerance to Fe deficiency in kiwifruit (*Actinidia deliciosa*). *Tree Physiol.,* 22:869-876.

Rombolá, AD; Brüggemann, W; Tagliavini, M; Marangoni, B and Moog, PR 2000. Iron source affects iron reduction and re-greening of kiwifruit (*Actinidia deliciosa*) leaves. *J. Plant Nutr.*, 23:1751-1765.

Römheld, V. and Marschner, H. (1986) Evidence for a specific uptake system for iron phytosiderophores in roots of grasses. *Plant Physiol.*, 80, 175–180.

Römheld, V 2000. The chlorosis paradox: Fe inactivation as a secondary event in chlorotic leaves of grapevine. *J. Plant Nutri.*, 23:1629-1643.

Römheld, V and Marschner, H 1981. Iron deficiency stress induced morphological and physiological changes in root tips of sunflower. *Physiol. Plant.*, 53:354-360.

Rosenfield, CL; Reed, DW and Kent, MW 1991. Dependency of iron reduction on development of unique root morphology in *Ficus benjamina* L. *Plant Physiol.*, 95:1120-1124.

Rutherford, PM; Dudas, MJ and Samek, RA 1994. Environmental impacts of phosphogypsum. *Sci. Total Environ.*, 149:1-38.

Salazar-García, S 1999. Iron nutrition and deficiency: a review with emphasis in avocado (*Persea americana* Mill.). *Revista Chapingo Serie Hort.*, 5:67-76.

Sanz, M; Pascual, J and Machin, J 1997. Prognosis and correction of iron chlorosis in peach trees: influence of fruit quality. *J. Plant Nutri.,* 20:1567-1572.

Schikora, A and Schmidt, W 2001. Iron stress-induced changes in root epidermal cell fate are regulated independently from physiological responses to low iron availability. *Plant Physiol.,* 125:1679-1687.

Schönherr, J 2001. Cuticular penetration of calcium salts: Effects of humidity, anions and adjuvants. *J. Plant Nutri. Soil Sci.*, 164:225-231.

Scimidt, W 2003. Iron solutions: acquisition strategies and signaling pathways in plants. *Trends Plant Sci.*, 8:188-193.

Sites, JW; Leonard, CD and Stewart, I 1953. Citrus fruit quality as affected by iron deficiency and its correction. Florida Agri. Exp. Sta. Ann. Rep.,pp.183-184.

Snowden, RED and Wheeler, BD 1993. Iron toxicity to fen plant-species. *J. Ecol.,* 81:35-46.

Socias-i-Company, R; Aparisi, G and Felipe, AJ 1995. A genetical approach to iron chlorosis in deciduous fruit trees. In: J. Abadía, (Ed.), Iron Nutrition in Soils and Plants. Kluwer Academic Publishers, Dordrecht, The Netherlands, pp.167-174

Sommer, K 1993. 'Cultan'-Fertilizers and Fertilization Technique: Fundamentals and Further Development. In: Proc. Dhalia Greidinger Memorial International Workshop on Controlled Slow Release Fertilizers; Technion- Israel Institute of Technology Haifa, Israel, 7–12 March 1993.

Spiller, S and Terry, N 1980. Limiting factors in photosynthesis. II. Iron stress diminishes photochemical capacity by reducing the number of photosynthetic units. *Plant Physiol.*, 65:121-125.

Syrgiannidis, G 1985. Control of iron chlorosis and replant diseases in peach by using the GF 677 rootstock. *Acta Hortic.*, 173:383-388.

Tagliavini, M and Marangoni, B 2002. Major nutritional issues in deciduous fruit orchards of Northern Italy. *HortTech.*, 12:26–31.

Tagliavini, M and Rombolà, AD 2001. Iron deficiency and chlorosis in orchard and vineyard ecosystems.: Review. *Europ. J. Agron.*, 15:71-92.

Tagliavini, M; Rombolá, AD and Marangoni, B 1995. Response to iron deficiency stress of pear and quince genotypes. *J. Plant Nutri.,* 18:2465-2482.

Taiz, L and Zeiger, E 1991. Plant Physiology. The Benjamin/Cummings Publ. Co., Inc. Redwood City, Calif., USA. 565 p.

Tamura, H; Knoche, M and Bukovac, MJ 2001. Evidence for surfactant solubilization of plant epicuticular wax. *J. Agric. Food Chem.*, 49:1809-1816.

Terry, N 1980. Limiting factors in photosynthesis. I. Use of iron stress to control photochemical capacity *in vivo*. *Plant Physiol.*, 65:114-120.

Terry, N and Abadía, J 1986. Function of iron in chloroplasts. *J. Plant Nutr.*, 9:609-646.

Thomine, S; Wang, R; Ward, JM; Crawford, NM and Schroeder, JI 2000. Cadmium and iron transport by members of a plant metal transporter family in Arabidopsis with homology to Nramp genes. In: *Proc. Nat. Acad. Sci.* USA, 97: 4991-4996.

Tognetti, R and Sebastiani, L 2004. Gas exchange and photosythetic efficiency in the combination of grafted peach/GF677 with iron deficiency. *Rivista Frutt. Ortoflori.*, 66: 58-62.

Udvardi, MK and Day, DA 1997. Metabolite transport across symbiotic membranes of legume nodules. *Ann. Rev. Plant Physiol. Plant Mol. Biol.,* 48:493–523.

Val, J; Sanz, M; Montanes, L and Monge, E 1995. Application of chlorophyll fluorescence to study iron and manganese deficiencies in peach tree. *Acta Hortic.*, 383:201-209.

Vasconcelos, M., Eckert, H., Arahana, V., Graef, G., Grusak, M.A. and Clemente, T. 2006. Molecular and phenotypic characterization of transgenic soybean expressing the Arabidopsis ferric chelate reductase gene, FRO2. *Planta,* 224: 1116–1128.

Vert, GA; Briat, JF and Curie, C 2001. *Arabidopsis* IRT2 gene encodes a root-periphery iron transporter. *Plant J.*, 26:181-189.

Vert, GA; Briat, JF and Curie, C 2003. Dual regulation of the Arabidopsis high-affinity root iron uptake system by local and long-distance signals. *Plant Physiol.*, 132:796-804.

Wallace, A 1991. Rational approaches to control of iron deficiency other than plant breeding and choice of resistant cultivars. *Plant Soil*, 130:281-288.

Wallace, A and Lunt, OR 1960. Iron chlorosis in horticultural plants. A review. *Proc. Amer. Soc. Hort. Sci.,* 75:819-841.

Wang, CJ and Liu, ZQ 2007. Foliar uptake of pesticides-present status and future challenge. *Pestic. Biochem. Physiol.*, 87:1-8.

Waters, BM; Blevins, DG and Eide, DJ 2002. Characterization of *FRO1*, a pea ferric-chelate reductase involved in root iron acquisition. *Plant Physiol.*, 129:85-94.

Westwood, MN and Lombard, PB 1983. Pear rootstocks: present and future. *Fruit Var. J.*, 36:24-28.

Wu, H., Li, L., Du, J., Yuan, Y., Cheng, X., and Ling, H.Q. 2005. Molecular and biochemical characterization of the Fe(III) chelate reductase gene family in Arabidopsis thaliana. *Plant Cell Physiol.,* 46: 1505–1514.

Yuan, Y.X. *et al.* 2005. AtbHLH29 of Arabidopsis thaliana is a functional ortholog of tomato FER involved in controlling iron acquisition in strategy I plants. *Cell Res.*, 15: 613–621.

Zaharieva, TB; Gogorcena, Y and Abadía, J 2004. Dynamics of metabolic responses to iron deficiency in sugar beet roots. *Plant Sci.*, 166:1045-1050.

Zheng, SJ; Tang, C; Arakawa, Y and Masaoka, Y 2003. The responses of red clover (*Trifolium pratense* L.) to iron deficiency: a root Fe (III) chelate reductase. *Plant Sci.*, 164:679-687.

Section IV:
Proteomic Research

12

SIGNIFICANCE OF POST-TRANSLATIONAL MODIFICATIONS FOR PROTEOMIC RESEARCH

A. K. Trivedi[1] and A. Hemantaranjan[2]

[1]*National Bureau of Plant Genetic Resources, Regional Station Bhowali, Nainital – 263132 (India)*
[2]*Department of Plant Physiology, Institute of Agricultural Sciences, Banaras Hindu University, Varanasi – 221005 (India)*

INTRODUCTION

Addition or removal of a functional group from an amino acid side chain, the modification of protein termini, cleavage of the synthesized polypeptide chain or the covalent cross-linking between separate protein domains is known as post-translational modification (PTM). Several proteins, particularly those derived from eukaryotic sources, undergo covalent modification either during their ribosomal synthesis or (more usually) after synthesis is complete, this is known as co-translational and post-translational modification. The stage during protein synthesis/ maturation at which the co- or post-translational modification occurs evidently differs but both are often referred as post-translational modifications (PTMs). These modifications invariably influence some structural aspect or functional role of the affected protein (Walsh 2009). Post-translational modification provides a dynamic mechanism for regulation of protein function. These well expand the chemical diversity of functional groups as well as regulatory potential ahead of those defined by the standard 20 amino acid side chains. Many cellular proteins undergo considerable amounts of Post-translational Modifications (PTMs) in response to certain stimuli and this dynamic process occurs in various cell compartments to dictate the fate and activity of the modified proteins (Cain *et al.,* 2014).

Some such modifications (e.g., glycosylation) are common in their occurrence, while others occur only rarely (e.g., selenoylation, the co-translational incorporation of selenium into some proteins). Some PTMs are reversible (e.g., phosphorylation in many cases), while others are irreversible (e.g., proteolysis). Most but not all have an evident function. Most of these modifications are introduced into target proteins via specific enzymatic steps/ pathways, while some (e.g., the oxidation of cysteine and methionine residues or the deamidation of asparagine or glutamine residues) arise non-enzymatically (Walsh 2009). Over 300 PTMs have been reported in the literature (Walsh, 2005) and new ones are regularly added to the list. They regulate all aspects of cellular homeostasis. Determination of the site of PTM and extent of modification is crucial for interpretation of cellular signaling networks. Despite the pivotal roles of PTMs in cellular functions, studies on PTMs have not really been feasible until recently.

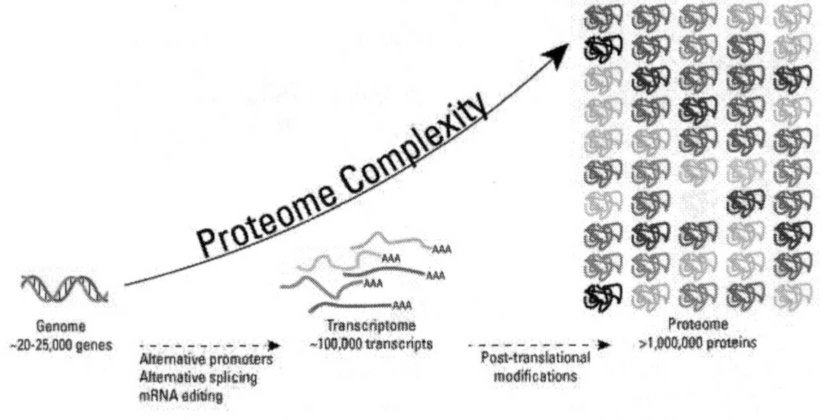

Fig. 1. A view of post translational modification for protein complexity (Source: www.thermofisher.com)

COVALENT MODIFICATIONS

The most commonly analyzed PTMs are covalent modifications, largely because of the close association between their dys- regulation and several environmental/ stress conditions. Moreover, several biologically important alterations in protein function occur as a result of their addition or removal. Covalent PTMs include both chemical moieties such as phosphate and carbohydrates and functional polypeptides such as ubiquitin and Small Ubiquitin-like Modifier (SUMO). The reversible addition of a chemical moiety such as phosphate, catalyzed by specific protein kinases, plays an essential role in regulating signaling in cells.

Incorporation of one or more phosphate groups on specific amino acid side chains within a protein, often induces important protein conformational change in protein. This has deep effects on protein activity and protein-protein interactions (Eyers *et al.,* 2008).

Protein glycosylation is more complex, with two main mechanisms for covalent binding of the glycan to the polypeptide: N-glycosylation (where the glycan is attached to an Asn residue within a tripeptide consensus sequence, Asn-X- Ser/Thr (where X represents any amino acid except Pro) and O-glycosylation (where the glycan is attached to a Ser or a Thr residue) (Morelle and Michalski 2007). Protein glycosylation is functionally important for modulating inter- and intracellular protein activities such as cell adhesion. It has a decisive role for mitosis and cell division.

The functional importance of protein ubiquitination is dependent on the number of ubiquitin monomers attached at a single site as well as the site of ubiquitin chain linkage. Ubiquitin is a 76-amino acid (~8 kDa) protein. It is covalently attached via its C-terminal carboxyl group to the ε-amino groups of lysine residues on target proteins (Weissman, 2001). Via Lys48, poly-ubiquitination of ubiquitin plays an essential role in proteasome-mediated induced protein degradation. Mono-ubiquitination regulates enzyme activity, protein-protein interactions and protein trafficking (Drews *et al.,* 2007). Several of the PTM events that occur on a single protein are known to be synergistic. Phosphorylation can promote protein ubiquitination, thus coupling an extracellular stimuli-initiated signal transduction event with consequent protein degradation (Weissman, 2001).

POST-TRANSLATIONAL MODIFICATIONS IN SECRETED PEPTIDE SIGNALS

Posttranslational modifications are known to alter the physico-chemical properties of peptides by changing net charge, hydrophilicity, and/or conformation, thereby modulating the binding ability and specificity of peptides for their target receptor proteins (Matsubayashi, 2014). In secreted peptide signals of plants so far three types of posttranslational modifications have been identified: tyrosine sulfation, proline hydroxylation and hydroxyproline arabinosylation (Matsubayashi, 2011).

Tyrosine Sulfation

Tyrosine sulfation is a posttranslational modification occasionally found in peptides and proteins synthesized through the secretory pathway in both plants and animals (Moore, 2003). This modification is mediated by TPST enzyme, which catalyzes the transfer of a sulfate from 3-phosphoadenosine 5-phosphosulfate (PAPS) to the phenolic group of tyrosine (Komori *et al.*, 2009). Although the tyrosine sulfation motif in peptides is not precise, the minimum requirement for tyrosine sulfation in plants is the presence of an aspartic acid residue N-terminally adjacent to a tyrosine residue. Multiple acidic amino acids near this tyrosine residue drastically improve sulfation (Hanai *et al.*, 2000).

Proline Hydroxylation

Hydroxyproline residues have been observed in almost all post translationally modified small peptide signals except for PSK, which has no proline residues. Proline hydroxylation is mediated by prolyl-4-hydroxylase (P4H), which belongs to family of 2-oxoglutarate-dependent dioxygenases that require 2-oxoglutarate and O_2 as co-substrates (Myllyharju, 2003). P4H is a trans membrane protein that localizes in both the endoplasmic reticulum and the Golgi complex.

Hydroxyproline Arabinosylation

Hydroxyproline residues in several secreted peptide signals are further modified with an O-linked L-arabinose chain (Okamoto *et al.*, 2013). This modification is also important in a large family of extra cellular proteins, the so-called hydroxyproline-rich glycoproteins (Kieliszewski *et al.*, 2010), these include extensins (Kieliszewski and Lamport 1994), proline-rich proteins (Kieliszewski *et al.*, 1992) and arabinogalactan proteins (Ellis *et al.*, 2010).

REDOX POST-TRANSLATIONAL MODIFICATIONS

Reactive oxygen species (ROS) and reactive nitrogen species (RNS) play an important role in redox signaling mainly through a set of reversible post-translational modifications (PTMs) of cysteine thiols on proteins (Zaffagnini *et al.*, 2012). The reaction between ROS and RNS with protein thiols involves cysteine residues which have unique physico-chemical properties. These residues are considered reactive when their ionization constant (p*K*a) is lower than that of free cysteines (8.3) (Klomsiri *et al.*, 2011), usually ionoization constant range between 3 and 7. This involves that at physiological pH, these residues are pre-dominantly found as thiolates (i.e., deprotonated,-S−). Thiolates are stronger nucleophiles than thiol counter-parts (-SH). Protein containing reactive cysteines can undergo different types of redox modifications in response to different ROS-and RNS-dependent redox signals. Physiologically, the most relevant ones are disulfide bond formation (S-S, intra- or inter-molecular), glutathionylation (S-SG), S-nitrosylation (S-NO) and oxidation to sulfenic acid (S-OH). Because of its unstable nature, sulfenic acids can generate disulfide bonds by reacting with a second cysteine nearby. These may evolve to irreversible oxidized forms such as sulfinic (S-O_2H) and sulfonic acids (S-O_3H). These redox PTMs are mostly controlled by oxidoreductases namely thioredoxins (TRXs) and glutaredoxins (GRXs) (Zaffagnini *et al.*, 2012).

TYPES OF POST-TRANSLATIONAL MODIFICATIONS

Although, more than 300 different types of PTMs have been identified and new ones are regularly added to the list as a consequence of recent researhes and advancement in analytical techniques. Among these common PTMs are as following:

I. **Acetylation:** Acetylation (i.e., the incorporation of a $COCH_3$ group) of the N-terminal alpha-amine of polypeptides is a prevalent modification in eukaryotes. This PTM is enzymatically introduced by N-alpha-acetyltransferases. Acetylation of actin is important in the formation of actin filaments.

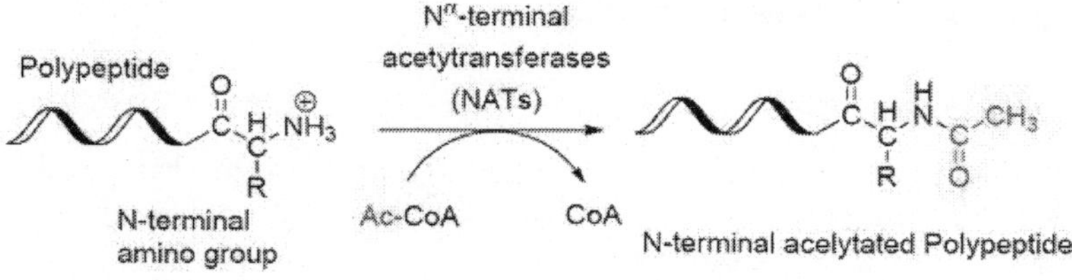

Fig. 2. Acetylation reaction (Source: https://en.wikipedia.org/wiki/Acetylation)

II. **Acylation:** Acylation [attachment of a (RCO-) group] helps some polypeptides to interact with or anchor in biological membranes.

III. **ADP-ribosylation:** Attachment of one or more ADP and ribose moieties to a polypeptide's backbone is known as ADP–ribosylation. This modification is achieved enzymatically by ADP-ribosyltransferase enzymes. It has several roles in cell signaling and control.

IV. **Amidation:** The enzymatic replacement of a polypeptide's C-terminal carboxyl group with an amide functional group is known as amidation. It is characteristic of many bioactive peptides/ short polypeptides. It influences biological activity/ stability of these polypeptides.

V. **γ – Carboxyglutamate formation:** γ – Carboxyglutamate formation is enzymatic conversion of target glutamate residues into γ – carboxyglutamate.

VI. **Disulfide bond formation:** Disulfide bonds are formed between the thiol (-SH) groups of cysteine residues. These usually assist in the folding and conformational stabilization of some proteins.

VII. **Glycosylation:** It is covalent attachment of a carbohydrate component to the polypeptide backbone. Glycosylation increases solubility, influence biological half-life and / or biological activity of some proteins.

VIII. **Hydroxylation:** The enzymatic conversion of target aspartate residues into ß-hydroxyaspartate or target asparagines residues into ß-hydroxy asparagine. It is important for the structural assembly of some proteins.

IX. **Phosphorylation:** Phosphorylation is covalent attachment of one or more phosphate groups to the polypeptide backbone. It influences or regulates biological activity of several polypeptide hormones.

X. **Sulfation:** The transfer of a SO_3^- groups to target tyrosine residues is known as sulfation. It influences proteolytic processing of some polypeptides.

ROLE OF PTMS IN CONTROLLING THE ANAPLEROTIC ACTIONS:

Phosphoenol pyruvate carboxylase (PEPC), a key enzyme responsible for replenishing components of the tricarboxylic acid (TCA) cycle is responsible for fixing atmospheric carbon. In addition to this, it has several roles in non-photosynthetic reactions (O'Leary *et al.,* 2011). One of these non-photosynthetic roles is its anaplerotic function. It replenishes TCA cycle intermediates according to plant physiologic and metabolic requirement. Strategies for controlling the anaplerotic actions of PEPC include post-translational phosphorylation and ubiquitination.

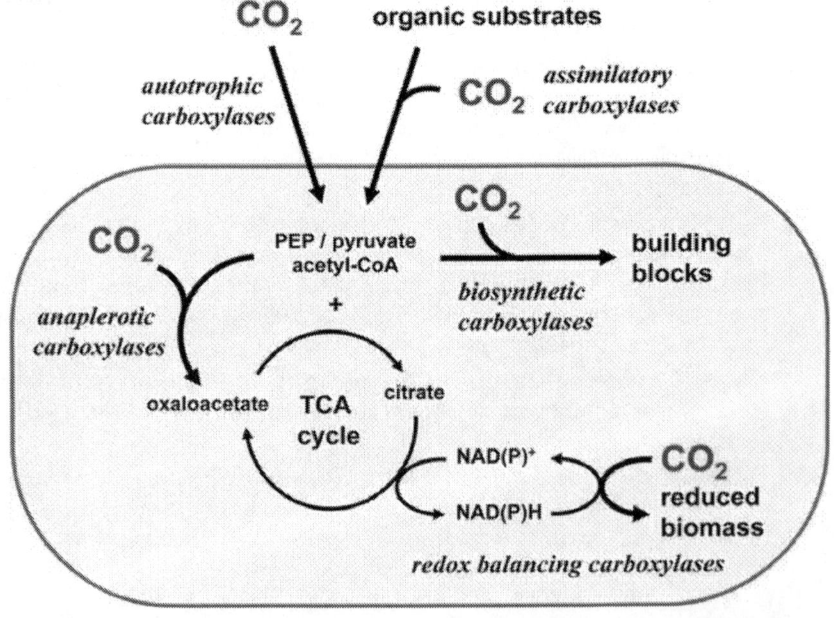

Fig. 3. Anaplerosis in TCA Cycle (Source: www.lamondlab.com)

STRATEGIES FOR CHARACTERIZATION OF PTMS

The identification of PTMs requires large amounts of proteins and a highly sensitive method for their detection. Two-dimensional gel electrophoresis (2DE) has been widely applied to proteomic analysis of PTMs. Based on the PTMs to be identified once the proteins have been prefractionated, it frequently provides sufficient amounts. Therefore, protein species in different modification states can be resolved in the gel and subjected to an analysis of the modifications. The modified proteins can be visualized in the gel by specific methods suitable for different PTMs (Patton, 2002).

Traditionally, PTMs have been identified by Edman degradation, amino acid analysis, isotopic labelling or immunochemistry. During recent years, mass spectrometry (MS) has proven to be extremely useful in PTM discovery. The presence of covalent modifications in proteins affects the molecular weight of the modified amino acids. The mass increment or deficit can be detected by MS. MS has several advantages for the study and characterization of PTMs, such as:

(*i*) Very high sensitivity,

(*ii*) Ability to identify the site of PTM,

(*iii*) Discovery of novel PTMs,

(*iv*) Capability to identify PTMs in complex mixtures of proteins,

(*v*) Ability to quantify the relative changes in PTM occupancy at distinct sites.

APPLICATION OF MASS SPECTROMETRY FOR CHARACTERIZATION OF PTMS

Mass spectrometry (MS) is an analytical strategy widely used for the identification and characterization of gene products within biological systems. A variety of MS instruments can be used for the identification and quantification of proteins present in complex biological mixtures. For characterization of PTMs, prior to MS analysis, proteins and/ or their constituent peptides should be separated using some form of chromatography. Mass analysis of a single purified protein can theoretically permit the identification of the type of PTM present. Yet, detailed analysis, at the peptide level, is usually required to identify the site(s) of modification. A modified peptide derived from such a sample can be detected on the basis of a mass increment or deficit relative to the unmodified form of the same peptide. Some modifications induce a steady characteristic mass change (Eyers *et al.,* 2008). Such as, study of mass differences representative of phosphorylation (+80 Da), methylation (+14 Da), or S-nitrosylation (+29 Da) is invariable. It is an indicator of modified peptide species. Presence of a specific mass difference does not realistically allow identification of the site of modification. For this purpose tandem MS is needed. Although, the site of modification can sometimes be directly inferred if only a single potential site is present in the presumed peptide sequence. Moreover, observation of a defined mass increment may be indicative of the number of diverse modifications. In usual analysis, mass change which arises as a result of trimethylation (+42.05 Da), acetylation (+42.01 Da) is usually indistinguishable when using mass spectrometers with limited mass accuracy. Similarly, phosphorylation (+79.9663 Da) and sulfation (+79.9568 Da) can only be distinguished with instruments capable of high mass accuracy or by using modification-specific analytical strategies such as affinity enrichment or enzymatic treatment (e.g., phosphatase).

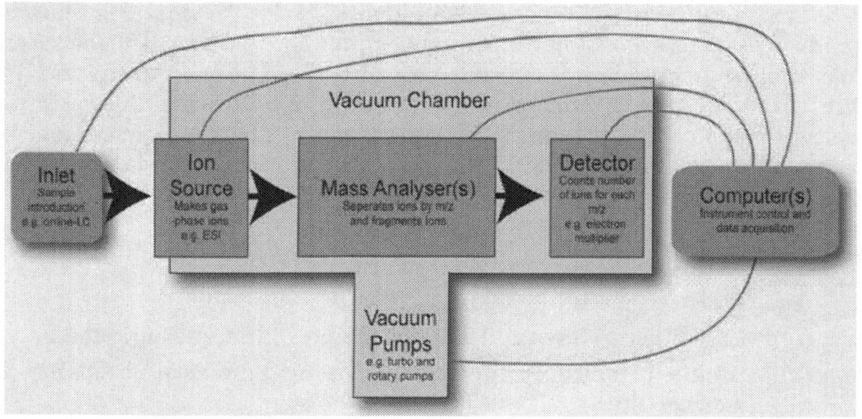

Fig. 4. An overview of mass spectrometer (Source: http://arm.asm.org)

1. Approaches to Characterize PTM Sites by LC-MS/MS

Characterization of PTM sites by LC-MS/MS can be achieved using a number of complementary approaches. These can be broadly divided into following four categories:

(i) Tandem MS using one or more of a number of available fragmentation mechanisms.

(ii) Removal of the modification between consecutive mass spectrometric analyses.

(iii) Selective enrichment of modified proteins or peptides based on the modified functional group prior to MS/MS.

(iv) PTM-specific multistage MS strategies. Tandem MS (MS/MS) usually used in a typical product ion analysis refers to the mass spectrometric analysis of all ions in a sample, prior to the subsequent selection and fragmentation of an ion of a specific *m/z*, and the subsequent MS analysis of the resultant fragment ions.

If applied to peptides, the amino acid sequence as well as type and site of modification can be determined. Large-scale characterization of the proteins present in a complex sample, such as a cell lysate, is performed using a standard "bottom-up" proteomics work- flow. In this protein is digested using a suitable protease (most commonly trypsin) generating smaller analytes, which are better suited to available MS analysis workflows. Prior to MS analysis these peptides are fractionated using liquid chromatography [strong cation exchange (SCX) or reverse phase (RP) or combination of both] (Washburn *et al.,* 2001). But collision-induced dissociation (CID) is the preferred method of the fragmentation for most proteomic applications. CID results in preferential fragmentation of the most superficial bonds. In the case of modified peptides, this is often the covalent bond linking the modification to the peptide backbone. As a consequence, when peptides containing a PTM undergo CID, neutral loss of the modification is often the most abundant fragmentation product. This leads to vagueness in investigating the site of modification.

Loss of the PTM is thought to be due to the large amount of energy transferred to the analyte during collisional activation. It is randomized over many vibrational degrees of freedom before dissociation takes place. Vibrational dissipation of this excess translational energy causes dissociation of the weakest bonds. This usually happens at the bulky side chains containing the modification. Phosphopeptides cause a major problem in this regard. CID has been found to be very useful in characterizing sites of phosphorylation on specific proteins (Haydon *et al.,* 2003) as well as in large-scale proteomic studies (Reinders and Sickmann 2005). CID results primarily in neutral loss of the phosphate group, which results as a loss either of 80 Da (HPO_3) or 98 Da (H_3PO_4, β-elimination) (Schweppe *et al.,* 2003). The primary fragmentation pathway is loss of this phosphate moiety. Therefore, the production of sequence-determining b- and y- fragment ions may be limited. This makes sequence analysis of the peptides difficult. Peptides containing many PTMs, including acetylation and methylation, can also be readily characterized using CID (Fraga *et al.,* 2005).

Alternate fragmentation mechanisms such as electron capture dissociation (ECD) and electron transfer dissociation (ETD) enable radical-initiated bond cleavage following the transfer of electrons. These do not promote analyte fragmentation by transfer of vibrational energy. Hence, labile PTMs are usually maintained on the peptide backbone ions during fragmentation. Specific sites of modification, such as glycosylation and phosphorylation, can thus be accurately identified using these alternate tandem MS strategies (Tyler *et al.,* 2009; Zubarev, 2004). Nevertheless, the usefulness of ETD decreases with a decrease in charge state. This shifts toward non-dissociative electron transfer for peptides with a precursor ion charge of less than 3, which is most often generated by tryptic peptides. This results in an intact electron transfer product species, $[M+2H]^{+\cdot}$, and limited informative fragment ion generation (Swaney *et al.,* 2007).

The observation of improved sequence coverage when highly charged, generally larger peptides are subjected to ETD has supported proteolytic digestion with enzymes other than trypsin. Lys-C, which cleaves C-terminal to lysine resides, typically generates larger peptides than trypsin (which cleaves C-terminal to both lysine and arginine). The application of Lys-C digestion in conjunction with ETD has enabled better fragmentation, and higher sequence coverage. This results increased PTM site discovery when compared to procedures using tryptic peptides (Zhang *et al.,* 2008). While ETD and ECD have proved really useful at overcoming some of the PTM-associated neutral loss issues observed during CID, but besides availability of these alternative fragmentation mechanisms some problems still remain. For example sulfopeptides are more susceptible to neutral loss as compared to phosphopeptides of identical amino acid sequence, both during ECD as well as ETD (Medzihradszky *et al.,* 2007). It seems that the amount of neutral sulfate loss can be reduced (and conversely site identification can be improved) by fragmenting the peptide-alkali metal adduct as opposed to the protonated peptide ions (Medzihradszky *et al.,* 2007).

APPLICATION OF SELECTIVE ENRICHMENT TECHNIQUES TO STUDY PTMS

Many PTMs are of dynamic nature and stoichiometry of modification is relatively low. It causes further complication in their characterization. Furthermore, the efficiency of ionization and/or detection of modified peptide species (particularly phosphopeptides) (Steen *et al.,* 2006), when compared to their non-modified counterparts, can be compromised. Therefore, strategies have been developed which target the explicit identification of peptides or proteins carrying specific post-translational modifications. Selective enrichment techniques can be grouped in the following three categories:

(i) Incorporation of a tagged modification for selective enrichment.
(ii) Affinity purification using specific biological agents.
(iii) Affinity enrichment based on chemical functionality of the modification.

Affinity Purification With Biological Agents

For the selective isolation of modified polypeptides biological molecules that bind with high specificity to the modification or the modified residue of interest, such as lectins are ideal. Lectins bind to carbohydrate therefore, they have been extensively used for the selective enrichment of glycoproteins, glycopeptides, and oligosaccharid (Drake *et al.,* 2006). Bioaffinity molecules such as lectins have an additional advantage due to the distinct carbohydrate-binding specificity of different families of lectins. Specific lectins can therefore be exploited to affinity purify defined oligosaccharide structures on glycoconjugates (Nawarak *et al.,* 2004).

Affinity Capture Technique

There are several solid-phase materials that can be used for the enrichment of modified peptide or protein species based on chemical functionality. Such enrichment of phosphorylated species is usually based on the principle of immobilized metal affinity chromatography (IMAC) which adsorbs phosphopeptides to chelated metal ions (e.g., Fe^{3+}, Ga^{3+}) through metal-phosphate ion-pair interactions (Posewitz Tempst 1999). In this regard, TiO_2- and ZrO_2- based columns also adsorb phosphorylated peptide species well (Kweon and Hakansson 2006). During IMAC or TiO_2 enrichment, phosphopeptides bound to the immobilized metal ions are eluted either by increasing the pH or by competitively eluting with phosphate ions. But concurrent enrichment of highly acidic peptides is a problem, which bind to the IMAC columns by the same chemical principles as the negatively charged phosphopeptides. Methyl esterification of acidic groups has been found to overcome these issues up to certain extent (Ficarro *et al.,* 2002), although secondary modifications that occur during this reaction may influence subsequent peptide characterization by MS (Haydon *et al.,* 2003). Binding of acidic peptides to TiO_2 can also be reduced by sample binding in the presence of 2,5-dihydroxybenzoic acid (DHB) or glutamate (Wu *et al.,* 2007).

Hydrophilic interaction chromatography (HILIC) is usually used for the semi-selective enrichment of glycopeptides (Calvano *et al.,* 2008; Picariello *et al.,* 2008). Enrichment of glycans has been demonstrated to occur via a number of mechanisms including hydrogen bonding and ionic interactions. It can therefore often be used in the separation of isobaric species, an extremely useful feature that complements MS/MS analysis (Wuhrer *et al.,* 2009).

MS/MS STRATEGIES TO STUDY PTMS

Modified peptides often behave characteristically during *MS/MS* analysis, generating unique PTM signatures due to either loss of functional groups or the formation of diagnostic fragment ions. Therefore, mass spectrometric methods can often be customized to allow the desired selection and analysis defined populations of modified peptides. Usually, peptide identification is performed using product ion analysis. In this all of the product ions generated following decomposition of a precursor ion of defined *m/ z* are recorded. In order to utilize the specific fragmentation patterns of modified peptide ions, analysis can be focused toward precursor ions. These precursor ions generate a defined product or diagnostic fragment ion, termed precursor ion scanning. Analysis can be directed toward ions exhibiting a characteristic difference between their precursor and product ions which demonstrate loss of functional group; this is known as neutral loss scanning.

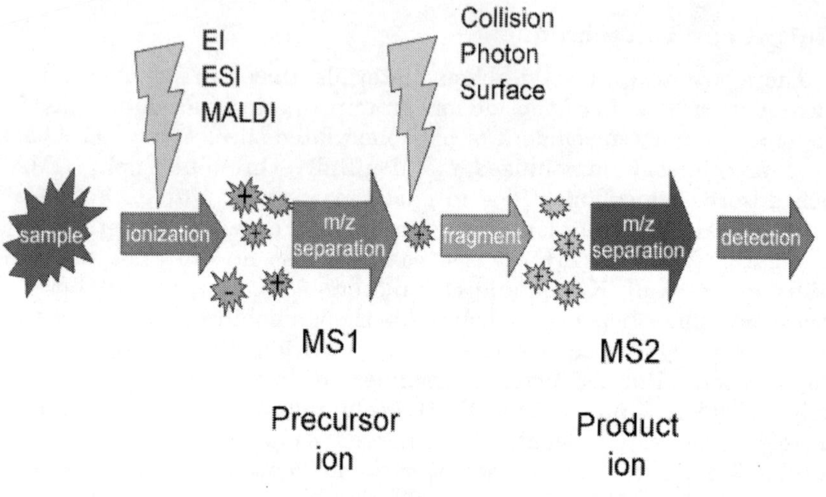

Fig. 5. MS/MS strategy to study PTMs (Source: www.en.wikipedia.org)

Application of Precursor Ion Scanning

In a single experiment, precursor ions (peptide ions) generate a specific product ion following fragmentation. A precursor ion scan determines the *m/z* of all of the precursor ions (IUPAC, 1997). During precursor ion analysis, the entire mass range is scanned in the first mass analyzer. This includes the precursor ions also whose fragmentation results in the selected product ion. The second mass analyzer is in the meantime fixed to monitor and transmit product ions of a specific *m/z* ratio only. The mass spectrum then details precursor ions whose fragmentation caused the formation of the product ion of interest. This filters out chemical 'noise' from the mass spectrum and increases sensitivity of analysis. During the analysis of phosphorylated peptides, a precursor ion scan of *m/z 79* (PO_3^-) or 63 (PO_2^-) in negative mode produces a mass spectrum containing only the phosphorylated precursor ions present within the sample (Car *et al.,* 1996). In the same way, phosphotyrosine generates analytical signal at *m/z* +216.042 (Steen *et al.,* 2002). These identified precursor ions can then be targeted for MS/MS to clarify their amino acid sequence and identify the site(s) of phosphorylation (Car *et al.,* 1996).

Application of Neutral Loss Scanning

Neutral loss scan determines the *m/ z* of parent ions which fragment to produce a product ion indicating loss of a specified (neutral) mass. The product ion generated following this neutral loss can then be selected for further fragmentation in an MS/MS/MS experiment. Neutral loss experiments of this type can be achieved only on those instruments which are able of multiple stages of tandem MS, such as ion-trapping instruments. Modified peptides which reveal such a PTM -specific loss during CID can easily be characterized using this technique. But like phosphorylated peptides, direct mapping of the site of modification is not easy. In addition to performing an MS/MS/MS CID experiment, the observation of a neutral loss during CID can also be used to trigger ECD or ETD of the precursor of interest. This is specially useful for those labile precursors for which sequence-determining fragment ions are difficult to generate (Sweet *et al.,* 2008). Combining ETD data generated in this manner with CID information found to be useful for improving confidence in the identification of sites of phosphorylation, glycosylation, and methylated arginine among others (Zubarev *et al.,* 2008).

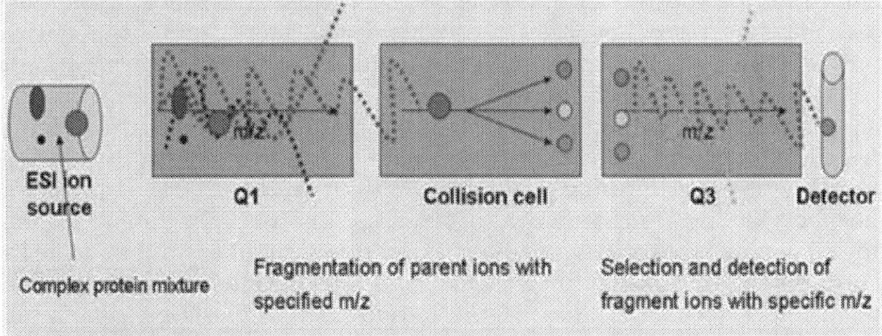

Fig. 6. Steps in selected reaction monitoring (Source: www.tankonyvtar.hu).

Application of Selected Reaction Monitoring / Multiple Reaction Monitoring

Selected reaction monitoring (SRM) and multiple reaction monitoring (MRM) begins with the fragmentation of a specific peptide ion to yield one or more defined product ions. Product ions are indicative for a specific series of amino acids. In SRM, a product ion of specified m/z derived from a particular precursor ion m/z window is usually used for the verification of the presence of a specific analyte. MRM further extend this, where a number of different fragment ion m/z values are used for the confirmation of the presence of desired peptide ions. This provides more confidence in the task. SRM and MRM can be used in the identification of modified peptides derived from a known post-translationally modified protein due to the increase in the sensitivity of identification and quantification achieved. This technique is known as "multiple reaction monitoring-initiated detection and sequencing" (MIDAS). This has been successfully used for the identification of sites of phosphorylation and ubiquitination from purified proteins which contains known modifications (Sahana *et al.*, 2007).

QUANTITATIVE ESTIMATION OF PTMS

Revelation of the stoichiometry of a given modification is important for correlation with the proportion of a protein exhibiting changed PTM-regulated functionality. This may be an enzyme activity or a binding affinity. Nevertheless, mass spectrometric signal intensity depends on the amount of analyte as well as on its chemical properties. Therefore, comparison of the signal intensities of a peptide and its modified counterpart is not exact reflection of the proportion of modified peptide. Avoiding the potential issues of changes in ionization efficiency of differentially modified peptides of the same primary sequence, relative quantification of stoichiometry between two or more systems can be achieved with the use of stable isotope labeling. Reference of peptides, particularly internal peptides, are chemically identical to the desired analyte, which can be differentiated by mass due to the incorporation of heavy-isotope forms of particular elements (^{13}C, ^{15}N), are usually used for the quantification of different peptide states. Absolute quantification of phosphorylated and ubiquitinated proteins has been successfully achieved by using isotope dilution methods (Mayor *et al.,* 2007). In case of phosphorylation stoichiometry determination, phosphrylated isotope-labeled surrogate peptides are synthesized, quantified as well as added to the sample of interest in known amounts. This permits calculation of the stoichiometry of modification of the native peptide based on comparable signal ion intensity. Stable isotope labeling with amino acids in cell culture (SILAC) permits to calculate the relative amounts of modified peptides present between multiple cell populations. Merit of this strategy is that it permits the combination of samples undergoing analysis during an early stage in the sample preparation workflow. This minimizes differences, and is capable of determining the relative stoichiometry of a large number of modified peptides simultaneously (Pan *et al.*, 2008). For specific assess of the stoichiometry of modified peptides, SILAC is best used in conjunction with one or more of the enrichment or MS-targeted strategies (Qiu and Regnier 2005).

Normalized ion currents can be effectively used as an alternative to stable isotope labeling, for the relative quantification of modified peptides between LC-MS/MS runs. The relative amount of modified peptides in two or more samples is then calculated using difference in the signal ion intensities of the desired peptide species (Steen *et al.*, 2005).

In conclusion, after completion of some genome sequencing projects current focus is to study transcriptome. But transcriptome is highly dynamic and changes considerably through time due to developmental progression and circadian rhythms, and in response to the environment. Post-translational control of transcription (co)activators has emerged as a crucial mechanism for reprogramming the transcriptome. This robust regulatory mechanism provides the cell with an on / off switch for transcription as well as a system to control the rate of this process in response to changing signals (Spoell *et al.*, 2010). Determination of posttranslational modifications (PTM) of proteins is fundamental in elucidation of the intricate processes that govern cellular events, like cell division, growth, and differentiation. PTMs are involved in most cellular processes including the maintenance of protein structure and integrity, regulation of metabolism and defense processes, and in cellular recognition events and morphology changes. Therefore, in order to precisely understand physiological processes it is important and essential to understand post translational modification. The ongoing research activities indicate that LC-MS/MS-based methodologies are constantly under development. It is expected that in the near future it will be finally possible to assess the true extent of post-translational modifications and their influence on cellular function.

References

Cain, J.A., Solis, N., and Cordwell, S. J. 2014. Beyond gene expression: The impact of protein post-translational modifications in bacteria. *J Proteomics,* **97**, 265–286.

Calvano, C. D., Zambonin, C. G., and Iensen, O. N. 2008. Assessment of lectin and HILIC based enrichment protocols for characterization of serum glycoproteins by mass spectrometry. *J. Proteomics,* **71**, 304-317.

Carr, S. A., Huddleston, M. J., and Annan, R. S. 1996. Selective detection and sequencing of phosphopeptides at the femtomole Analysis of Post-translational Modifications level by mass spectrometry. *Anal. Biochem.,* **239**, 180-192.

Drake, R. R., Schwegler, E. E., Malik Diaz, J., Block, T., Mehta, A., and Sem O. J. 2006. Lectin capture strategies combined with mass spectrometry for the discovery of serum glycoprotein biomark ers.10.1074/mcp.M600176-MCP200. *Cell. Proteomics,* **5,** 1957-1967.

Drews, Q., Zong, C., and Ping, P. 2007. Exploring proteasome complexes by proteomic approaches. *Proteomics,* **7,** 1047-1058.

Ellis, M., Egelund, J., Schultz, C.J., and Bacic, A. 2010. Arabinogalactan-proteins: key regulators at the cell surface? *Plant Physiol.*, **153,** 403 – 19.

Eyers, C. E., and Gaskell, S. J. 2008. Mass spectrometry to identify post-translational modifications. Wiley Encyclopedia of .Chemical Biology. doi:10.1002/ 9780470048672. wecb469.

Ficarro, S. B., McCleland, M. Stukenberg, P. T., Burke, D. J., Ross, M. Shabanowitz, J., Hunt, D. F., and Wt F. M. 2002. Phosphoproteome analysis mass spectrometry and its application to : *Saccharomyces cerevisiae. Nat. Biotechnol.,* **20,** 301-305.

Fraga, M. F., Ballestar, E., Villar-Garea, A., Boix-Chornet, M., Espada, J ., Schotta, G., Bonaldi, T., Haydon, C., Ropero, S., Petrie, K., Iyer, N. G., Perez-Rosado, A., Calvo, E., Lopez, J. A., Cano, A., Calasanz, M. J., Colomer, D., Piris, M. A., Ahn, N., Imhot~ A., Caldas, C., Jenuwein, T., and Esteller, M.

2005. Loss of acetylation at Lys16 and trimethylation at Lys20 of histone H4 is a common hallmark of human cancer. *Nat. Genet.,* **37**, 391-400.

Hanai, H., Nakayama, D., Yang, H., Matsubayashi, Y., Hirota, Y., and Sakagami, Y. 2000. Existence of a plant tyrosylprotein sulfotransferase: novel plant enzyme catalyzing tyrosineO-sulfation of preprophytosulfokine variants in vitro. *FEBS Lett.,* **470,** 97–101.

Haydon, C. E., Eyers, P. A., Aveline-Wolf, L. D., Resing, K. A., Maller, J. L., and Ahn, N. G. 2003. Identification of novel : phosphorylation sites on *Xenopus laevis* Aurora A and analysis of phosphopeptide enrichment by immobilized metal-affinity cromatography *Mol. Cell. Proteomics,* **2**, 1055-1067.

IUPAC 1997. Compendium of chemical I terminology, 2nd (The "Gold Book") ed, Blackwell Scientific Publications, Oxford.

Kieliszewski, M.J. and Lamport, D.T. 1994. Extensin: repetitive motifs, functional sites, post-translational codes, and phylogeny. *Plant J.*, **5,** 157–72.

Kieliszewski, M.J., de Zacks, R., Leykam, J.F., and Lamport, D.T. 1992. A repetitive proline-rich protein from the gymnosperm Douglas fir is a hydroxyproline-rich glycoprotein. *Plant Physiol.*, **98,** 919–26.

Kieliszewski, M.J., Lamport, D.T., Tan, L., and Cannon, M.C. 2010. Hydroxy-proline-rich glycoproteins: form and function. *Annu. Plant Rev.*, **41,** 321– 42.

Klomsiri, C., Karplus, P. A., and Poole, L. B. 2011. Cysteine-basedredoxswitches in enzymes. *Antioxid. Redox Signal.*, **14,** 1065–1077.doi:10.1089/ars.2010.3376.

Komori, R., Amano, Y., Ogawa-Ohnishi, M., and Matsubayashi, Y. 2009. Identification of tyrosylprotein sulfotransferase in *Arabidopsis*. *Proc. Natl. Acad. Sci. USA,* **106,** 15067–72.

Kweon, H. K, and Hakansson, K., 2006. Selective zirconium dioxide based enrichment of phosphorylated peptides for mass spectrometric analysis, doi:10.1021/ac0522355. *Anal. Chem,.* **78,** 1743-1749.

Matsubayashi, Y. 2011. Small post-translationally modified peptide signals in *Arabidopsis. Arabidopsis Book,* **9**, e0150.

Matsubayashi, Y. 2014. Posttranslationally Modified Small-Peptide Signals in Plants. *Annu. Rev. Plant Biol.*, **65**, 385-413.

Mayor, T., Graumann, J., Bryan, J., MacCoss, M. J., and Deshaies, R. J. 2007. Quantitative profiling of ubiquitylated proteins reveals proteasome substrates and the substrate repertoire influenced by the Rpn10 receptor pathway. *Mol. Cell. Proteomics,* **6**, 1885-1895.

Medzihradszky, K F., Guan, S., Malt by, D. A., and Burlingame, A. L. 2007. Sulfopeptide fragmentation in electron-capture and electron-transfer dissociation. *J. Am. Soc. Mass Spectrom.,* **18**, 1617-1624.

Moore K. L. 2003. The biology and enzymology of protein tyrosine O-sulfation. *J. Biol. Chem.,* **278,** 24243–46.

Morelle, W., and Michalski, J. C. 2007. Analysis of protein glycosylation by mass spectrometry. *Nat. Protoc.,* **2**, 1585-1602.

Myllyharju, J. 2003. Prolyl 4-hydroxylases, the key enzymes of collagen biosynthesis. *Matrix Biol.*, **22,** 15–24.

Nawarak, J., Phutrakul, S., and Chen, 2004. Analysis of lectin-bound glycproteins in snake venom from the Elapide and Viperidae families. *J. Proteome Res.,* **3,** 383-392.

O'Leary, B., *et al.* 2011. "The remarkable diversity of plant PEPC (phosphoenolpyruvate carboxylase): Recent insights into the physiological function and post-translational controls of non photosynthetic PEPCs," *Biochemical Journal*, **436**, 15–34.

Okamoto, S., Shinohara, H., Mori, T., Matsubayashi, Y., and Kawaguchi, M. 2013. Root-derived CLE glycopeptides control nodulation by direct binding to HAR1 receptor kinase. *Nat. Commun.*, **4,** 2191.

Pan, C., Gnad, F., Olsen, J. V., and Mann, M. 2008. Quantitative phosphoproteome analysis of a mouse liver cell line reveals specificity of phosphatase inhibitors. *Proteomics,* **8**, 4534-4546.

Patton, W.F. 2002. Detection technologies in proteome analysis. *J. Chromat. B: Analytical Technologies in the Biomedical and Life Sciences,* **771**, 3–31.

Picariello, G., Ferranti, P., Mamone, G., Roepstorff, P., and Addeo, F. 2008. Identification of N-linked glycoproteins in human milk by hydrophilic interaction liquid chromatography and mass spectrometry. *Proteomics,* **8**, 3833-3847.

Posewitz, M. C., and Tempst, P. 1999. Immobilized gallium(l11) affinity chromatography of phosphopeptides doi:10.1021/ac981409y. *Anal. Chem.,* **71,** 2883-2892.

Qiu, R., and Regnier, F. E. 2005. Use of Multidimensional lectin affinity Chromatography in differential glycoproteomics doi:10.1021/ac048751x. *Anal. Chem.,* **77,** 2802-2809.

Reinders, J., and Sickmann, A. 2005. State –of –the – art in phosphoproteomics. *Proteomics* **5,** 4052-4061.

Sahana Mollah, I. E. W., Phung, Q., Arnott, D., Dixit, V. M., Lill, J. R. 2007. Targeted mass spectrometric strategy for global mapping of ubiquitination on proteins. *Rapid Commun. Mass Spectrom,.* **21**, 3357-3364.

Schweppe, R.E., Haydon, C.E., Lewis, T, S., Resting, K.A., and Ahn, N.G. 2003. The characterization of protein post-translational modification by mass spectrometry. *Acc. Chem, Res.,* **36,** 453-461.

Spoel1, S. H., Tada, Y., and Loake, G. J. 2010. Post-translational protein modification as a tool for transcription reprogramming. *New Phytologist,* **186,**333-339.

Steen H, Kuster B, Fernandez M, Pandey A, Mann M. 2002. Tyrosine phosphorylation mapping of the epidermal growth factor receptor signaling pathway. *J. Biol. Chem.,* **277**, 1031–1039.

Steen, H., Iebanathirajah, J. A., Rush, J., Morrice, N., and Kirschner, M. W. 2006. Phosphorylation analysis by mass spectrometry: myths, facts, and the consequences for qualitative and quantitative measurements. *Mol. Cell. Proteomics,* **5,** 172-181.

Steen, H., Jebanathirajah, J. A., Springer M., and Kirschner, M. W. 2005. Stat isotope-free relative and absolute quantitation of protein phosphorylation stoichiometry by MS. *Proc. Natl. Acad. Sci. USA,* 102, 3948-3953.

Swaney, D. L., McAlister, G. C., Wirtala, M., Schwartz, J. C., Syka, J. E. P., and Coon, J. J. 2007. Supplemental activation method for high-efficiency electrontranster dissociation of doubly protonated peptide precursors. *Anal. Chem.*, **79**, 477-485.

Sweet, S. M., Mardakheh, F. K., Ryan, K. J., Langton, A. J., Heath, J. K., and Cooper, H. J. 2008. Targeted online liquid chromatography electron capture dissociation mass spectrometry for the localization of sites of in vivo phosphorylation in human Sprouty2. *Anal. Chem.*, **80**, 6650-6657.

Tyler, R. K., Chu, M. L., Johnson, H., McKenzie, E. A., Gaskell, S. J., and Eyers, P. A. 2009. Phosphoregulation of human Mpsl kinase. *Biochem. J.*, **417**, 173-181.

Walsh, C. 2005. Post translational modification of proteins: expanding nature's inventory. B. Roberts, Colorado.

Walsh, G. 2009. Post-translational modifications in the context of therapeutic proteins: An introductory overview. In: Post-translational Modification of Protein Biopharmaceuticals. (Ed. Walsh, G). WILEY-VCH Verlag GmbH & Co. KGaA, Weinheim, ISBN: 978-3-527-32074-5.

Washburn, M. P., Wolters, D., and Yates, J. R., 3rd 2001. Large-scale analysis of the yeast proteome by multidimensional protein identification technology. *Nat. Biotechnol.*, **19**, 242-247.

Weissman, A. M. 2001. Themes and variations on ubiquitylation. *Nat. Rev. Mol. Cell. Biol.*, **2**, 169-178.

Wu, J., Shakey, Q., Liu, W., Schuller, and Follettie, M. T. 2007. Global profiling of phosphopeptides by titania affinity enrichment. *J. Proteome Res.*, **6**, 4684-4689.

Wuhrer, M., de Boer, A. R., and Deelder, A. M. 2009. Structural glycomics using hydrophilic interaction chromatography (HILIC) with mass spectrometry. *Mass Spectrom. Rev.*, **28**, 192-206.

Zaffagnini, M., Bedhomme, M., Marchand, C. H., Morisse, S., Trost, P., and Lemaire, S. D. 2012. Redox regulation in photosynthetic organisms: focus on glutathionylation. *Antioxid. Redox Signal.*, **16,** 567–586.doi:10.1089/ ars. 2011.4255.

Zhang, Q., Schepmoes, A. A., Brock, J. W. C., Wu, S., Moore, R. J., Purvine, S. O., Baynes, J. W., Smith, R. D., and Metz, T. O. 2008. Improved methods for the enrichment and analysis of glycated peptides. doi:10.1021/ac801704j. *Anal. Chem.*, **80**, 9822-9829.

Zubarev, R. A. 2004. Electron-capture dissociation tandem mass spectrometry. *Curr. Opin. Biotechnol.*, **15**, 12-16.

Zubarev, R. A., Zubarev, A. R., and Savitski, M. M. 2008. Electron capture/transfer versus collisionally activated/induced dissociations: solo or duet? *J. Am. Soc. Mass Spectrom.*, **19**, 753-761.

Section V:

Medicinal Plants, In Vitro Regeneration and Natural Products

13

IN VITRO REGENERATION AND ACCLIMATIZATION STUDIES OF SOME OF THE IMPORTANT ENDANGERED MEDICINAL PLANTS OF ASIA

Kuldeep Yadav[1] and Narender Singh[1*]

Plant Tissue Culture Laboratory, Department of Botany, Kurukshetra University, Kurukshetra -136 119, India

**Corresponding author E-mail: nsheorankukbot11@gmail.com*

INTRODUCTION

Biodiversity provides to humankind enormous direct economic benefits in the form of timber, food, fiber, medicines, industrial enzymes, food flavors, fragrances, cosmetics, emulsifiers, dyes, plant growth regulators and pesticides. But its benefits to human health are incalculable and the plants, since time immemorial, have been used globally across the varied cultures throughout the known civilizations as a valuable and safe natural source of medicine and agents of therapeutic, industrial and environmental utilities.

Medicinal plants have the subject of man's curiosity since time immemorial (Constable, 1990). Almost every civilization has a history of use of medicinal plant (Ensminger *et al.*, 1983). The World Health Organization has listed 20,000 medicinal plants globally (Gupta and Chadha, 1995); India's contribution is 15-20 per cent (Singh, 2000). It has been found that one out of every 125 plants species studied at the Herb Research Foundation, Boulder, produced a major drug along with a market value of $ 200 million per year in the US. It has also been reported that of the 118 (out of the top 150) prescription drugs in the US, 75.4 percent are based on plants (Dobson, 1995). In India, about 2000 drugs used are of plant origin (Dikshit, 1999).

Human beings have been utilizing plants for basic preventive and curative health care. Approximately 80 per cent of the people in the developing countries still rely on traditional medicine for their primary health care needs. This usually involves the use of plant extracts (Viera and Skorupa, 1993). In India, the references to curative properties of some herbs in the Rig-Veda seem to be the earliest records of plants in medicine.

PROBLEM STATEMENT

Interest in phytomedicine has exploded in the last few years and about 500 different plant species are as key ingredients and many are still being collected from the wild (Mendelsohnm and Balick, 1994). The trend of using natural products is increasing steadly. Herbal medicines have attained much more popularity in the treatment of ailments due to increasing awareness of personal health maintenance through natural products. The resurgence of public interest in plant-based medicine coupled with rapid expansion of pharmaceutical industries has necessitated an increased demand for medicinal plants, leading to over exploitation that threatens the survival of many rare species. Also, many medicinal plants are disappearing at an alarming rate due to rapid agricultural and urban development, uncontrolled deforestation and indiscriminate collection. As a result, 20-25 percent of existing plant species in India has become endangered. The degree of threat to natural populations of medicinal plants has increased because more than 90 per cent of medicinal plant raw material for herbal industries in India and also for export is drawn from natural habitat (Dhar *et al.*, 2002). The IUCN Red list (2007) revealed that the number of threatened species is increasing gradually.

Most of the plants raised through seeds are highly heterozygous and show great variations in growth, habit and yield and many have to be discarded because of poor quality of products for their commercial release. Likewise, majority of the plants are not amenable to vegetative propagation through cutting and grafting, thus limiting multiplication of desired cultivars. Moreover many plants propagated by vegetative means contain systematic bacteria, fungi and viruses which may affect the quality and appearance of selected items. Most of the medicinal plants either do not produce seeds or seeds are too small and do not germinate in the soil. Thus mass-multiplication of disease free planting material is a general problem.

Biotechnology has emerged as a strong tool in mass multiplication and improvement of all plant species. Biotechnology involving modern tissue culture, cell biology and molecular biology offers the opportunity to develop new germplasm that are well adapted to changing demands (Rout *et al.,* 2000). The tissue culture technique has emerged as a potential method to meet the challenges in this direction.

HISTORY

This technique in fact is the outcome of work of Haberlandt (1902) who attempted first plant cell culture *in vitro* on a nutrient medium and laid the foundation stone of plant tissue culture. He gave the concept of totipotency by culturing palisade layer of leaf from a plant *Lamium purpureum*. Went (1926) discovered first plant growth hormone - Indole acetic acid (IAA). The first successful organ culture was achieved by White (1934) with the demonstration of potentially unlimited growth of excised tomato root tips. Van Overbeek (1941) was first to add coconut milk for cell division in *Datura*. Skoog and Miller (1957) gave the concept of hormonal control (auxin: cytokinin) of organ formation. Some other notable scientific discoveries for controlling plant regeneration and morphogenesis in culture included basic and applied aspect of organogenesis and somatic embryogenesis (Reinert, 1959), practical micropropagation and production of virus free plants (Morel, 1960), culture and regeneration of protoplasts (Cocking, 1960), test tube fertilization in flowering plants (Kanta *et al.,* 1962), haploid plant production (Guha and Maheshwari, 1964; Bourgin and Nitch, 1967), production of secondary metabolites (Kaul and Staba, 1965), introduction of term somaclonal variations (Larkin and Scowcroft, 1981), electrofusion of protoplasts (Zimmerman and Scheurich, 1981), use of immobilized cells for bio-transformation of digitoxin introdigoxin (Alfermann *et al.,* 1983), development of binary vector system for plant transformation (Fraley *et al.,* 1985), transformation of protoplast by direct DNA microinjection (Crossway *et al.,* 1986), first monocot transformation by *Agrobacterium tumefaciens* (Bytebier *et al.,* 1987), automated mass propagation with organogenesis and embryogenesis (Levi *et al.,* 1986), electroporation of intact plant tissues for direct delivery (Dekeyser *et al.,* 1990), production of first transgenic plant (Huang *et al.,* 1991), successful metabolic engineering for increased alkaloid production (Yun *et al.,* 1992), improved plant transformation through electrophoresis (Giesbach, 1994), development of a binary bacterial artificial chromosome (BBAC) vector for *Agrobacterium* mediated transformation (Hamilton *et al.,*1996).

PLANT TISSUE CULTURE

Plant tissue culture refers to growing and multiplication of cells, tissues and organs of plants on defined solid or liquid media under aseptic and controlled environmental condition. Micropropagation refers to *in vitro* plant propagation in an aseptic, controlled and artificial environment which includes axillary budding, induction of adventitious buds and somatic embryogenesis. Beside micropropagation, plant tissue culture is also employed in haploid production, production of disease free and resistant plants, elimination of breeding barriers, biosynthesis of secondary metabolites, generation of variability, germplasm conservation and selection of desirable traits (Pijut *et al.*, 1990; Karp, 1994; Roja and

Rao, 1988). The efficient regeneration of plants from cell, tissue, and organ culture is recognized as prerequisite for application of modern genetic and biotechnological approaches to crop improvement (Litz and Gray, 1992; Rai *et al.*, 2010). Tissue culture derived plant material provides industrial source of various necessary metabolic compounds which are of medicinal value (Kumar *et al.*, 2011).

The process of micropropagation is usually divided into several stages, i.e., initiation of explants, subculture of explants for proliferation, shooting, rooting and hardening. Success of *in vitro* regeneration depends on a series of stages, each with a specific set of requirements (Fernie and Willmitzer, 2001). Some of these include kinds of tissue or explants, composition of medium, plant growth regulators (PGRs), media additives, culture environment etc. (Forsyth and Van Staden, 1984; Vaillant *et al.*, 2005; Grison, 1991; Ondo Ovono *et al.*, 2009). The *in vitro* culturing is dependent mainly on the composition of the media. Among the different media, MS (Murashige and Skoog, 1962) medium was found to be the most common and frequently used media. The MS medium was used either as described originally or with little variation and combination of phytohormones. The ratio of auxin and cytokinin also play an important role in the rooting and shooting in plant tissue culture. It has been reported that low concentration of auxin and high concentration of cytokinin promotes shoot induction whether high concentration of auxin and low concentration of cytokinin promotes rooting but the intermediate concentration induces callus formation (Murashige, 1974). Explants selection also play a significant role in regeneration of plants. These involves selection of organs that serve as tissue source, physiological and ontogenic stage of organ, the season in which explant is being obtained; the size of explant and overall quality of plant from which explants are to be obtained.

Organized growth occurs either when as growing points (apical meristems) of shoots, roots, leaf initials etc. are transferred to the culture or when such structures are formed during the culture of previously unorganized tissue. Process of organ formation is called as organogenesis. This adopts two pathways one when the direct development of shoots from the axillary and lateral buds and another when shoot formation takes place via callus formation. Organogenesis has been reported in different plants involves utilization of leaf segments, axillary bud, shoot tips, nodal segments, cotyledonary nodes, hypocotyls, epicotyls, cotyledons, immature floral buds with inflorescence etc.

The process of embryo initiation and development from vegetative or non-gametic cells is known as somatic embryogenesis. This involves two pathways:

1. When the direct development of somatic embryos take place from the explant.
2. When somatic embryo formation takes place via callus.

It involves the formation of bipolar structure both shoot and root meristems, and develops in a manner similar to zygotic embryos. These embryoids can develop in to fully functional plants under appropriate condition.

Use of tissue culture for micropropagation was initiated by Morel (1963) for haploid propagation of orchid. Some commercially important rare and endangered medicinal plants have been successfully regenerated under *in vitro* conditions such as *Picrorhiza kurroa* (Lal *et al.,* 1988), *Saussurea lappa* (Arora and Bhojwani, 1989; Sudhakar *et al.*, 1997), *Chlorophytum borivilianum* (Purohit *et al.,* 1994; Sharma and Mohan, 2006), *Sterculia urens* (Purohit and Dave, 1996), *Rauwolfia micrantha* (Sudha and Seeni, 1996), *Rauvolfia serpentina* (Sarker *et al.,* 1996), *Vitex negundo* (Sahoo and Chand, 1998), *Terminalia arjuna* (Kumari *et al.,* 1998), *Eremurus persicus* (Verma *et al.,* 2002), *Rotula aquatic* (Martin, 2002), *Curculigo orchioides* (Wala and Jasrai, 2003), *Saussurea obvallata* (Joshi and Dhar, 2003), *Bacopa monnieri* (Mohapatra and Rath, 2005), *Celastrus paniculatus* (Rekha *et al.*, 2005), *Chlorophytum arundinaceum* (Lattoo *et al.,* 2006), *Santalum album* (Sanjaya *et al.,* 2006), *Wattakaka volubilis* (Chakradhar and Pullaiah, 2006), *Splachnum ampullaceum* (Gonzalez *et al.*, 2006), *Saussurea involucrata* (Guo *et al.,* 2007), *Tylophora indica* (Faisal *et al.,* 2007), *Citrus indica* (Lasker *et al.,* 2009), *Swertia chirata* (Balaraju *et al.,* 2009), *Clerodendrum serratum* (Sharma *et al.,* 2009), *Sarcostemma brevistigma* (Thomas and Shankar, 2009), *Picrorhiza kurroa* (Jan *et al.,* 2009), *Ilex khasiana* (Dang *et al.,* 2011), *Terminalia bellerica* (Mehta *et al.,* 2012), *Heliotropium keralense* (Sebastian and Hariharan, 2013), *Muscari muscarimi* (Uzan *et al.,* 2014) etc. (Table 1).

Table 1: *In vitro* multiplication of various rare and endangered medicinal plants

S.N.	Plant name	Explant used	Media type	Best response	Reference
1	*Picrorhiza kurroa*	shoot tip	MS	Kn (4.0 mg/L)+ IAA (1.0 mg/L)	Lal *et al.* (1988)
2	*Saussurea lappa*	cotyledon	MS	BAP (5.0 µM) + GA_3 (3.0 µM) + IAA (0.1µM)	Arora and Bhojwani (1989)
3	*Chlorophytum borivilianum*	shoot bud	MS	BA (22.2µM)	Purohit *et al.* (1994)
4	*Saussurea lappa*	cotyledonary shoot tip	MS	TDZ (0.45µM)	Sudhakar *et al.* (1997)
5	*Sterculia urens*	cotyledonary node	MS	BAP (2.0 mg/L)	Purohit and Dave (1996)
6	*Rauwolfia micrantha*	node	MS	(13.2 µM) BA + (2.68 µM) NAA	Sudha and Seeni (1996)

S.N.	Plant name	Explant used	Media type	Best response	Reference
7	*Rauvolfia serpentina*	node	MS	1.0 mg/L BA and 0.1 mg/L NAA	Sarker *et al.* (1996)
8	*Terminalia arjuna*	leaf	MS	2,4-D (5.0 mg/L) + Kn (0.01 mg/L)	Kumari *et al.* (1998)
9	*Eremurus persicus*	root tuber	MS	IAA (1.0mg/1) + BA (1.0 mg/L) + casein hydrolysate(1000 mg/L)	Verma *et al.* (2002)
10	*Rotula aquatic*	axillary bud	MS	BAP (1.0 mg/L) + IBA (0.5 mg/L)	Martin (2002)
11	*Curculigo orchioides*	meristem-tip	MS	BA (2.21 µM)	Wala and Jasrai (2003)
12	*Saussurea obvallata*	epicotyle	MS	Kn (1.0 µM) + NAA (0.25 µM)	Joshi and Dhar (2003)
13	*Bacopa monnieri*	leaf and node	MS	BAP (2.0 mg/L)	Mohapatra and Rath (2005)
14	*Chlorophytum arundinaceum*	shoot crown	MS	Kn (4 x 10^{-6} M) + IBA (2 x 10^{-6} M)	Lattoo *et al.* (2006)
15	*Chlorophytum borivilianum*	immature floral buds	MS	Kn (2.0 mg/L) + 2,4-D (1.0 mg/L)	Sharma and Mohan (2006)
16	*Vitex negundo*	node	MS	BA (2.0 mg/L)	Sahoo and Chand (1998)
17	*Santalum album*	node	MS	NAA (0.53µM) + BA (11.09µM)	Sanjaya *et al.* (2006)
18	*Wattakaka volubilis*	seedling node	MS	Kn (5.0 mg/L) + NAA (0.1 mg/L)	Chakradhar and Pullaiah (2006)
19	*Saussurea involucrata*	leaf	MS	BAP (10.0 µM) + NAA (2.5 µM)	Guo *et al.* (2007)
20	*Tylophora indica*	node	MS	BA (2.5 µM) + NAA (0.5 µM) + AA (100 mg/L)	Faisal *et al.* (2007)
21	*Citrus indica*	leaf	WPM	BAP (0.5 mg/L) + TDZ (0.25 mg/L) + NAA (0.25 mg/L)	Lasker *et al.* (2009)
22	*Swertia chirata*	seedling shoot tip	MS	BAP (1.0 mg/L) + Kn (0.1 mg/L)	Balaraju *et al.* (2009)

S.N.	Plant name	Explant used	Media type	Best response	Reference
23	*Clerodendrum serratum*	node	MS	BAP (0.25 mg/L) + IAA (0.25 mg/L) + AdS (15 mg/L)	Sharma *et al.* (2009)
24	*Sarcostemma brevistigma*	node	MS	BA (4.0 µM)	Thomas and Shankar (2009)
25	*Picrorhiza kurroa*	node	MS	NAA (0.6 mg/L)	Jan *et al.* (2009)
26	*Ilex khasiana*	seedling node	MS	BA (6.63 µM)	Dang *et al.* (2011)
27	*Terminalia bellerica*	seedling node	MS	BAP (3.5 mg/L) + IBA (1.0 mg/L)	Mehta *et al.* (2012)
28	*Heliotropium keralense*	root	MS	2,4-D (0.5 mg/L) + 2 iP (4.0 mg/L)	Sebastian and Hariharan (2013)
29	*Muscari muscarimi*	zygotic embryo	MS	BAP (4.0 mg/L) + NAA (0.5 mg/L)	Uzan *et al.* (2014)

Gloriosa superba L.

Gloriosa superba L. (family Colchicaceae) also known as Malabar glory lily or Kalihari is a perennial tuberous climbing herb, widely scattered in tropical and sub-tropical parts of Africa and Southeast Asia (Fig. 1a-b). In India, the plant is usually found in the Himalayan foot-hills, Central India, Tamil Nadu, Andhra Pradesh and Bengal (Sivakumar and Krishnamurthy, 2002). In Haryana, it is rarely found in Kurukshetra, Pinjoor and Morni regions (Jain *et al.*, 2000). Its attractive wavy edged orange-red flower is the national flower of Zimbabwe and also the state flower of Tamil Nadu in India. A postal stamp is also issued by the Postal Department of India for this flower.

The plant is used to cure cancer, arthritis, mediterranean fever, gout, rheumatism, inflammation, ulcers, bleeding piles, skin diseases, leprosy, impotency and snake-bites (Kapoor, 2001; Jana and Shekhawat, 2011). The tubers and seeds of this plant contain a highly demanding and very costly alkaloid, colchicine (Rajagopal and Kandhasamy, 2009) (Fig. 2a). Due to the action of colchicoside on spindle-fibre formation during cell-division, the plant has been identified as a potential anti-cancer drug. Colchicine is capable of inducing polyploidy and is used by a number of researchers in cytological and plant breeding research (Kumar, 1953). The colchicine content in the genera *Colchicum* has been reported to be lower than in *Gloriosa* (Bellet and Gaignault, 1985).

Usually *G. superba* is vegetatively propagated by tubers but the rate is very low as only two tubers are produced per year (Krause, 1986).

Moreover, the fungal contamination is also associated with the tubers such as *Curvularia lunata* and tuber-rot caused by *Sclerotium* species (Mrudul *et al.*, 2001). In nature, less seed germination with poor viability is responsible for its diminishing population size. The poor propagation coupled with over exploitation by the local population as well as pharmaceutical companies has put this plant into acutely threatened species (Sivakumar and Krishnamurthy, 2000).

Fig. 1. Mature plant and closer view of: **(A-B)**: *G. superba*; **(C-D)**: *C. paniculatus* (Lal *et al.*, 2010); **(E-F)**: *S. acmella*; **(G)**: *A. calamus*; **(H-I)**: *A. marmelos.*

Explant type and growth regulators

Ghosh *et al.* (2002) established the root cultures of *G. superba* from organogenic callus cultures initiated from petiole explants in SH medium (Schenk and Hildebrandt, 1972) containing 1.0 mg/L α-naphthalene acetic acid (NAA) and 1 mg/L kinetin (Kn). They also concluded that all precursors (except *trans*-cinnamic acid) enhanced colchicine content of root cultures without adversely affecting root growth and *p*-coumaric acid + tyramine treatment (each at 20 mg/L) increased colchicine content to 1.9 mg/g cell dry weight. Sivakumar *et al.* (2003) tested three different types of explants viz. dormant, non-dormant corm buds and *in vitro* derived multiple shoots and obtained excellent responses using MS medium supplemented with B5 vitamins, 6% sucrose, 2iP (N6-(2-Isopentenyl) adenine), AdS (Adenine sulphate) and ANL for corm formation using dormant corm buds as explants. Sivakumar and Krishnamurthy (2004) inoculated six categories of explants viz. roots, corm buds, leaves, stems, pedicels and shoot tips and noticed maximum callus induction (98%) in non dormant corm bud explants cultured on 4.52 μM 2,4-D and 2.32 μM Kn (Kinetin). The greatest number of multiple shoots (57) was observed in corm-derived calluses. Sivakumar *et al.* (2004) observed maximum colchicine (0.02%) production from the calluses induced from corm buds using B5 medium supplemented with tryptophan 5 mM + NAA 5.37 mM after twenty five days. Addition of 15% coconut water (CW) and 2 g/L activated charcoal (AC) to MS medium containing 1.5 mg/L BA + 0.2 mg/L NAA was found to increase the number of shoots from eight to fifteen per culture (Sayeed Hassan and Roy, 2005). Ghosh *et al.* (2007) inoculated the non dormant tubers on three different basal media, viz., MS (Murashige and Skoog, 1962) (3% sucrose), B5 (Gamborg *et al.*, 1968) (2% sucrose) and SH (Schenk and Hildebrandt, 1972) (3% sucrose) with or without different plant growth regulators and found that MS medium as most effective in inducing secondary tubers *in vitro*. Madhavan and Joseph (2010) observed callus induction from internodal explants cultured on MS medium supplemented with 2,4-D (3.0 mg/L) and BAP (0.5 mg/L) followed by emergence of shoot initials only in MS medium supplemented with Kn (2.0 mg/L) and 2,4-D (1.0 mg/L). Ade and Rai (2011) compared different growth media, viz., Murashige and Skoog (MS), B5 (Gamborg *et al.*, 1968), NN (Nitsch and Nitsch, 1969), WM (White, 1963) and Chu's N6 media having different combinations of auxins (2,4-D, IAA and IBA) and cytokinins (BAP and BA) and found MS medium as most suitable for the calli induction and multiple shoot formation followed by Gamborg's medium. Yadav *et al.* (2012) reported greatest *in vitro* tuberization (3.2) with the highest shoot regeneration frequency (90%) from the non-dormant tubers explants cultured on MS medium supplemented with BAP (2.0 mg/L) and NAA (0.5 mg/L) with 3% sucrose under16-h photoperiod (Fig. 3a).

(a) Colchicine

(b) Spilanthol

(c) Aegelinosides A and B

Fig. 2. Molecular structures of (a)Colchincine (b) Spilanthol (c) Aegelinosides A&B

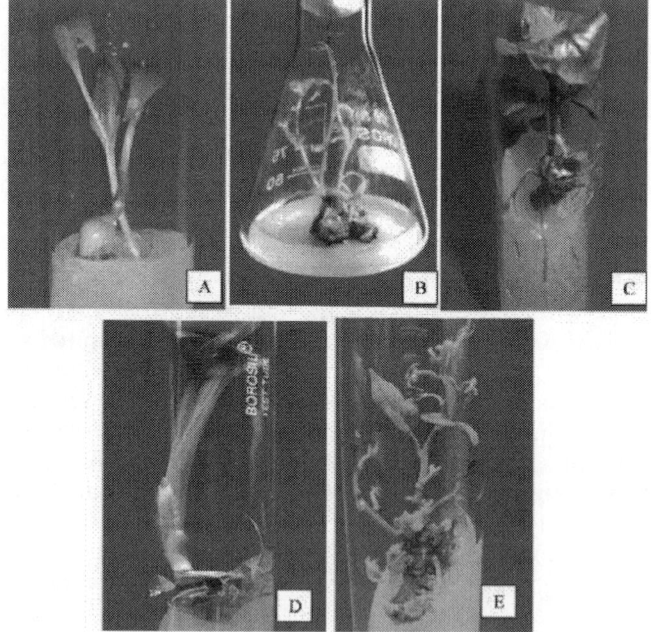

Fig. 3. ***In vitro* shoot regeneration in (A)**: *G. superba* (Yadav *et al.*, 2012); **(B)**: *C. paniculatus* (Lal *et al.,* 2010); **(C)**: *S. acmella* (Yadav and Singh, 2011a); **(D)**: *A. calamus*; **(E)**: *A. marmelos* (Yadav and Singh, 2011b).

Root development and acclimatization

Jadhav and Hegde (2001) transferred the micropropagated plantlets to pots containing soil, sand and farmyard manure (2:1:1) and obtained eighty percent survival rate. Sivakumar *et al.* (2003) noticed that, when mature embryoids obtained from leaf tissue of *G. superba* L. were transferred to BAP and IBA medium, gave rise to plantlets with single shoots and roots. Sayeed Hassan and Roy (2005) obtained highest percent rooting on half strength MS medium with 1.0 mg/L IBA + 0.5 mg/L IAA added as a supplement. Ghosh *et al.* (2007) noted best *in vitro* rooting in half strength liquid MS medium with one-fifth nitrates. Arumugam and Gopinath (2012) obtained ninety-six percent *in vitro* rooting by transferring the *in vitro* raised shootlets on MS medium containing BAP (8.0 mg/L) + GA_3 (1.0 mg/L) + Zen (0.5 mg/L) + NAA (1.0 mg/L) + AC (2.0 g/L) and transferred the rooted plantlets into small plastic nursery tray containing vermi-compost, sand and red soil (1:2:2). Yadav *et al.* (2012) transferred the *in vitro*-regenerated shoots to half-strength MS medium with 1.0 mg/L IBA + 0.5 mg/L NAA and obtained maximum root induction (66.6%). Later, these micropropagated plantlets were successfully acclimatized and transplanted to sterilized soil and sand mixture (3:1) in the glasshouse with seventy percent survival (Fig. 4a, d).

The amplification products of the regenerated plants showed similar banding patterns to that of the mother plant thus demonstrating the homogeneity of the micropropagated plants (Yadav *et al.,* 2013).

Celastrus paniculatus Willd.

Celastrus paniculatus Willd. (Celastraceae) commonly known as Malkangni, Jyotishmati, Bitter sweet is a woody climbing shrub with pendulous branches covered with small, white, wart-like excrescences (Fig. 1c-d). Leaves alternate, glabrous, membranous, ovate or orbilocular, toothed and abruptly pointed. Flowers are polygamous, pale yellow green in colour and present in terminal pendulous panicles (Rajesekharan and Ganeshan, 2002).

Seeds of this plant are the source of an Ayurvedic drug Jyothismati used in treating rheumatism, gout and neurological disorder. *Celastrus paniculatus* is well known for its ability to improve memory (Nadkarni, 1976). Pharmacological studies suggest that the oil obtained from the seeds possess sedative and anticonvulsant properties (Gatinode *et al.*, 1957). Seed oil has also been found to be beneficial to psychiatric patients (Hakim, 1964) and increased the intelligence quotient of mentally retarded children (Nalini *et al.*, 1986). The seed oil also reduces the cholesterol level by converting tryptophan into niacin which is responsible for depression in blood cholesterol level (Berdia and Verma, 1994).The seed oil is useful for treating abdominal disorders, beriberi and sores (Warrier *et al.*, 1994). Celapagin, celapanigin, celapanin, celastrine and paniculatine are the active constituents found in the seed oil of *C. paniculatus* (CSIR, 1992) (Fig. 2b).

Fig. 4. (A-C). *In vitro* rooting of: *G. superba*, *C. paniculatus* (Lal *et al.*, 2010) and *S. acmella* (Yadav and Singh, 2010) respectively; **(D-E)**: Acclimatized plants of *G. superba* and *S. acmella* (Yadav and Singh, 2010) respectively.

Leaf sap is an emmenagogue and antidote for opium poisoning (Warrier *et al.*, 1994). Flower possesses analgesic and anti-inflammatory activities. Bark is reported to be abortifacient, depurative and a brain tonic and taken internally for snake bite (Govil, 1993). Root-bark extract also shows antimalarial activity (Rastogi and Mehrotra, 1998). The powdered root is considered useful for the treatment of cancerous tumours (Parotta, 2001).

The conventional method of propagation of this medicinally important plant is through seeds. Poor seed viability and germination restricts the use of seeds in multiplication. Rekha *et al.* (2005) reported seed germination as low as 11.5%. Indiscriminate over exploitation from natural source to meet the growing demand by pharmaceutical industry coupled with low seed viability, lack of vegetative propagation methods and insufficient attempts for replenishment of wild stock of this medicinally important plant species have contributed to its threatened status.

Explant type and growth regulators

Arya *et al.* (2001) inoculated nodal explants on various culture media, viz., B5 (Gamborg *et al.*, 1968), MS (Murashige and Skoog, 1962) and SH (Schenk and Hildebrandt, 1972) medium and obtained highest number of shoots (4-5) on MS medium supplemented with 1.5 mg/L BAP + 0.1 mg/L NAA + additives (50 mg/L ascorbic acid and 25 mg/L each of adenine sulfate, arginine and citric acid). Nair and Seeni (2001) inoculated various explants (node, shoot tip, internode and leaf base) on MS medium and obtained maximum number of shoots (3.6) from nodal explants at 1.0 mg/L of BAP. Martin *et al.* (2006) achieved five shoots per nodal explants cultured on MS medium supplemented with 1.5 mg/L of benzyl adenine (BA) and 0.1 mg/L of NAA after two subculture cycles. De Silva and Senarath (2009) observed that MS medium supplemented with 5.0 µM BAP and 7.0 µM IAA, as the best medium for callus induction from different explants (nodal segments, shoot tips and leaf discs). Lal *et al.* (2010) noticed maximum callus growth from hypocotyls explants inoculated on MS medium fortified with 2.0 mg/L NAA in combination with 1.0 mg/L BAP (Fig. 3b). Lal and Singh (2010) also recorded maximum number of shoots (8.9) per nodal explant with hundred per cent bud break, when inoculated on MS medium supplemented with 1.0 mg/L BAP under 16 hrs photoperiod. Phulwaria *et al.* (2012) obtained maximum number of shoots on MS medium supplemented with BAP (0.5 mg/L) and IAA (0.1 mg/L) from nodal explants. Senapati *et al.* (2013) noticed maximum percentage of shoot multiplication (83.4%) with 8.2 shoots/explants on MS medium supplemented with 0.5 mg/L BAP + 0.1 mg/L NAA from nodal explants.

Root development and acclimatization

Arya *et al.* (2001) noticed poor rooting frequency (only 20–30%) under *in vitro* conditions, but by pulse treatment with 100 mg/L each of IBA + NOA (ß-naphthoxy acetic acid) + 10 mg/L of chlorogenic acid solution for 2-3 minutes, they noticed that eighty-three percent of regenerated shoots showed *ex vitro* rooting. More than 90% of *ex vitro* rooted plants survived after hardening and acclimation. De Silva and Senarath (2009) found MS medium with 5.6 µM IAA + 9.6 µM IBA as the best (73.3 %) medium for *in vitro* rooting of *C. paniculatus*. Further, the micropropagated plantlets were acclimatized using a mixture of river sand: top soil: compost (1:1:1) with highest survival rate (80.0%).While, Lal *et al.* (2010) noticed hundred percent *in vitro* rooting with half strength MS medium fortified with 0.5 mg/L NAA (Fig. 4b). Root formation with callusing at the shoot base occurred in all concentrations of auxins. The micropropagated plantlets were successfully acclimatized in pots containing sterilized soil and sand mixture (3:1) with seventy percent survival. Phulwaria *et al.* (2012) compared the two auxins, NAA and IBA for *ex vitro* root induction and observed hundred per cent rooting in the regenerated shoots treated with indole-3-butyric acid (300 mg/L) for 3 min. Senapati *et al.* (2013) achieved maximum rooting of 73.3% with 4.8 roots/shoot on half-strength MS media fortified with 0.5 mg/L indole-3-acetic acid (IAA).

Phulwaria *et al.* (2012) and Senapati *et al.* (2013) assessed the genetic fidelity of the regenerated plants using random amplified polymorphic DNA and detected no polymorphism among the regenerated plants and mother plant.

Spilanthes acmella Murr.

Spilanthes acmella Murr. commonly known as Akarkara or toothache plant is an important medicinal plant with rich source of therapeutic constituents. It belongs to family Asteraceae, the tribe Helianthae, and the sub-tribe Ecliptinae. The plants are annual herbs, or short-lived perennials, approximately half meter tall with prostrate or ascending branched cylindrical hairy stems and simple ovate opposite leaves with stipules. They have characteristic yellow flower heads (Anonymous, 1989) (Fig. 1e-f).

The herb is widely distributed in tropics and sub-tropics including tropical America, North Australia, Africa, Malaya, Borneo, India, Sri Lanka etc (Jansen, 1981). In India, it has been reported from South India, Chattishgarh and Jharkhand and from Morni hills in Haryana (Singh, 1995).

The flowers and leaves of this plant have been used as traditional medicine for stammering, toothache, stomatitis and throat complaints. It has potent diuretic activity and the ability to dissolve urinary calculi. It exhibits antimalarial properties as well (Singh, 1995; Ramsewak *et al.*, 1999; Pandey and Agarwal, 2009). The major pungent compound reported in *S. acmella* is alkaloid spilanthol (N-isobutyl-2,6,8- decatrienamide) (Fig. 2b). Spilanthol, the most active antiseptic alkaloid extracted from this plant, is found effective at extremely low concentrations against blood parasites, and indeed is a poison to most invertebrates while remaining harmless to warm-blooded creatures. It is further recommended as a cure for dysentery and rheumatism, and to enhance the immune system. It exhibits general immunomodulator properties when used internally, boosting production of leukocytes and antiviral interferon, as well as promoting phagocytosis. It stimulates wound healing, protects the individual from cold and flu (Anonymous, 1989). The leaves are also used to treat bacterial and fungal skin diseases.

With the increasing worldwide demand currently, the interest in utilization and conservation of medicinal plants is rapidly increasing, and mass micropropagation of plants via *in vitro* culture techniques has become an alternative approach to produce large-scale rapid plant multiplication to extract valuable chemical products rather than from plants grown and harvested in the field (Rao and Reddy 1983). A lot of research efforts are being made to develop and refine micropropagation methods for large-scale plant multiplication of *Spilanthes acmella.*

Explant type and growth regulators

Saritha *et al.* (2002) recorded highest shoot proliferation (98%) with maximum number of shoots per hypocotyl explants cultured on MS medium supplemented with 0.5 mg dm^{-3} benzyladenine (BA) in combination with 0.1 mg dm^{-3} NAA. Haw and Keng (2003) reported multiple shoots from axillary bud explants on MS medium supplemented with BAP (2.0 mg/L). Leng *et al.* (2004) studied the effect of explant type, explant orientation, culture temperature and culture vessel type on *in vitro* regeneration of *Spilanthes acmella.* They found that nodal explants cultured vertically in Erlenmeyer flask (250, 500, or 1000 mL) on MS medium supplemented with 0.5 mg/L BA produced normal multiple shoots upto the temperature of 28°C. Pandey and Agrawal (2009) reported best green and compact callus on MS medium supplemented with 1.0 µM NAA and 10.0 µM BA using seedling leaf explants. Singh *et al.* (2009a) achieved best gel complexation for encapsulation of shoot tips using 3% sodium alginate and 100 mM calcium chloride. Maximum percent response for the conversion of encapsulated shoot tips into plantlets was obtained on growth regulator-free full-strength liquid MS medium. Singh *et al.* (2009b) established an efficient and improved *in vitro* propagation system for *Spilanthes acmella* L. using transverse thin cell layer (tTCL) culture method. Yadav and Singh (2010) recorded maximum number of shoots (4.3) per nodal explant in the MS medium fortified with 1.0 mg/L of BAP. Yadav and Singh (2011) obtained maximum *in vitro* flowering percent (50%) of cultured nodal explants on MS medium containing BAP (2.0 mg/L) + IAA (0.5 mg/L) under photoperiod of 16 hrs photoperiod after 7 weeks of culture (Fig. 3c). Joshi and Jadhav (2013) cultured nodal segments on different basal media (MS, B5, N and N, WM and LS) and obtained maximum *in vitro* shoot proliferation on MS media supplemented with 3% glucose. Singh *et al.* (2013) found that MS medium supplemented with 5.0 µM N6-benzylaminopurine, 1.0 µM 2,4-dichlorophenoxyacetic acid and 1.0 µM α-naphthaleneacetic acid as the best medium to induce dedifferentiation from the leaf cells and for further maintenance of the callus cultures. Sharma and Shahzad (2013) found that MS medium supplemented with 0.25 µM TDZ induced a maximum of 30.0 shoots per explant.

Root development and acclimatization

Haw and Keng (2003) observed that MS medium supplemented with 0.5 mg/L BA alone was sufficient for proliferating *in vitro* rooting of multiple shoots of *S. acmella.* Pandey and Agrawal (2009) noted hundred percent *in vitro* rooting in all the excised shoots inoculated in half strength MS medium supplemented with IBA (0.1 µM) within two weeks. Saritha and Naidu (2008) and Yadav and Singh (2010) also achieved *in vitro* rooting using MS medium containing IBA (Fig. 4c). Singh *et al.* (2009b) observed hundred percent rooting after 4 weeks on growth regulator free MS medium. Singh and Chaturvedi (2010) successfully rooted *in vitro* excised shoots on half-strength MS medium with 50 g/L sucrose, with a frequency of hundred percent. Supplement of either 6% sucrose or 6% glucose in MS medium also produced *in vitro* rooting in the plantlets which may be due to the increased osmotic stress (Joshi and Jadhav, 2013). All of the above researchers noticed that more than 60% of the plantlets were successfully acclimatized and established in soil (Fig. 4e). Singh and Chaturvedi (2010) noticed comparable amount of scopoletin content (0.10 µg/g DW) in the leaves of micropropagated *S. acmella* plants to those of field-grown mother plants, as estimated by HPLC technique.

***Acorus calamus* Linn.**

Acorus calamus Linn. (family Araceae) is a semiaquatic herb with creeping rhizomes and sword shaped long leaves commonly known as "sweet flag" or "Bach" is an important endangered medicinal plant having great medicinal properties (The Wealth of India, 2003). It is found in marshy land, shallow water and pond edges of the Northern temperate, sub-tropical and warm regions of Indian sub-continent. It grows upto a height of 3-4 feet with sword-shaped leaves, small yellow/ green flowers and branched rhizomes (Fig. 1g) (Verma and Singh, 2012).

Various bioactive molecules, *viz.*, acorin, α- and β–asarone, asaryldehyde, caryophylene, isoasarone, methylisoeugenol, and safrol have been isolated from this plant (Fig. 2d). This plant has been used in traditional medicine since time immemorial for its antispasmodic, carminative, antidepressant and antianthelmintic properties (McGaw *et al.* 2002). It is also used for treatment of epilepsy, mental ailments, chronic diarrhea, dysentery, bronchial catarrh, intermittent fevers and tumors (Anonymous, 2000). It has also been used in various products like shampoo, bath soap, skin care, perfume and air fresheners.

Due to its commercial importance and extensive use in medicines, there has been a growing demand for this plant. Conventionally, the plant can be propagated only by vegetative means. The plant rarely produces flowers or set seed (Pai, 2005). The conventional propagation techniques of this plant are inadequate to provide the planting material required for large scale cultivation and to meet the growing need of pharmaceutical industry.

Explant type and growth regulators

Rani *et al.* (2000) standardized an efficient micropropagation protocol using rhizome segment cultured on MS medium containing BAP (4.0 mg/L) + IAA (0.5 mg/L) and noticed maximum shoot proliferation. Anu *et al.* (2001) cultured rhizome buds in MS medium fortified with BAP (8.87μM) + NAA (5.37μM) under 12 hrs photoperiod of 30 μE m^{-2} s^{-1} light intensity and obtained 8-10 multiple shoots by the eighth week of inoculation. Ahmed *et al.* (2007) observed best shoot proliferation (98.99%) from rhizome tip explants cultured on MS medium containing Kinetin (2.0 mg/L) + NAA (0.05mg/L). Sandhyarani *et al.* (2011) established an efficient and improved the dual-phase culture system consisting of agar-solidified MS medium overlaid by liquid fraction of the same medium and obtained maximum multiple shoot induction by supplementation of NAA (0.5 mg/L) + BAP (2.0 mg/L). Lee and Han (2011) subcultured the seedlings of *A. calamus* on MS medium supplemented with TDZ (*N* - phenyl-*N'*-[(1,2,3-thidiazol-5-yl) urea]) and obtained a maximum of 11.0 shoots per explant. Verma and Singh (2012) observed maximum number of shoots (4.4) per rhizome explant on MS medium fortified with BAP (2.0 mg/L) in combination with NAA (0.5 mg/L) where 90% of explants showed proliferation (Fig. 3d).

Root development and acclimatization

Rani *et al.* (2000) obtained highest rooting frequency (43.7%) from regenerated shoots inoculated on MS medium with 2.0 mg/L IAA. They transferred the regenerated plantlets to sterilize soil with ninety percent survival rate. Ahmed *et al.* (2007) noticed the maximum of 40.51% of root formation in the micro shoots implanted on MS medium supplemented with 1.0 mg/L IBA. The plant lets were then transplanted into small plastic pot containing garden soil and humus (3:1) with eighty percent survival. Lee and Han (2011) obtained best root formation on MS media containing 1.0 mg/L NAA in *A. calamus.* Sandhyarani *et al.* (2011) and Verma and Singh (2012) also cultured maximum rooting percent with IBA. They successfully acclimatized the regenerated plantlets in pots containing sterilized soil and sand mixture (3:1) with 75% survival rate under the field conditions.

***Aegle marmelos* (L.) Corr.**

Aegle marmelos (L.) Corr., (Rutaceae) is a popular medicinal plant in the Ayurvedic and Siddha systems of medicines used to treat a wide variety of ailments. In India, this plant is known as “Bael Tree”. It is mostly found in tropical and subtropical region. The tree grows wild in dry forests on hills and plains of central and southern India, Burma, Pakistan, Bangladesh, Sri Lanka, Northern Malaya, Java and Philippine Islands (Islam *et al.*, 1995). It is medium sized tree having profuse dimorphic branched, alternate, trifoliate, deep green leaves, membranous leaflets, large sweet scented, greenish white flowers, large, globose fruits (Fig. 1h-i) (Purohit and Vyas, 2005). It flowers from May to July and yields an average of 350-400 fruits (200-250 kg) per tree (Mazumder *et al.*, 2006).

Almost all parts of the tree are used in preparing herbal medicine (Kala, 2006).The roots are useful for treating diarrhea, dysentery, and dyspepsia (Mazumder *et al.*, 2006).The aqueous extracts of the stem and root bark are used to treat malaria, fever, jaundice, and skin diseases such as ulcers, urticaria, and eczema (Nadkarni, 1954). In pharmacological trails, both the fruit and root showed showed antiamoebic and hypoglycaemic activities (Ponnachan *et al.*, 1993). The plant is rich in alkaloids, among which aegline, marmesin, marmin, and marmelosin are the major ones.

A series of phenylethyl cinnamides, named anhydromarmeline, aegelinosides A and B, were isolated from the leaves of *Aegle marmelos* as a-glucosidase inhibitors (Fig. 2c). The compounds luvangetin and pyranocoumarin, isolated from seeds showed significant antiulcer activity (Goel *et al.*, 1997). Essential oil isolated from the leaf has antifungal activity (Rana *et al.*, 1997). It is an out breeder, and is routinely propagated by seeds for cultivation in northern India (Singh *et al.*, 1976).

As seed progeny are not uniform, no standard varieties are available. Seeds have short viability and are prone to insect attack. Vegetative propagation through root suckers is slow, difficult and cumbersome (Ajithkumar and Seeni, 1998).

Explant type and growth regulators

Islam *et al.* (1995) achieved *in vitro* regeneration of *A. marmelos* from callus cultures of embryonic tissues. Ajithkumar and Seeni (1998) reported *in vitro* mass multiplication of *A. marmelos* using young nodal explant and concluded the dependency of bud break on cytokinin supply, but the synergistic combination of 2.5 mg/L BAP and 1.0 mg/L IAA induced maximum shoot proliferation. Nayak *et al.* (2007) achieved highest frequency plantlet regeneration (86.6%) from cotyledonary nodes cultured on MS medium supplemented with 6.6 µM BAP + 1.14 µM IAA. Hazeena and Sulekha (2008) recorded highest callus induction and proliferation from cotyledon explants inoculated on MS medium supplemented with benzyl adenine (2.2 µM) and 2,4-dichlorophenoxy acetic acid (2.26 µM). According to Pati *et al.* (2008), nodal explants excised from mature tree during the months of September to October were found to be most ideal with maximum (85.7%) *in vitro* bud burst. Warrier *et al.* (2010) noticed the effectiveness of MS medium and Woody Plant Medium (WPM) over Gamborgs B5 medium for *in vitro* shoot multiplication using mature nodal segments. Yadav and Singh (2011) noticed creamish friable compact callus accompanying multiple shoots (8.0) from nodal segment of mature tree cultured on MS medium fortified with 2.0 mg/L BAP + 0.5 mg/L 2,4-D (Fig. 3e). Abirami and Kumar (2013) noticed maximum percent callus induction (86%) from nodal explants cultured on 2,4-D (1.5 mg/L) and later obtained maximum number of shoots on MS + BA (1.5 mg/L) + 2,4-D (0.2 mg/L). Behera *et al.* (2013) achieved *in vitro* plant regeneration from the diverse explants such as epicotyl, cotyledon, hypocotyl and root explants of *A. marmelos*. Epicotyl explants cultured on MS medium supplemented with BAP (2.2 µM) + IAA (1.425 µM) produced highest average number of shoots/explant.

Root development and acclimatization

Nayak *et al.* (2007) transferred the *in vitro* regenerated shoots to half strength MS supplemented with auxins IAA, IBA and NAA and noticed best rooting in the medium supplemented with 14.7 µM IBA. Later, they transferred the regenerated plantlets in small plastic pots containing presoaked vermiculite and noticed eighty percent survival under field conditions. Pati *et al.* (2008) obtained hundred percent rooting percent on half strength MS medium supplemented with IBA 49.0 + IAA 5.7 µM. They transferred the rooted plantlets to glass bottles containing different carrier substrates viz. autoclaved soil, soil + sand + FYM (1:1:1) and coconut husk. Yadav and Singh (2011) implanted the *in vitro* regenerated shoots in half strength MS medium fortified with IBA, IAA and NAA and noticed best rooting with 1.0 mg/L IAA. Hazeena and Sulekha (2008), Warrier *et al.* (2010) and Abirami and Kumar (2013) also observed best *in vitro* rooting with IBA. Hazeena and Sulekha (2008) observed the superiority of sand potting media and recorded good survival rate (75%). On the other hand, Behera *et al.* (2013) noted highest rooting frequency in the medium augmented with 2.85 µM IAA.

Pati *et al.* (2008) also tested the genetic fidelity of micropropagated plants and mother tree using 13 RAPD, 3 ISSR and 2 DAMD primers and noticed the genetic integrity in the result.

Conclusion and Future Prospects

Plant tissue culture refers to growing and multiplication of cells, tissues and organs of plants on defined solid or liquid media under aseptic and controlled environment. Micropropagation allows rapid production of high quality, disease-free and uniform planting material. The cultured cells and tissue can take several pathways. The pathways that lead to the production of true-to-type plants in large numbers are the preferred ones for commercial multiplication.

Acknowledgement

Authors are grateful to Kurukshetra University, Kurukshetra for providing laboratory facilities and also to University Grants Commission,

New Delhi for providing research support in the form of major research project.

References

Abirami, H. and Kumar, P.S. 2013. *In vitro* regeneration and extraction of secondary metabolites in *Aegle marmelos* (L.) Correa. *Asian Journal of Plant Science and Research,* 3: 99-106.

Ade, R. and Rai, M. 2011. Multiple shoot formation in Gloriosa superba: A rare and endangered Indian medicinal plant. *Nusantara Bioscience* 3: 68-72.

Ahmed, M.B., S. Ahmed, M. Salahin, R. Sultana, M. Khatun, M.A. Razvy, M.M. Hannan, R. Islam and M.M. Hossain, 2007. Standardization of a suitable protocol for *in vitro* clonal propagation of *Acorus calamus L.* - An important medicinal plant in Bangladesh. *American-Eurasian Journal of Scientific Research*, 2(2): 136-140.

Ajithkumar, D. and Seeni, S. 1998. Rapid clonal multiplication through *in vitro* axillary shoot proliferation of *Aegle marmelos* (L.) Corr. a medicinal tree. *Plant Cell Rep.*, 17: 422-426.

Alfermann, A.W., Bergmann, W., Figur, C., Helmbold, U., Schwantag, D., Schuller, I. and Reinhard, E. 1983. In: Mantell SH, Smith H (eds) Plant Bictechnology, Cambridge University Press, Cambridge London New York New Rochelle Melbourne Syndey, pp. 67–74.

Anonymous. 1989. The Wealth of India: A Dictionary of Indian Raw Materials and Industrial Products. Vol. 10. Publication and Information Directorate. CSIR, New Delhi: 11–12.

Anonymous. 2000. Thai Herbal Pharmacopoeia Volume II, Department of Medical Sciences, Ministry of Public Health, Thailand, Prachachon Co., Ltd., Bangkok

Anu, A., Babu, K.N., John, C.Z. and Peter, K.V. 2001. *In vitro* clonal multiplication of Acorus calamus. *Journal of Biochemistry and Biotechnology*, 10: 53-55.

Arora, R. and Bhojwani, S.S. 1989. *In vitro* propagation and low temperature storage of *Saussurea lappa* C.B. Clarke - an endangered, medicinal plant. *Plant Cell Rep.*, 8: 44-47.

Arumugam, A. and Gopinath, G. 2012. *In vitro* micropropagation using corm bud explants: an endangered medicinal plant of *Gloriosa superba* L. *Asian Journal of Biotechnology,* DOI: 10.3923/aibkar.2012.

Arya, V., Singh, R.P. and Shekhawat, N.S. 2001. A micropropagation protocol for mass multiplication and off-site conservation of *Celastrus paniculatus*- a vulnerable medicinal plant of India. *Journal of Sustainable Forestry*, 14: 107-120.

Balaraju, K., Agastian, P. and Ignacimuthu, S. 2009. Micropropagation of *Swertia chirata* Buch.- Hams. ex Wall: a critically endangered medicinal herb. *Acta Physiol. Plant.*, 31: 487-494.

Behera, P.R., Thirunavoukkarasu, M. and Chand, P.K. 2013. Adventitious plantlet regeneration from different explants of *Aegle marmelos* (L.) Corr. *J. Medicinal Plant Research*, 7: 2761- 2768.

Bellet, P. and Gaignault, J.C. 1985. Le *Gloriosa superba* L. et la production de substances colchiciniques. *Ann. Pharm. Fr.*, 43: 345–347.

Bourgin, J.P. and Nitch, J.P. (1967) Obtention de Nicotiana haploides àpartir de'étamines cultivées in vitro. *Ann. Physiol. Vég.*, 9: 377–382.

Bytebier, B., Deboeck, F., De Greve, H., Van Montagu, M. and Hernalsteens, J.P. 1987. T-DNA organization in tumor cultures and transgenic plants of the monocotyledon Asparagus officinalis. *Proc. Natl. Acad. Sci. USA*, 84: 5345-5349.

Chakradhar, T. and Pullaiah, T. 2006. Effect of explant source on axillary shoot multiplication during micropropagation of a rare medicinal plant - *Wattakaka volubilis* (L) Stapf. *J. Plant Biochem. Biotechnol.*, 15: 43-45.

Cocking, E.C. 1960. A method for the isolation of plant protoplasts and vacuole. *Nature* (London), 187: 927-929.

Constable, F. 1990. Medicinal plant biotechnology. *Planta,* 56: 421-425.

Crossway, A., Oakes, J. V., Irvine, J. M., Ward, B., Knauf, V. C. and Shewmaker, C. K, 1986: Integration of foreign DNA following microinjection of tobacco mesophyll protoplasts. *Mol. Gen. Genet.*, 202: 179–185.

CSIR.1992. Wealth India-raw material Vol.3. Council for Scientific and Industrial Research (CSIR), New Delhi, pp. 412-413.

Dang, J.C., Kumaria, S., Kumar, S. and Tandon, P. 2011. Micropropagation of *Ilex khasiana,* a critically endangered and endemic holly of Northeast India. *AoB PLANTS*, DOI:10.1093/aobpla/plr012.

De Silva, M.A.N. and Senarath, W.T.P.S.K. 2009. Development of a Successful Protocol for *in vitro* Mass Propagation of *Celastrus paniculatus* Willd. – A Valuable Medicinal Plant. *Tropical Agricultural Research,* 21: 21 – 29.

Dekeyser, R.A,. Claes, B., De Rycke, R.M.U., Habets, M.E., Van Montagu, M.C. and Caplan, A.B. 1990. Transient gene expression in intact and organized rice tissues. *Plant Cell,* 2: 591–602.

Dhar, U., Manjkhola, S., Joshi, M., Bhatt, A., Bisht, A.K. and Joshi, M. 2002. Current status and future strategy for development of medicinal plant sector in Uttaranchal, India. *Current Sci,* 83: 956-964.

Dikshit, V.K. 1999. Export of medicinal plants from India: need for resource management. In Biodiversity-North-east India Prespectives: People's Participation in Biodiversity Conservation. Kharbuli, B.; Shyem, D. and Kayang, H. (eds.). NEBRC, North-Eastern Hill University, Shillong, pp. 85-88.

Dobson, A. 1995. *TREE,* 10: 390-391.

Ensminger, A.H., Ensminger, M.E., Konlande, J.E. and Robson, J.R.K. 1983. *Food and Nutrition Encyclopedia,* 2: 1427-1441. Pegus Press, Clovis, California, U.S.A.

Faisal, M. Ahmad, N. and Anis, M. 2007. An efficient micropropagation system for *Tylophora indica*: an endangered, medicinally important plant. *Plant Biotechnology Reports,* 1: 155- 161.

Fernie, A.R., Riesmeier, J.W., Martiny, A., Ramalingam, S., Willmitzer, L. and Trethewey, R.N. 2000 Consequences of the expression of a bacterial glucokinase in potato tubers, both in combination with and independently of a yeast-derived invertase. *Aus. J. Plant Physiol.,* 27: 827–833.

Forsyth, C. and Van Staden, J. 1984. Tuberization of Dioscorea bulbifera stem nodes in culture. *J. Plant Physiol.,* 115: 79-83.

Fraley, R.T., Rogers, S.G., Horsch, R.B., Eichholtz, D.A., Flick, J., S., Fink, C.L., Hoffmann, N.L., and Sanders, P.R. 1985. The SEV system: A new disarmed Ti plasmid vector system for plant transformation. *Bio-Technology* (NY), 3: 629-635.

Gamborg, O.L., Miller, R.A. and Ojima, K. 1968. Nutrient requirements of suspension cultures of soybean root cells. *Exp. Cell Res.,* 50, 151–158.

Gatinode, B.B., Raiker, K.P., Shroff, F.N. and Patel, J.R.1957. Pharmacological studies with malkanguni, an indigenous tranquilizing drug (preliminary report). *Current Practice,* 1: 619-621.

Ghosh, B., Mukherjee, S, Jha, T.B. and Jha, S. 2002. Enhanced colchicine production in root cultures of *G. superba* by direct and indirect precursors of the biosynthetic pathway. *Biotechnol. Lett.,* 24: 231– 4.

Ghosh, S., Ghosh, B. and Jha, S. 2007. In vitro tuberisation of Gloriosa superba L. on basal medium. *Scientia Horticulturae,* 114: 220–223.

Giesbach, R.J. 1994. Marketing of genetically engineered tomato: 'Flavr-Savr' improved plant transformation through electrophoresis. *Plant Sci.,* 102: 81-89.

Goel, R.K., Maiti, R.N., Manickam, M. and Ray, A.B. 1997. Antiulcer activity of naturally occurring pyranocoumarin and isocoumarins and their effect on prostanoid synthesis using human colonic mucosa. *Indian Journal of Experimental Biology,* 35: 1080–1083.

Gonzalez, M.L., Mallon, R., Reinoso, J. and Rodriguez-Oubina, J. 2006. *In vitro* micropropagation and long-term conservation of the endangered moss *Splachnum ampullaceum. Biologia Plantarum,* 50: 339-345.

Govil, J.N. 1993. Medicinal plants: New Vistas of Research, Part-2. p. 393.

Grison, C. 1991. Influence des facteurs d'environnement sur le cycle ve´ge´tatif de la pomme de terre. *La pomme de terre fran.caise.,* 462: 7-15.

Guha, S. and Maheshwari, S.C. 1964. *In vitro* production of embryos from anthers of *Datura. Nature,* 204: 497.

Guo, B., Gao, M. and Liu, C.Z. 2007. *In vitro* propagation of an endangered medicinal plant *Saussurea involucrata* Kar. et Kir. *Plant Cell Report.* 26: 261-265.

Gupta, R. and Chadha, K.L. 1995. Medicines and aromatic plants in India. In: Advances in Horticulture, Medicinal and Aromatic Plants. Chadha, K.L. and Gupta, R. (eds.). Malhotra Publishing Housing, New Delhi, pp. 1-44.

Haberlandt, G. 1902. Kulturversuche mit isolierten Pflanzenzellen. *Sber. Akad. Wiss. Wien,* 111: 69-92.

Hamilton, C.M., Frary, A., Lewis, C. and Tanksley, S.D. 1996. Stable transfer of intact high molecular weight DNA into plant chromosomes. *Proc Natl Acad Sci. USA,* 93: 9975–9979.

Haw, A.B. and Keng, C.L. 2003. Micropropagation of *Spilanthes acmella* L., a bio-insecticide plant, through proliferation of multiple shoots. *J. Appl. Hort.,* 5: 65-68.

Hazeena, M.S. and Sulekha, G.R. 2008. Callus induction and regeneration in *Aegle marmelos* (L.) Corr. using cotyledon explant. J. Trop. Agric. 46:79-84.

Huang, Y., Diner, A.M. and Karnosky, D.F. 1991. *Agrobacterium rhizogenes*-mediated genetic transformation and regeneration of a conifer:*Larix deciduas. In Vitro* Cell. *Dev. Biol. Plant.*, 27: 201-207.

Islam, R., Hossain, M., Karim, M.R. and Joarder, O.I. 1995. Regeneration of *Aegle marmelos* (L.) Corr. Plants *in vitro* from callus cultures of embryonic tissues. *Curr. Sci.*, 69: 494- 495.

IUCN 2007. *The IUCN Red List of Threatened Species*. http://www.iucnredlist.org.

Jadhav, S.J. and Hegde, B.A. 2001. Somatic embryogenesis and plant regeneration in *Gloriosa* L. *Indian Journal of Experimental Biology,* 39: 943- 946.

Jain, S.P., Verma, D.N., Singh, S.C., Singh, J.S. and Kumar, S. 2000. Flora of Haryana, CIMAP, Lucknow.

Jan, A., Phalisteen, S., Thomas, G. and Shawl, A.S. 2009. Auxin prompted improved micropropagation protocol of *Picrorhiza kurroa*: an endangered medicinal plant. *Plant Tissue Cult. and Biotech.*, 19: 161-167.

Jana, S. and Shekhawat, G.S. 2011. Critical review on medicinally potent plant species: *Gloriosa superba*. *Fitoterapia*, 82: 293-301.

Jansen, R.K. 1981. Systematics of *Spilanthes* (Compositae Heliantheae). *Systematic Botany,* 6: 231.

Joshi, M. and Dhar, U. 2003. In vitro propagation of *Saussurea obvallata* (DC.) Edgew. – an endangered ethnoreligious medicinal herb of Himalaya. *Plant Cell Rep.*, 21: 933-939.

Joshi, V. and Jadhav, S.K. 2013. Effect of Media Types and Carbon Sources on *In vitro* Shoot Proliferation of *Spilanthes acmella* (L.) Murray. *Plant Tissue Cult. and Biotech.,* 23: 283-288.

Kala, C.P. 2006. Ethnobotany and ethnoconservation of *Aegle marmelos* (L) Correa. *Indian Journal of Traditional Knowledge,* 5: 541-550.

Kanta, K., Rangaswamy, N.S. and Maheshwari, P. 1962. Test-tube fertilization in a flowering plant. *Nature*, 194: 1214-1217.

Kapoor, L.D. 2001. Traditional uses of medicinal plant In: Ayurvedic Medicinal Plant. CRC Press, New Delhi.

Karp, A. 1994. Origin, causes and uses of variation in plant tissue cultures. In: Plant Cell and Tissue Culture. Vasil, I.K. and Thorpe, T.A. (eds.). Dordrecht, Kluwer Academic Publ, pp. 139-150.

Kaul, B. and Staba, E.J. 1967. Visnagin: Biosynthesis and isolation from *Ammi visnaga* suspension cultures. *Science*, 150: 1731-1732.

Larkin, P.J. and Scowcroft, W.R. 1981. Somaclonal variation: a novel source of variability from cell cultures for plant improvement. *Theor. Appl. Genet.,* 60: 197-214.

Krause, J. 1986. Production of *Gloriosa* tubers from seeds. Acta. Hortic. 177: 353–360.

Kumar, L.S. 1953. Doubling of chromosomes induced by gloriosine isolated from *Gloriosa superba* Linn. *Nature*, 171: 791-792.

Kumar, S., Mangal, M., Dhawan, A.K. and Singh, N. 2011. Assessment of genetic fidelity of micropropagated plants of *Simmondsia chinensis* (Link) Schneider using RAPD and ISSR markers. *Acta Physiol. Plant.*, 33: 2541-2545.

Kumari, N.; Jaiswal, V. and Jaiswal V.S. 1998. Induction of somatic embryogenesis and plant regeneration from leaf callus of *Terminalia arjuna* Bedd. *Curr. Sci.*, 75: 1052-1055.

Lal N., Ahuja PS, Kukreja AK and Pandey B. 1988. Clonal propagation of *Picrorhiza kurroa* Royle ex Benth. by shoot tip culture. *Plant Cell Rep.*, 7: 202-205.

Lal, D. and Singh, N. 2010. Mass multiplication of *Celastrus paniculatus* Willd – an important medicinal plant under *in vitro* conditions using nodal segments. *Journal of American Science,* 6: 55-61.

Lal, D., Singh, N. and Yadav, K. 2010. In vitro studies on Celastrus paniculatus. Journal of Tropical Medicinal Plants, 11:169-174.

Laskar, M.A., Hynniewta, M. and Rao, C.S. 2009. *In vitro* propagation of *Citrus indica* tanaka- An endangered progenitor species. *Indian J. Biotechnol.*, 8: 311-316.

Lattoo, S.K., Bamotra, S., Sapru Dhar, R., Khan, S. and Dhar, A.K. 2006. Rapid plant regeneration and analysis of genetic fidelity of *in vitro* derived plants of *Chlorophytum arundinaceum* Baker - an endangered medicinal herb. *Plant Cell Rep.*, 25: 499-506.

Lee, J.H. and Han, T.H. 2011. Micropropagation of the Plantlets Derived from Seeds in the Genus *Acorus* (*A. calamus* and *A. gramineus*). *Hort. Environ. Biotechnol.*, 52: 89-94.

Leng, T., Haw, A. and Keng, C.L. 2004. Effect of reduced N6-benzyladenine, explant type, explant orientation, culture temperature and culture vessel type on regeneration of adventitious shoot and *in vitro* plantlets of *Spilanthes acmella*. *J. Plant Biol.*, 47:15-20.

Levi, M., Tarquini, F., Sgorbati, S. and Sparvoli, E. 1986. Determination of DNA content by static cytofluorimetry in nuclei released from fixed plant tissue. *Protoplasma*, 132: 64–68.

Litz, R.E. and Gray, D.J. 1992. Organogenesis and somatic embryogenesis. In: Biotechnology of perennial fruit crops. (Eds. Hammerschlag FA and Litz RE) p. 3-34, CAB International, Wallingford. U.K.

Madhavan, M. and Joseph, J.P. 2010. *In vitro* root tuber industrial from leaf and nodal explants of *Gloriosa superba* L: An endangered medicinal plant of India. *Plant Archiv.,* 10: 611-615.

Martin, G., Geetha, S.P., Raja, S.S., Raghu, A.V., Balachandran, I. and Ravindran, P.V. 2006. An efficient micropropagation system for *Celastrus paniculatus* Willd.: a vulnerable medicinal plant. *J. For. Res.*, 11: 461-465.

Martin, K.P. 2002. Rapid *in vitro* multiplication and *ex vitro* rooting of *Rotula aquatica* Lour., a rare rhoeophytic woody medicinal plant. *Plant Cell Rep.*, 67: 547-548.

Mazumder, R., Bhattacharya, S., Mazumder, A., Pattnaik, A.K., Tiwary, P.M. and Chaudhary, S. 2006. Antidiarrhoeal evaluation of *Aegle marmelos* (Correa) Linn. root extract. *Phytotherapy Research*, 20: 82–84.

Mehta, J., Sain, M., Mathuriya, B.L., Naruka, R., Kavia, A. and Sharma, D.R. 2012. Rapid micropropagation and callus induction of *Terminalia bellerica* Roxb. - An endangered plant. *Asian Journal of Plant Science and Research*, 2: 364-368.

Mendelsohn, R. and Balick, M. 1994. The value of undiscovered pharmaceuticals in tropical forests. Econ. Bot. 49: 223-228.

Mohapatra, H.P. and Rath, S.P. 2005. *In vitro* studies of *Bacopa monnieri* - An important medicinal plant with reference to its biochemical variations. *Indian J. Exp. Biol.*, 43: 373-376.

Morel, G. 1960. Producing virus free Cymbidiums. *Amer. Orchid Soc. Bull.*, 19: 495-497.

Morel, G. 1963. La culture in vitro du meristème apical de certaines orchidées. CR Hebd Séances Acad Sci. 256: 4955- 4957.

Mrudul, V., Shirgurkar, C.K. and John, R.S. 2001. Factors affecting *in vitro* microrhizome production in turmeric. Plant Cell Tiss. Org. Cult. 64: 5-11.

Murashige, T. 1974. Plant propagation through tissue cultures. *Ann. Rev. Plant Physiol.*, 25: 135-166.

Murashige, T. and Skoog, F. 1962. A revised medium for rapid growth and bioassays with tobacco tissue cultures. *Physiol. Plant.,* 15: 473-497.

Nadkarni, A.K. 1976. In: Nadkarni's Indian Materia Medica, pp. 1: 296. Popular Prakashan, Bombay, India.

Nadkarni, K.M. 1954. Indian material Medica, 3rd edn. Popular Book Depot, Bombay, India, pp. 45-49.

Nair, L.G. and Seeni, S. 2001. Rapid *in vitro* multiplication and restoration of *Celastrus paniculatus* (Celastraceae), a medicinal woody climber. *Indian J. Exp. Biol.,* 39: 697-704.

Nalini, K., Aroor, A.R., Kumar, K.B. and Rao, A. 1986. Studies on biogenic amines and metabolites in mentally retarded children on *Celastrus* oil therapy. *Altern. Med.*, 1: 355-360.

Nayak, P., Behera, P.R. and Manikkannan, T. 2007. High frequency plantlet regeneration from cotyledonary node cultures of *Aegle marmelos* (L.) Corr. *In vitro Cell. Dev. Biol. Plant,* 43:231–236.

Nitsch, J. and Nitsch, C. 1969. Haploid plants from pollen grains. *Sci.,* 163: 85-87.

Ondo Ovono, P., Kevers, C. and Dommes, J. 2009. Effects of reducing sugar concentration on in vitro tuber formation and sprouting in yam (*Dioscorea cayenensis– D. rotundata* complex). *Plant Cell, Tissue and Organ Culture*, 99: 55-59.

Pai, A. 2005. The population ecology of a perennial clonal herb *Acorus calamus* L. (Acoraceae) in Southeast Ohio, USA. Dissertation College Arts Sci., Ohio Univ., USA. p. 1-153.

Pandey, V. and Agrawal, V. 2009. Efficient micropropagation protocol of *Spilanthes acmella* L. possessing strong antimalarial activity. *In vitro Cell. Dev. Biol. Plant.* 45: 491-499.

Parotta, J.A. 2001. Healing plants of peninsular India. CABI, New York.

Pati, R., Chandra, R., Chauhan, U.K., Mishra, M., Srivastava, N., (2008): *In vitro* clonal propagation of Bael (*Aegle marmelos* Corr.) CV. CISH-B1 through enhanced axillary branching. *Physiol. Mol. Biol. Plant*, 14: 337-346.

Phulwaria, M., Rai, M.K., Patel, A.K., Kataria, V. and Shekhawat NS. 2013. A genetically stable rooting protocol for propagating a threatened medicinal plant- *Celastrus paniculatus. AoB PLANTS*, 5: pls054; doi:10.1093/aobpla/pls054

Pijut, P.M., Domir, S.C., Lineberger, R.D. and Schreiber, L.R. 1990. Use of culture filterate of *Ceratocystis ulmi* as a bioassay to screen for disease tolerant *Ulmus Americana. Plant Sci.*, 70: 191-196.

Ponnachan, P.T.C., Paulose, C.S. and Panikar, K.R. 1993. Effect of the leaf extract of *Aegle marmelos* (L.) Corr. in diabetic rats. *Indian Journal of Experimental Biology*, 31:345–7.

Purohit, S.D. and Dave, A. 1996. Micropropagation of *Sterculia urens* Roxb. - An endangered tree species. *Plant Cell Rep.*, 15: 704-706.

Purohit, S.D., Dave, A. and Gotam, K. 1994. Micropropagation of safed musli (*Chlorophytum borivilianum*), a rare Indian medicinal herb. *Plant Cell Tiss. Org. Cult.* 39: 93-96.

Purohit, S.S. and Vyas, S.P. 2005. Medicinal Plant Cultivation - A Scientific Approach. Agrobion. India), pp. 282.

Rai, M.K., Asthana, P., Jaiswal, V.S. and Jaiswal, U. 2010. Biotechnological advances in guava (*Psidium guajava* L.): recent development and prospects for future research. *Trees,* 24: 1-12.

Rajagopal, C., and Kandhasamy, R. 2009. Genetic variability of *kazhappai kizhangu* (*Gloriosa superba* L.) in Tamil Nadu assessed using morphological and biochemical traits. *J. Trop. Agr.*, 47: 77-79.

Rajesekharan, P.E. and Ganeshan, S. 2002. Conservation of medicinal plant biodiversity - an Indian perspective. *J. Med. Arom. Plant. Sci.,* 24:132-147.

Ramsewak, R.S., Erickson, A.J. and Nair, M.G. 1999. Bioactive Nisobutylamides from the flower buds of *Spilanthes acmella. Phytochemistry,* 51: 729-732.

Rana, B.K., Sing, U.P. and Taneja, V. 1997. Antifungal activity and kinetics of inhibition by essential oil isolated from leaves of *Aegle marmelos* (L.) Corr. *Journal of Ethnopharmacology,* 57: 29–34.

Rani, A.S., Subhadra, V.V. and Reddy, V.D. 2000. *In vitro* propagation of *Acorus calamus* Linn. – a medicinal plant. *Indian Journal of Experimental Biology,* 38: 730-732.

Rao, N.K. and Reddy, R.K. 1983. Threatened plants of Tirupati and its environs. In: Jain S.K., Rao P.R. (eds): An Assessment of Threatened Plants of India. Department of Environment, Howrah: 167–168.

Rastogi, R.P. and Mehrotra, B.N. 1998. Compendium of Indian medicinal plants, vol.5. National Institute of Science Communication (NISCOM), New Delhi.

Reinert, J. 1959. Uber die Kontrolle der Morphogenese und die Induktion von Adventivembryonen an Gewebekulturen aus Karotten. *Planta,* 53: 318-333.

Rekha, K., Bhan, M.K., Balyan, S.S. and Dhar, A.K. 2005. Cultivation prospects of endangered species *Celastrus paniculatus* Willd. *Nat. Prod. Rad.*, 4: 483-486.

Roja, G. and Rao, P.S. 1998. Biotechnological investigation in medicinal plants for the production of secondary metabolites. In: Role of biotechnology in medicinal and aromatic plants. Khan, I. and Khanum, A. (eds.). Ukaaj Publ., Hyderabad, pp. 95-125.

Rout, G.R.; Samantaray, S.and Das, P. 2000. *In vitro* manipulation and propagation of medicinal plants. *Biotechnol. Advances,* 18: 91-120.

Sandhyarani, N., Kishor, R. and Sharma, G.J. 2011. Clonal propagation of triploid *Acorus calamus* Linn. using dual-phase culture system. *J. Crop Sci. Biotech.*, 14: 213-217.

Sanjaya, Muthan, B., Rathore, T.S. and Rai, V.R. 2006. Micropropagation of an endangered Indian sandalwood (*Santalum album* L.). *J. of Forest Research,* 11: 203-20.

Saritha, K.V. and Naidu, C.V. 2008. Direct shoot regeneration from leaf explants of *Spilanthes acmella. Biol Plant.,* 52: 334–338.

Saritha, K.V., Prakash, E., Ramamurthy, M. and Naidu, C.V. 2002. Micropropagation of *Spilanthes acmella* Murr. *Biol. Plantarum,* 45: 581-584.

Sarker, K.P., Islam, A., Islam, R., Hoque, A. Joarder, O.I. 1996. In vitro propagation of *Rauvolfia serpentina* through tissue culture. *Planta Med.*, 62: 358-359.

Sayeed Hassan, A.K.M. and Roy, S.K. 2005. Micropropagation of *Gloriosa superba* L. through high frequency shoots proliferation. *Plant Tiss. Cult. Biotech,*15: 67-74.

Schenk, R.U. and Hildebrandt, A.C. 1972. Medium and techniques for induction and growth of monocotyledons and dicotyledons plant cell cultures. *Can. J. Bot.,* 50, 199–204.

Sebastian, D.P. and Hariharan, M. 2013. Micropropagation of *Heliotropium keralense* - a globally endangered and endemic medicinal plant through indirect organogenesis from roots. Int. J. Curr. Microbiol. *App. Sci.,* 2: 56-62.

Senapati, S.K., Aparajita, S. and Rout, G.R. 2013. Micropropagation and assessment of genetic stability in *Celastrus paniculatus*: An endangered medicinal plant. *Biologia,* 68: 627-632.

Sharma, A. and Shahzad, A. 2013. Efficient Micropropagation of *Spilanthes acmella* (L.) Murr.: A Threatened Medicinal Herb. *British Biotechnology Journal,* 3(3):

Sharma, M., Rai, S.K., Purshottam, D.K., Jain, M., Chakrabarty, D., Awasthi, A., Nair, K.N. and Kumar, A. 2009. *In vitro* clonal propagation of *Clerodendrum serratum* (Linn.) Moon (barangi): a rare and threatened medicinal plant. *Acta Physiol Plant.,* 31: 379-383.

Sharma, U. and Mohan, J.S.S. 2006. *In vitro* clonal propagation of *Chlorophytum borivilianum* Sant. et Fernand., a rare medicinal herb from mature floral buds along with inflorescence axis. *Indian J. Exp. Biol.*, 44: 77-82.

Singh, H.B. 2000. Alternative sources for some conventional drug plants of India. In Ethnobotany and Medicinal Plants of Indian Sub-continents. Maheshwari, J.K. (ed.). Scientific Publishers, Jodhpur, India, pp. 63-78.

Singh, M. and Chaturvedi, R. 2010. Improved clonal propagation of *Spilanthes acmella* Murr. for production of scopoletin. *Plant Cell Tiss. Organ Cult.*,103: 243-53.

Singh, M., Roy, B., Tandon, V. and Chaturvedi, R. 2013. Extracts of dedifferentiated cultures of *Spilanthes acmella* Murr. possess antioxidant and anthelmintic properties and hold promise as an alternative source of herbal medicine. *Plant Biosystems*, DOI: 10.1080/11263504.2013.766278.

Singh, S.K., Rai, M.K., Asthana, P. and Sahoo, L. 2009b. An improved micropropagation of *Spilanthes acmella* L. through transverse thin cell layer culture. *Acta Physiol. Plant.*, 31: 693-98.

Singh, S.K., Rai, M.K., Asthana, P., Pandey, S., Jaiswal, V.S. and jaiswal, U. 2009a. Plant regeneration from alginate-encapsulated shoot tips of *Spilanthes acmella* (L.) Murr., a medicinally important and herbal pesticidal plant species. *Acta Physiol Plant.,* 31: 649-653.

Singh, U.R., Panday, I.C. and Prasad, R.S. 1976. Propagation of bael (*Aegle marmelos* Corr.). *Punjab Horticulture Journal,* 16: 57-59.

Singh, V. 1995. Herbal folk remedies of Morni hills (Haryana), India. *Fitoterapia,* 66: 425–430.

Sivakumar, G. and Krishnamurthy, K.V. 2000. Micropropagation of *Gloriosa superba* L-an endangered species of Asia and Africa. *Curr. Sci.,* 78: 31-32

Sivakumar, G. and Krishnamurthy, K.V. 2004. *In vitro* organogenetic responses of *Gloriosa superba. Russ. J. Plant Physiol.,* 51: 790- 798.

Sivakumar, G., Krishnamurthy, K.V. and Rajendran, T.D. 2003. *In vitro* corm production in *Gloriosa superba* L., an Ayurvedic medicinal plant. *J. Horticult. Sci. Biotech.*, 78: 450-453.

Sivakumar, G., Krishnamurthy, K.V., Hahn, E.J. and Paek, K.Y. 2004. Enhanced *in vitro* production of colchicines in *Gloriosa superba* L. - an emerging industrial medicinal crop in south. *India. J. Horticul. Sci. Biotech.*, 79: 602-605.

Skoog, F. and Miller, C.O. 1957. Chemical regulation of growth and organ formation in plant tissues cultured *in vitro. Symp. Soc. Exp. Biol.*, 11: 118-131.

Sudha, C.G. and Seeni, S. 1996. *In vitro* propagation of *Rauwolfia micrantha*, a rare medicinal plant. *Plant Cell Tiss. Org. Cult.*, 44: 243-248.

Sudhakar, J.T., Badari Narayan, S. and Narayana, D.B.A. 1997. Rapid *in vitro* propagation of *Saussurea lappa*, an endangered medicinal plant, through multiple shoot cultures. *In vitro Cell. Dev. Biol. Plant.*, 33: 128-130.

The Wealth of India. 2003. NISCAIR, CSIR, New Delhi, I: A, 63-65.

Thomas, T.D. and Shankar, S. 2009. Multiple shoots induction and callus regeneration in *Sarcostemma brevistigma* Wight and Arnott, a rare medicinal plant. *Plant Biotechnol. Rep.*, 3: 67-74.

Uzun, S., Parmaksiz, I., Uranbey, S., Mirici, S., Sarihan, E.O., Ipek, A., Kaya, M.D., Gürbüz, B., Arslan, N., Sancak, C., Khawar, K.M. and Özcan, S. 2014. In vitro micropropagation from immature embryos of the endemic and endangered *Muscari muscarimi* Medik. *Turk. J. Biol.*, 38: 83-88.

Vaillant, V., Bade, P. and Constant, C. 2005. Photoperiod affects the growth and development of yam plantlets obtained by *in vitro* propagation. *Biol. Plant.*, 49: 355-359.

van Overbeek, J., Conklin, M. E., and Blakeslee, A. F. 1941. Factors in coconut milk essential for growth and development of Datura embryos. *Science*, 94: 350.

Verma, S. and Singh, N. 2012. *In vitro* Mass Multiplication of *Acorus calamus* L. - An Endangered Medicinal Plant. *American-Eurasian J. Agric. Environ. Sci.*, 12: 1514-1521.

Verma, S., Magotra, R. and Koul, A.K. 2002. *In vitro* multiplication of *Eremurus persicus* Boiss. (Liliaceae) – an endangered species. *Phytomorphology*, 52: 315-321.

Viera, R.F. and Skorupa, C.A. 1993. Brazilian medicinal plants gene bank. *Acta. Hort.*, 330: 51-88.

Wala, B.B. and Jasrai, Y.T. 2003. Micropropagation of an endangered medicinal plant: *Curculigo orchioides* Gaertn. *Plant Tiss. Cult.*, 13: 13-19.

Warrier, P.K., Nambiar, V.P.K. and Ramankutty, C. 1994. Indian medicinal plants – a compendium of 500 species, vol 2. Orient Longman, Madras.

Warrier, R., Viji, J. and Priyadharshini, P. 2010. *In vitro* propagation of *Aegle marmelos* L. (Corr.) from mature trees through enhanced axilliary brancing. *Asian J. Exp. Biol. Sci.*, 1: 671- 676.

Went, F.W. 1926. On growth accelerating substances in the coleoptile of *Avena sativa. Proc K Akad Wet Amsterdam.*, 30: 10–19

White, P.R. 1934. Potentially unlimited growth of excised tomato root tips in a liquid medium. *Plant Physiol.*, 9: 585-600.

White, P.R. 1963. The cultivation of animal and plant cells. Ronald press, New York.

Yadav, K. and Singh, N. 2010. Micropropagation of *Spilanthes acmella* Murr. – An Important Medicinal Plant. *Nature and Science,* 8: 5-11.

Yadav, K. and Singh, N. 2011a. *In vitro* flowering of shoots regenerated from cultured nodal explants of *Spilanthes acmella* Murr. - an ornamental cum medicinal herb. *Analele Universităţii din Oradea - Fascicula Biologie,* 18: 60-64.

Yadav, K. and Singh, N. 2011b. *In vitro* propagation and biochemical analysis of field established wood apple (*Aegle marmelos* L.). *Analele Universităţii din Oradea - Fascicula Biologie.*, 18: 23-28.

Yadav, K., Aggarwal, A. and Singh, N. 2012. Actions for *ex situ* conservation of *Gloriosa superba* L. - an endangered ornamental cum medicinal plant. *J. Crop Sci. Biotechnol.,* 15: 297- 303.

Yadav, K., Aggarwal, A. and Singh, N. 2013. Evaluation of genetic fidelity among micropropagated plants of *Gloriosa superba* L. using DNA based markers- a potential medicinal plant. *Fitoterapia*, 89: 265-270.

Yun, D.J., Hashimoto, T. and Yamada, Y. 1992. Metabolic engineering of medicinal plants: transgenic *Atropa belladonna* with an improved alkaloid composition. *Proceedings of the National Academy of Sciences.* 24:11799-803.

Zimmerman, U. and Scheurich, P. 1981 High frequency fusion of plant protoplasts by electric fields. *Planta,* 151: 26-32.

14

BIOASSAY-DIRECTED DISCOVERY OF NATURAL PRODUCTS FOR THEIR ANTIMICROBIAL POTENTIAL

Renu Singh* and Ancy Fernandes

Department of Botany, University School of Sciences, Gujarat University, Ahmedabad-380009, Gujarat, India;
**Corresponding author. Tel.: +91 9426849407;*
E-mail address: drrenusingh2012@gmail.com

INTRODUCTION

Natural products are a proven source of compounds for most therapeutic areas including oncology, cardiovascular disease, infective disease, disease of the Central Nervous System (CNS), metabolic diseases and inflammation. Bioactivity information is useful to design new scaffold based (focused) libraries and to explore the chemical space around validated structure elements with confirmed activity. Initial steps in new drug discovery involve identification of new chemical entities (NCE's) which can be either sourced through chemical synthesis or can be isolated from natural products through biological activity guided fractionation.

Bioactive natural products can be used as drugs, lead compounds, biological or pharmacological tools, feedstock products (raw materials for the production of drugs), excepients and nutraceuticals. They are present in herbs, dietary supplements, spices and foods. Some of them are important flavors, fragrances, dyes and cosmetics, and others are used as insecticides, anti-feedants, pesticides and rodenticides. Bioassay offers a special advantage in the standardization and quality control of heterogenous botanical products. Such products can be "heterogenous" due to the presence of mixtures of bioactive components either from the same or from purposefully mixed botanical sources. Extracts must be screened for biological activity, the active extracts selected, fractionations directed with bioassays, and the bioactive compound identified and then exploited. Nature only uses a few building blocks for its biosynthetic pathway, but the great diversity of the end products results from various "decorating" enzymes found in each species, that introduce new functionalities. When compared with libraries of synthetic substances, natural products offer the prospects of discovering a greater number of compounds, with strictly more complex structures (Pieters and Vlietinck, 2005). Since only 5-15% of the higher plants have been systematically investigated for the presence of bioactive compounds, nature's biodiversity remains largely unexplored. This is even truer for other sources from nature such as marine organisms, microorganisms and animals (Cragg *et al.,* 1997).

Much of the exploration and utilization of natural products as antimicrobials arises from microbial sources. It was discovery of penicillin that led to later discoveries of antibiotics such as streptomycin, aureomycin and chloromycein. Though most of the clinically used antibiotics are produced by soil micro-organisms or fungi, higher plants have also been a source of antibiotics (Trease, 1972). Examples of these are the bacteriostatic and antifungal properties of Lichens, the antibiotic action of allinine in *Allium sativum* L. (Garlic), or the antimicrobial action berberines in goldenseal (*Hydrastis canadensis*). Plants based antimicrobials have enormous therapeutic potentials. They are effective in the treatment of infectious diseases while simultaneously mitigating many of the side effects that are often associated with synthetic antimicrobials. The primary botanical used as an antimicrobial is *Hydrastis* with sales of 4.7% in 1995 (Gruenwald, 1997). While antiinfectives agents make up 24% of the pharmaceutical market (Iwu *et al.*, 1999).

Natural pharmaceuticals (Naturaceuticals), nutraceuticals and cosmecueticals are of great importance as a reservoir of chemical diversity aimed at new drug discovery and are explored for antimicrobial, cardiovascular, immunosuppressive and anticancer drugs. Around 80% of all such products are of plant origin. Examples of plant products and derivatives used by the pharmaceutical industry include paclitaxel, vincristine, artemisinin, camptothecin, podophyllotoxin, etc.

Different approaches to drug discovery using higher plants can be distinguished: random selection followed by chemical screening; random selection followed by one or more biological assays; follow-up of biological activity reports (e.g., ecology based); follow-up of ethnomedical (traditional medicine) use of plants. The later approach includes plants used in organized traditional medical systems; herbalism and folklore; the use of databases (Fabricant and Farnsworth, 2001). Biological activity guided fractionation has been the process deployed to identify the lead druggable candidate from any given phytochemical matrix. Two approaches might be followed as the design of extraction for bioactive guided fractionation leading to compound isolation to act as a lead compound.

I. **Parallel approach**: This approach may be applied when the biological activity of the plant is known by its traditional use. The objective of this approach is to isolate compounds responsible for the activity based on their biological activity. In parallel extraction approach, three types of extracts are obtained viz. 100% methanolic extract, 50% methanolic extract, and 100% aqueous extract from crude plant. The most active fraction based on the primary screening for bioactivity is chosen for further extraction and evaluation.

II. **Sequential Approach**: This approach may be useful when the subject plant is not known and random selection strategy is adopted for plants. Extraction is done on the polarity of the solvent and fractions are obtained in a sequential process using hexane, chloroform, ethyl acetate, and butanol as solvents. Further extraction involves purification stage where, structural elucidation is done for different compounds.

Bioactivity-guided fractionation of any crude extract from natural sources in any case would lead to a wide array of possible outcomes at different stages. Also, these outcomes might provide unforeseen opportunities for modulating the discovery design during subsequent stages depicts the possible outcomes of a typical bioactivity-guided fractionation (Katiyar *et al.,* 2012).

The classical drugs were all discovered more or less by accident. In recent, years, a more systemic approach to the discovery of medicaments has been developed and it's in the area of directed screening and bioassay directed fractionation. Whereas, it is useful to discover a new and unsuspected use for a known compound it is more exciting to the chemist to find a new structural entity with such activity. The primary screening test must be tolerant of the many impurities present in such extracts and yet sensitive enough to reveal the presence of a low level of potentially interesting substances. Typically such first screens are characterized by having throughout and relatively low information content. An example would be screening against robust intact bacterial pathogens and noting survival in the presence of various concentrations of extract. Even amongst the antibiotics less than two dozen structural entities proceeded in their natural state directly into medicinal use from among the more than 10,000 presently known antibiotics found in nature.

This review discusses the different approaches, advantages and challenges and finally shows some practical example of bioassay directed separation of compounds to portray the discovery and development of medicine from bioassay-directed separation of compounds.

CURRENT AND POTENTIAL APPLICATIONS

The primary benefits of using plant derived medicines are that they are relatively safer than synthetic alternatives, offering profound therapeutic benefits and more affordable treatment. Plant based antimicrobials represent a vast untapped source of medicine continued and further exploration of plant antimicrobials needs to occur. Plant based antimicrobials have enormous therapeutic potential. They are effective in the treatment of infectious diseases while simultaneously mitigating many of the side effects that are often associated with synthetic antimicrobials. Phytomedicines usually have multiple effects on the body. Their actions often act beyond the symptomatic treatment of disease. An example of this is *Hydrastis canadensis. Hydrastis* not only has antimicrobial activity, but also increases blood supply to the spleen promoting optimal activity of the spleen to release mediating compounds (Murray, 1995). The potential for developing antimicrobials into medicines appears rewarding, from both the perspective of drug development and the perspective of phytomedicines. The immediate source of financial benefit from plants based antimicrobials is from the herbal products market. This market offers many opportunities for those cultivating new crops, as many of the plants that are wildcrafted today must be cultivated to match the demands of this market. Again *Hydrastis*, one of the top selling antimicrobials in the US herbal market, represent an example of a herb that has undergone domestication. Originally this plant, native to eastern North America, was wild crafted. *Hydrastis*, has been used by Native Americans for many conditions, including as an antimicrobial for infections. Efforts to cultivate this plant were undertaken in order to supply the demands of the herbal products market and to battle it's threatened extinction.

Plants have provided a source of inspiration for novel drug compounds, as plant derived medicines have made large contributions to human health and well-being (Table 1). Their role is twofold in the development of new drugs, first: they may become the base for the development of a medicine, a natural blueprint for the development of new drugs, or second: a phytomedicine to be used for the treatment of the disease. There are numerous illustrations of plant derived drugs. Some selected examples, including those classified antiinfective, are presented below. The isoquinoline alkaloid emetine obtained from the underground part of *Cephaelis ipecacuanha*, and related species, has been used for many years as and amoebicidal drug as well as for the treatment of abscesses due to the spread of *Escherichia histolytica* infections. Another important drug of plant origin with a long history of use is quinine. This alkaloid occurs naturally in the bark of *Cinchona* tree. Apart from its continued usefulness in the treatment of malaria, it can be also used to relieve nocturnal leg cramps. Currently, the widely prescribed drugs are analogs of quinine such as chloroquine. Some strains of malarial parasites have become resistant to the quinines, therefore antimalarial drugs with novel mode of action are required. Similarly, higher plants have made important contributions in the areas beyond antiinfectives, such as cancer therapies. Early examples include the antileukaemic alkaloids, vinblatine and vincristine, which were both obtained from the Madagascar periwinkle (*Catharanthus roseus* syn. *Vinca roseus*).The essential ingredient which contrasts bioassay-directed studies with the more classical phytochemical approaches is the design and use of appropriate testing cascades to ensure that the biological features sought will have a high probability of being embodied in the compounds isolated (Fig. 1 and Fig. 2).

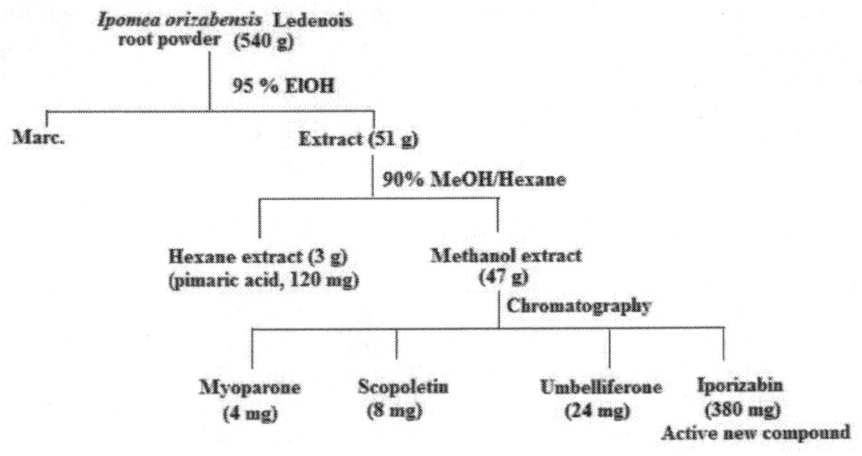

Fig. 1. Isolation scheme for iporizabin from *Ipomoea orizabensis.* (Source: Authors)

CHALLENGES

The objective is the targeted isolation of new bioactive plant products, i.e. lead structures with novel structures and novel mechanisms of action. Several "drug discovery hurdles" have to be taken. When an active extract has been identified the first hurdle to be taken is dereplication (rapid identification of known compounds from a partially purified mixture and identification of enough of an unknown structure to prioritize or conduct isolation). This can mean either a full identification of a natural product after only partial purification or partial identification to the level of a family of known compounds. The second hurdle is the purification of the new compound by chromatographic methods, the third hurdle is structure determination by spectroscopic methods and when the novel structure has been established the fourth hurdle is scale up of the production. The fifth and final hurdle is the medicinal chemistry phase, and when an optimized lead compound has been characterized the drug development process can start.

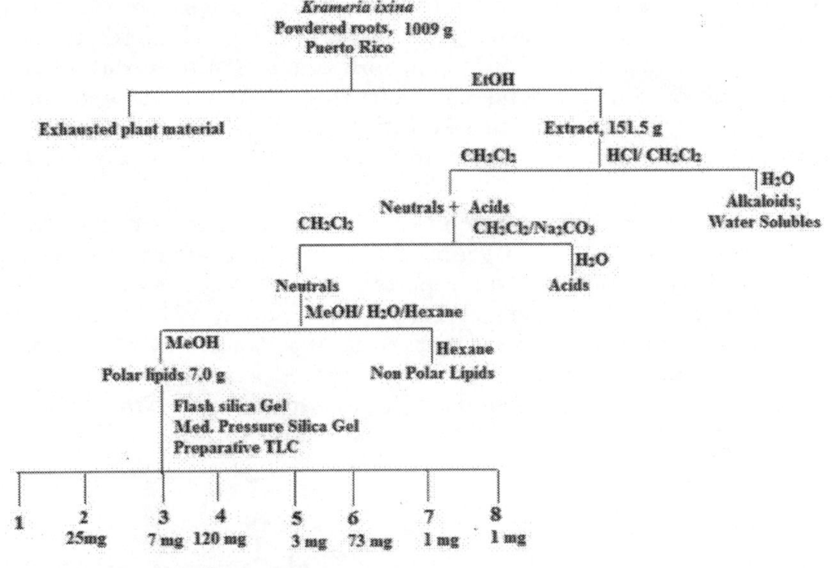

Fig. 2. Bioassay-directed fractionation of antifungal agents from *Krameria ixina*. (Source: Authors)

There are some general requirement for screening bioassay of plant extracts; validity lack of ambiguity accuracy reproducibility, simplicity and a reasonable cost. Some special considerations include a high selectivity (to limit the number of leads for follow-up evaluation), a high sensitivity (to detect low concentration of active compounds) and a high specificity (to be insensitive to a wide variety of inactive compounds).

Isolation of compounds in their pure form and evaluating them for their pharmacological properties leading to drug discovery is a long, tedious, time-consuming and expensive path. Drug discovery from natural sources has become very expensive (currently estimated at US$500 million) and time consuming (5 to 6 years in the 1980s to 15 to 22 years in the 21st century). Two-thirds of the cost goes to leads that fail during the clinical trials. Fifty percent of all potential drugs fail because of adsorption, distribution, metabolism, excretion or toxicity problems. Due to the long periods involved in discovery and preclinical and clinical trials, which require large numbers of volunteers (1000 to 5000 for long-term effects), many companies have abandoned the search for natural products and turned their attention towards combinatorial chemistry and modification/ analogue synthesis of existing drugs. Therefore, the new approaches like metabolomics and high-throughput screening (HTS) are used to screen and evaluate several metabolites in a short time (Ramawat *et al.*, 2009).

Microbially derived natural products are major sources of antibiotics and other medicines, but discovering new antibiotic scaffolds and increasing the chemical diversity of existing ones are formidable challenges. New drug discovery is facing serious challenges due to reduction in number of new drug approvals coupled with exorbitant rising cost. Advent of combinatorial chemistry provided new hope of higher success rates of new chemical entities (NCEs); however, even this scientific development has failed to improve the success rate in drug discovery.

The time required for development of pharmaceuticals ranges from a few years to as many as 20 years. Example, the chemical structure of paclitaxel (Taxol®) (Fig. 3) was reported and identified as the cytotoxic active constituent of extracts of *Taxus brevifolia* in 1971 (Wani *et al.*, 1971). Taxol® was approved for marketing as a cancer chemotherapeutic agent at the end of 1992, 20 years later. On average new pharmaceuticals require a decade for development and commercialization. This timeframe has not changed and appreciably in the last quarter century.

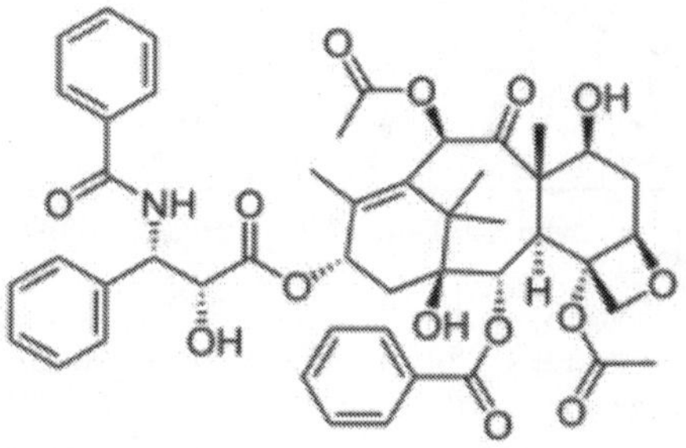

Fig. 3. Chemical structure of Taxol

Concern over the availability of the quantity of a chemical entity required for development and market needs has been the one most limiting factor for the pharmaceutical industries interest in natural products. A realistic assessment of quantities of plant material required for the preparation the amount of chemical substance necessary for the entire development process will address this concern. Under this scenario, the presence of an active product would be at a concentration of only 0.001% of dry weight of biomass. Current technologies are capable of identifying potential utility in the bioassay and of isolation and characterization of the natural product that occurs in plant biomass at this low concentration. Identification of biological activity isolation of active product and determination of chemical structure require at the most 50 mg of the chemical substance (Cremin and Zeng, 2002; Eldridge *et al.*, 2002). Isolation of this amount of chemical substance requires about 5 kg of dry plant material.

At this point with the chemical substance isolated and characterized and its biological properties determined a decision point reached: Is the chemical structure novel? Does this substance represent a potential new prototype? If the answers to these questions are yes we now have a natural product hit and must be decide whether to carry it forward into development.

Proceeding to the next step means assessing the real potential of the substance. Confirmatory bioassays must be carried out to make sure that the suspected biological activity is actually present. There must be followed with secondary biological assays to gain a full understanding of the breadth and selectivity of the biological activity and preliminary toxicology tests (if it cures a particular disease but kills the patient, then it is not really likely to be a drug substance).To carry out these assessments approximately 400-500mg of pure active chemical substance is needed. That represents a 10 fold increase in the plant material required as much as 100 kg of biomass is needed for processing to gain this additional information.

The quantity of pure chemical substance required for full preclinical development and a subsequent clinical trial is roughly 2 kg of pure product. Therefore, in the "worst case" scenario for active product, (isolation yield from biomass of only 0.001%) 200,000 kg of dry plant biomass would be required to produce the required pure active product to this point. The challenges must be identified and addressed if we are to return natural products to their preeminent position as the foundation of new pharmaceutical discovery and development.

MODERN METHODS FOR FINDING OUT NEW PHARMACOLOGICALLY ACTIVE PLANT CONSTITUENTS

When searching for new pharmacologically active plant constituents in the classic way one usually proceeds in three steps which are the-

Extraction and Isolation

Extraction is the crucial first step in the analysis of medicinal plants, because it is necessary to extract the desired chemical components from the plant materials for further separation and characterization. The basic operation included steps, such as pre-washing, drying of plant materials or freeze drying, grinding to obtain a homogenous sample and often improving the kinetics of analytic extraction and also increasing the contact of sample surface with the solvent system. Proper actions must be taken to assure that potential active constituents are not lost, distorted or destroyed during the preparation of the extract from plant samples. If the plant was selected on the basis of traditional uses (Fabricant and Farnsworth, 2001), then it is needed to prepare the extract as described by the traditional healer in order to mimic as closely as possible the traditional 'herbal' drug. The selection of solvent system largely depends on the specific nature of the bioactive compound being targeted. Different solvent systems are available to extract the bioactive compound from natural products. The extraction of hydrophilic compounds uses polar solvents such as methanol, ethanol or ethyl-acetate. For extraction of more lipophilic compounds, dichloromethane or a mixture of dichloromethane/ methanol in ratio of 1:1 are used. In some instances, extraction with hexane is used to remove chlorophyll (Cos *et al.,* 2006). As the target compounds may be non-polar to polar and thermally labile, the suitability of the methods of extraction must be considered. Various methods, such as sonification, heating under reflux, soxhlet extraction and others are commonly used (United States Pharmacopeia and National Formulary, 2002; Pharmacopoeia of the People's Republic of China, 2000; The Japanese Pharmacopeia, 2001) for the plant samples extraction. In addition, plant extracts are also prepared by maceration or percolation of fresh green plants or dried powdered plant material in water and/or organic solvent systems. A brief summary of the experimental conditions for the various methods of extraction is shown in Fig. 4 (Harborne, 1998). The other modern extraction techniques include solid-phase micro-extraction, supercritical-fluid extraction, pressurized-liquid extraction, microwave-assisted extraction, solid-phase extraction, and surfactant-mediated techniques, which possess certain advantages. These are the reduction in organic solvent consumption and in sample degradation, elimination of additional sample clean-up and concentration steps before chromatographic analysis, improvement in extraction efficiency, selectivity, and/ kinetics of extraction. The ease of automation for these techniques also favors their usage for the extraction of plants materials (Huie, 2002).

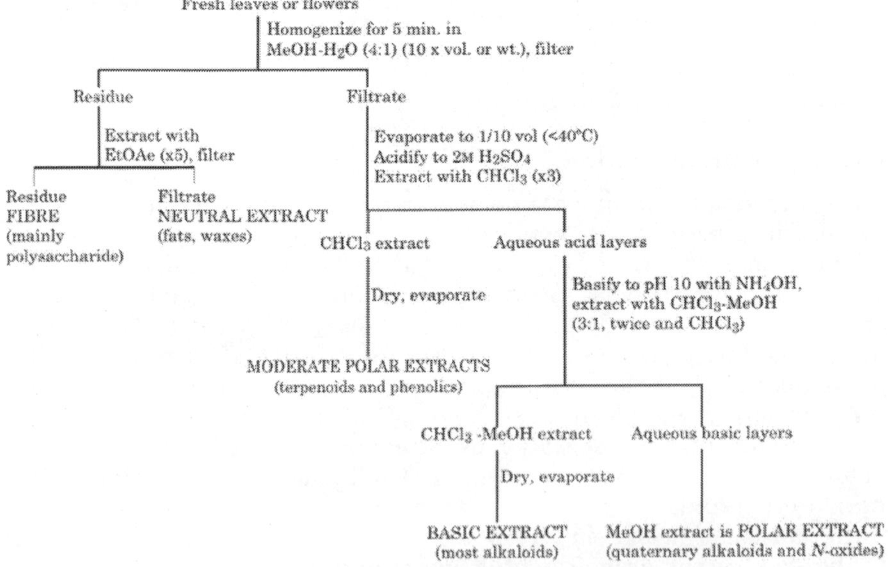

Fig. 4. A general procedure for extracting fresh plant tissues and fractionating into different classes according to polarity (Source: Authors).

Pharmacological Screening and Separation of Plant Constituents

Bioassay-guided fractionation has been the state-of-the art method for identifying bioactive natural products for many years. This approach involves repetitive preparative-scale fractionation and assessment of biological activity up to the isolation of pure constituents with the selected biological activity. The purified plant extract is fractioned and the obtained fractions submitted to a primary pharmacological screening. Only those fractions that showed some efficiency are exploited chemically (Fig. 5). The separation and purification of plant constituents is mainly carried out using one or other, or a combination, of four chromatographic techniques: paper chromatography (PC), thin layer chromatography (TLC), gas liquid chromatography (GC) and high performance liquid chromatography (HPLC). The choice of technique depends largely on the solubility properties and volatilities of the compounds to be separated. PC is particularly applicable to water-soluble plant constituents, namely the carbohydrates, amino acids, nucleic acid bases, organic acids and phenolic compounds. TLC is the method of choice for separating all lipid-soluble components, i.e. the lipids, steroids, carotenoids, simple quinines and chlorophylls. By contrast, the third technique GLC finds its main application with volatile compounds, fatty acids, mono- and sesquiterpenes, hydrocarbons and sulphur compounds. Alternatively, the less volatile constituents can be separated by HPLC, a method which combines column efficiency with speed of analysis. Additionally, it may be pointed out that there is considerable overlap in the use of above techniques and often a combination of PC and TLC, TLC and HPLC or TLC and GLC may be the best approach for separating a particular class of plant compound. Besides the above techniques, a few others viz. electrophoresis, droplet countercurrent chromatography (DCCC), affinity chromatography and differential ultracentrifugation etc. are used occasionally in phytochemical research (Harborne, 1998).

Structure Elucidation of Isolated Plant Constituents

In identifying a plant constituent, once it has been isolated and purified, it is necessary first to determine the class of compounds and then to find out which particular substance it is within that class. Its homogeneity must be checked carefully beforehand, i.e. it should travel as a single spot in several TLC and or/PC systems. The class of compound is usually clear from its response to colour tests, its solubility and R_F properties and its UV spectral characteristics. /biochemical tests may also be invaluable: presence of a glucoside can be confirmed by hydrolysis with ß-glucosidase, of a mustard oil glycoside by hydrolysis with myrosinase and so on. Complete identification within the class depends on measuring other properties and then comparing these data with those in the literature. These properties include melting point (for solids), boiling point (for liquids), optical rotation (for optically active compounds) and R_F or RR_t

(under standard conditions). However, equally informative data on a plant substance are its spectral characteristics: these include ultraviolet (UV), infrared (IR), nuclear magnetic resonance (NMR) and mass spectral (MS) measurements. A known plant compound can usually be identified on the above basis. Direct comparison with authentic material (if available), should be carried out as final confirmation. If authentic material is not available, careful comparison with literature data may suffice for its identification. If a new compound is present, all the above data should be sufficient to characterize it. With new compound, however, it is preferable to confirm the identification through chemical degradation or by preparing the compound by laboratory synthesis. Identification of new plant compounds by X-ray crystallography is now routine, and can be applied whenever the substance is obtained in sufficient amount and in crystalline form (Harborne, 1998).

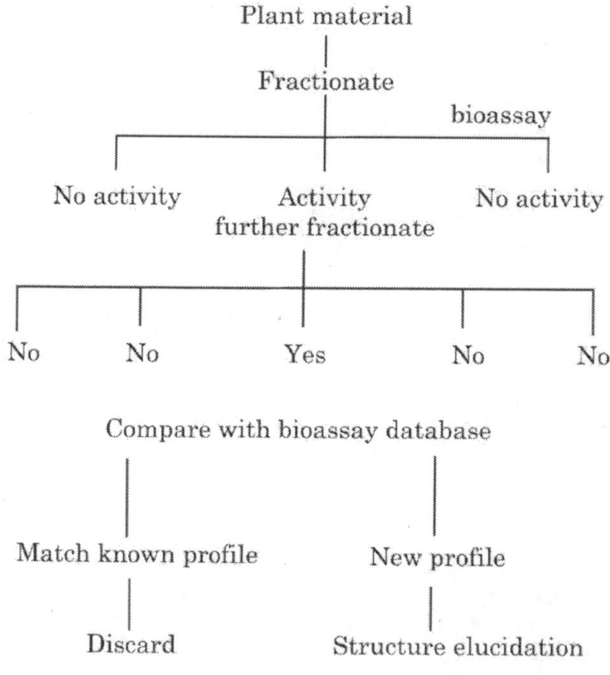

Fig. 5. Bioassay-driven discovery strategy for new phytochemicals (Source: Authors)

Future Prospects and Conclusions

Many companies in the pharmaceutical industry have foregone natural products as sources of new compounds because of the long lead times and quality of compounds for drug development available by conventional screening and fractionation approaches. However, as drug discovery pipelines grow thin for new leads, and current blockbusters go off patent, there is a growing need for compounds to be developed in the fight against disease. One source that should be included in the drug discovery paradigms of today is natural products. Advances in biotechnology and in analytical and in separations technology at the nano-scale level are providing solutions for the complex source and potentially leading to new compounds and scaffolds from which to derive new medicines. Affinity screening of natural products and industrializing the natural product drug discovery process provide access to the rich source of compounds. Therefore the research on natural products and compounds derived from natural products has been accelerated in recent years due to their importance in drug discovery.

Plants have provided humans with many of their essential needs, including life-saving pharmaceutical agents. In the last 6 years, five new plant derived drugs have been launched onto the market and many more are currently undergoing clinical trials. As a vast proportion of the available higher plant species have not yet been screened for biologically active compounds, drug discovery from plants should remain essential component in the search for new medicines, particularly with the development of highly sensitive and versatile analytical methods.

The past few decades have clearly demonstrated that natural products represent an unparallel reservoir of molecular diversity to drug discovery and development and are complementary to combinatorial libraries. For the developing countries the approval as drug of standardization and formulated plant extracts might be the starting point of an innovative and successful local pharmaceutical industry, which can complete with the western pharmaceutical companies not only for the treatment of major diseases but also for the treatment of severe and life threatening diseases.

Scientists from divergent fields are investigating plants anew with an eye to their antimicrobial usefulness. A sense of urgency accompanies the search as the pace of species extinction continues. Laboratories of the world have found literally thousands of phytochemicals which have inhibitory effects on all types of microorganisms *in vitro*. More of these compounds should be subjected to animal and human studies to determine their effectiveness in whole-organism systems, including in particular toxicity studies as well as an examination of their effects on beneficial normal microbiota. Attention to these issues could usher in a badly needed new era of chemotherapeutic treatment of infection by using plant-derived principles.

Acknowledgements

Authors thank Department of science & Technology (DST), New Delhi, India for support.

References

Afolayan, AJ and Meyer, JJM. 1997. The antimicrobial activity of 3,5,7-trihydroxyflavone isolated from the shoots of *Helichrysum aureonitens*. *J. Ethnopharmacol.,* **57:** 177-81.

Al-Bayati, FA. 2009. Isolation and identification of antimicrobial compound from *Mentha longifolia* L. leaves grown wild in Iraq. *Ann. Clin. Microbiol. Antimicrob.,* **8**: 20-6.

Almeida, LS; Murata, RM; Yatsuda, R; Dos Santos, MH; Nagem, TJ; Alencar, SM; Koo, H and Rosalen, PL. 2008. Novel Antimicrobial agents and Strategies. *Phytomed.,* **15:** 886-91.

Ambrosio, SR; Furtado, NA; de Oliveria, DC; da Costa, FB; Martinis, CH; de Carvalho, TC; Porto, TS and Veneziani, RC. 2008. Antimicrobial activity of kaurane diterpenes against oral pathogens. *Z. Naturforsch C.*, 63: 326-30.

Becker, H; Scher, JM; Speakman, JB and Zapp, JB. 2005. Bioactivity guided isolation of antimicrobial compounds from *Lythrum salicaria. Fitoterapia,* **76:** 580-84.

Boonphong, S; Puangsombat, P; Baramee, A; Mahidol, C; Ruchirawat, S and Kittakoop, S. 2007. Bioactive compounds from *Bauhinia purpurea* possessing antimalarial, antimycobacterial, antifungal, anti-inflammatory and cytotoxic activities. *J. Nat. Prod.,* **70:** 795-801.

Boonsri, S; Karalai, C; Ponglimanont, C; Kanjana-opas, A and Chantrapromma, K. 2006. Antibacterial and cytotoxic xanthones from the roots of *Cratoxylum formosum. Phytochem.,* **67:** 723-27.

Chou, SC; Everngam, MC; Sturtz, G and Beck, JJ. 2006. Antibacterial activity of components from *Lomatium californicum. Phytother. Res.*, **20:** 153-56.

Cos, P; Vlietinck, AJ; Berghe, DV and Maes, L. 2006. Anti-infective potential of natural products: How to develop a stronger *in vitro* 'proof-of-concept'. *J. Ethnopharmacol.*, **106:** 290-302.

Cragg, GM; Newman, DJ and Snader, KM. 1997. Natural products in drug discovery and development. *J. Nat. Prod.,* **60:** 52-60.

Cremin, PA and Zeng, L. 2002. High-throughput analysis of natural product compound libraries by parallel LC-MS evaporative light scattering detection. *Anal. Chem.,* **74:** 5492-500.

da Silva, LL; Nascimento, MS; Cavalheiro, AJ; Silva, DH; Castro-Gamboa, I; Furlan, M; Bolzani and Vda, S. 2008. Antibacterial activity of labdane diterpenoids from *Stemodia foliosa*. J. *J. Nat. Prod.*, **71:** 1291-93.

Drewes, SE and van Vuuren, SF. 2008. Antimicrobial acylphloroglucinols and dibenzyloxyflavonoids from flowers of *Helichrysum gymnocomum* DC. *Phytochem.,* **69:** 1745-49.

Ejele, AE; Iwu, IC; Enenebeaku, CK; Ukiwe, LN and Okolue, BN. 2012. Bioassay-guided isolation, purification and partial characterization of antimicrobial compound from basic metabolite of *Garcinia* kola. *JETEAS,* **3:** 668-72.

Eldeen, IMS; Elgorashi, EE; Mulholland, DA and van Staden, J. 2006. Anolignan B: A bioactive compound from the roots of *Terminalia sericea. J. Ethnopharmacol.*, **103:** 135-38.

Eldridge, GR; Vervoot, HC; Lee, CM; Cremin, PA; Williams, CT; Hart, SM; Goering MG; O'Neil-Jhonson, M and Zeng, L. 2002. High-throughput method for the production and analysis of large natural product libraries for drug discovery. *Anal. Chem.*, **74:** 3963-71.

Erasto, P; Grierson, DS and Afolayan, AJ. 2006. Bioactive sesquiterpene lactones from the leaves of *Vernonia amygdalina. J. Ethnopharmacol.*, **106:** 117-20.

Eyong, KO; Folefoc, GN; Kuete, V; Beng, VP; Krohn, K; Hussain, H; Nkengfack, AE; Saeftel, M; Sarite, SR and Hoerauf, A. 2006. Newbouldiaquinone A: A naphthoquinone-anthraquinone ether coupled pigment, as a potential antimicrobial and antimalarial agent from *Newbouldia laevis. Phytochem.,* 67: 605-9.

Fabricant, DS and Farmsworth, NR. 2001. The value of plants used in traditional medicine for drug discovery. *Environ. Health Perspect.*, **109:** 69-75.

Gruenwald, J. 1997. The herbal remedies market in the US, market development, consumer, legislation and organizations. Phyotpharm Consulting, Communique.

Harborne, JB .1998. Phytochemical Methods: A Guide to Modern Techniques of Plant Analysis, 3rd edn. Chapman and Hall, New York.

Heilmann, J; Brun, R; Mayr, S; Rali, T and Sticher, O. 2001. Minor cytotoxic and antibacterialcompounds from the rhizomes of *Amomum aculeatum. Phyotchem.,* **57:** 1281-85.

Huie, CW. 2002. A review of modern sample-preparation techniques for the extraction and analysis of medicinal plants. *Anal. Bioanal. Chem.,* **373:** 23-30.

Iwu, MW; Duncan, AR and Okunji, CO. 1999. New antimicrobials of the plant origin. *In: J. Janick (ed), Perspectives on New Crops and New Uses. ASHS Press, Alexandria, V.A.* p.457-462.

Jovel, EM; Zhou Xi L.; Ming, DS; Wahbe TR and Towers, GHN. 2007. Bioactivity-guided isolation of the active compounds from *Rosa nutkana* and quantitative analysis of ascorbic acid by HPLC. *Can. J. Physiol. Pharmacol.,* **85:** 865-71.

Kambizi, L., Sultana, L. and Afolayan, AJ. 2005. Bioactive Compounds Isolated from *Aloe ferox*: A Plant Traditionally Used for the Treatment of Sexually Transmitted Infections in the Eastern Cape, South Africa. *Pharm. Biol.,* **42:** 636-39.

Katiyar, CA; Gupta, SK and Katiyar, S. 2012. Drug discovery from plant sources: An integrated approach. *AYU,* **33:** 10-19.

Kim, YS; Kim, H; Jung, E; Kim, JH; Hwang, W; Kang, EJ; Lee, S; Ha, BJ; Lee, J and Park, D. 2012. A novel antibacterial compound from *Siegesbeckia glabrescens. Molecules*, 17: 12469-77.

Kuete, V; Wabo, GF; Ngameni, B; Mbaveng, AT; Metuno, R; Etoa, FX; Ngadjui, BT; Beng, VP; Meyer, JI and Lall, N. 2007a. Antimicrobial activity of the methanolic extract, fractions and compounds from the stem bark of *Irvingia gabonensis* (Ixonanthaceae). *J. Ethnophamacol.*, 114: 54-60.

Kuete, V; Eyong, KO; Folefoc, GN; Beng, VP; Hussain, H; Krohn, K and Nkengfack, AE. 2007b. Antimicrobial activity of the methanolic extract and of the chemical constituents isolated from *Newbouldia laevis. Pharmazie*, 62: 552-56.

Kuete, V; Nguemeving, JR and Penlap, BV. *et al.*, 2007c. Antimicrobial activity of the methanolic extracts and compounds from *Vismia laurentii* De Wild (Guttiferae). *J. Ethnopharmacol.,* 109: 372-79.

Kuete, V; Ngameni, B; Fotso, SCC; Kengap, TR, Tchaleu, NB, Meyer, DN; Lall, JJM and Kuiate, JR. 2008. Antimicrobial activity of the crude extracts and compounds from *Ficus chlamydocarpa* and *Ficus cordata* (Moraceae). *J. Ethnopharmcol.,* **120:** 17-24.

Lee, DG; Jung, HJ and Woo, ER. 2005. Antimicrobial property of (+)-lyoniresinol-alpha-O-beta-D-glucopyranoside isolated from the root bark of *Lycium chinense* Miller against human pathogenic microorganisms. *Arch. Pharmacal. Res.*, 28: 1031-36.

Martini, ND; Katerere, DRP and Eloff, JN. 2004. Biological activity of five antibacterial flavonoids from *Combretum erythrophyllum* (Combretaceae). *J. Ethnopharmacol.*, **93:** 207-12.

Mathabe, CM; Hussein, AA; Nikolova, RV; Basson, AE; Meyer, JJ and Lall, N. 2008. Antibacterial activities and cytotoxicity of terpenoids isolated from *Spirostachys africana. J. Ethnopharmacol.*, **116:** 194-97.

Mbaveng, AT; Ngameni, B; Kuete, V; Simo, IK; Ambassa, P; Roy, R; Bezabih, M; Etoa, FX; Ngadjui, BT; Abegaz, BM; Meyer, JI; Lall, N and Beng, VP. 2008. Antimicrobial activity of the crude extracts and five flavonoids from the twigs of *Dorstenia barteri* (Moraceae). *J. Ethnopharmacol.*, **116:** 483-89.

Mskhiladze, L; Kutchukhidze, J; Chincharadze, D; Delmas, F; Elias, R and Favel, A. 2008. *In vitro* antifungal and antileishmanial activities of steroidal saponins from *Allium leucanthum C. Koch* - a Caucasian endemic species. *Georgian Med. News.*, 154: 39-43.

Murray, M. 1995. *The Healing Power of Herbs.* Prima Publishing. Rocklin, C.A. p. 162-171.

Naz, S; Ahmad, S; Rasool, SA; Sayeed, SA and Siddiqi, R. 2006. Antibacterial activity directed isolation of compounds from *Onosma hispidum. Microb. Res.*, **16:** 43-8.

Pathong, K; Pongcharoen, W; Phongpaichit, S and Taylor, WC. 2006. Tetraoxygenated xanthones from the fruits of *Garcinia cowa. Phytochem.,* **67:** 999-1004.

Peng, SC; Cheng, CY; Sheu, F and Su, CH. 2008. The antimicrobial activity of heyneanol A extracted from the root of Taiwanese wild grape. *J. Appl. Microbio.*, **105:** 485-91.

Pereira, AP; Ferreira, ICFR; Marcelino, F; Valentão, P; Andrade, PB; Seabra, R; Estevinho, L; Albino Bento, A; and Pereira, JA. 2007. Phenolic compounds and antimicrobial activity of Olive (*Olea europaea* L. *Cv.* Cobrançosa) Leaves. *Molecules,* 12: 1153-62.

Pessini, GL; Filho, BPD; Nakamura, CV and Cortez, DAG. 2003. Antibacterial activity of extracts and neolignans from *Piper regnellii* (Miq.) C. DC. var. *pallescens* (C. DC.) Yunck. *Mem. Inst. Oswaldo Cruz.*, **98:** 1115-20.

Pharmacopoeia of the People's Republic of China. 2000. English ed., The Pharmacopeia Commission of PRC, Beijing.

Pieters, L and Vlietinck, AJ. 2005. Biological isolation of pharmacologically active plant components, still a valuable strategy for finding of new lead compounds? *J. Ethnopharmacol.*, **100:** 57-60.

Prachayasittikul, S; Buraparuangsang, P; Worachartcheewan, A; Isarankura-Na-Ayudhya, C; Ruchirawat, S and Prachayasittikul V. 2008. Antimicrobial and antioxidative activities of bioactive constituents from *Hydnophytum formicarum* Jack. *Molecules*, **13:** 904-21.

Qi, SH; Zhang, S; Qian, PY and Wang, BG. 2008. Antifeedant, antibacterial, and antilarval compounds from the South China Sea seagrass *Enhalus acoroides. Botanica Marina.*, **51:** 441-47.

Ramawat, KG; Dass, S and Mathur, M. 2009. The chemical diversity of bioactive molecules and therapeutic potential of medicinal plants. In *Herbal Drugs: Ethnomedicine to Modern Medicine.* Edited by Ramawat KG. Springer-Verlag Berlin Heidelberg, Chapter 2.

Renault, S; De Lucca, AJ; Boue, S; Bland, JM; Vigo, CB and Selitrennikoff, CP. 2003. CAY-1, a novel antifungal compound from cayenne pepper. *Med. Mycol.*, 41: 75-81.

Rukachaisirikul, T; Innok, P; Aroonrerk, N; Boonamnuaylap, W; Limrangsun, S; Boonyon, C; Woonjina, U and Suksamrarn, A. 2007. Antibacterial pterocarpans from *Erythrina subumbrans. J. Ethnopharmacol.*, **110:** 171-75.

Singh, DN; Verma N; Raghuwanshi, S; Shukla, PK and Kulshreshtha, DK. 2006. Antifungal anthraquinones from *Saprosma fragrans. Bioorg. Med. Chem. Lett.,* 16: 4512-14.

The Japanese Pharmacopeia. 2001. Fourteenth ed., JP XIII, The Society of Japanese Pharmacopeia, Japan.

Trease, G and Evans, W. 1972. *Pharmacognosy.* Univ. Press, Aberdeen, Great Britain. p. 161-163.

United States Pharmacopeia and National Formulary, USP 25, NF 19. 2002. United States Pharmacopeial Convention Inc., Rockville.

Wani, MC; Taylor, HL; Wall, ME; Coggon, P and McPhail, AT. 1971. Plant antitumor agents. VI. The isolation and structure of taxol, a novel antileu-kamia and antitumor agent from *Taxus brevifolia. J. Am. Chem. Soc.,* **93:** 2325-27.

Watt, E. Van der and Pretorius, JC. 2001. Purification and identification of active antibacterial components in *Carpobrotus edulis* L. *J. Ethnopharmacol.*, **76:** 87-91.

Yang, X; Summerhurst, DK; Koval SF; Ficker, C; Smith ML; and Bernards, MA. 2001. Isolation of an antimicrobial compound from *Impatiens balsamina* L. using bioassay-guided fractionation. *Phytother. Res.,* **15:** 676-80.

Yin, S; Fan, CQ; Dong, L and Yue, JM. 2006. Psoracorylifols A-E, five novel compounds with activity against *Helicobacter pylori* from seeds of *Psoralea corylifolia Tetrahedron,* **62:** 2569-75.

Yoshida, H; Iwata, N; Katsuzaki, H; Naganawa, R; Ishikawa, K; Fukuda, H; Fujino, T and Suzuki, A. 1998. Antimicrobial Activity of a Compound Isolated from an Oil-Macerated Garlic Extract. *Biosci. Biotech. Biochem.,* **62:** 1014-17.

Zhou, L; Li, D; Wang, J; Liu, Y; and Wu, J. 2007. Antibacterial phenolic compounds from the spines of *Gleditsia sinensis* Lam. *Nat. Prod. Res.,* **21:** 283-91.

15

SWERTIA CHIRAYITA: ENDANGERED MEDICINAL HERB OF TEMPERATE HIMALAYA

Dhiman Mukherjee*

Regional Research Station, Uttar Banga Krishi Viswavidyalaya, Kalimpong (Darjelling)-734301 (W.B)
**Incharge, All India coordinated Research Project on Medicinal and Aromatic Plants, RRS, Kalimpong.*
E-mail : dhiman_mukherjee@yahoo.co.in

INTRODUCTION

Swertia chirayita is an important medicinal plant found in Uttarakhand. *Swertia chirayita* is considered the most important plant for its bitterness, antihelminthic, hypoglycemic, hepatoprotective, and antiviral properties. Out of many genuine approaches for deriving relevant pieces of information, the novel techniques of plant tissue culture provide a viable alternative for managing these valuable resources in a sustainable manner. There are few interesting reports about the tissue culture of *Swertia chirayita*. In addition, several others have been also made with this medicinal plant. However, this plant obviously leaves scope for several physiological studies, which are needed for proper understanding of the drug originating plant. This review aims to discuss vital aspects of the endangered medicinal herb of temperate Himalaya.

Swertia chirayita

The following is the brief description of the plant:

Scientific name: *Swertia chirayita* (Robex. ex Flem) Karsten

Family: Gentianaceae

Local name: Chiraita, Tite, or Lektite or Khalu, Pothe, Chirota

English Name: Chiretta

Swertia chirayita especially needs to be explored thoroughly through different disciplines of plant sciences, since today's healthcare systems mainly rely on drugs derived from plant materials. Most of the world's populations depend on traditional medicine to meet daily health requirements, especially in developing countries, where plants are the main source of medicine. China uses about 5000 species, while India uses 7000 species of medicinal plants. The WHO has estimated that over 80 per cent of the world population meets its primary health care needs through traditional medicine. About 47.2 percent of the surface of the earth can be classified as dry land wherein 600 m ha of land is devoted to dry land agriculture worldwide and more than 700 million people live in these regions (Mukherjee, 2009). Medicinal plants have higher demand in the market and are quite suitable to our soils and weather conditions, and highly remunerative business under temperate Himalayan zone. Thus as an alternate cropping system medicinal plants could be included in the cropping system and as well cultivated as sole crop. There are records of about 400 plants of therapeutic value in the Darjeeling - Sikkim Himalaya. Indiscriminate and non-systematic collection of medicinal plants has led to severe pressure on the availability of these plants, many of which are now rare, threatened or endangered. Six species, which come in category of threatened plants in the Darjeeling Himalaya are *Aconitum heterophyllum* (Wall), *Podophyllum hexandrum* (Royle), *Nardostachys jatamansi* (DC), *Picrorhiza kurrooa* (Benth), *Swertia chirata* (Ham) and *Bergenia ciliata* (Har.) Stenb (Rai *et al.*, 2000).

Table 1: Medicinal plant distribution under Indian traditional system.

World	3,60,000 species
India	40 % plants- grown in16 agro- climatic zones
Ayurveda, siddha, unani	7195 plants
Folk medicines	4720 plants
Ayurveda	1773 plants
Siddha	1122 plants
Unani	751 plants

(Source : Kalaichelvi and Swaminathan, 2009)

Himalaya is well known for its rich biodiversity in medicinal and aromatic plants. Owing to its unique geographical set-up, topography and undulant landscape, the climatic conditions varied along an altitudinal gradient which attributes diversified ecological habitat ranging from tropical forest, grassland to alpine meadow with vast and diverse natural resources. Unfortunately, some of the natural resources are unscientifically exploited for various purposes, particularly medicinal plants

(Mukherjee, 2014 a). Most of the medicinal plants are being extracted for drug and pharmaceutical industries, traditional use and research purposes from wild populations. These activities adversely affect the existence of a number of plants, particularly those of high commercial value. Further, with the developmental activities in the higher Himalayan region, the original populations of many of these valuable resources have been vanishing. *Swertia chirayita* is among the several medicinal plant species growing in the Himalayan region (Mukherjee, 2014).

Swertia chirayita (Roxb. ex Flem.) Karst is an important medicinal plant endemic to the Himalayan regions (IUCN, 2009). It is over exploited due to faulty collection methods, habitat disturbance and less information about its biology. It is found growing in moist hillsides of temperate forests between the altitudes of 1200 - 3000 m asl and is indiscriminately collected from its wild habitat for trade and local medicinal use. *Swertia chirayita* belongs to the Gentianaceae family and contains many of the compounds that are responsible for its therapeutic properties such as xanthones, flavonoids, terpenoids, iridoids and secoiridoid glycosides (Pant *et al.* 2000). This plant is used locally in India and other part of world as an infusion prepared by grinding the plant and steeping it in water overnight. In addition, *Swertia chirayita* has only recently been brought into cultivation with limited success.

In lower altitude *Alnus nepalennsis* (alder) was found to be the most abundant and preferred tree, which is non-leguminous nitrogen fixing tree and help to grow chiratyita and other high value medicinal plants. Survey revealed that medicinal plants viz. *Panax pseudo-ginseng Picrorhiza kurrooa, Podophylum hexandrum* prefer to grow under *Acer oblongum, Pinus wallichiana, Betula alnoides, Mahornia acanthifolia* and *Swertia chirata, Valeriana hardwicki* prefer to grow with *Betula utilis, Pyrus pashia, Mahornia acanthifolia* and *Dichroa febrifuga* in high hill. Few plant of *Coptis teeta* was also observed in association with *Taxus baccata* in some field (Mukherjee, 2013 b).

MEDICINAL USES

S. chirayita is utilized extensively in Eastern traditional medicine such as Ayurveda, Unani, Siddha and also in traditional Chinese and Tibetan medicine as well as in local healing in Darjeeling – Sikkim himalaya. Its medicinal usage is reported in Indian pharmaceutical codex, the British and the American pharmacopoeias. This species was first introduced in the Edinburgh Pharmacopoeia in 1839 and is reported in British and American Pharmacopeias to be used as an infusion or a tincture (Mukherjee *et al.*, 2013). An astringent stimulant prepared with the plant is an outstanding medication for weak stomach, particularly when it results in indigestion, bloating and nausea. In addition, this bitter tonic is also said to be effective in protecting the liver. It is laxative and an appetizer. It also corrects the disordered process of nutrition and restores the normal function of the system. It is an effective herb for reducing fevers and beneficial in the treatment of malarial fevers. It is also effective in hysteria and convulsions. It also shows hypoglycemic activity. Its importance also observed in the field of anemia, bronchial asthma, liver disorders, hepatitis, gastritis, constipation, dyspepsia, skin diseases, worms, epilepsy, ulcer, scanty urine, hypertension, melancholia and certain type of mental disorder, secretion of bile, blood purification and diabetes (Rai 2003). In Ayurveda, *S. chirayita* is described as bitter (*tikta*) in taste and its thermal action defined as cooling (*shita*), easily digestible (*laghu*) and dry (*ruksha*) (Joshi and Dhawan 2005).

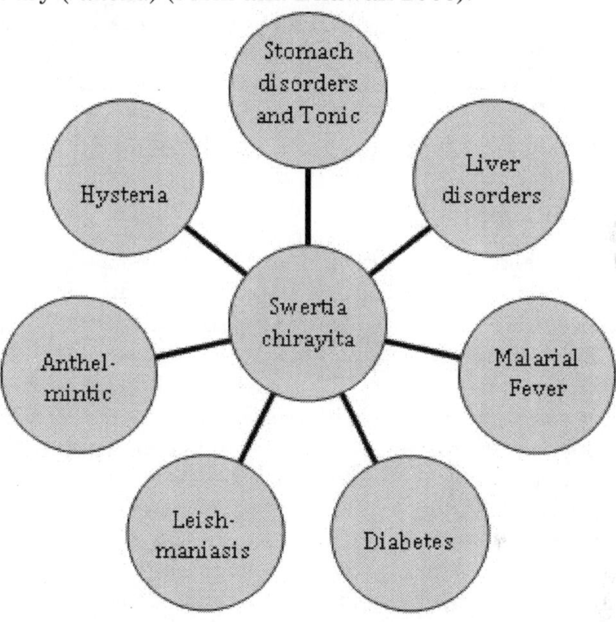

Fig. 1 : Role of *S.chirayita* in human body (Source: Author)

Pharmacological activity

The ethanolic extract of *S.chirayita* exhibits hypoglycemic activity. This herb is an excellent drug for strengthening the stomach and promoting its action. It is used in the treatment of dyspepsia and diarrhoea (Mukherjee, 2013). This plant possesses anthelmintic that is, worms destroying, properties and is used in killing intestinal worms, due to this property children used it in daily drinking puprose. An infusion of the herb is taken for this purpose. It serves as an effective tonic in case of general weakness and during convalescence. The infusion of the plant can be taken in doses of 60 ml or 40 tablespoons twice a day, before meals. The herb is used in the form of an infusion or tincture. The infusion is prepared in the hot water with aromatics likes cloves and cinnamon. It is generally taken in doses of 15 to 30 ml or 1 to 2 tablespoon. The root of the plant is useful in checking hiccups and vomiting. It is taken in doses of 0.5 to 2 grams with honey. The chirayata herb is remarkable for its properties in the removal of all kinds of intestinal worms. An infusion of the herb is prepared and taken regularly during parasitic infestations. The *Swertia chiretta* is a therapeutic plant and its remedial usage has been recognized in the Indian pharmaceutical codex as well as the British and American Pharmacopoeias. In addition, the curative value of the herb has also been recorded by the ancient Indian herbal medicine system Ayurveda and other conventional medical systems. The herb as well as its extracts is used as a bitter stimulant to treat fever as well as curing several skin problems. It is much employed in urinary complaints with uneasiness in the region of the kidneys, frequent urging to urinate, which is accomplished with difficulty, and in cases of uric aciddeposits. It is a remedy also for convalescence from exhausting sickness, and for a tonic and nervous forms of dyspepsia.

PLANT DESCRIPTION

This plant belongs to family *Gentianceae*, found in India (Darjeeling hill, Kashmir, Meghalaya and Khasi hills), Nepal and Bhutan. This is a biennial or perennial herb of seasonal growth. The plant is found at an altitude of 1200–3000 m, from Kashmir to Bhutan, and in the Khasi hills at 1200 –1500 m.

It usually has a single, stout, elongated stem, about 2–5 ft long stem, the middle portion is round, while the upper is four-angled, with a prominent decurrent line at each angle. The stems are orange brown or purplish in colour. The stem is cylindrical at the base and quadrangular upwards. Transverse section of Stem shows, single layered epidermis; parenchymatous cortical cells with mucilage, minute acicular crystals, resin (as dark brown mass), and oil droplets; endodermis; single thin walled pericycle cells layer; rounded and isodiametric pith cells with prominent intercellular spaces. Ovate or lanceolate leafs with entire margin curly in nature. Transverse section of leafs shows single epidermis layer covered with a thick, striated cuticle, more strongly developed on the upper surface than the lower; anisocytic stomata; single layered palisade tissue, spongy messophyll cells with minute acicular crystal and mucilage (Bhardwaj *et al.*, 2011). The root is simple, tapering and stout, short, almost 7 cm long and usually half an inch thick. However, in a few cases, branching from the root has also been recorded. The roots are generally small, 5-10 cm in length, light brown, somewhat twisted and gradually tapering, and have a few rootlets or remnants. Transverse section of root shows, 2-4 layers of cork, thick-walled parenchymataous secondary cortex cells with mucilage, minute acicular crystals (also in present phloem region), and resin (as dark brown mass); secondary phloem composed of thin-walled strands of sieve tubes, companion cells and phloem parenchyma; lignified and thick-walled scalariform, simple and bordered pitted secondary xylem vessels, tracheids parenchyma and xylem fibers. Flowering in chirayita is in the form of numerous small, axillary, opposite, lax cymes arranged as short branches and the whole inflorescence is 2 ft long. Flowers are small, stalked, green-yellow, tinged with purple colour. The flower-tube is twice as long as the sepal-cup and divided near the base into four ovate–lancelike segments. The upper surface of the petal has a pair of nectaries covered with oblong scales and ending as fringes. Flowering pods are egg-shaped, and the seeds are minute, smooth, and many angled. Flowers are monocious. The seeds are light brown to dark brown in colour.

a. Vegetative phase of plant

b. Flowering phase (Inflorescence)

Fig. 2. Swertia chirayita plant (Source: Author)

Table 2 : Soil environment suitable for *Swertia chirayita* cultivation.

Sl.no.	Particulars	Values
1	pH	4.8 -5.6 (Acidic)
2.	E.C.E. (inch/cm)	0.09-.13 (Neutral)
3	Organic carbon (%)	0.93-1.62 (High)
4	Available N (Kg/ha)	250-321 (Medium)
5	Available P_2O_5 (Kg/ha)	70-90 (High)
6	Available K_2O (Kg/ha)	125-260 (Medium)

a. Purple leaf (beneath surface) b. Dark brown-purple stem node

Fig. 3. Purple tinge below leaf surface (a) and on stem node (b)(unique character) of *S. chirayita* (Source: Author)

PRINCIPLE CONSTITUENTS

Swertia chirata contains bitter yellow acid known as ophelic acid, two bitter glycosides chiratin (not a pure substance) and amarogentin (phenol carbonic acid ester of sweroside, a substance related to gentiopicrin), two alkaloids gentianine and gentiocrucine. This plant is chemically characterised by the presence of taxonomically important groups of secondary metabolites such as, xanthones, iridoids, mangiferin and C-lucoflavones. Among the different seco-iridoids present in *S. chirayita*, swertiamarin is the one of the most predominant and have well-established, important therapeutic applications.

Chiretta is reported to contain a yellow bitter acid, **ophelic acid**, two bitter glucosides, chiratin and amarogentin, gentiopicrin, two yellow crystalline phenols, a neutral, yellow crystalline compound, and a new **xanthone,** swerchirin. The herb contains swertanone, swertenol, episwertinol, chiratenol, gammacer-16-en-3ß-ol, 21-a-H-hop-22(29)-en-3ß-ol, taraxerol, oleanolic acid, ursolic acid, swerta-7,9(11)-dien-3ß-ol, pichierenol. It also yields, 1,3,6,7-tetrahydroxyxanthone-C-2-ß-D-glucoside (mangiferin) having a significant immunomodulatory potential.A new xanthone, 1,5-dihydroxy-3,8-dimethoxyxanthone besides swerchirin and **7-O-Me swertiarin** and monohydroxy terephthalic acid and 2,5-dihydroxy terephthalic acid were isolated from the herb. The herb also yields 1,5,8-trihydroxy-3-methoxyxanthone, 1-hydroxy-3,5,8-trimethoxyxanthone,1-hydroxy-3,7,8-trimethoxy xanthone, 1,8-dihydroxy-3,5-dimethoxy-xanthone, 1,8-dihydroxy-3,7-dimethoxy xanthone and 1,3,6,7-tetrahydroxy xanthone C-2-ß-D-glucoside (mengiferin), **1,3,8-trihydroxy-5- methoxy xanthone**, 1,3,5,8-tetrahydroxy xanthone and 1,3,7,8-tetrahydroxy xanthone, a novel dimeric xanthone the alkaloids gentianine, gentiocrucine and enicoflavine (Khan *et al.*, 2014).

Table 3: list of important bioactive constituents isolated from *Swertia chirayita*

Active constituents	Biological activities
Amarogentin (chirantin)	Topoisomerase inhibition, chemo-preventive and antileishmanial effects.
Amaroswerin	Gastro-shielding.
Gentianine	Anti-inflammatory, anesthetic, antihistaminic, anticonvulsant properties, hypotensive, antipsychotic, lenitive, diuretic, antimalarial, antiamoebic and antibacterial properties.
Swerchirin	Antimalarial, hypoglycemic, hepatoprotective, pro-heamatopoitic, and weak chemo preventive pharmacological effects.
Swertiamarin	Analgesic property

(Source : Tabassum *et al.*, 2012)

QUANTIFICATION OF MAJOR PHYTOCHEMICALS

Chiretta stimulates the digestion and helps to normalize blood sugar, which makes it useful for diabetics. Studies with animals suggest that this herb reduces the sugar levels only when they are high, which lowers the risk of hypoglycaemia (Mukherjee, 2013). Also 95% ethanol extract and four fractions of *Swertia chirayita* were tested for blood sugar lowering activity in rats. The hexane fraction caused maximum lowering although the ethanol extract was clearly active. Swertiamarin is main constituents of plant given the anti-diabetic effect due to an active metabolite, gentianine, that upregulates PPAR-γ gene expression in 3T3-L1 cells. A xanthone compound, mangiferin effect on the atherogenic potential of streptozotocin (STZ)-diabetes was investigated. Swertia chirayita possesses in vitro and in vivo antioxidant activity, the liver and kidney of CCl_4-intoxicated animals exhibited decrease in superoxide dismutase (SOD), catalase (CAT), and glutathione (GSH) levels. Additionally, these organs exhibited increase in malondialdehyde (MDA) level. Out of seven compounds extracted from this plant, five components exhibited antioxidant activity at different magnitude of potency (Patel and Patel, 2014).

The main phytochemicals in crude aqueous and ethanolic extracts of different plant parts of *Swertia chirayita* collected from nine different districts of Nepal representing West, East and Central Nepal were quantified using HPLC/DAD (High Performance Liquid Chromatography-Diode Array Detection). The quantities of these phytochemicals were also compared between wild and cultivated plant parts of *Swertia chirayita*. Amarogentin, mangiferin, swertiamarin were the main phytochemicals in all extracts. The highest quantity of all the three phytochemicals was found in inflorescence and leaf mixture of all the collected plants samples. There was no significant difference in the amounts of these three phytochemicals between extracts from wild and cultivated plants. The result from this study substantiates the validity of cultivated *Swertia chirayita* for medicinal purposes and trade. All three phytochemicals i.e., mangiferin, swertiamarin and amarogentin were present in highest quantities in IL aqueous and ethanolic extracts. Therefore, this plant part could be used for medicinal preparations. Aqueous and ethanolic root extracts had the lowest amounts of mangiferin and amarogentin whereas amount of swertiamarin were found to be lowest in aqueous and ethanolic stem extracts (Phoboo *et al.*, 2010).

Chemical Compounds Residing in Chirayita

Amarogentin (chirantin)

It is secoiridoid glycoside, and is the most acerbic substance found. It tastes bitter even at a dilution of 1:58,000,000 and can be procured from, *Swertia chirayita* (Roxb ex. Flem) Karst (Singh, 2008). It acquires Topoisomerase inhibition, chemo-preventive (Saha and Dass, 2005) and antileishmanial effects.

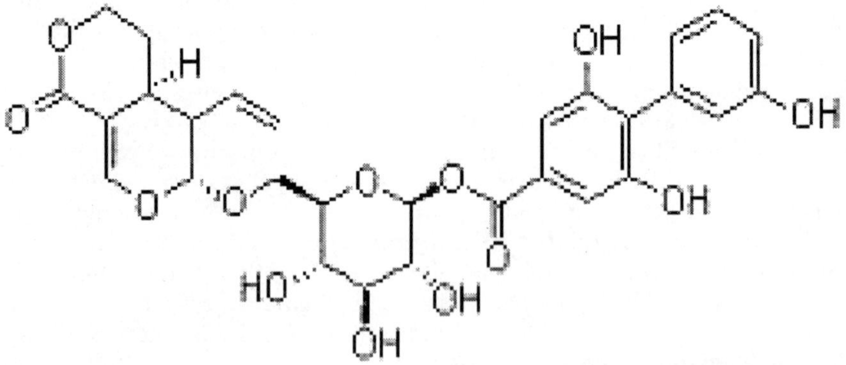

Fig. 4. A biphenylcarboxylic acid moiety; biosynthesized by a polyketide-type pathway, with three units of acetyl-CoA and one unit of 3-hydroxybenzoyl- CoA (Source: chemicalbook.com)

Amarogentin content in Swertia chirayita

Observation revealed that amarogentin content of S.chirayita was quite higher in true type of chirota. Study conducted by scientist from RRS, Kalimpong center (2006-07) that in leaf concentration of amarogentin vary from 0.238 to 0.391 and in stem 0.077 to 0.096 % (Plant sample analysis was done by Directorate of Medicinal and Aromatic Plant, Gujrat, Anand).

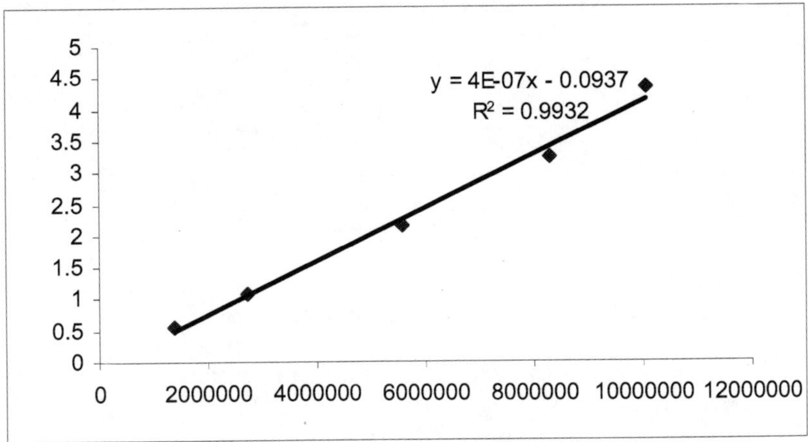

Fig. 5. Calibration curve of *S.chirayita* sample analysis

Table 4: Analysis report of *Swertia chirayita* plant (Sample from Darjeeling hill)

Plant sample parts	sample wt(g)	Dissolve ml	Area	ug/ 20ul	mg/g	%	Mean	St dev
Leaf -1	0.5008	10	6201894	2.3805	2.3767	0.238	0.238	0.0013
	0.5008	10	6259112	2.4033	2.3995	0.240		
	0.5008	10	6201894	2.3805	2.3767	0.238		
Stem -1	0.5008	10	2192533	0.7767	0.7755	0.078	0.077	0.0005
	0.5008	10	2171521	0.7683	0.7671	0.077		
	0.5008	10	2170223	0.7678	0.7666	0.077		
Leaf - 2	0.5000	10	10012093	3.9045	3.9045	0.390	0.391	0.0010
	0.5000	10	10057640	3.9228	3.9228	0.392		
	0.5000	10	10016740	3.9064	3.9064	0.391		
Stem - 2	0.5006	10	2640542	0.9559	0.9548	0.095	0.096	0.0003
	0.5006	10	2654262	0.9614	0.9603	0.096		
	0.5006	10	2654282	0.9614	0.9603	0.096		

1. **Amaroswerin:** It is a Secoiridoid glycoside collected from *Swertia chirayita* and found to be gastro-shielding (Niiho, 2005).

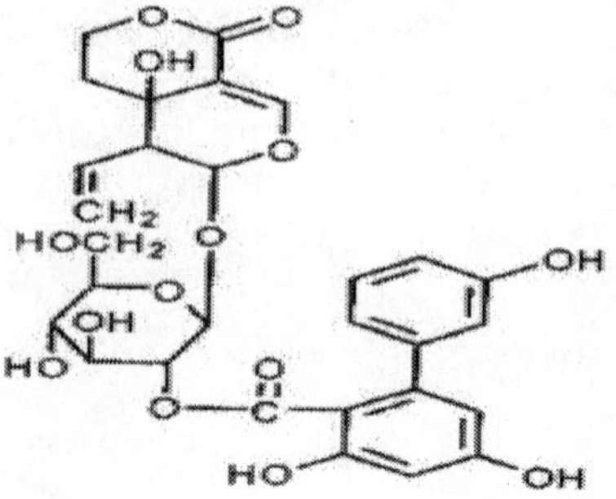

Fig. 6. An iridoidal glycoside having molecular formula $C_{29}H_{30}O_{14}$ (Source: Plant -expert.com)

2. **Swerchirin:** A medicinally foremost xanthone, obtained from several plants of family Gentianaceae including *Swertia chirayita;* having antimalarial, hypoglycemic, hepatoprotective, pro-heamatopoitic and weak chemo preventive pharmacological effects.

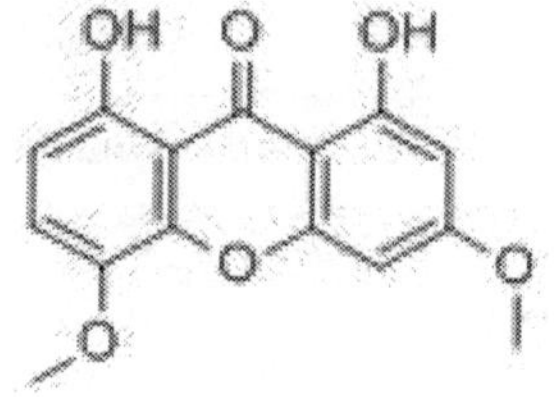

Fig. 7. Methylbellidifolin; 1,8- Dihyroxy- 3,5-dimethoxy-9H-xanthen-9one (Source: Chemicalbook.com)

3. **Swertiamarin** A Secoiridoid glycoside obtained from this plant having analgesic property. Swertiamarin, has been reported to possess a number of pharmacological properties such as hepato-protective and antiedematogenic/anti-inflammatory, free radical scavenging and antispastic (Vaijanathappa and Badami, 2009).

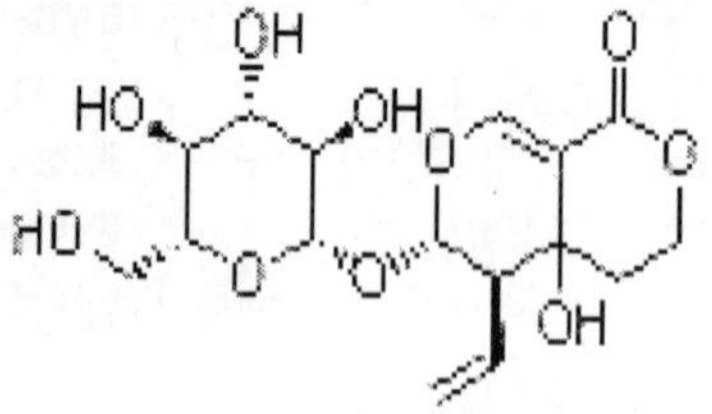

Fig. 8. Swertiamaroside having molecular formula $C_{16}H_{22}O_{10}$ (Source: Chemicalbook.com)

*Swertia*marin content in different species of *Swertia* and substitutes of chiretta

Different species of *Swertia* are used for various medicinal purposes. In India, among the fourteen species of Swertia used for medicinal purposes, *S. chirayita* is considered as medicinally most important. Reportedly, some of the allied species of S. chirayita are also used as common substitutes of the herb. In this particular experiment analysis of the content of swertiamarin was attempted in different species of Swertia, namely *S. cordata*, S. purpurascens and *S. paniculata*, which were collected during their flowering stages. Swertiamarin content was also analysed in the chiretta samples available in the market. It is evident that the profiles vary not only in terms of the swertiamarin content but also in terms of the different peaks obtained for each of the samples. Among the allied species of chiretta, *S. paniculata* was found to contain maximum amount of swertiamarin (1.2%), whereas the lowest was found in market sample I (0.003%).

Table 5: Variation in swertiamarin content in different species of Swertia and samples procured from herbal market (Khari baoli, New Delhi).

Sl. No.	Swertia species	Swertiamarin content (%)
1.	*S. chirayita*	0.16±0.16b
2.	*S. cordata*	0.29±0.17b
3.	*S. paniculata*	1.2±1.43a
4.	*S. purpurascens*	0.32±0.62b
5.	Market sample I	0.003±0.00b
6.	Market sample II	0.37±0.00b
	LSD	0.76

(Source :http://shodhganga.inflibnet.ac.in/bitstream/10603/6307/11/11_chapter% 206.pdf)

A significant variation was observed in swertiamarin content within the market samples that were analysed. Swertiamarin was almost negligible in sample I whereas in sample II only 0.37% of swertiamarin was detected. These observations highlight the variability in the chemotypic profiles of the samples available in the market.

4. **Xanthones :** Over all Xanthones are important bioactive constituent present in the drug which shows CNS down regulation in mice and rats.

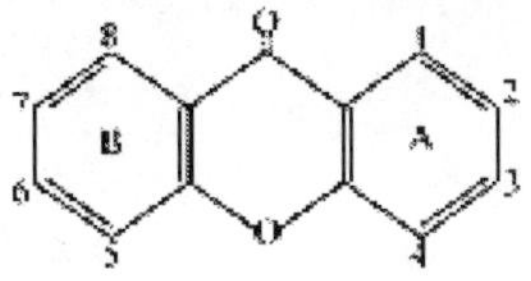

Fig. 9. An organic compound having molecular formula $C_{13}H_8O_2$ (Source: Chemicalbook.com)

Different metabolites viz., protein, free amino acids and chlorophyll-*a* and chlorophyll-*b* contents decreased with the plant maturity. However, the soluble and insoluble carbohydrates, nucleic acids (both RNA and DNA) and lipid contents showed high in advancement of plant growth and development stages. Enzymes activities like catalase, peroxidase, α-amylase were more recorded in seedling stage than sapling. The appreciable activity of acid/alkaline phosphatases supports the necessity of free inorganic phosphate metabolites for growth and development. However, the gradual decrease in enzyme activities like IAA-oxidase, protease, RNase recorded from seedling to sapling stage (Pradhan and Lama, 2012).

PHYSIOLOGY OF SWERTIA CHIRATA

Due to its high demand in the pharmaceutical market and poor seed germination, the plant is categorized as critically endangered. Seed germination is one of the biggest problems for its proper conservation. As per various report concerned, germination of this endangered plant was only 12-20% (Mukherjee and Chakraborty, 2008). Understanding seed germination is a pre-requisite to ensure species conservation. The germination of seeds collected from six microhabitats was studied at 20°C, 25°C, and 30°C, both under a 14/10 h light/dark photoperiod and in continuous darkness. Microhabitat and temperature significantly affect seed germination, germination rate, germination recovery (GR), and GR rate. Overall, the seeds collected from under canopy showed a significantly ($p < 0.05$) higher germination than those from open habitats, at 20°C, 25°C, and 30°C (14/10 h light/dark photoperiod). Germination was negligible in continuous darkness but after transfer to a 14/10 h light/dark photoperiod, the seeds from under canopy significantly recovered at 20°C and at 25°C ($p < 0.05$), and showed the highest germination percentage compared to seeds collected from tree base, stump base, shrubberies, and grassy slope. Similarly, at 30°C, seeds from under canopy recorded the highest GR percentage. In general, seed germination, mean germination rate, seed GR, and GR rate were significantly greater ($p < 0.05$) at 25°C. Among the microhabitats tested, variation in GR rate was significant ($p < 0.05$). Seeds were confirmed to be positively photoblastic (Pradhan and Badola, 2012).

Table 6 : Effect of different hormonal treatments on seed germination of *Swertia chirayita.*

Treatments	Germination (%)	Days require for onset of germination	Days required for completion germination
Control	36.66±2.11	32.00	57.33
GA_{50}	40.00±4.31	28.00	57.00
GA_{100}	48.00±2.13	29.33	56.33
GA_{200}	50.15±3.40	28.00	56.00

GA_{300}	54.00±4.11	23.00	54.33
GA_{400}	63.66±3.92	22.00	53.66
GA_{500}	37.66±3.22	26.00	51.00
IAA_{50}	43.33±3.32	33.00	59.33
IAA_{100}	40.00±3.26	30.33	57.00
IAA_{200}	46.15±5.02	30.33	55.66
IAA_{300}	46.66±3.91	28.33	54.33
IAA_{400}	47.15±5.92	29.66	53.33
IAA_{500}	55.00±3.96	27.66	52.33
LSD (P=0.05)	10.60	3.07	N.S

(Source : Mukherjee, 2008)

Table 7 : Seed germination as influenced by different sowing depth and soil composition.

Treatments	Germination (%)	Days require for onset of germination	Days required for maximum germination
Depth of soil (cm)			
0.50	56.25 ±3.9	26.31	44.39
1.00	44.41±2.33	32.66	48.66
LSD (P=0.05)	4.68	3.28	NS
Proportion of soil:sand:FYM			
1:1:1	40.66± 3.92	31.33	48.50
2:1:1	28.33±4.66	35.83	51.21
1:2:1	57.66±3.31	29.99	45.16
1:1:2	61.00±2.99	28.75	44.50
LSD (P=0.05)	9.54	2.07	2.02

(Source : Mukherjee, 2008)

In general, Gentianaceae is reported to have physiological dormancy (Nikoleaeva, 1977). Seed germination study revealed significant improvement in germination under GA_3 treatment compared to control (Table 2). However, the extent of improvement varied among different concentration. Observation on germination in *Swertia chirayita*, shows epigeal type of germination. Amongst all hormonal treatments maximum seed germination percentage was observed with $GA_{3\ 400}$ (63.66±3.92) and significantly superior to rest of the hormonal treatments. This recorded 73.64% more seed germination over the control. This treatment took least time to start germination and this was within 22 days after seed sowing and was at par with GA_3 300. Days required for completion of germination was least with GA_3 400 and was followed by with GA_3 300. Further, soil depth also influence plant physiological system of this endangered plant species. Sowing at 0.50 cm showed highest germination with less time required for onset of germination and significantly superior to higher depth. (Table 3). With various soil composition germination percentage and onset of germination were almost similar in soil, sand and FYM in 1:2:1 and 1:1:2 proportion, and this were at par with each other. Further table showed that amongst all soil media maximum germination goes with compositions of 1:1:2 (soil : sand : FYM) and was followed by 1:2:1 and, significantly superior to rest of soil composition.

The result of *Swertia* species, with improved germination under GA_3 and IAA treatments correspond well with these generalization, as both these substances are considered best for breaking physiological dormancy. The present study recommends GA_3 as the best treatment for improving seed germination. Higher percentage of germination by GA_3 400, indicates that it is the best source. Sowing at 0.50 cm depth sowed highest germination with less time required for onset and completion of germination. Amongst all soil media maximum germination goes with compositions of 1:1:2 (soil : sand : FYM). Low germination percentage under natural conditions may be responsible poor distribution of the species in nature.

CONSERVATION ASPECT

Owing to its medicinal importance, the plant has been harvested from the wild for a long time and is now considered endangered in its natural habitats. This requires immediate action for conservation and sustainable management of the herb. Keeping in mind the diverse ecogeographical zones occupied by the plant species, it is important to assess the existing diversity among populations occupying different niches (Mukherjee *et al.*, 2009). In order to devise pragmatic conservation strategies for endangered plant species such as *S. chirayita,* identification of intra-population and inter-population diversity is a preliminary step. This enables identification of regions representing maximum genetic variations and therefore population from these areas can be conserved at utmost priority. Further, as for most of the medicinal herbs, in case of *S. chirayita* also, authentic material is not cultivated and collected, leading to substitution. This is due to sheer ignorance on the part of the collector or unavailability of the genuine germplasm (Mukherjee *et al.,* 2013). In such cases, diagnostic tools become essential for proper identification of authentic samples. Enormous variations at morphological and biochemical levels have been observed among different species of *Swertia.* Also, the use of morphological markers as was done in the above-mentioned study is not appropriate for identification of dried samples (the manner in which, product is sold in the market), as the plants floral and other identifying characteristics cannot be recognised (Mukherjee and Sharma, 2014).

So far all the medicinal plants that reach the common market are collected from the wild state. The majority of traders collect wild medicinal plants through untrained and unskilled labourers. This is generally in the form of extraction of the entire plants, roots, rhi- zomes, tubers and bulbs before seed set, making low regeneration of the species in the habitat. Over-exploitation of these species, as well as trampling at particular places during collection has changed the habitat conditions causing a gradual loss of other associated species. The market stock of medicinal plants comes from private collectors and suppliers with legal permits for collection of herbs from specific collection sites. These sites are demarcated and collection permits issued by the Forest Department, Government of India. Major medicinal herbs are scattered throughout the region, but some of the sites are identified for portable collection. A collection cycle is maintained every year under rotation so that the same sites are not exploited repeatedly. Although this system has definite benefit to the conservation of these herbals it is not strictly followed. Absence of forest officials at remote sites allows collectors to enter sites other than the demarcated ones and there is a great deal of illegal collecting that cannot be measured from market investigations (Mukherjee, 2013). The collection season for the major species is shown in Fig. 8. After collection all the materials are brought to the respective road-heads and then transported to the markets at Gangtok, Siliguri, Kalimpong and other destinations such as Calcutta and New Delhi. The pricing and transaction of amount involved is kept strictly secret. Even the actual destination of the shipment is not known as dealers maintain a guarded attitude towards relevant inquiries.

Fig. 10. Conservation of *S. chirayita* under natural habitat at Lava (2200 m asl), Darjeeling (Source: Author, D. Mukherjee)

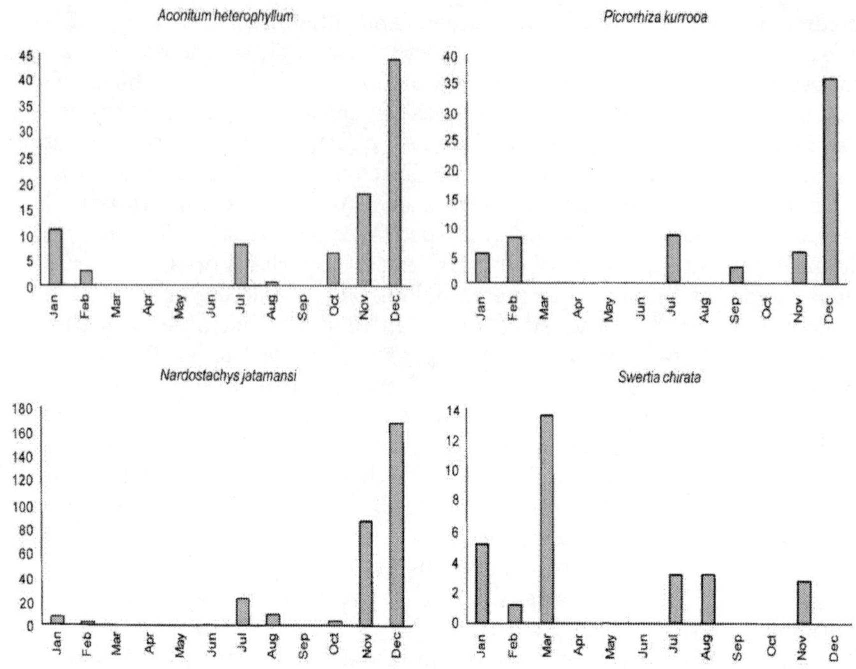

Fig. 11. Seasonal extraction (x100 kg) of medicinal plants from Sikkim (Rai *et al.*, 2000)

Role of Biotechnology in Conservation

Biotechnology can play an indispensable role in conservation of *chirayita* species. It can engross different modes of conservation. Farming of *chirayita* at low heights is prohibited because of several environmental factors like fertility and textures of soil, pH, humidity etc. Tissue culture technique has found to be very useful for conservancy of intimidate *Swertia chirayita* as little plant material can produce a big number of disease free propagules which can be revive in their innate environment. Systematized cultivation of the plant is important to guarantee continuous supplementation and affirmation of drug. Exercise in quality control of raw material of chiratiya is labeling of DNA markers that associate DNA finger printing data with quantitiy of selected markers. Research is still going on molecular investigation of chirayita for protection of this wild endangered species, establishing methods for breeding *in-vitro* and making efforts for assuring continuous supply of its raw material. Micropropogation is one of the most established techniques till date.

In vitro culture is an efficient method for *ex situ* conservation of plant biodiversity and multiplication of the endangered species. Micropropagation/Clonal propagation techniques using shoot tip and nodal segments are must for mass-scale multiplication and conservation of an endangered or threatened and medicinally important species within short period and limited space (Balaraju *et al.*, 2009). It enables the propagation of the endangered species from minimum plant material that is available for propagation. *S. chirayita,* though a widely used medicinal plant, is now a rare and endangered species, a consequence of its ruthless and destructive harvesting. Further, low germination percentage and poor viability of the seeds, long gestation periods and delicate field handling are some of the factors that discourage commercial cultivation and agro-technology development of *S. chirayita.* In view of considering the problems beset with conventional propagation of the species and its endangered status, efforts have been made to propagate the plant by *in vitro* techniques. Wawrosch *et al.* (1999) from different part of world described adventitious shoot formation from root explants of *in vitro* germinated seedlings of *S. chirata.* The regenerated shoots were 3 mm in length. Further, a two-step regeneration protocol was optimised in order to reduce the problems of callus formation, hyperhydricity and small shoot size. Ahuja *et al.* (2003) filed patent for medium composition for faster propagation of *S. chirayita* from mature explants. The patent report cited a shoot multiplication fold that ranged from 11.4 to 26.2, while the rooting frequency reportedly ranged between 50-80% in time duration of about 8 weeks, and a survival of 70% was reported.

A procedure for regeneration of complete plantlets of *Swertia chirata* via indirect organogenesis is described. Callus was obtained from in vitro regenerated roots on Murashige and Skoog (MS) medium supplemented with varying concentrations of 6, benzylaminopurine (BAP) and 2,4-dichlorophenoxyacetic acid (2,4-D). 13.32 μM BAP in combination with 0.90 μM 2,4-D proved to be the most effective concentration for callus induction, multiplication and adventitious shoot regeneration from callus surface. The optimal hormone combination for shoot multiplication was shown to be 8.88 μM BAP, 2.85 μM indole-3 acetic acid (IAA) and 271.45 μM adenine sulphate (Ads) giving an average of 10.70 shoots after 4 weeks and 17.50 shoots after 8 weeks. Individual elongated shoots were rooted on half-strength MS medium supplemented with varying concentrations of auxins. Best rooting was obtained with 4.90 μM indole-3 butyric acid (IBA) where an average of 14.40 and 21.50 roots per shoot could be obtained after 4 and 8 weeks, respectively. In vitro raised plantlets with well developed shoots and roots were acclimatized successfully (Pant *et al.*, 2012).

Strategies for Conservation

Using available information based on field survey, it is assumed that the causes of its degradation are largely overexploitation, low regeneration in the natural habitats and clearing of forests for developmental activities. For example, construction of roads and canals in many natural habitats has been responsible for the major loss of biodiversity. The government has imposed total ban on collection or removal of planting materials of this important species from their natural populations, but the question of declining populations still remains. Therefore, its long-term survival is at risk in the Himalayan region. Keeping in view the importance and conservation value of *S. chirayita*, some conservation measures need to be taken, i.e. notify the natural populations for the protection of its habitats, detailed study on phenology with understanding of whole life cycle, sustainable collection of germplasm from natural populations for developing elite generations in future, rehabitation of species before taking any developmental activities, establishment of nurseries in nearby areas of natural habitats, development of large-scale seedlings and their plantation, restrictions in field survey and collection in sensitive areas having natural populations and creation of awareness among the local healers to control over-exploitation in natural habitats to maintain its posterity for future. Since policy-makers and scientist cannot conserve the species without wider support from the local communities and government agencies, particularly the Forest Department, a multifarious effort is required to conserve the species. The R&D institutions and universities working on conservation issues need to develop appropriate strategies for sharing their experiences and scientific information for the conservation of this important species. Loss of species can have important consequences for the well-being and prosperity of the human population, because it reduces the strength and capacity of ecosystems to provide essential goods and services. This includes socio-cultural, religious, historical, scientific and educational resources (Purohit *et al*., 2013)

References

Ahuja, A., Koul, S., Kaul, B.L., Verma, N.K., Kaul, M.K., Raina, R.K., Qazi, G.N. 2003. Media compositions for faster propagation of Swertia chirayita. WO 03/045132 AL, US Patent 7238527.

Balaraju, K., Agastin, P. and Ignacimuthu, S. 2009. Micropropagation of *Swertia chirata* Buch- Hams. Ex Wall: a critically endangered medicinal herb. *Acta Physiologia Plantarum*, 31(3): 487-494.

Pradhan, S and Lama, P.C. 2012. Biochemical analysis during early seedling growth of *Swertia chirata* Buch. Ham. in Darjeeling Hill of Eastern Himalayas. *Environment and Ecology*, 30(4):. 1395-1398

Bharat, K., Pradhan, B.K. and Badola, H.K. 2012. Effects of microhabitat, light and temperature on seed germination of a critically endangered Himalayan medicinal herb, *Swertia chirayita* : Conservation implications. *Plant Biosystem*, 146 (2) : 345-351.

Bhardwaj, A., Khatri, P., Soni, M.L. 2011. Potent herbal hepatoprotective drugs – A review. *Journal of Advanced Scientific Research*, 2(2): 15-20

Chandra, S., Kumar, V., Bandopadhyay, R. and Sharma, M.M. 2009. SEM and Elemental Studies of *Swertia chirayita*: A Critically Endangered Medicinal Herb of Temperate Himalayas. *Current Trends in Biotechnology and Pharmacy*, 6 (3): 381-388.

Joshi, P. and V. Dhawan. 2005. *Swertia chirayita* - an overview. *Current Science,* 89: 635-640.

Kalaichelvi, K and Swaminathan, A.A. 2009. Alternate land use through cultivation of medicinal and aromatic plants - A Review. *Agriculture Review*, 30 (3) : 176-183

Khan, A., Kumari,S., Ishaq, F. 2014. Sardar Bhagwan Singh Post Graduate Institute of Biomedical Sciences & Research, Dehradun, India

Mukherjee, D and Chakraborty, S. 2008. Effect of growth hormones on seed germination and growth pattern of *Swertia chirayita* L. *Environment and Ecology,* 26 (4B) : 2116-18.

Mukherjee, D. 2008. Germination improvement in *Swertia chirayita* : an endangered medicinal herb. *Journal of Medicinal and Aromatic Plant Sciences,* 30 (2) :136-138.

Mukherjee, Dhiman 2013 b. Prospect and need of agro-forestry in the hills of Darjeeling. *Indian Agriculturist*, 57 (1) : 21-27.

Mukherjee, Dhiman 2013. Improved agricultural practices for high value medicinal plants under mid to high altitude situation. *Journal of Crop and Weed,* 9 (1): 201-06.

Mukherjee, Dhiman 2014. Effect of forest microhabitat on growth of high altitude plants in Darjeeling Himalaya. *Journal of Interacademicia*, 18(1): 20-30.

Mukherjee, Dhiman and Sharma, B.R. 2014. Conservation and scope of medicinal plant in Eastern Himalaya. *The Himalayan Review*, 1 (1): 102-110.

Mukherjee, Dhiman. 2009. Medicinal plant in Darjeeling hills. In : *KRISHI SANDESH* (eds. Rai, S., Mokthan, M.W. and Ali, S). Miazik International Volunteer Center, Japan. pp 118-121.

Mukherjee, Dhiman. 2013. Secondary forest wealth and its sustainable conservation of eastern himalaya. *MFP News letter*, 23 (3) : 11-16.

Mukherjee, Dhiman. 2014a. Lead lecture on "Impact of shifting climate on Darjeeling Himalaya agro-ecosystem". *In* National Seminar *on "Adaptation and mitigation strategies of climate change for sustainable livelihood". Cooch Behar Association for Cultivation of Agricultural Sciences* (COBACAS), March, 05 – 07, 2014. *Uttar Banga Krishi Viswavidyalaya*, Pundibari, Coochbehar, West Bengal, India. pp 12-13.

Mukherjee, Dhiman., Chakraborty, S. and Baskey, S. 2013. Near to extinct medicinal plants of Darjeeling hills: Protect for future. *Himalayan Times*, 4 (7): 17-18.

Mukherjee, Dhiman., Chakraborty, Soumendra., Roy, Amar and Mokthan, M.W. 2009. Differential approach of germplasm conservation of high value of medicinal plants in north eastern himalayan region. *International Journal of Agriculture Environment & Biotechnology,* 2 (4) : 332-340.

Niiho Y. 2005. Gastro protective effects of bitter principles isolated from Gentian root and *Swertia* herb on experimentally induced gastric lesions in rats. *Journal of Natural Medicine*, 60:74-88.

Nikoleaeva, M.G. 1977. Factors controlling the seed dormancy pattern. In The *Physiology and Biochemistry of Seed Dormancy and Germination* (ed. Kahan, A.A), Elsevier pp 51-54.

Pant, M., Bisht, P. and Manju P. Gusain. 2012. *In vitro* propagation through root-derived callus culture. *African Journal of Biotechnology*, 11(29): 7408-7416.

Pant, N., Jain, D.C and Bhakuni, R.S. 2000. Phytochemicals from genus *Swertia* and their biological activities. *Indian Journal of Chemical Sciences,* 39(B): 565-586.

Patel, B.C and Patel, N.M. 2014. A review on antidiabetic and antioxidant activity of bhunimbadi churna. *International Journal of Pharmaceutical Research and Bioscience*, 3 (2) : 223-237.

Phoboo, S., Pinto, De Silva., Bhowmik, C., Jha, P.K. and Shetty, K. 2010. Quantification of major photochemical of *Swertia chirayita*, A medicinal plant from Nepal. *Ecoprint*, 17 59-68.

Purohit, V.K., Bahuguna, Y.M., Tiwari, D., Tiwari, A., Andola, H.C and Negi, S.R. 2013. *Swertia chirayita* (Roxb. ex Fleming) Karsten on the verge of extinction in the Himalayan region. *Current Science*, 104 (2): 161-162.

Rai, L., Prasad, P and Sharma, E. 2000. Conservation threats to some important medicinal plants of the Sikkim Himalaya. *Biological Conservation*, 93 (1): 27–33.

Rai, L.K., Prasad, P. and Sharma, E. 2000. Conservation threats to some important medicinal plants of the Sikkim Himalaya. *Biological Conservation*, 93: 27-33.

Rai, M.B. 2003. Medicinal plants of Tehrathum district, Eastern Nepal. *Our Nature,* 1:42-48.

Saha, P. and Das, S. 2005. Fraction of *Swertia chirata* prevent carcinogenic risk due to DMBA Exposure. *Indian Journal of Medical Research*, 10: 22-27.

Singh, A. 2008. Phytochemicals of the Gentianaceae: A review of pharmacological properties. *International Journal of Pharmacology Science,* 1: 33-36.

Tabassum, S., Mahmood, S., Hanif, J., Hina, M. And Uzair, B. 2012. An Overview of Medicinal importance of *Swertia chirayita. International Journal of Applied Science and Technolog,* 2 (1)*:* 298 - 300

Vaijanathappa, J. and Badami. S. 2009. Antiedematogenic and free radical scavenging activity of swertiamarin isolated from *Enicostemma axillare. Plant Medicine,* 75: 12-17.

Wawrosch, C., Maskay, N. and Kopp, B. (1999). Micropropagation of the threatened Nepalese medicinal plant *S. chirata* Buch.-Ham.ex Wall. *Plant Cell Reproduction*, 18: 997-1001.

Website: IUCN (2009). IUCN IUCN Red list of threatened Species www. Iucnredlist.org.

Section VI:

Plant Physiology in Sustainability of Agriculture

16

PHYSIOLOGICAL AND BIOCHEMICAL PROPERTIES OF *GLIRICIDIA* : ITS CULTIVATION A SCOPE FOR REMUNERATIVE VENTURE FOR FARMERS

Prasann Kumar*, Padmanabh Dwivedi and A. Hemantaranjan

Department of Plant Physiology, Institute of Agricultural Sciences, Banaras Hindu University, Varanasi

**Corresponding address: prasann0659@gmail.com*

INTRODUCTION

Gliricidia sepium adapts very well in a wide range of soils ranging from eroded acidic (pH 4.5-6.2) soils, fertile sandy soils, heavy clay, calcareous limestone and alkaline soils. *Gliricidia* tolerates fire and the tree; quickly re-sprout with the onset of rains. Soils in the tropics are low in organic matter and have low nutrient supplying capacity. Soil organic matter plays an important role in crop production. Continuous application of organic matter as farm compost, farmyard manure and plant residues is needed to maintain or increase soil organic matter content. However, short supply of organic manures and competitive uses of farm residues as feed and fuel make it difficult to apply these to soil at desirable levels. Green leaf manuring is one of the important farming practices for increasing organic matter content in the soil. In highly eroded soils, especially in the tropics region, soil lacks adequate amount of nitrogen. Green leaf manure plants such as *Gliricidia* can play an important role in tropical farming systems for increasing the soil fertility. Growing *Gliricidia* plants on farm bunds serves dual purpose of producing green leaf manure rich in nitrogen, under field conditions and also helps in conserving soil through reduced regular top soil erosion.

MORPHOLOGICAL AND BIOCHEMICAL FEATURES OF *GLIRICIDIA*

1. The leaves contain nutrients: N (2.4%), Phosphorus (P) (0.1%), Potassium (K) (1.8%), Calcium (Ca) and Magnesium (Mg).
2. It grows fast and is tolerant to pruning.
3. The foliage can be used as green manure.
4. It can thrive in dry, moist, acidic soils or even poor degraded, infertile soils under rainfed conditions.
5. *Gliricidia* is a root nodulating, N-fixing and multipurpose legume. Many scientists conclude that *Gliricidia* leaves decompose quickly and release maximum nitrogen within 15 days. It increases mineralization and mineral N pool in the soil. Alley cropping system was started in the late twentieth century to improve/remove crop rotation with an idea to replenish nutrients and cultivate cereals simultaneously on the same piece of land (Kang *et al.*, 1981). The alley cropping involves hedgerows intercropping of food crops with fast growing shrubs/ tree species. This is leguminous species because they contribute better to soil (Tables 1 and 2). Quick decomposition of *Gliricidia* leaves may be attributed to high concentration of nitrogen (high quality) (Pandey and Rai, 2007).

Table 1: Nitrogen uptake (kg ha^{-1}) by maize crop in *Gliricidia* based alley-cropping system under two methods of leaves application (incorporated into soil; denoted with I and surface application; denoted with S) at different times at South Andaman, India.

Treatments of leaves application	N uptake (kg ha^{-1})							
	15 days		30 days		45 days		60 days	
	I	S	I	S	I	S	I	S
0 WAS	1.56	1.06	2.57	2.30	5.13	4.74	14.29	13.21
2 WAS	0.80	0.88	2.46	2.19	7.11	6.33	16.75	15.95
4 WAS	0.42	0.44	1.34	1.13	5.23	4.42	11.95	10.91
Control	0.34	0.34	0.45	0.45	2.21	2.21	5.44	5.44
Urea	1.84	1.84	3.33	3.33	7.00	7.00	19.20	19.20
LSD(P<0.05)	0.22		0.34		0.89		1.14	

(Source: Pandey and Rai, 2007) Where; WAS denotes "Week after sowing"

6. Anti-nutritional factors: Available data indicate that some potentially active toxic substance is found in *Gliricidia sepium*. HCN content has been reported up to 4 mg/kg and cyanogens may be present (Andrew Speedy). High level of nitrates (during the rainy season) is suspected of causing "Cattle Fall Syndrome" in Colombia but levels declined to negligible in winter. *Gliricidia* may be a nitrate accumulator (Andrew Speedy).

Evidences of toxicity under practical feeding conditions have been rare. The balance of evidence suggests that the plant could be toxic to non-ruminants but conclusive evidence of toxicity of ruminants under normal feeding is lacking.

Table 2: Nitrogen recovery (kg a^{-1}) by maize in a *Gliricidia* based alley-cropping system under two methods of leaves application (incorporation into soil; denoted with I and surface application; denoted with S) at different times at South Andaman, India.

Treatments of leaves application	N uptake (kg ha^{-1})							
	15 days		30 days		45 days		60 days	
	I	S	I	S	I	S	I	S
0 WAS	1.00	0.59	1.75	1.53	2.41	2.09	7.29	6.39
2 WAS	0.38	0.44	1.66	1.44	4.04	3.40	9.31	8.66
4 WAS	0.06	0.08	0.74	0.57	2.49	1.82	5.36	4.50
Urea	1.23	1.23	2.37	2.37	3.95	3.95	11.33	11.33
LSD(P<0.05)	0.22		0.34		0.89		1.14	

(Source: Pandey and Rai, 2007) where; WAS denotes "Week after sowing"

CULTIVATION AND MANAGEMENT PRACTICES

PROPAGATION

Gliricidia can be propagated through stem cuttings or seed.

Stem cutting or stake method

Gliricidia cutting are taken from stems of at least one –year-old plants. These should be from brownish-green mature branches and should measure 2-6 cm in diameter and 30-100 cm in length. The stem cutting is normally cut obliquely at both ends, discarding the younger tips and the base is inserted 20-50 cm into the soil. The cuttings should be planted on bunds in the rainy season immediately after these are cut from the stems. The plants grow quickly from cuttings. Propagation from stakes is simple but suitable mainly for situations where only a few trees are to be established. For hedges, cuttings are planted closely at 50 cm spacing. The hedge can be periodically pruned to provide fodder, green manure, firewood or stakes for new fences.

Seed propagation

Gliricidia seeds are soaked in water for 8-10 h, preferably overnight. The soaked seeds are sown in small polythene bags filled with a mixture of red soil, sand and farmyard manure (1:1:1) and watered regularly. Direct sowing of seeds requires good land preparation and regular weeding. It is not necessary to treat the fresh seeds before planting. However, when the seeds are not fresh, they need to be soaked overnight in hot water and planted immediately. Nearly all the seeds will germinate within a week. Generally, 3-4 month old seedlings can be planted on bunds in the rainy season. Seed propagation method is more convenient for establishing a large number of plants. Seeds or seedlings inoculated with suitable strains of *Rhizobium* are necessary where *Gliricidia* is not naturalized. In countries where *Gliricidia* is native or naturalized, local bacteria will already live in association with the roots of *Gliricidia*. In such case, using local soils in the nursery will automatically provide the right bacteria strains (Growing *Gliricidia*, NFT Highlights). The macromolecule contents in seeds cotyledons were significantly affected by variations in storage time and imbibition of seeds (Table 3). From the Table 1 it is clear that starch and proteins are not affected by imbibition time. Increasing storage time significantly altered enzyme activity but no significant difference was observed for total amylase and α-amylase during the period of seed imbibition, although there was a significant interaction between the two parameters (Table 3).

Table 3: Biochemical analysis of macromolecule contents in *Gliricidia sepium* seeds during storage and imbibition period.

Source of Variations	G%	ST	SUC	TSS	RS	TP	AA	AT	α-A
Storage	141.24*	72.324*	331.40*	178.86*	106.25*	7.15*	33.64*	25.49*	3.994*
Imbibitions	238.14*	1.424^{ns}	108.61*	81.99*	10.868*	1.755^{ns}	3.233*	0.592^{ns}	0.512^{ns}
Storage* Imbibitions	12.481*	18.658*	15.264*	7.337*	3.680*	3.311*	5.950*	3.316*	6.483*
CV%	14.87	25.34	21.02	14.96	18.18	18.71	11.30	9.68	14.10

(Source: Reis *et al.*, 2011)

Where; G%= Germination; ST= Starch; SUC=sucrose; TSS= Total Soluble Sugar; RS= Reducing Sugar; TP= Total Proteins; AA= Amino acids; AT= Total Amylase; α-A=α-amylase; ns= Not Significant; * Significant at 5% respectively by F test

Planting

Gliricidia is commonly grown from cutting, although for establishment in poor soils, growing from seed will lead to better root development. *Gliricidia* stem cutting with 2-6 cm diameters and 30-100 cm length or 3-4 month old seedlings can be planted on buds at 50 cm spacing during the rainy season or when there is enough moisture in the soil. For steep slopes, closer plant spacing of <20 cm is recommended for better soil erosion control. *Gliricidia* plants from stem cuttings grow faster than those grown from seeds.

Pruning

One year after planting, harvesting can be started by lopping the plants at 75 cm above the ground. For good management, plants should be pruned at appropriate time. Pruning is done at least thrice during the year i.e., in June (before sowing of the rainy season crop), in November (before sowing of the post rainy season crop) and in March (before sowing of the summer crops).

Chopping

These plants can be cut into small pieces. The chopped material is applied to the soil surface as mulch or incorporated into the soil as green manure. The management of *Gliricidia* in the field depends on what one plans to use it for. If it is being grown for firewood, then one can coppice it cut it close to base so it resprouts new growth. If it is for fodder or green manure, pruning at 0.3-1.5 m will make more leaves grow. If we want to use *Gliricidia* in the field of maize or increasing the fertility then we plant 6-8 week-old seedlings from the nursery at spacing of 1m x 1m, in between where one plans to plant rows of maize. Weed all unwanted plants which might interrupt the growth of the young *Gliricidia* plants. Free the plants from all material that might catch fire and create a firebreak around the field. After 18-24 months of growth, the field of *Gliricidia* is often ready for cutting. The *Gliricidia* is cut when they reach 30 cm in height. Leave the cutting in the field for 2 weeks until them drop leaves. Two weeks after cutting, the *Gliricidia* biomass is ready to be mixed into the soil (Growing *Gliricidia*, NFT Highlights)

Pest and Disease

As we know it is widely grown throughout the tropics and remained free of many diseases. Recent advancement in surveys in Central America, however, noted the frequent occurrence of serious disease "little leaf disease" in natural population of plants. Besides these, *Cercosporidium gliricidiasis* and brown leaf spot are widely recorded. Under humid conditions, it causes defoliation. In some areas in Bali, Indonesia *Gliricidia* is infested with an aphid (*Aphid craccivora*) particularly at the onset of the rains, which causes blackening of the leaf surface and in severe cases the death of the primordia and shedding of young leaves (Suttie, 2005). Evaluation of 16 provenances of *Gliricidia sepium* showed that three provenances (G14, G17 and N14) were quite resistant to aphid infestation (Suttie, 2005). *Colletotrichum gloeosporioides*, expressed as small, dark, rounded leaf spots, is more common than *Colletotrichum gliricidiasis* in Nigeria. Survey reports have also found leafy scorch and powdery black spot at several sites. Many investigations are in progress for determination of causal agents for this disease.

Germplasm improvements

There are limited numbers of germplasm of this crop because this genus has not been collected extensively. According to Polhill and Sousa (1981), there are only four species considered of which *G. sepium* is the only species with real agronomic potential; it is only recently that an effort has been made to collect a representative range of its germplasm. In *Gliricidia*, seed supply is often a problem as seed set is sporadic and uncertain in environments without a marked dry season. It is a cross pollinated crop and there is one limitation in vegetative propagation existence of large monogenotypic stands, with no inherent variation to combated new pests and disease. Hughes (1991) reported that Oxford Forestry Institute collected seeds of many provenances of *G. sepium* from Latin American countries. Similarly, Simons and Dunsdon (1992), reported that tests were conducted over 147 select sites across the tropics which showed that Retalhuleu of Guatemala and Belen Rivas of Nicaragua performed better than other provenances.

BENEFITS OF *GLIRICIDIA*

Role in soil management

1. Increased organic matter content in the soil. Many agroforestry trees besides *Gliricidia* can be used for increasing organic matter content in soil and especially leguminous trees can be used in the fields to solve efficiently the low organic matter content in soil. These plants can be used as single source for both organic matter and nitrogen nutrients. [a] *Fedherbia albidia*: The fall of leaves at the beginning of the rainy season naturally enriches the soil and improves the micro-nutrient. The other uses of these trees are, this produces abundant, quality fodder during the dry season, mainly in the form of pods. It also provides good fuelwood and building material. [b] *Tephrosia vogelli*: Cut the trees down at ground level just before the second season, after sun drying for 2-3 days, shake the leaves off and remove woody branches for fuelwood or other uses. The leaves may be left on the surface, buried in the ridge or lightly covered with soil.
2. Improves soil physical properties. Trees with deep roots help to recycle nutrients from the deep soil to the top soil. The nutrients go from the roots to the leaves. When the leaves fall to the soil, then the nutrients are available for the crops.
3. Restores and improves the soil fertility. This plant fixes nitrogen from the air due to bacteria which live on the roots. Those bacteria process the nitrogen from the air to a form of nitrogen that can be absorbed by the plant. Then, nitrogen is available for the crops. It is a drought- resistant and valuable water-conserving species, because in the dry season, it sheds most of its leaves, hence reducing water loss through transpirat-ion. It also improves soil aeration and reduces soil temperature.
4. Increases crop yields.

5. Allows the water to infiltrate into the soil more quickly rather than run off the surface.
6. Increase water-holding capacity of the soil.
7. Reduces soil erosion. Hedgerows in alley cropping control soil erosion.
8. Increases nutrient availability in the soil due to production of carbon dioxide and organic acids during decomposition of the plant material.
9. It adds valuable nutrients such as N, P, K, Ca and Mg to the soil. *Gliricidia* plants grown on 700-m long bunds can provide about 30 kg N $ha^{-1}yr^{-1}$ under rainfed systems with 700-800 mm annual rainfall.
10. Reduced environmental risks associated with chemical fertilizers. Use of *Gliricidia* as green manure minimizes the usage of chemical fertilizers that are very expensive and also environmentally unfriendly.
11. Acts as a barrier and filter to the rainwater running down the surface of a slope. *Gliricidia* roots stabilize lands with high slopes. This can be used to reclaim denuded land or land infested with *Imperata cylindrica.*
12. Leaves of *Gliricidia* contain more than 20% crude protein, essential oils and these crude proteins are nutritious for cattle. But livestock not conditioned to it may not readily eat it. In such situations, wilting the leaf may be helpful. Essential oils from the leaf of this plant are as effective as modern medicine to combat pathogenic microorganisms.

Table 4: Chemical composition of the leaf essential oil of *Gliricidia sepium*

Sl.No.	Identified components	Percentage
1.	Propylene glycol	25.1
2.	Coumarin	18.2
3.	[Z]-3-Hexenol	17.7
4.	β-Farnesene	14.2
5.	[E]-2- Hexenol	6.5
6.	Thymol	3.6
7.	Benzyl alcohol	3.5
8.	Caryophyllene	2.3
9.	α- Farnesene	2.0
10.	2-Pentene-1-ol	<1
11.	Isovanillin	<1
12.	Isobutyl alcohol	<1
13.	Phenyl alcohol	<1

Sl.No.	Identified components	Percentage
14.	Phenylethyl alcohol	<1
15.	Phenol	<1
16.	Crotonic aldehyde	<1
17.	5,6-Dihydro-4H-cyclopenta-(6)-furan	<1

(Source: Jose and Reddy, 2010)

Table 5: Chemical composition of the flower essential oil of *Gliricidia sepium*

Sl.No.	Identified compounds	Percentage
1.	Benzyl alcohol	0.35
2.	Nonanol	0.62
3.	Maltol	4.42
4.	2-Butyl-2-hexanol	1.03
5.	Myrtenol	12.73
6.	Eucarvone	0.88
7.	Geraniol	0.72
8.	Nonanoic acid	0.55
9.	Myrtenal	0.78
10.	Hydroquinone	21.64
11.	p-Mentha-1,8-dien-1-ol	1.83
12.	p-Mentha-1,4-dien-2-ol	0.73
13.	p-Mentha-1,4-dien-7-ol	0.71
14.	Decanoic acid	0.31
15.	γ-Nonalactone	1.31
16.	Coumarin	43.07
17.	Dodecanoic acid	0.64
18.	Tetradecanoic acid	0.46

(Source: Jose and Reddy, 2010)

It is found that the essential oils of leaf and flower of *Gliricidia sepium* showed varying degree of antibacterial activities, it indicates self defence mechanism within the *Gliricidia sepium* (Table 4 and 5). Due to the emergence of the antibiotic resistant pathogens, plants are being looked upon as an excellent alternate to combat the spread of multi drug resistant microorganism. The variation of the susceptibility of micro-organism towards the essential oils could be attributed to their intrinsic properties that are related to the permeability of their cell surface to the essential oils. Its leaves are normally measured to be one of the most palatable among the tropical leguminous forages (Glover, 1989). The worth of this species for feed and feed supplement purposes (Whetton *et al.*, 1997) was resolute through the tissue culture method of degradation products of both crude leaf and soluble protein extracts of the leaf by rumen microbes (Rastrelli luca, *et al.*, 1999). Occurrence of soluble alcoholic sugar pinitol (Calle *et al.*, 1986), coumarins, hydrocarbons, and flavonoid glycosides in the leaves has been surveyed by several scientist earlier (Herath *et al.*, 1998). Most of the previous chemical investigations on *G. sepium* have focused on the isolation of potential allelopathic and toxic compounds from the heartwood (Rastrelli luca, *et al.*, 1999) (Figure 1 and 2).

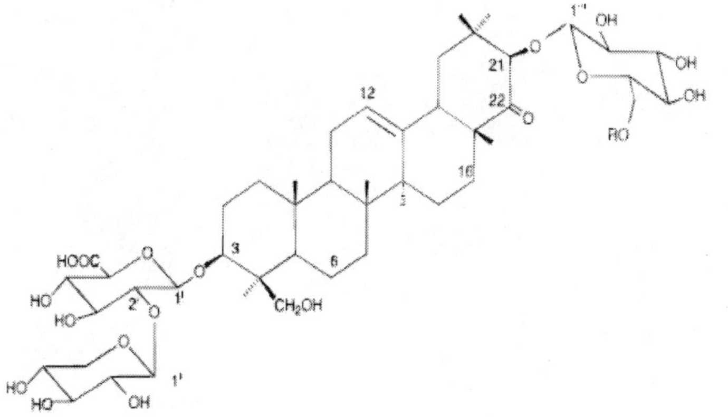

Fig. 1. Structure of triterpene saponins 1 and 2 isolated from *G. sepium* roots. (Source: Rastrelli luca, *et al.*, 1999)

13. Impact of *Gliricidia sepium* intercropping on soil organic matter fractions in a maize-based cropping system. According to Wander (2004) soil organic matter is often the most important source of plant nutrient for farming system as well store house of available moisture, providing energy for fauna residing in the soil and increasing soil aggregate structure. The intercropping system based on *Gliricidia*/Maize holds promises for increasing productivity in maize–based cropping systems on depleted soil in many countries. Beedy *et al.* (2010) reported that the *Gliricidia*/Maize intercrop had significant and positive effects on soil organic matter, particulate organic matter-carbon and soil organic matter fractions. He estimated that soil organic matter was 12% higher, particulate organic matter 40% higher, particulate organic matter-carbon 62% higher and particulate organic matter-nitrogen 86% higher in the *Gliricidia* intercrop compared to sole maize, indicating nitrogen enrichment of particulate organic matter (Table 6).

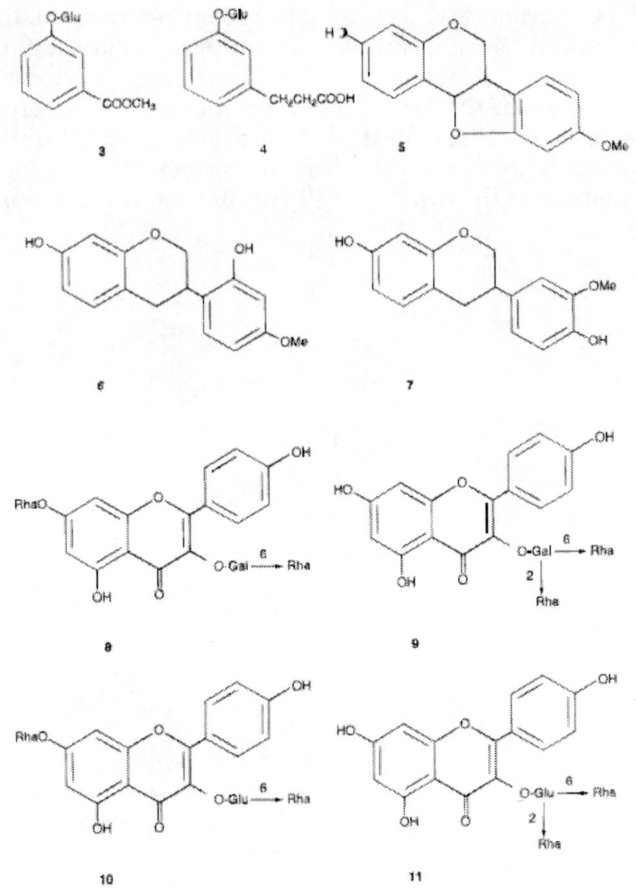

Fig. 2. Aromatic compounds 3-11 isolated from *G. sepium* leaves
(Source: Rastrelli luca *et al.*, 1999)

Table 6: The effect of cropping system and added N or added P on concentration of plant nutrient in study soils

	Inorganic N (mg kg^{-1})	Exch. K^+ (mg kg^{-1})	CEC (c mol kg^{-1})
Sole maize			
N= 0	6.2(3.1)	166(39.8)	7.75(0.99)
N=48	6.6(3.1)	132(40.1)	8.82(0.99)
N=96	11.0(2.1)	81.0(24.9)	8.22(0.65)
Gliricidia/Maize			
N=0	11.1(2.3)	96.4(27.5)	8.01 (0.71)
N=48	6.6(2.3)	130(27.8)	9.08 (0.71)

N=96	15.8(1.2)	181 (12.6)	8.48 (0.36)
		Bray P (mg kg^{-1})	
Sole maize			
P=0	40.4 (12.5)		
P=20	68.6 (12.5)		
P=40	87.3 (8.2)		
Gliricidia/Maize			
P=0	31.7(9.1)		
P=20	61.6(9.0)		
P=40	78.8(4.7)		

(Source: Beedy *et al.*, 2010)

where; N= Inorganic nitrogen, Exch. K^+= Exchangeable potassium, CEC= Cation-exchange capacity, Bray P= Available phosphorous with the Bray P1 extract

Table 7: Significance of intercropping with *Gliricidia*

	SOM	POM	POM-C	POM-N	P-C/N	Inorg. N	CEC
Statistical significance			Non-significant (NS) (P=0.05)				
Cropping system	<0.0001	<0.0001	<0.0001	<0.0001	<0.0001	0.0054	0.0017
N	NS	0.0034	0.0026	<0.0001	0.0013	<0.0001	NS
P	NS	NS	NS	NS	0.0399	0.0334	NS
Clay	0.0161	NS	NS	NS	NS	NS	0.0091
N x P	NS	0.0118	NS	NS	NS	NS	NS
Cropping system x Sand	NS	NS	NS	NS	NS	0.0232	NS
P x Sand	NS	NS	NS	NS	NS	0.0416	NS
Cropping system x Clay	NS	NS	NS	NS	NS	NS	0.0049
Planned contrast: (Intercrop + N=0) versus (Sole maize + N=48 kg ha^{-1})			<0.0001	<0.0001			

(Source: Beedy *et al.*, 2010)

Where, SOM= Soli organic matter, POM= particulate Organic Matter, POM-C= Particulate organic matter- Carbon, POM-N= Particulate organic matter-Nitrogen, P-C/N= Carbon–Nitrogen Ratio of POM, Inorg. N= Inorganic nitrogen, CEC= Cation Exchange Capacity.

Table 8: *Gliricidia* and Maize biomass inputs

Sl. No.	Plant Type	Treatments	Biomass added	References
1.	Maize stover	SM+ 0	2.6	Kang *et al.* (1999)
2.	*Gliricidia* pruning	*Gliricidia* intercrop	5.7	Kang *et al.* (1999)
3.	*Gliricidia* pruning	*Gliricidia* intercrop	4.3	Vanlauwe *et al.* (2005)
4.	*Gliricidia* pruning	*Gliricidia* intercrop	5.5	Vanlauwe *et al.* (2005)
5.	*Gliricidia* pruning	*Gliricidia* intercrop	2.5	Vanlauwe *et al.* (2005)
6.	Maize stover	GM + 48	8.2	Makumba *et al.* 2006
7.	Maize stover	SM +48	5.4	Makumba *et al.* 2006
8.	Maize stover	SM+ 0	1.7	Makumba *et al.* 2006
9.	*Gliricidia* pruning	*Gliricidia* intercrop	4.6	Makumba *et al.* 2006

(Source: Beedy *et al.*, 2010)

where, SM+0= Sole maize+0 kg ha^{-1} N, SM+48=Sole maize + 48 kg ha^{-1} N, GM + 48= *Gliricidia*/Maize + 48 kg ha^{-1} N.

From Tables 6-8, it is clear that use of the *Gliricidia* intercrop with maize had linear positive impacts on combating soil fertility. Particulate organic matter pools, which act as store house of plant nutrients, were tremendously enhanced by *Gliricidia*/Maize intercropping, which was also related with decrease in particulate organic matter C/N ratio significantly. It also increases nitrogen content in soil significantly. If we gradually increase the *Gliricidia*/Maize intercropping then nitrogen fixing tree intercrops may play an important role in preventing or slowing soil degradation and in maintaining the source of plant nutrient to maize in this intercropping system. The effect of the intercrop did not interact with the effects of the fertilizers rates, but it did interact with textural covariates affecting inorganic nitrogen and cation exchange capacity. In coarse textured soil, *Gliricidia*/Maize intercropping maintains cation exchange capacity.

ADDITIONAL USES

1. Plants are grown a live fence around cattle pastures and for delineating boundaries. Hedgerows in alley cropping slow weed growth and have been shown to reduce the incident of disease in groundnut crops.
2. Stems are used as fuel wood. This is very good for charcoal production. The special features of woods are the wood burns slowly without sparking and with little smoke. Timber of *Gliricidia* is very durable and termite resistant so it is used frequently for making railways sleepers, farm implements, furniture, house construction and as mother posts in live-fence establishments.

3. Plants serve as support for other plant species. Flowers can be fried and eaten. Leaves are rich in protein and highly digestible for ruminants like cattle and goat, as they are low in fiber and tannin. Flowers attracts honeybees, hence it is good source for improving pollination in plants and production of honey. It is considered that this plant is a traditional remedy for hair loss, boils, bruises, burns, colds, cough, debility, eruptions, erysipelas, fever, fractures, gangrene, headache, itch, prickly heat, rheumatism, skin tumors, ulcers, urticaria and wounds.
4. A mixture of ground leaves or bark and cooked seeds of *Gliricidia* is used as poisonous bait for mice and rats and, in fact, the Latin name *Gliricidia* means rodent poison. The leaves, seeds or powdered bark are poisonous to humans when mixed with cooked rice or maize and fermented.
5. Leaves are used as insectides and insect repellent.

Nazli *et al.* (2008) reported that repellent activity of *G. sepium* (ethanol extract) was studied by testing it against adult Mosquito *Aedes aegypti* (L.). Results were compared with the *Citronella* oil. The maximum repellency percentage afforded by the repellent *G. sepium* was 78%, whereas with *citronella* oil, the repellency percentage was 74%, (Tables 9-11). As a result of many studies it may be concluded that the ethanolic leaves extract of *Gliricidia sepium* is non-irritating and longer lasting than the standard repellent. In this context, it is worth mentioning that research in this field would open door for future exploitation of our natural heritage and its commercialization in modern era.

Table 9: Efficacy of *G. sepium* (ethanol extract) as repellent against mosquitos *Aedes egypti*

Sl.No.	Name of repellants	Dose (ml)	Replicates	No. of insects released in each cage	No. of insects repelled in each cage	Mean	Repellency percentage
1.	*Citronella* Oil	0.2	1	50	41		
			2	50	42	37	74
			3	50	39		
2.	*G. sepium*	0.2	1	50	40		
			2	50	38	39	78
			3	50	39		
3.	Control	0	1	50	0	-	
			2	50	0		0
			3	50	0		

(Source: Nazli *et al.*, 2008)

6. The tree produces dense masses of attractive white or pink flowers. The flowers are a good source of forage for bees. The flowers appear during January -March when the tree is leafless.

It can be grown along paths of parks, gardens and avenues. It can also be pruned to a height of 2.0 m and maintained as a hedge.

Table 10: Antibacterial activity of ethanol extract of leaves of *Gliricidia sepium*

Sl. No.	Strains	Plant part used	Inhibition zone (cm)
1.	*E. Coli*	Leaf extract	1.5
2.	*S. aureus*	Leaf extract	1.1
3.	*Pseudomonas spp.*	Leaf extract	1.2
4.	*S. typhi*	Leaf extract	1.0
5.	*Klebsilla spp.*	Leaf extract	1.3

(Source: Nazli *et al.*, 2008)

Table 11: Effect of *G. sepium* extracts on nematode (meloidogyne) mortality number of mortality (initial number of nematode 50)

Sl. No.	Concentration of extract	Plate1	Plate2	Plate3	Palte4	Mean	Percentage
1.	G/0	02	01	04	01	02	4
2.	G/1	50	48	46	50	48.5	97
3.	G/2	30	33	28	29	30	60
4.	G/3	17	21	10	13	15.25	30

(Source: Nazli *et al.*, 2008) (where, G/0, G/1, G/2, G/3 are the dilution of the culture, respectively).

Table 12: Proximal analysis of the rations and *Gliricidia sepium* flour

1.	Ration/Feed	DM (%)	CP (%)	EE (%)	CF (%)	NFE(%)	TDN(%)	AS (%)
2.	TA	88.17	15.64	6.61	10.94	57.33	74.03	9.48
3.	TR	82.53	19.55	1.38	21.73	48.87	71.78	8.47
4.	GS	90.53	22.7	2.20	20.81	43.66	70.26	10.63

(Source: Gonzalez-Villalobos *et al.*, 2002)

where, TA: Supplementation provided through a commercial concentrate,
TR: 60% cow meal + 40% GS flour + 50 g/animal/day of complete mineral mixture,
GS: *Gliricidia sepium* plant, DM: Dry matter, CP: Crude protein, EE: Ether extract, CF: Crude fiber, NFE: Nitrogen free extract, TDN: Total digestible nutrient, AS: Ashes.

7. *Gliricidia sepium* can be used as supplementary feeding of cross breads female calves.

Gliricidia sepium is a leguminous shrub with great potential and its introduction could represent an alternative means of minimizing nutritional deficiencies that are presented in young animals, still in the process of growth, during periods of fodder scarcity. In this sense, Gonzalez-Villalobos *et al.* (2002), evaluated growth (partial and final body weight and average daily gain) of predominantly Holstein crossbreed female calves fed with two different supplementation sources, under the same grazing management by the farm (Table 12 and 13).

Table 13: The effect of supplementation on partial and final body weight and average daily gain

Sl.No.	Parameters	TA	TR
1.	Number of female calves	7	7
2.	Weight at start of trial (Kg)	86.8±1.05	87.00±1.10
3.	Weight at 14 days (Kg)	93.93±1.19	91.49±1.19
4.	Weight at 28 days (Kg)	101.91±1.77	93.65±1.77
5.	Weight at 42 days (kg)	104.89±2.34	94.52±2.34
6.	Final weight, 56 days (kg)	106.65±2.51	96.06±2.51
7.	Average daily gain (g/d.)	334.61±42.0	157.0±42.0

(Source: Gonzalez-Villalobos *et al.*, 2002)

where, TA: Supplementation provided through a commercial concentrate, TR: 60% cow meal + 40% GS flour + 50 g/animal/day of complete mineral mixture

Few results also indicate that when laying chickens were fed 4.5% sun-cured *Gliricidia* in diets they gave good egg production, egg weight and yolk colour. It has been found that diet containing up to 10% *Gliricidia* can be fed to growing chicks without affecting performance and survival.

BENEFITS TO FARMERS

1. Enhancement of soil productivity.
2. Increase in crop yield.

BENEFITS TO ENVIRONMENT

1. Improved soil quality.
2. Reduced soil erosion
3. Reduced pollution as the use of chemical fertilizers is minimized.

BENEFITS TO NATIONAL ECONOMY

1. Sustainable agricultural productivity.
2. Reduced reliance on fossil fuels for producing fertilizers.
3. Increased carbon sequestration: *Gliricidia sepium* can help in the battle to control global warming by sequestering atmospheric CO_2. Most attention so far has been on the carbon sequestration in trees, but soils

can also contain considerable amounts of carbon, some of which is released up to harvest. Harvest has significant effects on soil carbon and harvest biomass, with very small effects on standing biomass; in contrast, pruning has small effects on soil carbon and harvested biomass and a more pronounced effect on standing biomass (Table 14).

Table 14: Average carbon stocks in soil and biomass and average annual harvest of firewood

Pruning	High initial soil carbon				Medium initial soil carbon				Low initial soil carbon			
	Harvest				Harvest				Harvest			
	100%	50%	25%	Mean	100%	50%	25%	Mean	100%	50%	25%	Mean
					Soil carbon (t C/ha)							
75%	34.79	43.78	48.24	42.27	21.21	30.19	34.66	28.69	12.08	20.98	25.47	19.51
50%	34.80	43.55	47.91	42.09	21.19	31.11	34.33	28.88	12.08	20.72	25.08	19.29
25%	34.81	43.20	47.37	41.79	21.23	29.62	33.79	28.21	12.09	20.34	24.55	18.99
Mean	34.80	43.51	47.84	42.05	21.21	30.31	34.26	28.59	12.08	20.68	25.03	19.26
					Standing biomass carbon(t/ha)							
75%	23.68	24.68	24.68	24.35	23.36	24.68	24.68	24.24	14.25	24.53	24.63	21.14
50%	26.14	27.51	27.51	27.05	28.35	29.68	27.51	28.51	15.07	27.24	27.35	23.22
25%	32.91	34.41	34.41	33.91	32.40	34.40	34.41	33.74	19.15	34.05	34.18	29.12
Mean	27.58	28.87	28.87	28.44	28.04	29.59	28.87	28.83	16.16	28.61	28.72	24.49
					Harvested biomass (t DM/ha/yr)							
75%	25.42	13.25	6.62	15.10	25.02	13.25	6.62	14.97	15.79	13.16	6.60	11.85
50%	25.13	13.08	6.54	14.92	24.88	13.08	6.54	14.84	15.73	12.96	6.51	11.73
25%	24.44	12.71	6.36	14.50	24.06	12.71	6.36	14.37	15.57	12.54	6.31	11.48
Mean	25.00	13.01	6.51	14.84	24.66	13.01	6.51	14.73	15.70	12.89	6.47	11.69

(Source: Wise and Cacho, 1999)

In Table 14, figures represent the average amount of carbon in soil and standing biomass (t C/ha) and the average amount of biomass harvested as firewood (kg DM/ha) per year. These figures were estimated as (Table 15)

Table 15: Formula for calculation of average amount of carbon in soil and standing crop biomass

$\sum_{t=1}^{25} Yijt$	$Yij = \dfrac{\quad}{25}$

(Source: Wise and Cacho, 1999)

where, the Y*ijt* represents annual output *I* under treatment *j*, where *i*=soil carbon (St), biomass carbon (Bt) or firewood harvested. Hence these results measure average annual stocks over the planning horizon and do not reflect any difference in the time paths of biomass accumulation.

Conclusions

Gliricidia sepium is known as mother of cocoa. *Gliricidia sepium* (Jacq.) Steud. (Syn. *Gliricidia maculate* H.B.K.) is a fast growing, tropical, leguminous tree, known commonly as *Gliricidia* or *madre* tree, *Gliricidia sepium* is well known in plantation agriculture as a shade tree. This is a drought tolerant species. It is one of the commonest and best–known multipurpose trees in many parts of Central America, Mexico, West Africa, West Indies, South Asia and tropical Americas. It is a nitrogen-fixing tree that is native to Central and South America. *Gliricidia* is a small fast growing, deciduous, thorn less tree. Its short trunk is covered with soft gray bark marked by longitudinal cracks. It grows up to 10 m height. It has an open crown and a trunk of less than 30 cm diameter. The trees are used for timber, firewood, hedges, medicinal purpose, charcoal, live fences, plantation shade, poles, soil stabilization and as green manure. The toxic property of the seeds and bark has given rise to the generic epithet. *Gliricidia* was introduced to Sri Lanka from West Indies and later to India. Farmers should be encouraged to grow *Gliricidia* on farm bunds and borders of field. Use of *Gliricidia* as green manure in crop production systems provides a sustainable mean of soil fertility along with inorganic sources of nutrients.

References

Beedy T.L., Snapp S.S., Akinnifesi F.K. and Sileshi G.W. 2010. Impact of *Gliricidia sepium* intercropping on soil organic matter fractions in maize-based cropping system. *Agri., Ecosys. & Environ.*, **138(3-4):** 139-146.

Calle, J., Rivera, A., Joseph-Nathan, P. 1987. Pinitol from the leaves of *Gliricidia sepium. Planta Medica.*, **25:** 303-306.

Gonzalez-Vilalobos D., Palomares-Naveda R., Navarro E., Razz R., Soto-Castillo G. and Quintero Moreno A. 2002. The use of *Gliricidia sepium* in the supplementary feeding of crossbred female calves. *Revista Cientifica,* **XII (5):** 384-387.

Growing *Gliricidia,* NFT Highlights- A quick guide to useful nitrogen fixing trees around the world, A publication of the Forest, Farm and Community Trees Network (FACT Net), Winrock International.

Glover N. 1989. *Gliricidia*: production and use. Nitrogen Fixing Tree Association: Honolulu, HI, 44pp.

Herath H.M.T.B., Dassanayaka R.S., Priyadarshani A.M.A., De Silava S., Wannigama G.P., Jamie J., 1998. Isoflavonoids and petrocarpan from *Gliricidia sepium. Phytochem.*, **47:** 117-119.

Hughes C.E. 1991. *Leucaena* improvement through use of lesser known species. *Leucaena Res. Rep.,* **12:** 98-100.

Jose B. and Reddy L.J. 2010. Evaluation of antibacterial activity of the leaf and flower essential oils of *Gliricidia sepium* from South India. *Intl. J. App. Pharma.*, **2(2):** 20-22.

Kang B.T., Caveness F.E., Tian G. and Kolawole G.O. 1999. Long term alley cropping with four hedgerow species on an Alfisol in southernwestern Nilgiriaeffect on crop performance, soil chemical properties and nematode population. *Nutr. Cycl. Agroeco.*, **54:** 145-155.

Kang B.T., Wilson G.F. and Sipkens L. 1981. Alley cropping maize (*Zea mays* L.) and Leucaena (*Leucaena leucocephala* Lam) in southern Nigeria. *Plant and Soil*, **63:**165-179.

Makumba W., Janseen B., Oenema O., Akinnifest F.K., Mweta D. and Kwesiga F. 2006. The long-term effects of a *Gliricidia* –Maize intercropping system in Suthern Malawi. *Agric. Ecosyst. Environ.*, **118:** 237-243.

Nazli R., Akhter M., Ambreen S., Solangi A.H. and Sultanan N. 2008. Insecticidal, nematicidal and antibacterial activities of *Gliricidia sepium. Pak. J. Bot.* **40(6):** 262-2629.

Pandey C.B. and Rai R.B. 2007. Nitrogen cycling in *Gliricidia sepium* alley cropping in humid tropics. *Trop. Eco.*, **48(1):** 87-97.

Polhill R.M. and Sousa M.S. 1981. Robinieae. In: Polhill, R.M. and Raven, P.H. (eds), *Adv. Leg. Syst.* Royal Botanic Gardens, Kew, p. 283.

Resis R.C.R., Dantas B.F., Castro R.D.D., Antunes C.G.C., Silva F.F.S.D. and Pelacani C.R. 2011. Reserve Mobilization During Imbibitions of Stored *Gliricidia sepium* (Jacq.) Steud. (Leguminosae-Papilionoideae) Seeds. *Revista Brasileira de Sementes*, **33(3):** 549-560.

Rastrelli L., Caceres A., De Simone F., Aquino R. 1999. Studies on the constitution of *Gliricidia sepium* (Leguminosae) leaves and roots: isolation and structural elucidation of new triterpenoids saponins and aromatic compounds. *J. Agric. Food. Chem.*, **47:** 1537-1540.

Suttie J. M. 2005. http://www.fao.org/ag/AGP/AGPC/doc/Gbase/DATA/ Pf000156. htm (3 of 7) 23/06/2005 10:38:15

Speedy A. (Editor). *Gliricidia sepium.* Tropical Feeds and Feeding Systems.

Simons A.J. and Dunsdon A.J. 1992. Evaluation of the potential for genetic improvement of *Gliricidia sepium.* Report to ODA on Forestry Research Project R-4525. Oxford Forestry Institute, 176 pp.

Vanlauwe B., Aihou K., Tossah B., Diesls J., Sanginga N. and Merckx R. 2005. *Senna siamea* trees recycle Ca from Ca-rich subsoil and increase the topsoil pH in Agroforestry system in the West African derived savanna zone. *Plant Soil*, **269:** 285-296.

Wani S.P. (Principle Scientist) (Watersheds and Regional Project Coordinator) ICRISAT, Hyderabad.

Wander M.M., Traina S.J., Stinner B.R. and Peters S.E. 1994. Organic and conventional management effects on biologically active soil organic matter pools. *Soil Sci. Soc. Am. J.*, **58:** 1130-1139.

Wise R. and Cacho O. 1999. A Bioeconomic Analysis of Soil Carbon Sequestration in Agroforestry. Working paper CO_2. ACIAR project ASEM 1999/093, http://www.une.edu.au/feb/Econ/carbon/.

Whetton M., Rossiter J.T. and Wood C.D. 1997. Nutritive evaluation of nitrogenous fractions in leaves of *Gliricidia sepium Caliandra calothyrsus* in relation to tannin content and protein degradation by rumen microbes *in vivo. J. Agric. Food Chem.*, **45:** 3570-3576.

17

THIOUREA IMPROVES THE PERFORMANCE OF CROPS

R.S. Meena* and A. Hemantaranjan[1]

**Department of Agronomy,*
[1]Department of Plant Physiology, Institute of Agricultural Sciences, Banaras Hindu University, Varanasi -221 005 (UP), India
**E-mail: rsmeenaagro@gmail.com*

INTRODUCTION

In the current scenario increasing demand for sustainable agriculture, understanding and improving crop plants for environmental stress tolerance, has become necessary. Abiotic and biotic stress factors cause extensive losses to agricultural production worldwide; there has been extensive research concern to achieve stress tolerance in crop plants without compromising their yield. The technique of developing stress tolerant plants began with the transfer of individual functional genes either through plant breeding or genetic engineering. Both the approaches have so far generated wealth of information resource and germplasm base, but could not give much impact as far as consistency, reliability and visible effect at the field level are concerned. An alternative approach in this direction has been the application of some external effectors molecules, which can induce the expression of critical signaling molecules involved in stress tolerance (Srivastava *et al.*, 2008).

Bio-regulators are the organic compounds, other than plant nutrients, that modify physiological processes in plant. These bio-regulators acts as chemical catalyst in the plant and improve physiology and reproductive efficiency in the plant. Bio-regulators use is an emerging plant biotechnology approach which can modify plant gene expression, can affect levels of DNA, RNA, enzymes and finally their products such as protein, carbohydrates, lipids, allelochemicals for enhancing yield and phytonutrients in food crops (Olaiya *et al.*, 2013). Bio-regulators possibly improve the sucrose transport and increase dry matter partitioning for grain production. Application of bio- regulator like thiourea increase crop growth, yield and quality parameters. Use of thiourea is one of the effective means of delaying the senescence of leaves as well as retarding the abscission of reproductive organs. Seed treatment along with foliar spray of thiourea applied at initiation of branches and flowering stages brought about significant improvement in growth parameters of crops (Kumar and Kaushik, 2014; Meena *et al.*, 2014). Heat stress due to increased global atmospheric temperature could be one of the major constraints for limiting crop yield and quality in many crops (Wahid *et al.*, 2007). Heat and drought stress in field crops can be managed by applying bio-regulators are able to induce long term thermo-tolerance in plants and can be helpful in mitigating the yield reduction threats as well as are helpful in producing good quality grains (Wahid *et al.*, 2007; Kumar *et al.*, 2013). Of a variety of growth substances, thiourea is believed to play an important role in oxidative stress defense system (Kaya1 *et al.*, 2015).

Thiourea is a sulphydryl compound (Jocelyn, 1972), which contains one –SH group besides containing nitrogen in the form of –NH2 as evident from its structural formula given below

Molecular Formula	CH4N2S
Structure	H_2NCSNH_2
Molar mass	76.12 g/mol
Melting point	182 °C
Density	1.41 g/cm^3
Origin of Thiourea	New Latin
First Known Use	1884

$$NH_4SCN \xrightarrow{180°C} H_2NC(S)NH_2 \xleftarrow{H_2S} H_2NCN$$

(a) (b)

$$H_2NC(S)NH_2 + RX \longrightarrow [H_2N \overset{\ldots}{=} \underset{SR}{C} \overset{\ldots}{=} NH_2]^+X^- \quad (I)$$

$$\xrightarrow{HO^-} RSH + H_2NC(O)NH_2 \quad (II)$$

where X is a halogen and R is an alkyl.

Thiourea (-SH compounds) is not only the fertilizer but is an important plant growth bio-regulator with Sulphydryl group chemical having 36.8 per cent nitrogen and 42.1 per cent sulphur which helps in photosynthesis, synthesis of chlorophyll (Garg *et al.*, 2006), transportation of food, assimilate partitioning, altered pattern of protein synthesis, changes in the activities of antioxidants and so many other bio-chemical activities in the plant. It is highly water soluble and easily absorbed in the living tissues. The plant stress tolerance can be improved up to a certain extent with the exogenous application of stress alleviating chemicals. Among stress alleviating compounds, thiourea is an important molecule with two functional groups, "thiol" is important to oxidative stress response and "imino" partly fulfils the N requirement (Farooq *et al.*, 2009).

It is interesting to note that thiourea promoted growth in cytokinin requiring soybean and tobacco callus tissues in the absence of kinetin (Erez, 1978). The increase in fresh and dry weights obtained was a result of cell multiplication. Callous tissues grew well when repeatedly sub-cultured in kinetin-free medium containing thiourea. Best growth was obtained with 10^{-4} to 10^{-3} M thiourea. At these concentrations synergism was demonstrated between thiourea and cytokinins, kinetin, BA and zeatin. Thiourea caused greater growth than diphenyl urea and phenyl thiourea. The possibility that a breakdown product of thiourea was responsible for the enhanced growth was noted excluded.

Thiourea is a non-physiological thiol and has been employed by various researchers to impart stress tolerance and improve yield of crops like wheat, and maize. Thiourea application improved the plant growth potential and photosynthetic efficiency. This was concomitant with the onset of early maturity and increased crop yield (Shau *et al.*, 2006; Amin *et al.*, 2013). Thiourea also plays a vital role in the physiology of plants mainly as a sulphydryl compound and to some extent as an amino compound like urea. The stimulating action of thiourea in various physiological activities of plants is well known. Thiourea has been proved effective in mitigating drought in crops (DOW, 2004). It promotes growth in cytokinin requiring callus tissues in absence of kinetin in various crops (Erez, 1978). Kolyada (1970) noticed significantly increased germination and seedling vigour due to soaking of seeds in thiourea solution. Soaking of seeds and foliar spray of thiourea have been reported growth and development of plants (Solanki, 2002). It has also been reported that thiourea regulated the plant growth by maintaining higher photosynthetic rate upto the reproductive stage and increased the yield by improving carbon partitioning towards sink. Application of thiourea also increase flower, fruit setting, grain filling and test weight in different crops (Patel and Singh, 1980).

Several reports are available showing beneficial effects of thiourea on yield in many crops under such conditions, application of sulphydral compounds like thiourea can be a potential alternative to improve photosynthetic efficiency, assimilate partitioning and increased growth and yield (Sahu *et al.*, 1993; Anitha *et al.*, 2004). Foliar applied thiourea at critical crop stages improved photosynthetic efficiency, assimilate partitioning and increased chlorophyll contents, shoot dry matter and seed weight had been found in many crops (Jagetiya and Kaur, 2006). These increases in yield were associated with beneficial effects of foliar applied thiourea on total dry matter production, pod number, pod and seed weight, on total chlorophyll contents, starch, reducing sugars and soluble proteins (Mathur *et al.*, 2006).

ROLE OF THIOUREA IN *KHARIF* CROPS

PEARL MILLET

Thiourea play an important role in pearlmillet, if seed soaking and foliar spray of 1000 ppm, increase the number and weight of grains per spike, length of spike, grain (21.5 q ha^{-1}) and stover yield (37.4 q ha^{-1}). However, higher plant height, number of tillers and dry matter production of pearlmillet. Furthermore, an improvement in dry matter distribution to spike head of pearlmillet was also observed by Choudhary (2003). Similarly, results of a field study on pearl millet grown on loamy sand soils revealed that application of thiourea in terms of seed soaking, foliar spray and seed soaking + foliar spray (500 ppm) increased the chlorophyll content, nitrogen concentration in leaf and protein content in grains (Lakhana, 2001).

MAIZE

Foliar spray of thiourea 500 ppm increased the dry matter accumulation per plant, LAI and number of leaves per plant at physiological maturity of maize Sachan (1991). While, Sahu *et al.* (1993) and Rathore (1996) in maize reported that foliar spray of thiourea improved the LAI, dry matter accumulation, LAD and chlorophyll content in leaves. Sharma (1997) observed that foliar spray of thiourea at 1000 ppm on maize increase of 18.8, 17.4 and 15.8 per cent in cob weight, LAI and chlorophyll content, respectively. In another, Sahu and Solanki (1991) reported that foliar spray of 0.1 per cent thiourea at grain formation stage significantly increased the grain yield of maize by 34.1 per cent over water sprayed control. However, foliar spray of 0.1 per cent urea did not improve grain yield, it could be inferred that the biological activity of thiourea was primarily on account of its -SH group. Sachan (1991) while, working on maize and observed that foliar spray of 500 ppm thiourea increased the yield components *viz.*, length of cob, diameter of cob, number of grains per cob and grain weight per cob. A significant increase of 25.4 per cent in grain yield was also noted over control. Sahu *et al.* (1993) reported that soaking of maize seeds in 500 ppm thiourea solution increased the grain and biological yield by 13.4 and 14.7 per cent, respectively over control. However, foliar spray of 1000 ppm thiourea at 30 and 45 days after sowing increased the grain yield of maize by 40.6 per cent over control. Rathore (1996) observed that thiourea infusion at 1000 ppm in maize increased the number of seeds and seed weight per cob over water control. Overall, salt stress induced inhibitory effects on growth attributes of maize plants. However, exogenously applied thiourea enhanced to some extent stress tolerance in maize plants by enhancing shoot fresh weights, chlorophyll content, leaf water potential, N and P content and by decreasing electrolyte leakage and Na content in the maize plants under saline conditions (Kaya1 *et al.*, 2015).

CLUSTER BEAN

A field study was conducted on cluster bean to find out the effect of foliar spray of thiourea and results indicated that seed soaking + foliar spray of 500 ppm thiourea at branching significantly increased plant height, dry matter accumulation, LAI, CGR, RGR, NAR and chlorophyll content Jharia (2002). Similarly, Solanki (2002) was observed on cluster bean with foliar application of thiourea (500 ppm) proved effective in increasing dry matter distribution in leave and pod, total chlorophyll content, and LAI over control. Seed treatment + foliar spray of thiourea 500 ppm at initiation of branches and flowering stages brought improvement in all growth parameters of cluster bean. It may also be noted that not only accumulation of dry matter was increased due to the effects of thiourea spray; translocation of dry matter as well as its efficiency were also found to be higher in crop plants sprayed with thiourea as compared to un sprayed crop. Thiourea brought significant improvement in days to maturity. The earliest maturity was recorded in treatment seed + foliar application with thiourea 500 ppm (Meena *et al.*, 2014; Meena *et al.*, 2014).

GREEN GRAM AND BLACK GRAM

In a field study under semi-arid climate, Singh (1999) found that seed soaking and foliar spray of thiourea (0.05%) in green gram at pre flowering stage significantly increased the total chlorophyll content, dry matter accumulation and number of pods per plant over control. Singh (2002) also reported that seed soaking + foliar spray of thiourea at branching gave significantly higher pods per plant, number of grains per pod, test weight, straw yield, biological yield and harvest index in black gram compared to control.

GROUNDNUT

In an experiment (Ramesh *et al.*, 2008) observed that increase in number of pods per plant, number of kernel per pod and yield of groundnut due to foliar spray of 0.1 per cent thiourea. Moezi (1993) observed that foliar spray of thiourea on groundnut increased the number of pods per plant, number of kernels per pod, pod yield per plant and pod yield per hectare.

ROLE OF THIOUREA IN *RABI* CROPS

WHEAT

Foliar application of 500 ppm thiourea at tillering and flowering stage observed higher grain yield of wheat (DOW, 2004). Similarly, Yadav (2005) observed that 500 ppm thiourea (seed soaking + foliar spray) significantly enhanced the yield attributes, *viz.*, number of spikes per meter row length, number of grains per spike, spike length and test weight and grain, straw and biological yield and net returns of wheat as compared to control (seed soaking in water + water spray). Application of 500 ppm thiourea (seed soaking + foliar spray) increased the growth parameters (plant height, number of tillers m^{-1} row length, total dry matter accumulation, dry matter accumulation in leaf and panicle (90 DAS and at harvest), LAI (60 and 90 DAS), CGR, RGR (60-90 days and 90 days to harvest), LAD, NAR and chlorophyll content in leaves.

OAT

A study was conducted on oat under semiarid climate and reported that foliar application of 500 ppm thiourea at tillering and flowering significantly increased the dry matter accumulation in stem, leaf and panicle. The increase in LAI and LAD was 9.5 and 8.8 per cent over control, respectively. Contrary to this, foliar spray of 1000 ppm thiourea at tillering and flowering proved largely ineffective in this respect. Similar response of oat to foliar spray of thiourea and noted that foliar application of 500 ppm thiourea at tillering and flowering produced significantly higher grain yield, biological yield and harvest index. The yield attributes like number of tillers, number of penicle, length of penicle, number of grains per spike and test weight also increased Yadav (2000).

MUSTARD

There are several reasons of low productivity of mustard and one of them as improper sink-source relationship and internal hormonal imbalance which are of primary importance. Several studies on different crops have shown that plant growth substances have potential to enhance the productivity of crops. The low productivity of mustard has been attributed to a coefficient of effectiveness and the lack of efficient partitioning of photosynthates between reproductive and vegetative plant parts. Since different growth substances like thiourea used for the increase water and crop productivity, they can more effectively improve assimilate partitioning and yield of crops under water sufficient and deficit conditions (Hayat *et al.*, 2007). Thiourea foliar spray play important role to increase water and crop productivity. Thiourea acts as chemical catalyst in the plant and improve physiological and reproductive efficiency in the plant and also improve the sucrose transport and increase dry matter partitioning and seed yield (Dadhic *et al.*, 2014). Similarly, Garg *et al.* 2006 reported that S is involved in synthesis of fatty acids and increases protein contents through synthesis of sulphur containing amino acids such as cysteine, cystine and methionine. Increased leaf area index with foliage thiourea was due to stimulatory effects of sulphur on cell division which helped in leaf expansion and increased light interception improved the vegetative growth of mustard. Improved leaf area by sulphur nutrition provided greater surface area for production and partitioning of photo-assimilates towards growing sinks and produced high crop growth rate and net assimilation rate. However, improved crop growth by foliar applied thiourea can be attributed to increased photosynthetic efficiency (Anjum *et al.* 2011; Garg *et al.* 2006). Improvement in number of branches, dry matter accumulation, crop growth and relative growth rate were recorded favorably influenced by thiourea spray. The growth attributes increase under foliar spray of thiourea at both stages may be attributed to the better availability of nutrients. Dry matter accumulation is a result of photosynthesis, which mainly depends upon the utilization of solar energy through crop canopy along with other essentials. The increase in crop dry matter with thiourea might be attributed due to considerable increase in number of branches and effective utilization of nutrient, moisture and light in mustard (Dadhic *et al.*, 2014).

CHICKPEA

Narang (1988) found out in a study conducted on chickpea and observed that soaking of seeds in 0.1 per cent thiourea solution and foliar spray of 0.1 per cent thiourea significantly increased the grain yield by 14.1 and 47.08 per cent over control, respectively. In same experiment also observed taht increase in chlorophyll, water soluble carbohydrate and N and S content of chickpea leaves by 0.1 per cent thiourea seed treatment. Further, foliar application of 0.1 per cent thiourea also significantly increased the N, K, S, Ca+Mg and Zn content of chickpea leaves over control. Patel (1991) found that 0.1 per cent thiourea spray at flowering significantly increased the number of pods per plant, number of grains per pod, seed index and yield of chickpea when compared with water sprayed control.

OPIUM POPPY

A field experiment was conducted on opium poppy by Saini (1991) and reported that soaking of seed with 150, 300 and 450 ppm thiourea solution did not bring about a significant increase in plant height and number of leaves per plant. However, significant improvement was observed in LAI, CGR and dry matter accumulation per plant as compared to water spray. However, observed that seeds soaked in 150, 300 and 450 ppm thiourea did not influence the number of capsules per plant, girth of capsule, capsule weight and seed yield but brought a significant impact on the latex yield, representing an increase of 12.3, 11.3 and 18.1 per cent, respectively. The foliar spray of 0.1 per cent thiourea significantly increased the latex and seed yield by 24.6 and 36.0 per cent, respectively over control. Similarly, Intodia (1992) also observed that foliar spray of 500 ppm thiourea at capsule development stage in opium poppy significantly enhanced the seed and latex yield by 9.9 and 11.5 per cent, respectively over control. While, capsule development improved the dry matter accumulation, LAI and LAD.

CORIANDER

Balai (2005) conducted a field experiment and found that 500 ppm thiourea spray at vegetative and flowering stages recorded higher magnitude of growth parameters *viz.*, plant height and number of branches per plant in coriander. However, higher number of umbels per plant, umbellets per umbel, seeds per umbel and test weight in coriander.

TOBACCO

Khalil *et al.* (1987) observed that soaking of freshly harvested seeds of tobacco in 0.3 per cent solution of thiourea significantly increased the germination per cent and seedling growth as compared to seed soaking in water or no seed treatment.

POTATO

A seven years study conducted in green house and field condition on loamy sand soils of Shimla revealed that in green house study, the application of thiourea increased the phosphorous uptake by 44.7 per cent over control. Continuous application of thiourea and phosphorus significantly increased the soil phosphorus by 15 ppm when compared with control. The results further revealed that application of 5.0 kg ha^{-1} thiourea alone was found to save 22 kg phosphorus ha^{-1} for potato crop (Sud *et al.*, 1991).

REFERENCES

Amin, A.A., Abd El-Kader, A.A., Shalaby, M.A.F., Gharib, F.A., Rashad, E.S.M. (2013). Physiological effects of salicylic acid and thiourea on growth and productivity of maize plants in sandy soil. *Commun. Soil Sciece and Plant Anal.* **44**: 1141–1155.

Anitha, S., Sreenivasan, E., Purushothaman, S.M. (2004). Effect of thiourea application on cowpea productivity under rainfed conditions. *Trop. Agric.* **42**:53–54.

Anjum, F., Wahid, A., Javed, F., Arshad, M. (2011). Potential of foliar applied thiourea in improving salt and high temperature tolerance of bread wheat (*Triticum aestivum* L.). *Int. J. Agric. Biol.* **13**:251-256.

Balai, L.R. (2005). Effect of thiourea on growth and productivity of coriander [*Coriandrum sativum]* varieties under normal and late sowing conditions. *Ph.D. Thesis, Rajasthan Agricultural University, Bikaner.*

Chaudhary, B.R. (2003). Effect of thiourea and zinc on the productivity of pearlmillet. *M.Sc. (Ag.) Thesis,* Rajasthan Agricultural University, Bikaner.

Dadhich, R.K., Reager, M.L., Kansotia, B.C., Meena, R.S. (2014). Efficacy of Growth Substances on Mustard (*Brassica juncea* L.) under Hyper Arid Environmental Condition of Rajasthan. *The Eco.* **3&4**: 269-272.

DOW. (2004). Report of Project Director (Wheat), *Directorate of Wheat Research,* Karnal. pp76-81.

Erez, A. (1978). Thiourea, a growth promoter of callus tissue. *J. Exp. Bot.* **29**: 159-165.

Farooq, M., Wahid, A., Kobayashi, N., Fujita, D., Basra, S.M.A. (2009). Plant droughts stress: effects, mechanisms and management. *Agron. Sust. Dev.* **28**: 185- 212.

Garg, B.K., Burman,U., Kathju, S. (2006). Influence of thiourea on photosynthesis, nitrogen metabolism and yield of clusterbean (*Cymopsis tetragonoloba*) under rain fed conditions of Indian arid zone. *Plant Grow. Reg.* **48**:237-245.

Hayat, S., Ahmad, A. (2007). A Plant Hormone. Springer Publishers, Dordrecht, the Netherlands. pp. 1-371.

Intodia, S.K. (1992). Effect of levels and sources of sulphur and foliar applied chemicals on yield and quality of opium poppy (*Papaver somniferum* L.). *Ph. D. Thesis, Rajasthan Agricultural University, Bikaner.*

Jagetiya, B.L., Kaur, M.J. (2006). Role of thiourea in improving productivity of soybean. *Int. J. Plant Sci.* **1**:308–310

Jharia, A. (2002). Effect of iron fertilization and thiourea spray on the productivity of cluster bean (*Cyamopsis tetragonoloba* (L.) Taub.). *M.Sc. (Ag.) Thesis, Rajasthan Agricultural University, Bikaner.*

Jocelyn, P.C. (1972). Biochemistry of -SH group: The Occurrence, Chemical Properties, Metabolism and Biological Function of Thiols and Di - sulphides. Academic Press, London.

Kaya., Sonmez, O., Ashraf, M., Polat, T., Levent, T., Salih, A. (2015) Exogenous application of nitric oxide and thiourea regulates on growth and some key physiological processes in maize (*Zea mays* L.) plants under saline stress. *Soil-Water J.*, Special Issue, 61-66.

Khalil, S.K.H., Mordalen, V.M., Cherkason, S.V., Juk Yaneko, G.Y.L. (1987). Effect of physiologically active substance on sowing qualities and seedling growth of tobacco, *Tabak.* **3**(1) : 39-40.

Kolyada, T.I. (1970). Effect of thiourea on seed germination and plant growth. *Byul uses. Inst. Sel.* Khoz Mikrolbiol. *Field crop Abst.*, 3 : 74-81

Kumar, B., Singh, Y., Ram, H., Sarlach, R.S. (2013). Enhancing seed yield and quality of Egyptian clover (*Trifloium alexandrinum* L.) with foliar application of bio-regulators. *Field Crops Res.* **146**: 25-30.

Kumar, V., Kaushik, M.K. (2014). Growth substances response to clusterbean [*Cyamopsis tetragonoloba* L.] Growth parameters. *Agric. Sust. Devel.* **2**(1):11-13

Lakhana, R.C. (2001). Effect of nitrogen and thiourea on growth and productivity of pearlmi llet (*Pennisetum glaucum* (L.) R.Br. Emend Stuntz). *M.Sc. (Ag.) Thesis,* Rajasthan Agricultural University, Bikaner.

Mathur, N., Singh, J., Bohra, S., Bohra, A., Vyas, A. (2006) Improving productivity of mungbean by application of thiourea under arid conditions. *World J. Agric. Sci.* **2**:185–187

Meena, V.K., Kaushik, M.K., Meena, R.S., Meena, V. S., Meena, B.P. (2014). Effect of growth regulators on clusterbean [*Cyamopsis tetragonoloba* (L.)]

Growth under Aravali hills environment in Rajasthan. *The Bio.* **9**(2): 547-550, 2014.

Moezi, A.A. (1993). Phosphorus, sulphur and iron status of soil of southern Rajasthan and the effect of their application on chlorosis, Nutrient availability and yield of groundnut (*Archis hypogea* L.). *Ph.D. Thesis, Raj. Agril. Univ., Bikaner.*

Narang, S.K. (1988). Effect of sulphur, sulphur derivatives and ascorbic acid on cryoprotection, growth and productivity by gram *(Cicer arietinum* L.*) M.Sc. (Ag.) Thesis. Dept. of Agron. Raj. Agril. Univ. Bikaner, Campus Udaipur.*

Olaiya, C.O., Gbadegesin, M.A., Nwauzoma, A.B. (2013). Bioregulators as tools for plant growth, development, defense and improvement. *African J. Biot.* **12**: 4987–4999

Patel, J.C. Singh, R.M.(1980). Effect of irrigation, mulching and bio regulators on the production of sunflowers. *Indian J. of Agron.* **25**:122-128.

Patel, J.S. (1991). Effect of elemental sulphur and foliar applied chemicals on growth, nutrition and productivity of gram (*Cicer arietinum* L.) on calcareous soil. *Ph. D. Thesis*, Rajasthan Agricultural University, Bikaner.

Ramesh, A., Hasegawa, H., Sugimoto, W., Maki, T., Ueda, K. (2008). Adsorption of gold (III), platinum(IV) and palladium(II) onto glycine modified crosslinked chitosan resin. *Bioresour. Technol.* **99**: 3801–3809.

Rathore, S.S. (1996). Organ in susion mediated effect of free radical quenchers on upoxygonase activity and grain development in maize (*Zea mays* L.). *M.Sc. (Ag.) Thesis, Rajasthan Agricultural University, Bikaner.*

Sachan, A.S. (1991). Effect of sulphur fertilization and foliar applied chemicals on yield and quality of maize (*Zea mays* L.) on calcareous soils. *M.Sc. (Ag.) Thesis, Rajasthan Agricultural University, Bikaner.*

Sahu, M.P., Solanki, N.S., Dashora, L.N. (1993) Effects of thiourea, thiamine and ascorbic acid on growth and yield of maize (*Zea mays* L.). *J. Agron. Crop Scies.* **171**:65–69

Sahu, M.P., Solanki, N.S. (1991). Role of sulphydryl compounds in improving dry matter partitioning and grain production of maize (*Zea mays* L.). *J. Agron. Crop Sci.* **167**: 356-359.

Sahu, M.P., S.M. Kumawat, N. K. Ramaswamy, S. F. D'Souza, Singh, G. (2006). Sulphydryl bioregulator technology for increasing wheat productivity. *Res. Bull. RAU-BARC*, 1–56.

Saini, L.R. (1991). Effect of thiourea treatments, foliar applied chemicals and elemental sulphur on mitigation of chilling injury and productivity of opium poppy (*Papaver somniferum* L.). *M.Sc. (Ag.) Thesis, Rajasthan Agricultural University, Bikaner.*

Sharma, D.D. (1997). Effect of zinc, iron and bioregulators on growth, dry matter partitioning and grain yield of maize (*Zea mays* L.). *Ph. D. Thesis, Rajasthan Agricultural University, Bikaner.*

Singh, J.P. (1999). Fodder and fodder seed production in Rajasthan. An approach, paper submitted by the author to Rajasthan Co-operative Dairy Federation, Bikaner.

Singh, V. (2002). Influence of zinc and thiourea on mothbean [*Vigna aconitifolia* (Jacq.) Marcehal] in arid regions of Rajathan. *M.Sc. (Ag.) Thesis, Rajasthan Agricultural University, Bikaner.*

Solanki, N.S. (2002). Effect of thiourea and demethyl sulphoxide on phosphorus use efficiency, dry matter partitioning and productivity of clusterbean [*Cyamopsis tetragonoloba* (L.) Taub.]. *Ph. D. Thesis, Maharana Pratap University of Agriculture and Technology, Udaipur.*

Srivastava, A.K., Ramaswamy, N.K. and D'Souza S F. (2008). Bioregulatory Role of "Thiourea" In Multistress Tolerance of The Indian Mustard (*Brassica juncea*). *Founder's Day Special Issue.* pp 113-120.

Sud, K.C., Sharma, R.C.; Trehan, S.P. and Bisi, B.S. (1991). Effect of thiourea on phosphorus use efficiency by potato *(Solanum tuberosum)* in Shimla hills. *Indian J. Agrc. Sci.* **61** (10) : 731-735.

Wahid, A., S. Gelani, M., Ashraf, Foolad, M.R. (2007). Heat tolerance in plants: An overview. *Env. Expe.Bot.* **61**: 199-223

Yadav, L.R. (2005). Effect of bioregulators on productivity of wheat (*Triticum aestivum* L.) emend fiori and paol varieties under normal and late sown conditions. *Ph.D. Thesis, Rajasthan Agricultural University, Bikaner.*

Yadav, N.S. (2000). Response of oat (*Avena sativa)* to levels of nitrogen, cutting management and foliar spray of thiourea. *Ph. D. Thesis, Rajasthan Agricultural University, Bikaner.*

Section VII:

Comprehensive Review

18

ROLE OF ALLELOPATHY IN SUSTAINABLE AGRICULTURE-WITH SPECIAL REFERENCE TO PARTHENIUM

Nisha Raghava, Ravindra P. Raghava, Lina Singh and Jayanti Srivastava

FIST-DST PG Dept., of Botany, S.M.M.T. (PG) College, Ballia- 277001, India.
E.mail-nisharaghava@yahoo.co.in / raghavarp@rediffmail.com

INTRODUCTION

We are in 21st century and there are a number of challenges to be faced to maintain the necessary food production. The increase in population could be 8 billion by 2020, which in turn increases the consumption of crops. The physiology and biochemistry have developed as powerful discipline during the 20th century, but only a limited number of instances have led to crop improvement. It is now widely agreed that new technologies must be not only economically viable, but also environmentally and socially sustainable. The challenge now is, to add the ecological dimension to crop productivity. Thus, there has to be a paradigm shift in research strategies, from commodity-centered approach to an integrated natural resources management procedure covering the entire cropping system (Mifflin, 2000; Swaminathan, 2006). Several significant steps have been taken during the last two years to remedy the situation, but their implementation has been slow. The time has come for the leaders of all major political parties to re-visit the advice given by **Pt. Jawaharlal Nehru** in 1974, when he said, **"Everything else can wait, but not agriculture,"** and adopt the ending of Farmers' distress as a National Common Minimum Programme (Swaminathan, 2007).

Plants are the key to life of the earth as they directly supply 90% of human calories nutrients and 80% of the protein intake the reminder being derived from animal produced. India's population is growing at the rate of above 1.8 percent a year. Besides, the population increase improved purchasing power associated with economic growth will enhance the demand for agricultural and horticultural crops. In predominantly rural and agricultural countries like India, agricultural progress will be the most effective social safety net against hunger and poverty. Hence, the ongoing fatigue for the "green revolution", in major crops should be converted into an "evergreen revolution" designed to promote productivity improvement, without ecological harm (Swaminathan, 2006, 2007). The experience from green revolution that sustainable productivity requires strict adherence to agricultural practices that ensure integrity of the ecological foundations of agriculture (Gahukar, 2010; Kesavan and Malarvannan, 2010). Evergreen revolution is achieving productivity in perpetuity involves less dependence on chemicals and more on biological inputs. The good researches in India greatly contributed in this area (Swaminathan, 2007, 2008).

Recent researches have demonstrated possibilities of such prospects in reality, especially in weed control. As a demand increase for sustainable agriculture and concern grows regarding the extensive use of synthetic chemicals (e.g. contamination of environment, herbicide resistance, increasing cost, etc.), attention is focused on reducing reliance upon synthetic herbicides and finding alternative strategies for weed management. Allelopathy holds great prospect for meeting some of those demands. Allelopathic novel weapons have emerged as a prominent mechanism to explain why some plant invaders can rapidly expand in habitats where they lack evolutionary familiarity (Inderjit *et al.*, 2000; Bais *et al.*, 2003). These novel weapons are powerful phytochemicals to which native competitors or herbivores are assumed to have no evolved tolerance (Harnden *et al.*, 2011).

Some authors have used the term in more restricted senses to describe only the harmful effects of one higher plant upon another. Allelopathy may be part of whole network of chemical communication between plants, between plant and other organisms and such communication may contribute to plant defence (Lovett and Ryuntyu, 1992). Chemicals that impose allelopathic influences are called allelochemicals or allelochemics. They may be largely classified as secondary plant metabolites, which are generally considered to be those compounds (such as alkaloids, phenolics, flavonoids, terpenoids, and glucosinolates), which do not play a role in primary metabolic processes essential for a plant's survival, and are produced as offshoots of primary metabolic pathways. Allelopathic potential can be used in several ways in agroecosystems. Allelopathy occurs through the release by one plant species of chemicals, which affect other species in its vicinity, usually to their detriment. This phenomenon has been observed for over 2000 years.

Reports as early as 300 BC documented that many crop plants (e.g. chick pea, barley, bitter vetch) destroyed weeds and inhibited the growth of other crop plants. Allelochemicals are present in virtually all plant tissues, including leaves, flowers, fruits, stems, roots, rhizomes, seeds and pollen. They may be released from plants into the environment by means of four ecological processes: volatilization, leaching, root exudation, and decomposition of plant residues.

Biological activities of receiver plants to allelochemicals are known to be concentration dependent with a response threshold. Responses are characteristically, stimulation or attraction at low concentrations of allelochemicals, and inhibition or repellence as the concentration increases (Lovett, 1989). Plants are exposed to allelochemicals and their growth and development are affected. The readily visible effects include inhibited or retarded germination rate; seeds darkened and swollen; reduced root or radicle and shoot or coleoptile extension; swelling or necrosis of root tips; curling of the root axis; discolouration, lack of root hairs; increased number of seminal roots; reduced dry weight accumulation; and lowered reproductive capacity. Early research grew out of observations of poor regeneration of forest species, crop damage, yield reductions, replant problems for tree crops, occurrence of weed-free zones, and other related changes in patterns of vegetation.

The basic approach used in allelopathic research for agricultural crops has been to screen both crop plants and natural vegetation for their capacity to suppress weeds. To demonstrate allelopathy, plant origin, production, and identification of allelochemicals must be established as well as persistence in the environment over time in concentrations sufficient to affect plant species. In the laboratory, plant extracts and leachates are commonly screened for their effects on seed germination with further isolation and identification of allelochemicals from greenhouse tests and field soil, confirming laboratory results. Interactions among allelopathic plants, host crops and other non-target organisms must also be considered. Furthermore, allelochemistry may provide basic structures or templates for developing new synthetic herbicides (Popovici *et al.*, 2011). Incorporation of allelopathic traits from wild or cultivated plants into crop plants through traditional breeding or genetic engineering methods could also enhance the biosynthesis and release of allelochemicals.

Allelopathy could be potentially used for weed control by producing and releasing allelochemicals from leaves, flowers, seeds, stems, and roots of living or decomposing plant materials (Weston, 1996). "Allelopathy can also be defined as an important mechanism of plant interference mediated by the addition of plant produced secondary products to the soil rhizosphere" (Weston 2005). These products are typically exuded into the soil profile and impact the growth of plants that are growing near the allelopathic species. Allelopathic species produce these compounds to reduce populations of other plant species that are directly competing for similar resources such as nutrients, sunlight, or water. Allelopathic compounds can show selectivity towards different plant species and can alter the composition of plant populations. Allelochemicals can be released in the soil by decomposition, root exudation, and volatilization of compounds by the allelopathic species. Allelopathic species can also suppress weed or crop growth by direct competition for limited resources, thus cover crops are normally killed off before crop establishment to avoid competition with the crop. Recently, agricultural research has looked at weed control using allelopathic crop residues. There is also interest in developing cultivars that produce higher amounts of these toxic compounds that suppress weed growth. Allelopathic cover crops would provide other benefits besides weed control in minimum or no tillage practices such as erosion prevention, moisture conservation (mulch), and improving soil productivity (Akemo *et al.,* 2000).

Using a living allelopathic crop that is incorporated into crop rotations, and also using it as mulch have looked at allelopathic weed suppression. Studies have shown that when compared to other leguminous cover crops such as peas or hairy vetch, rye provides much better weed suppression which suggests that it is caused by a chemical interaction and is not solely due to direct competition. Allelopathic species weed suppression could provide us with another possible approach to integrated weed management especially in organic farming or minor use crops. Although, allelopathic species provide weed suppression, replacing herbicide use with allelopathic species is not a viable option for complete weed control in large-scale operations. If allelopathic species could be incorporated in certain cropping systems to provide weed suppression, this could reduce dependency on synthetic herbicides that are potentially hazardous to our environment. This comprehensive review article focuses on conspicuous and indispensable aspects of allelopathy from the past to the present scenario, which have been understood to be vital for the sustainability of agriculture.

HISTORICAL BACKGROUND OF ALLELOPATHY

The earliest recorded observations of weed and crop allelopathy were by Theophrastus (300 BC). He first noticed that it is due to odour, the common problem in both Greek and Roman times were so-called soil-sickness, the decline yield of field. The term 'allelopathy' is derived from two Greek words 'Allelon' means each other and 'Pathos' means to suffer. Molisch (1937) coined this term, which refers to all biochemical interaction (stimulatory and inhibitory) among plants, including microorganism. Rice (1974) defined **'allelopathy as any direct or indirect effect by one plant, including microorganism, on another through the production of chemical compounds that escape into the environment and subsequently influence the growth and development of neighbouring plants'**. The complete review on allelopathy by Rice (1984) describes almost all aspects related to it. The use of the term "**allelopathy**" may therefore be somewhat controversial. Chemicals found to inhibit the growth of a species at a certain concentrations may help stimulate the growth of the same species or another at a lower concentration (Rice, 1984; Wills, 1985; Putnam and Tang, 1986). Aldrich (1984) describes two types of allelopathy:

- **True type:** The release into the environment of compounds that are toxic in the form in which they are produced.
- **Functional Type:** The release into the environment of a substance that is toxic as the result of transformation by micro-organisms.

The word allelopathy hence means: the injurious effect of one upon another (Rizvi *et al.*, 1992). These compounds then come in contact with neighbouring plants, which can then stimulate, or more typically negatively impact the growth of these neighbouring plants. Although these compounds are not restricted to plant-plant interactions, they can also interact with other organisms. Lovett looked at allelopathy as the complex of subtle communications between plants and also between plants and other organisms (Einhellig, 1995). Allelopathic species can indirectly affect their growth rate by manipulating microbes in their surrounding rhizosphere, and altering the plant-microbe interaction. Biorational alternatives are gaining increased attention for weed control because of concerns related to pesticide usage and dwindling numbers of labeled products, particularly for minor use crops (Weston, 1996). In certain cropping situations allelopathy may have the potential to be integrated into a weed management plan in order to reduce the use of synthetic herbicides, as well as provide other added benefits from the allelopathic crop.

Genetic engineering could be a useful tool in order to create cultivars that produce greater amounts of allelopathic compounds for weed suppression in agriculture. Pliny stated the nature of some plants though not actually deadly is injurious owing to its blend of scents or of juice (Weston, 2005). Although allelopathy does not always cause negative effects to neighbouring plants, throughout history it was probably easier to observe negative plant responses. de Candolle in the 1800's was one of the first to perform research on toxic root exudates from allelopathic species (Weston, 2005).

Willis (1985) suggested that six criteria be met, and even when maximally relaxed to just three, proving allelopathy is rarely, if ever accomplished (Blun *et al.*, 1999):

- Pattern of inhibition of one species by another.
- Putative aggressor must produce a toxin.
- Known mode of release of this toxin.
- Toxin transport or accumulation in the environment.

- Afflicted plants have means of uptake of toxin.
- Observed pattern of inhibition cannot be solely explained by physical competition or other biotic factors.

Proof of Allelopathy

Many field studies implicate allelopathy, but isolation and identification of the chemical agents require a rigorous laboratory effort. It is extremely difficult to prove that any deleterious effect is due to allelopathy rather than to competition for essential products. Numerous studies have provided evidence, but seldom have a specific protocol be followed to achieve convincing proof (Putnam and Tang, 1986). These authors pointed out those shortcomings of the discipline make it hard to differentiate between allelopathy and competition. These shortcomings include: -

- A general lack of nomenclature to adequately describe the plant responses that occur in this manner.
- A dearth of techniques to separate allelopathic interactions from competition.
- A failure to prove the existence of direct compared with indirect influences via other organisms/micro environmental modification.
- A considerable body of information has accumulated implicating allelopathy as an important form of plant interference.

According to Willis (1985), Putnam and Tang (1986) and Cheng (1992) the methodology dictates certain points for allelopathic research to be established to suggest that is operative. A pattern of inhibition one species by another must be shown using suitable controls, describing the symptoms and quantitative growth reductions; the putative aggressor plant must produce a toxin; there must be a mode of toxin release from the environment and thus the target plan; mode of toxin transport or accumulation in the environment must be evident; the afflicted plan must have some means of toxin uptake, be exposed to the chemical in sufficient quantities and time to cause damage, and to notice similar symptoms. The observed pattern of inhibition should not be solely by physical factors or other biotic factors, especially competition. It is important to stress that the above points do not prove that allelopathy is operative, only that it offers the most reasonable explanation for the observed pattern. According to Cheng (1992), once the chemical enters the environment, a number of interacting processes will take place. These processes have identified as:

- **Retention:** The related movement of the chemical from one location to another, through soil, water.
- **Transformation:** The change in form or structure of the chemical, leading to partial change or total decomposition of the molecule;

- **Transport:** Defines how the chemicals move in the environment.

Cheng (1992) also pointed out that these processes are influenced by the nature of the chemical, the organisms present, the properties of the soil, and environment conditions. The fate of the chemicals depends on the kinetics and interactions of individual processes with time, at a particular site under a particular site under a particular condition.

Allelochemicals

According to Putnam and Tang (1986) all alleged cases of allelopathy that have been studied appear to involve a complex of chemicals. Few studies on allelopathy, concentrate on the mechanism and processes, involved in the production of allelochemicals. Putnam and Tang (1986) and Einhellig (1987), raised question whether alleged biochemical agents were in sufficient concentrations and with enough persistence in the environment to affect a neighbouring or succeeding plant. These chemicals could be transformed during the course of extraction. According to Cheng (1992), allelopathic symptoms may not be manifested at the time or site of where plant damage has actually occurred. No single phytotoxin was solely responsible for or produced as a result of interference by a neighbouring plant, Rizvi *et al.* (1992) pointed out that the subject not only deals with the gross biochemical interactions and their effects on the physiological processes but also with the mechanism of action of allelochemicals at specific sites of action at the molecular level.

Allelopathy also may be one of several attributes, which enable a plant to establish in a new ecosystem (Callaway and Aschehoug, 2000; Callaway and Ridenour, 2004). From the emanation of allelochemicals, plants can regulate the soil microbial community in their immediate vicinity, affect herbivory, encourage beneficial symbiosis, change the chemical and physical properties of the surrounding environment, and directly inhibit the growth of competing plant species (Pedrol *et al.,* 2006). Allelopathic compounds play important roles in the determination of plant diversity, dominance, succession, and climax of natural vegetation and in the plant productivity of agro ecosystems.

Sources of Allelochemicals

According to Aldrich (1984), allelochemicals must be concentrated in the leaves, stem or roots than in the fruit or flowers. If it is concentrated in these organs it is likely that it could be available in time to interference with neighbouring plants. Rodosevich and Holt (1984) stated that the primary effect of allelopathy seems to result from an association with plant litter in or on the soil. Rice (1984) and Putnam (1985) reported that allelochemicals are present in virtually all plant tissue, i.e. leaves, fruit, stems, and roots. Leaves may be the most consistent source, while roots are considered to contain fewer and less potent toxins. According to them, there are four ways in which the chemicals are released:

a) Volatilization

Release into the atmosphere It is only significant under arid or semi-arid conditions. The compounds may be absorbed in vapour by surrounding plants, be absorbed from condensate in dew or may reach the soil and be taken up by the roots (Muller, 1966). Kanchan (1975) reported that *Parthenium hysterophorus* also released the allelochemicals through volatiles. The genera, which release volatiles, are *Artemisia, Salvia, Parthenium* and *Eucalyptus* (Rice, 1984).

b) Leaching

Rainfall, dew or irrigation may pats of plants that are subsequently deposited on other plants; on the soil leaching also may occur through plant residues. Their solubility will affect their mobility in soil water.

c) Root exudation

It takes place from plant roots into the soil environment. Whether these compounds are actively exuded, leaked or arise from the cells sloughing off the roots is not clearly understood at this time.

d) Decomposition of plant residues

It is difficult to determine whether toxic substance are contained in residues and simply released upon decomposition, or produced instead by micro-organisms utilizing the residues.

Natural Products Identified as Allelopathic Agents

Alleged allelochemicals represent a myriad of chemicals compounds from simple hydrocarbons and aliphatic acid to complex polycyclic structures. More than 10,000 secondary metabolites in plants are known to play an important role in allelopathy, which includes positive and negative effects (Bell and Koeppe, 1972; Rice, 1984; Waller, 1987; Liebl *et al.*, 1992; Inderjit and Dakshini, 1995; Olofsdotter *et al.*, 1995; Inderjit, 1996; Inderjit and Keating, 1999; Olofsdotter, 1998, 2001; Chung *et al.*, 2003; Inderjit and Duke, 2003; Vasilakoglou *et al.*, 2005, 2006; Dhima *et al.*, 2006a, b; Vaitor *et al.*, 2006; Machado, 2007). The secondary products could be classified in the following categories, but it is not possible to enumerate each and every chemical identified as an allelochemical. Whittaker and Feeney (1971), Rice (1984), Putnam and Tang (1986) divided allelochemicals into various major chemical groups:

- Simple water -soluble organic acids
- Simple unsaturated lactones
- Long -chain fatty acids and poly acetylenes
- Napthoquinone, anthroquinones and complex quinines
- Simple phenols

- Benzoic acid and derivatives
- Cinnamic acid and derivatives
- Flavonoids
- Tannins
- Terpenoids and steroids
- Amino acids polypeptides
- Alkaloids and cyanohydrins
- Sulphides and glucosides
- Coumarins
- Thiocynates
- Lactones
- Acetogenins

Mechanisms of Action of Allelopathic Compounds

Allelopathy involves the release of secondary metabolites that are termed allelochemicals. These compounds are biosynthetically produced but are not needed for the plants primary growth. Although, these compounds can affect plant growth indirectly by modifying the competition from other organisms including plants. "Many secondary metabolites found in plants have a role in defense against herbivores, pests and pathogens. The role of secondary metabolites in defense may involve deterrence/anti-feedant activity, toxicity or acting as precursors to physical defenses" (Bennett, 1994; Li *et al.,* 2010).

The amount of different allelopathic compounds produced by plants is enormous. Like synthetic herbicides, these allelochemicals can show species selectivity and can be applied to neighbouring plants by several mechanisms. Root exudation produces allelopathic compounds that are actively secreted directly into the soil rhizosphere by living root systems. The allelochemicals then move through the soil by diffusion and come into contact neighbouring plants. This creates a radius effect, where proximity to the allelopathic species results in greater concentrations of the allelochemical, which in turn typically decreases the growth of neighbouring plants. Compounds can also be leached off the roots and leaves into the soil profile by sufficient water application. Allelochemicals can also be given off though the air as volatile compounds, which can come into contact with neighbouring plants and cause different biological and physiological alterations.

"The diversity of allelochemicals produced by plants is vast, and the chemicals can range in structure from simple hydrocarbons to complex polycyclic aromatics" (Weston, 1996). This variation in sizes and structures of the allelochemicals produced for allelopathy allow for these compounds to have varying modes of action or attack strategies, which dictates what type of physiological impact they will have on neighbouring plant species. How allelopathic compounds interact in the soil depends on the turnover rate of allelochemicals and how they interact with the clay, organic matter and other soil properties (Weston and Duke, 2003; Weston, 2005).

Soil properties impact the movement of allelochemicals in the rhizosphere, which can alter their impact to neighbouring plants. These naturally produced secondary compounds can have chemical structures as complex as synthetic herbicides, they can also have the same wide range of selectivity and control for weeds. Knowing what species produce what type of allelopathic compounds is needed in order evaluate what pest species they will control. "Because all plant species produce a variety of organic compounds that may be released into the environment as exudates, leachates, or volatiles, allelopathy may be a widespread phenomenon" (Lau, 2007). With the wide range of secondary metabolites that plants produce, as well as the variety of release mechanisms used to distribute these compounds, these combinations lead to numerous defense or attack strategies that allelopathic species can use.

Using tobacco Inderjit *et al.* (2009) showed the first report demonstrating that the natural release of a volatile chemical inhibits the growth of neighbouring siblings. Last, allelopathic crop residues can be used as mulch or a cover crop, which can release allelochemicals from their decaying plant tissues. These compounds are released from decaying plant matter and end up in the soil rhizosphere where they move by diffusion. With these few mechanisms presented above you can see the great variations that exist for dispersing allelopathic chemicals in different species. Having knowledge of these release mechanisms is crucial if we are to effectively utilize allelopathic compounds for weed control in agriculture. Along with having a variety of chemical release mechanisms, the number of different compounds that allelopathy can include is enormous.

The mode of action of a chemical can broadly be divided into a direct and an indirect action (Rizvi *et al.,* 1992). Effects through the alternation of soil properties, nutritional status and an altered population or activity of microorganisms and nematodes represent the indirect action. The indirect action involves the biochemical on various important processes of plant growth and metabolism. Processes influenced by allelochemicals -

- **Mineral uptake**: Allelochemicals can alter the rate at which ions are absorbed by plants. A reduction in both macro –and micronutrients are uncounted in the presence of phenolic acids (Rice, 1974).
- **Cytology and ultrastructure**: A variety of allelochemicals have been shown to inhibit mitosis in plant roots (Rice, 1974).
- **Phytohormones and balance**: The plant growth hormones indole acetic acid (IAA) and gibberellins. IAA is present in both active and inactive forms, and is inactivated by IAA-oxidase.

IAA-oxidase is inhibited by various allelochemicals (Rice, 1974) other inhibitors block GA- induced extension growth.

- **Membranes and membrane permeability**: Many biological compounds exert their action through change in permeability of membranes. Exudation of compounds from roots on root slices have been used as an index of permeability because plant membranes are difficult to study (Harper and Balke, 1981).
- **Photosynthesis**: Photosynthetic inhibitors may be electron inhibitors or uncouples, energy-transfer inhibitors electron acceptors or a combination of the above (Einhellig and Rasmussen, 1979; Patterson, 1981).
- **Respiration**: Allelochemicals can stimulate or inhibit respiration both of which can be harmful to the energy producing process (Rice, 1974).
- **Protein synthesis**: Studies utilizing radio labeled C14 sugars or amino acids and traced incorporation of the label into protein, found that allelochemicals inhibit protein synthesis (Rice, 1974).
- **Specific enzyme activity**: Rice (1984) reported on a number of allelochemicals that inhibit the function of enzymes in the plant.
- Through conducting tissue (Rice, 1974).
- Maintaining Water relations (Rice, 1974).
- Through Genetic material (Aldrich, 1984).

Under natural conditions the action of allelochemicals seems to revolve round a fine tuned regulatory process in which many such compounds may act together on one or more of the above processes (Muller, 1966; Rizvi *et al.,* 1992). Factors affecting production of allelochemicals according to the environmental conditions to which they are exposed. Stress has a marked effect on the production of allelochemicals. Allelopathy is strongly coupled with other stresses of the crop environment, including insects and diseases, temperature extremes, nutrient and moisture variables, radiation and herbicides. These stress conditions often enhance allelochemicals production, thus increasing the potential for allelopathic interference (Einhellig, 1996).

According to Aldrich (1984) and Rice (1984), a variety of environmental conditions influence the variety of chemical produced:

- **Light stress:** Some allelochemicals are influenced by the amount, intensity and duration of light. The greatest quantities are produced during exposure to ultraviolet and long day photoperiods. Thus, under story plants filter out the
- **Minerals stress:** More allelochemicals are produced under conditions of mineral deficiency.
- **Drought stress:** Under these conditions more allelochemicals are produced.

- **Temperature stress:** In cooler temperatures, greater quantities are produce. The location within the plant and effect in specific allelochemicals seems to be variable.

There are also numerous other factors influencing the production of allelochemicals. The type and age of plant tissue during extraction is important since compounds are not uniformly distributed in plants. Production differs between species as well as within species.

Aldrich (1984) stated that environmental conditions that restrict growth tend to increase the production of allelochemicals. One could postulate that allelopathy may frequently be an accentuation of competition although not part of competition. It stress from competition increases the quantities of allelochemicals produced, it is conceivable that allelochemicals will inhibit the growth of some species and not others, thereby reducing the ability of the affected species to complete. The allelopathy plant and those affected by them are part of the ecosystems. If one factor changes, the change will also occur in some other factor. For example, light can be expected to interact with temperature and indirectly with soil moisture and other factors. Much of evidence indicates that several chemicals are released together and chemicals are released together and may exert toxicities in an additive or synergistic manner. Sometimes the allelopathic effect will be obvious and asserting, but in the majority of cases the effects are stable and thus more difficult to access.

Problems Studying Allelopathy

When monitoring the influence of allelopathic compounds on other plants it is difficult to separate the impact caused from competition versus the impact due to allelopathic interactions. This predicament was known as early as 1983, when Fuerst and Putnam proposed that "proof of competitive interference" includes:

(1) Identification of the symptoms of interference;

(2) Demonstration that the presence of the agent is correlated with reduced utilization of resources by the suspect;

(3) Demonstration of which resource(s) depleted by the agent are limiting resources; and

(4) Simulation of that interference (in the absence of the agent) by reduction of the supply of resources to levels that occur during interference

(5) Isolation, assay, characterization, and synthesis of the toxin;

(6) Simulation of the interference by supplying the toxin as it is supplied in nature; and

(7) Quantification of the release, movement, and uptake of the toxin.

It would be desirable but not essential to show that "the selectivity of the toxin to various species corresponds to the range of species affected by the allelopathic agent" (Fuerst and Putnam, 1983). From these lists of proofs for competition versus allelopathic interference you can see the problem with separating out these effects when conducting allelopathic research. "Allelopathy can play an important role in structuring plant communities, but allelopathic effects are often difficult to detect because many methods used to test for allelopathy can be confounded by experimental artifacts" Lau and others describe how in allelopathic studies, the use of activated carbon for neutralizing allelochemicals can actually affect nutrient availability and therefore impact plant growth. Finding proper techniques that can separate out chemical interference versus competition is the main dilemma when studying and evaluating the impact of allelopathic species (Lau *et al*, 2008).

The action of allelochemicals in target plant is diverse and affects a large number of biochemical reactions resulting in modifications of different physiological functions. Thus the results of allelochemical action can be detected at different levels of plant organization: molecular, structural, biochemical, physiological and ecological. Enzyme activities, cell division and ultrastructure, membrane permeability, ion uptake and as a consequence plant growth and development are modified by allelochemicals. Significant effects on photosynthesis and respiration are the best-characterized results of allelopathic interactions. Moreover allelopathic compounds seem to induce a secondary oxidative stress expressed as enhanced free radical production and induction of cellular antioxidant system. Plant survival under allelopathy stress conditions depends on plant defense leading to allelochemical detoxication, the process which may go on in parallel to cell defense reaction to oxidative stress (Gniazdowska, and Bogatek, 2005). Allelochemicals can also lead to increased cell membrane permeability. Consequently, cell contents spill and there is increased lipid peroxidation. Finally, there is slow growth or death of plant tissue. In addition, allelochemicals can also inhibit plants from absorbing nutrients from surroundings and affect the normal growth of plants.

A series of physiological and biochemical changes in plants induced by these compounds are shown in the figure on next page.

The impact of allelochemicals on the respiration of plants has mainly been shown to involve weakened oxygen absorption capacity, while the impact on photosynthesis has mainly been to reduce the chlorophyll content and photosynthetic rate (Patterson, 1981; Yu *et al.*, 2003). The allelochemicals amended the soil by inducing the stress conditions affecting the plant growth in general. These allelochemicals may be phenolics, as ferulic, coumaric or syringic acids (Blum and Shafer, 1988; Blum, 1998; Blum *et al.,* 1999; Li *et al*, 2010). These are reported in general to reduce the crop growth and yield by affecting the uptake of Ammonium ions by the developing plants (Bergmark *et al.,* 1995).

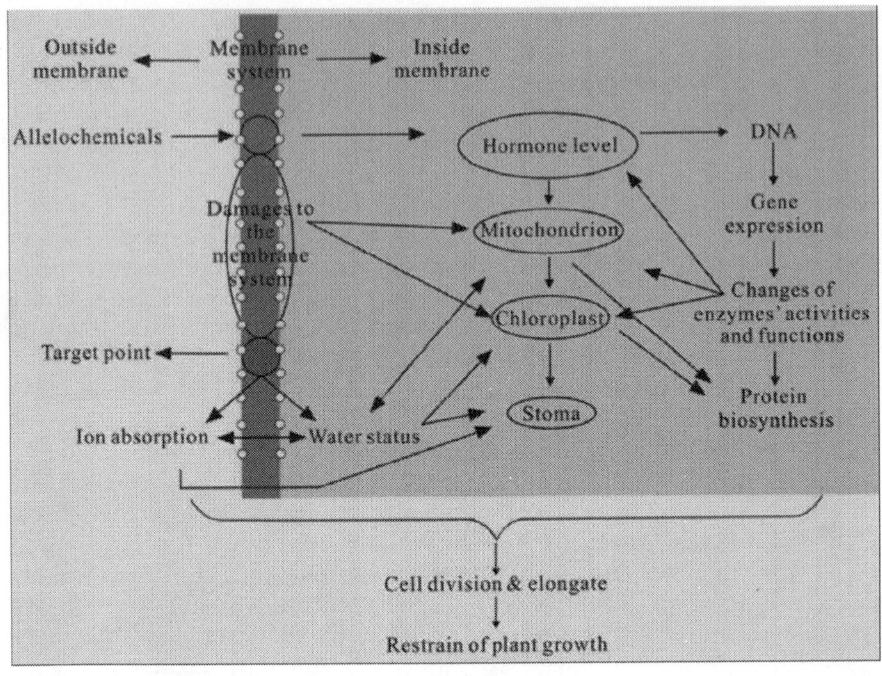

Fig. 1. Mechanism of action of allelochemicals (Wang *et al.,* 2006)

Techniques and Methodology in Allelopathic Research

Kohli *et al.* (1993) developed a mathematical model to predict tissue response to parthenin, an allelochemicals produced by Parthenium weed. Thijs *et al.* (1994) developed an elegant experimental design to separate competition and allelopathic interactions. They grew a target plant with varying densities of neighbouring plants in a pot. They then applied a toxin, atrazine to the pots and harvested the plants after a month. Interestingly they found that target plants with increasing neighbor densities actually increased growth. Finally, great progress has been identified. Flux rates of allelochemical production in the field can now be studied by using ion exchange resins to extract nutrient from soil (Weidenhammer, 1996). Perhaps unsurprisingly, there have been coincidental changes in the flora of intensively managed arable farmland (Marshall, 2001; Marshall *et al.,* 2001).

In association with the advances in agriculture, there is also evidence with farmland have shown marked reduction in range and population size over the past 30 years. Weeds are an important constraint on yield in most crops across the world. The global pesticide market was valued at $ 29 billion in 2000 (CPA, 2002) divided approximately between herbicides 48%, insecticides 27%, fungicides 19% and other products 6%. Improved crop management techniques including herbicides, have resulted in good control of weeds and have facilitated different cropping patterns and steadily increasing crop yields.

It is clear that some broad changes have occurred over the past 75 years in the arable plant communities. Allelopathy also plays an eminent role in the intra-specific and inter-specific competition and may determine the type of inter-specific association. Allelopathic potential has been reported for a number of weed species (Einhellig and Rasmussen, 1973; Shaukat *et al.,* 1985; Ahmed and Wardle, 1994; Seigler, 1996; Burhan and Shaukat, 1999; Rebaz *et al.*, 2001; Shaukat and Siddiqui, 2002; Kong *et al.,* 2002a, b; Tajuddin *et al.,* 2002; Marshall *et al.,* 2003; Shaukat *et al.*, 2003; Golisz *et al.,* 2008; Sharma *et al.*, 2009; Sangeetha and Baskar, 2015). A preliminary study has been made on systems engineering-based method for the evaluation of allelopathic potential in crops and its application by Sheng-peng and Yong-qing (2007).

Weeds Allelopathy and Agricultural Uses

Weeds are always considered as unwanted plants (Harlan and deWet, 1965). They compete with agricultural crops for nutrient, light and moisture and harbour many diseases and insects. There are many weed species that are allelopathic in nature. It is a viable weed management strategy but needs to be extensively studied under laboratory as well as in the field conditions. It is a natural and environment friendly technique, which may prove an effective strategy for weed management and thereby increase crop yields. In ancient Indian literatures, it is mentioned that every plant on this earth is useful for human beings, crops and animals. Weeds are one of the major problems in crop production around the world, and we are trending toward controlling these weeds with herbicides, which comes with increased environmental impact. Cultural practices and chemicals control the weeds in wheat crop. Cultural practices are weather dependent, while chemicals are not environmentally safe (Kassasion, 1971) and may affect the nutritive value of some crops (Saghir and Bhatti, 1970). Putnam and Duke (1974) first explored the possibility of utilizing allelopathic crops to dominate or inhibit weed growth in agricultural sites. The effect of weeds on crops, crops on weeds and crops on crops has invariably been emphasized (Klein and Miller, 1980). Weeds are enemies to the crop plants and have harmful effects on agricultural crops due to several factors such as competition for space, light and nutrients. The production of allelochemicals in crop plants and their release into the soil could influence the germination and growth of plant species (Rice, 1984).

According to Aldrich (1984), weeds interfere with crops in two ways:

- Inhibiting germination and seedling establishment.
- Inhibiting growth of the crop.

Organic chemicals released as leaf leachates, affect the crop plants. Weeds species are considered as rich source of secondary metabolites (Allelochemicals) and these chemicals improve a certain kind of environmental system on other plants growing in their vicinity and the phenomenon known as allelopathy (Nandal *et al.*, 1994). At this point they even had the idea of incorporating allelopathic potential into crops as well as using them as rotational crops, cover crops, or intercropping them for weed control in agriculture. If we could find a way successfully integrate allelopathy into agriculture we could reduce our herbicide use as well as gaining other benefits from an allelopathic crop. "*One common way that allelopathy may be utilized in weed management systems is through the manipulation of allelopathic cover crop residues in annual and perennial cropping systems*"(Weston, 1996).

Narwal (1994), the negative allelopathic effects on crops is used to develop green growth promoters. Most plants exhibit allelopathic effects on seed germination, growth and development of other plants by releasing allelochemicals into the soil, either as exudates from living organs or by decomposition of plant residues (Peters and Zam, 1981; Bansal, 1988; Smith and Martin, 1994; Narwal, 1999; Ahn and Chung, 2000; Sasikumar *et al.,* 2001; Singh *et al.,* 2002; Singh *et al.,* 2003a,b; Tawaha and Turk, 2003; Nouguchi and Ino, 2005; Singh *et al.,* 2005a,b; Sampietro and Vattuone, 2006; Bogatek *et al.,* 2006; Batish *et al.,* 2001, 2002a, b, 2006, 2007; Machado, 2007; Meksawat and Pornprom, 2010).

Multiple physiological effects have commonly been observed from treatments with many allelochemicals. The effects include decrease in plant height, absorption of water and mineral nutrients, ion uptake, leaf water potential, shoot turgor pressure, and osmotic potential, dry matter production, leaf area expansion, photosynthesis, chlorophyll accumulation, protein synthesis, reduction in stomatal aperture etc., in different plants (Einhellig and Kaun, 1971; Datta and Sinha-Roy, 1973, 1975; Patterson, 1981; Bhowmik and Doll, 1982, 1984; Einhellig *et al.,* 1970, 1985; Gerald *et al.,* 1992; Inderjit, 2001; Singh and Thaper, 2002; Inderjit and Duke, 2003; Verma and Rao, 2006; Sisodia and Siddiqui, 2009). Results obtained so far clearly demonstrate that some of the findings on allelopathic control of weeds, elimination of deleterious allelopathic effects of crops on crops, or exploitation of beneficial interactions in a rotation or mixed cropping system have an effect on seed germination, seedling growth and a direct bearing on crop production (Stoller, *et al.,* 1979; Rizvi, *et al.,* 1992; Qasem, 1993, 1995, 2001; Mehboob *et al.*, 2000; Ebana *et al.*, 2001; Gniazdowska and Bogatek, 2005; Hegab *et al.,* 2008; Kamal and Bano, 2008).

The only need is to identify of its uses. We can consider the weeds are useful plants. Many studies conducted at **Department Agronomy, Indira Gandhi Agricultural University, Raipur (India)** have clearly revealed that stimulatory allelopathic effects of weeds on crops can be utilized successfully for higher production (Oudhia, 1996; Lal and Oudhia, 1999). Weeds compete for light, nutrients, moisture and space with the crop and thus cause severe losses to yield. Losses in wheat due to range 17-25% (Shad, 1987) and in monetary terms it may be as high as Rs.28 billion (Hassan and Marwat, 2001). These effects are selective, depending upon the concentrations and residue type, either inhibitory or stimulatory to the growth of companion or subsequent crops or weeds (Kimber, 1973; Hall *et al.,* 1982; Bhowmik and Doll, 1984; Einhelling *et al.,* 1985; Purvis *et al.,* 1985; Cheema, 1988; Khan and Vaishya, 1992; Naseem, 1997; Naseem *et al.,* 2003; Cheema *et al.,* 2004; Jalili *et al.,* 2007). Sorghum is known to exude Sorgoleone an allelochemical from its living root system, which helps it out to compete with neighbouring weed species (Bertin *et al.,* 2007). One mechanism explaining the success of invasive weeds may be the production and release of allelopathic compounds by the invader that due to a lack of co-evolutionary history, have effects on plant neighbours in the introduced range (Abhilasha *et al.*, 2008).

In ancient Indian literatures, it is mentioned that every plant on this earth is useful for human beings, crops and animals. Allelopathy may be used as a tool in weed management by applying the residues of allelopathic weeds or crop plants as mulches, growing them in successions and leaving their residues in the field (Altieri and Doll, 1978; Drost and Doll, 1980; Putnam and DeFrank, 1979). Allelopathy is a novel approach to keep the environment safe and to develop sustainable agriculture (Yongqing, 2005). Among the various alternatives, use of natural plant products, offers a new approach for the management of noxious weeds and pests in a sustainable manner. Many members of Family Asteraceae (Compositae) have allelopathic potentials and the only need is to exploit them (Chon and Nelson, 2010; Ilori *et al.,* 2010). Differential effects of aqueous extracts of many members of Asteraceae have been studied by many workers (Stachon and Zindel, 1980; Schon and Einhellig, 1982; Eze and Gill, 1992; Inderjit and Dakshini, 1995; Ratwat *et al.,* 2002; Chon *et al.*, 2003; Bogatek *et al.,* 2006; Mulatu *et al.,* 2006; Batish *et al.*, 2007; Ilori *et al.*, 2007; Javed and Asghari, 2008; Javaid *et al.*, 2009). One of these plants well known for its allelopathic properties is ***Parthenium***.

Parthenium is of particular concern because of its invasive and allelopathic properties:

1. It is an aggressive colonizer that gets established in natural and manmade ecosystems, grassland habitats, open woodlands, river banks, flood plains, wildlife parks, open fields of settlements areas, bare areas along roadsides, crop fields, gardens, heavily stocked areas around yards and watering points etc. (McFadyen, 1992; Chippendale and Panetta, 1994; Tamado and Milberg, 2000; Pandey *et al.,* 2003; Tiwari *et al.,* 2005);
2. It produces large number of seeds and thus, it has the potential to become widespread;
3. The allelochemicals released directly from the weed or from seed leaching inhibit germination of other plants and the growth of pasture grasses, legumes, cereals, vegetables, other weeds, and even trees (Oudhia, 2000a,b; Batish *et al.,* 2002a,b; Singh *et al.,* 2003a, 2005a);

4. It induces changes in the physical and chemical properties of soil such as soil texture, soil pH, soil organic matter, soil nitrogen, soil potassium, soil phosphorus etc. (Batish *et al.,* 2002c; Bhowmik *et al.,* 2007);
5. It is unpalatable to herbivores, it replaces rangelands palatable grasses and adversely effecting animal's health damaging milk and meat quality (Tudor *et al.,* 1982; Kohli and Rani, 1994; Anonymous, 2000; Ayele, 2007; Wegari 2008);
6. Manual removal is difficult because it may cause dermatitis, hay fever, asthma, and allergic and even death in humans (Navie *et al.,* 1996; Evans, 1997; Oudhia, 2001; Bhowmik *et al.,* 2007).
7. It is a very strong contender to be recognized as phytohormones and awaits elucidation of its molecular level.

PARTHENIUM

Classification

Kingdom :	Plantae
(unranked) :	Angiosperms
(unranked) :	Eudicots
(unranked):	Asterids
Order :	Asterales
Family :	Asteraceae
Subfamily :	Asteroideae
Tribe :	Heliantheae
Genus :	***Parthenium***
Species :	***hysterophorus***

Parthenium hysterophorus L. is one of the ten worst weeds in the world. It is an obnoxious weed. It is an annual herb that aggressively colonizes and disturbs the sites. This herbaceous weed species is a native of Mexico, and has become widespread in a number of tropical and subtropical countries. The origin of this serious weed is South of USA, Central America and West Indies (Hind *et al.*, 1993). Since its first reported occurrence in India (Pune) in the early 1950s, and a little later in Australia, much has been written about this weed. Indeed, the CAB ABSTRACTS Database contains more than 400 references on Parthenium from 1985 till present. There have been a number of recent reviews on this weed in both Australia and India. One of the most comprehensive, being that the service by Navie *et al.* (1996) and this should be consulted for basic details of the biology and ecology of Parthenium. It spreads easily through trade as contaminants of grain and other crop products and by means of farm machineries (Mack and Lansdale, 2001).

Weed Status

The weed has achieved major weed status in India and Australia only within the last few decades (Navie *et al.,* 1996; Evans, 1997; Mahadevappa, 1997; Kaushik *et al.,* 2005; Bhowmik *et al.,* 2007). It is prevalent in fertile and marshy land and shady places. It is also observed on the roadsides and wasteland. It can tolerate drought condition (Dhawan and Dhawan, 1996). Parthenium weed, as it is commonly known as congress grass, as it is uniquely called in India, where the flower shape has been fancifully compared to the congress building in New Delhi. It has achieved major weed status in both these countries within a relatively short period. For example, it is not included in the weeds of Australia (Lamb and Collet, 1976) and, indeed, not mentioned in the classic treatise on the World's Worst Weeds.

Present Distribution in the World

Parthenium is native to subtropical areas in South and North America, but currently widely distributed in other tropical and subtropical countries such as Australia, India, China, Kenya West Indies, Australia, Ethiopia, Israel, Taiwan, India, Nepal, (Hara *et al.,* 1982; Picman and Picman, 1984; Peng *et al.,* 1988; Mishra, 1991; McFadyen, 1992; Medhin, 1992; Navie *et al.,* 1996; Evans, 1997; Tiwari *et al.,* 2005; Shabbir and Bajwa, 2006; Bhowmik *et al.,* 2007). It was first introduced in the southeast Queensland of Australia in 1955 by aircrafts and machinery parts during the Second World War. The weed was introduced a second time in central Queensland in 1958 by contaminated pasture seeds (Navie *et al.,* 1996). It has changed native habitat in Australian grasslands, open woodlands, riverbanks and floodplains (McFadyen, 1992; Chippendale and Penetta, 1994). It is considered as one of the noxious weeds of Australia (Haseler, 1996; Persons and Cuthbertson, 1992; Stephen and Sowerby,

1996; Adkins *et al.*, 1997). In South Africa, Parthenium was reported during the 1880s, but it becomes troublesome only since 1980s (Henderson, 2001). Parthenium weed was imported to Ethiopia from North America during the 1988 famine in Ethiopia as contaminants of grain food aid and distributed with grain (Murphy and Cheesman, 2006). In Alamata of Ethiopia alone, about 10,000 hectares of land has been invaded with Parthenium (Bezabih and Araya, 2002).

Parthenium (*P. hysterophorus* L.) is an exotic weed, accidentally introduced in India in 1955, through the imported food grains (Rao, 1956; Bhowmik and Sarkar, 2005), but the earliest record of this species goes back to 1814 by William Roxburgh, 'The Father of Indian Botany' (Paul, 2010). Since 1956, the weed has spread like wildfire throughout India. Parthenium due to its invasive capacity destroyed its natural ecosystems in National Wildlife Parks southern of India its invasiveness have been observed (Evans, 1997). Apart from its competitive ability, the invasiveness of this noxious weed is thought to be due to an ability to displace other species by means of allelopathy. Parthenium thought to be initially introduced in India before 1950s with food grains imported from the USA, but heavily introduced after 1956 through the transport of Milo (red wheat) from Mexico (Wise *et al.,* 2007). After 1956, the weed spread throughout India, in all kinds of disturbed habitats including roadsides, pasturelands, cultivated areas, etc. Parthenium could grow in all States of Australia, but is unlikely to become a major problem in winter-rainfall areas. The plant grows on a wide range of soil types, ranging from sand to heavy clays, but favours the later.

Species Identification

Parthenium (locally known with Kanike jhar/Paati jhar/Bahudal jhar) identification was conducted using plant characteristics thoroughly described by Grierson and Long (2001) and Tiwari *et al.* (2005). It is known as false ragweed, star weed, bitter weed, white top, santa maria, feverfew, white head, chatak chandani, ramphool, gajarghas, congress weed, broom-bush, or carrot weed (Oudhia, 2001; Tiwari *et al.,* 2005). The weed looks similar to other introduced weeds from the same family (Anonymous, 2000), including seedlings of cobbler's pegs, leaves of *Artemisia* spp, flowers of bishop's weed (*Ammi majus*), hemlock (*Conium maculatum*) and the ragweed (*Ambrosia* sp.). Parthenium resembles a young *Chrysanthemum* in its early (rosette) stage. It has a deep tap root and erect stem that becomes woody with age. The long tap root system of this weed allows it to survive in adverse environment (dry conditions) and to achieve high resource exploitation in short time period (Navie *et al.,* 1996). Its leaves resemble those of *Artemisia* sp. they are grayish-green ranging in size from 5-15 x 0.4-5 cm, deeply lobed, pubescent, and acute with decurrent base. The petiole is 3-4 cm long. The radial spread leaves encroach more area preventing other seedlings to grow (Kumari and Kohli, 1987). Stems are hairy, with creamy-white flower heads about 4 mm in diameter 34, produced in clusters at the tip of each branch. Each flower head produces about 4 black seeds, which are approximately 2 mm in length with two thin, white, spoon-shaped appendages.

The inflorescence of the plant is a lax panicle of numerous small heads and peduncles 4 to 10 mm long. The heads of the inflorescence are radiate, 4-8 mm across. The receptacle is convex and small. There are five campanulate involucres, green in color, elliptic to ovate in shape, acute, 3-4 x 1-2 mm in size and pubescent. There are five ray florets with 2 seriate disc florets on either side. The corolla is white to light yellow, ovate, orbicular, blobbed, and obtuse. The style is 1 mm long with a stout, and the stigma is obtuse. Disc florets are numerous, with light yellow corolla, which is narrowly campanulate, 1.5-2 mm long, five-lobe, and ovate-acute in shape. The disc florets have five stamens each, with linear anthers. The style of the disc floret is 1-5 mm and is puberulous. Achenes are flattened in shape and 2 mm long. Pappus consists in two broad, strongly reflected awns, 0.3-0.5 mm long, and puberulous along the side.

Life Cycles

P. hysterophorus L. has two stages (Narwal *et al.*, 2003) :-

a) Juvenile or rosette stage

It has a rosette with large, dark green, simple, radicle, pinnetisect small leaves and flowering is absent. The large lower leaves are spread on the ground like a carpet, without allowing vegetation underneath it.

b) Adult Stage

It is procumbent (trailing along the ground but not rooting), profusely branched, leafy herb resembling a bush or shrub because of its height (1-2.5m). The stem becomes tough and woody as the plant matures into a hardy bush. Enormous numbers of pollen grains (624 million per plant) are produced by anemophilous (by wind) pollination. It is an extremely prolific seed producer with up to 25,000 seeds (achenes) per plant. The plant is thermo- and photoinsensitive; hence, it grows round the year except in severe winters; in other words, it survives environmental extremes. It is a rapid colonizer and competes out other vegetation in its vicinity within two growing seasons. It grows in almost all types of soil except near the seashore, as the saline soil is not conducive to Parthenium flowering.

Reproductive Efficiency

The growth and reproduction cycle of Parthenium depends on climatic conditions such precipitation, available moisture, soil and air temperature (Pandey and Dubey, 1988). Pollination to achene maturation takes about 14 days. It takes 30 days to mature the achene from floral bud

(Lewis *et al.*, 1988). A single plant of Parthenium can produce as much as 25,000 achenes and it is capable of establishing four successive generations at same site in a year under favourable temperature and moisture conditions (Pandey and Dubey, 1988). Life cycle is short and pollen dispersal is done mainly by wind (Lewis *et al.*, 1991).

Phytochemical Properties of *P. hysterophorus*

The allelopathic nature has been well-documented and water-soluble phenolics and sesquiterpene lactones have been reported from the roots, stems, leaves, inflorescence, pollen and seeds (Evans, 1997). **Parthenin (1)** the major sesquiterpene lactone of *P. hysterophorus* was isolated as early as 1959 (Herz and Watanabe, 1959) and its structure was provisionally assessed as a guainolide. Subsequently (Herz *et al.*, 1962), the structure was revised to that of a pseudoguaianolide (methyl group is at C-5 in pseudoguaianolides, while in guaianolides, it is at position C-4). In a follow-up study of a collection tentatively identified as *P. hysterophorus* from the highlands of central Mexico, hysterin (a C-15 oxygenated ambrosano-lide) was identified as the major lactone (Romo, *et al*, 1996). It has been suggested that this species is derived from an extant population of *P. bipinnatifidum* by way of long-range dispersal. In a biochemical study of the genus Parthenium, Rodriguez (1975) found that three sesquiterpene lactone structure types are common to the genus; the herbaceous members of the section Argyrochaeta, of which Parthenium is a representative species, elaborate amorosanolides.

Although the "parthenin race" is predominant throughout the above-mentioned area, a population of Parthenium from southern Texas was found to contain large quantities of **hymenin** (2) instead of parthenin (Fig. 2). It should be added that populations of *P. confertum* var. *microcephalum,* a taxon suggested to be of hybrid origin between *P. hysterophorus* from southern Bolivia and central Argentina contain hymenin as the major lactone; two novel and unidentified minor lactones are also present. The possibility that other chemical races exist in Central and South America, Asia or Africa.

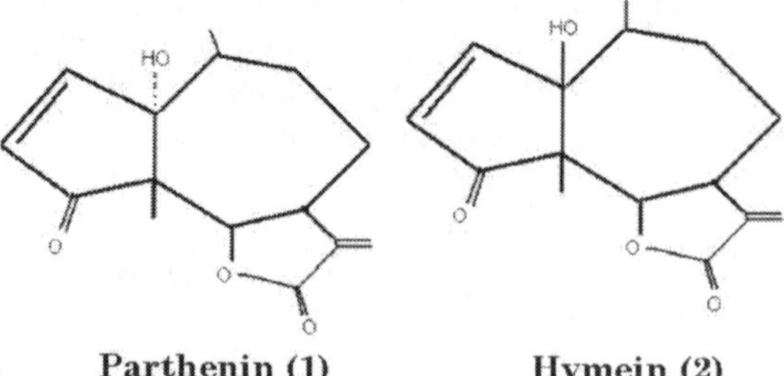

Fig. 2. Sesquiterpene lactones known from *P. hysterophorus* (Source: Rodriquez, 1975).

The distribution of parthenin (1) and hymenin (2), the C-1 hydroxy stereoisomer of parthenin, in populations of *P. hysterophorus* from the southeastern and southern United States, Mexico, Jamaica, Cuba, Trinidad, India, Bolivia and central Argentina is well studied. The presence of parthenin in the Indian populations suggests that *P. hysterophorus* was introduced from North America. Flavonoid and phenolic aglycone chemistry- A detailed investigation of flavonoids of *P. hysterophorus* from central Texas (Shen *et al.*, 1976) showed that quercetagetin 3, 7-dimethyl ether is a major flavonol. It is also a characteristic chemical marker for the genus Parthenium (Rodriguez, 1975). Small amounts of 6-hydroxyl kaempferol 3,7-dimethyl ether and glycosides, quercetin 3-O-glucoside, kaempferol 3-O-glucoside and kaempferol 3-O-arabinoglucoside are also present (Fig. 2a).

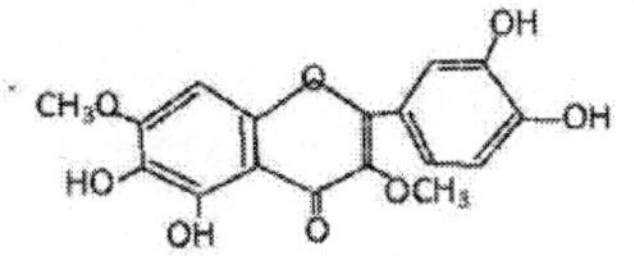

Quercetagetin 3, 7-dimethyl ether

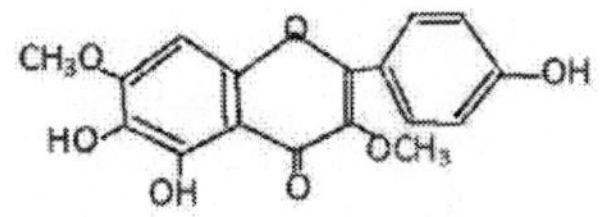

6- Hydroxyl kaempferol 3, 7-dimethyl ether

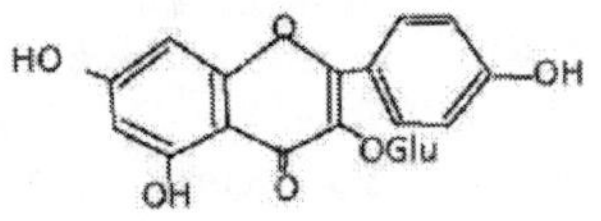

Kaempferol 3- O- Glucoside

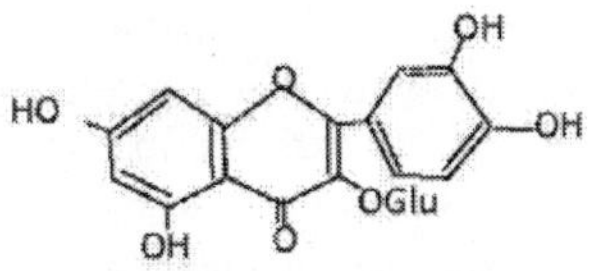

Quercetin 3- O- Glucoside

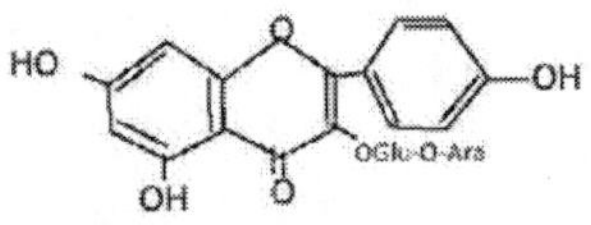

Kaempferol 3-O-Glucoarabinoside

Fig. 2a. Flavonoid and phenolic aglycone chemistry (Source: Shen *et al.,* 1976)

Chromatographic analysis of *P. hysterophorus* from various localities in Mexico and South America showed no qualitative difference in flavonoid constituents. Phenolic glycosides showed fluctuations in number and amounts depending on the collection site, but no consistent pattern was discernible (Towers *et al.*, 1977). Kanchan (1975) has found that fumaric acid accumulates in the stems and leaves and *p-anisic, p-hydroxybenzoic* and vanillic acids in the root while caffeic acid is present in all plant parts except the pollen. Paper chromatograms of hydrolyzed phenolic acid concentrates of Indian collection in our laboratories showed the presence of *p-coumaric acid* and ferulic acids in addition to caffeic acid.

Whole plant: Two types of sesquiterpene lactones, hysterin and dihydroisoparthenin have been isolated from plant *P. hysterophorus* (Picman *et al.*, 1982). Histamine (0.585%) is present in aerial parts of the plant (Kamal and Mathur, 1991). Syringaresinal has also been isolated from this weed (Das *et al.*, 1999). Three ambrosanolides: *α*-epoxymethylacrylyloxy parthenin, its 11*α*13-dihydro derivatives and 8*α*-eposeymethylacrylylos-ambrosin, have been isolated from chloroform extract of the aerial parts of the *P. hysterophorus* (Chhabra *et al.*, 1999). A normal sesquiterpenoid, charminarone (the first seco-pseudoguaianolide) has been isolated from the whole plant (Venkataiah *et al.*, 2003). Twenty-three compounds, representing 90.1% or more of the volatile oils, have been identified from *P. hysterophorus* (Pareek *et al.*, 2011). The toxic and inhibitory constituents contained by all parts (stem, leaves, leaf hair, flower, pollen grain) of *P. hysterophorus* are summarized in Table 1.

Table 1: Chemical constituents of *P. hysterophorus*

Main Group	Constituents	References
Terpenoids	**Sesquiterpene lactones**: germacranolides (including parthenolide, artemorin and chrysanthemonin) guaianolides (including chrysartemin A, partholide and chrysanthemolide) and eudesmanolides (including santamarin, reynosin and magnolialide), parthenin, cornopolin, artecanin, balchanin, costunolide, epoxyartemorin	Picman *et al.*, 1982; Patil and Hedge, 1988; Parsons and Cuthbertson, 2001; Pareek *et al.*, 2011; Batish *et al.*, 2007
	α-unsaturated γ-lactones: 3-β-hydroxyparthenolide, costunolide, 3-β-hydroxycostunolide, 8-α-hydroxyestafiatin, artecanin, two chlorine containing sesquiterpene lactones, 1-β-hydroxyarbusculin and 5-β-hydroxyreynosin.	Barnes *et al.*,2007
Flavonoids	Luteolin, apigenin, 6-hydroxykaempferol 3,6-dimethyl ether, 6-hydroxy kaempferol 3,6,4′-trimethyl ether (tanetin), quercetagetin 3,6-dimethyl ether, quercetagetin 3,6,3′-trimethyl ether (accompanied by isomeric 3,6,4′-trimethyl ether), quercetin, chrysoeriol, santin, jaceidin and centaureidin	Shen *et al.*, 1976; Pareek *et al.*, 2011;
Phenolic Derivatives	Caffeic, vanillic, ferfulic, chlorogenic and anisic acids	Parsons and Cuthbertson, 2001; Kanchan, 1975; Reinhardt *et al.*, 2009
Volatile oils (0.02–0.07%)	**Various monoterpene and sesquiterpene components**:e.g. camphor (56.9%), camphene (12.7%), p-cymene (5.2%), bornyl acetate (4.6%), tricylene, α-thujene, α-pinene, β-pinene, α-phellandrene, α-terpinene, γ-terpinene, chrysantheone, pinocarvone, borneol, terpinen-	Barnes *et al.*,2007; Pareek *et al.*,2011; Miranda *et al.*, 2014

Main Group	Constituents	References
	4-ol, ρ-cymen-8-ol, α-terpineol, myrtenal, carvacrol, eugenol, trans-myrtenol acetate, isobornyl 2-methyl butanoate, caryophyllene oxide, germacrene, farnesene and their esters, *trans*-ß-ocimene and ß-myrcene	
Amino acids	Rich in Glycine and proline and moderate amount with alanine and lysine	Gupta *et al.,* 1996
Amino sugars	N-acetylgalactosamine and N-acetylglucosamine	Gupta *et al.,* 1996
Others	8-*β*-Acetoxyhysterone C, Charminarone, 8*α*-Epoxymethylacrylyloxyambrosin, 8*α*-Epoxymethylacrylyloxy-11, 13-dihydroparthenin, 8*α*-Epoxymethylacrylyloxyparthenin, 2*β*-Hydroxycoronopilin, Hysterone (A, B, C, D), 1*α*, 2*β*, 4*β*-Trihydroxypseudoguaian-6*β*, 12-olide, Pyrethrin, tannins (type unspecified), melatonin, potassium chloride, protein.	Parsons and Cuthbertson, 2001; Barnes *et al.,* 2007; Zhou *et al.,* 2011

Leaves: Parthenin, hexacosanol, myricyl alcohol, â–sitisterol, campesterol, stigmasterol, betulin, ursolic acid, â-D-glucoside of â-sitosterol and saponin have been isolated from leaves of *P. hysterophorus*. The saponin on hydrolysis yield oleanolic acid and glucose. The aqueous extract of *P. hysterophorus* contains free amino acids, glucose, galactose and potassium choloride (4.8% Gupta *et al.*, 1977). Methoxypseudoguaia-nolides viz. 13-ethoxy-dihydroambrosin, 13-methoxydihydroparthenin and 2α, 13*α*-dimethoxy- dihydroparthenin have been isolated from leaves of this plant (Bhullar *et al.*, 1997). The leaves also contain parthenin, caffeic, chlorogenic, p-hydroxybenzoic, vanilic, salicylic, gentisic, neo-chlorogenic and proto-catechuic acids (The wealth of India, 2003). The allelochemical parthenin is sequestered at high level in capitate-sessile tirchomes on leaf surfaces of *P. hysterophorous* (Rainhardt *et al.*, 2005; Ambiye and Golatkar, 2010). Leaves contain about 5% parthenin (The wealth of India, 2003).

Flower: Methanolic extract of flower of *P. hysterophorous* contains several constituents such as 2â- hydroxycoronopilin, 8â-hydroxycoronopilin, 11-H,13-hydroxyparthenin, parthenin and coronopilin. Parthenin up to 8% is present in capitulum (The wealth of India, 2003). A new highly oxygenated pseudoguaianolides (8-â-acetoxyhysterone C), parthenin, coronopilin and hysterone C have also been isolated from the flowers (Ramesh *et al.*, 2003; Das *et al.*, 2005, 2007).

Root: Histamine (0.35%) is present in the roots of plant *P. hysterophorus* (Kamal and Mathur, 1991). The roots contain parthenin, caffeic, chlorogenic, phydroxybenzoic, p-anisic, vanilic, salicylic, gentisic, neochlorogenic and proto-catechuic acids (The wealth of India, 2003).

Parthenium is an aggressive weed and therefore poses a serious threat to the environment and biodiversity owing to its high invasion and allelopathic effect, which has the capacity to rapidly replace the native vegetation (Pandey *et al.,* 1993). Many efforts are being made to control the growth or to eradicate this weed by different researches (Jayachandra, 1971; Rajulu *et al.,* 1976; Francis, 1978; Rajendrudu and Rama Das, 1981; Singh, 1983; Syamasundar and Mahadevappa, 1986; Jayanth, 1986; McClay, 1987; Joshi, 1991, 2006; Dhawan and Dhawan, 1996; Mahadevappa, 1997; Dhileepan and McFadyen, 2001; Goyal and Brahma, 2001; Tomado *et al.*,2002; Tamado and Milberg, 2000, 2004; Dhileepan, 2009; Kushwaha *and* Maurya, 2012; Adkins and Shabbir, 2014; Guyana, 2014; Saini *et al.*, 2014; Shrestha, *et al.*, 2014; Tanveer *et al.*, 2015).

The aqueous foliar extracts of some trees and other herbs were used to check the germination and seedling growth of Parthenium (Singh and Rizvi, 1992; Dhawan and Dhawan, 1995; Sonkaran *et al.,* 1996; Sharma, 2003; Singh *et al.*, 2004). The allelopathic effects of *Amaranthus viridis* to suppress the growth of Parthenium have been studied (Athanassova, 1996; Thapar and Singh, 2003). Seed germination of Parthenium was inhibited by leaf leachates of *Azadirachta indica* and latex of *Calotropis procera* (Goyal and Singh, 2003). The survey conducted in India by Wahab (2005) has shown that species like *Cassia sericea, C. tora, C. auriculata, Croton bonplandianum, Amaranthus spinosus, Tephrosia purpurea, Hyptis suaveolens, Sida spinosa* and *Mirabilis jalapa* could suppresses Parthenium in natural habitats. The fresh weight and biomass of *Parthenium hysterophorus* decreased significantly by *Cassia occidentalis, Calotropis procera* and *Azadirachta indica* (Knox, 2013). Three C_4 grasses (*Setaria incrassata, Astrebla squarrosa,* and *Bothriochloa decipiens*) and one C_3 legume (*Clitoria ternatea*) have also been shown to suppress the growth of C_3 species Parthenium (Khan *et al.,* 2014).

Parthenium Allelopathy

The effects of Parthenium on different plants can be discussed under the following heads:

1. Regulation of growth with Parthenium extracts
2. Regulation of productivity with Parthenium extracts
3. Biochemical regulation with Parthenium extracts
4. Response of crops to Parthenium compost
5. Biocontrol value
6. Green manure value
7. Soil amendment value
8. Biocidal value
9. Human Sensitivity
10. Phytoremediation
11. Pharmaceutical Uses

REGULATION OF GROWTH WITH PARTHENIUM EXTRACT

Seed Germination and Seedling Growth

Further, Rice (1974) observed that many species of weeds produce toxins that are inhibitory to other weeds and often to themselves. It has been reported that the forage production of Indian croplands and grasslands was reduced by 40% and 90%, respectively due to the invasion of this weed (Khosla and Sobti, 1981). Pandey *et al.* (1993) reported that Parthenium suppress the germination of seeds of associated species, through the release of allelochemicals from decomposing biomass and root exudates. Parthenium competes directly with pasture species, reducing pasture vigour and seed set leading to habitat and ecosystem change (Evans 1997; O'Donnel and Adkins 2005; Shabbir and Bajwa 2006). Parthenium have allelopathic effect on neighbouring flora by which it is capable to replace most of the associated herbaceous species (Rajan, 1973; Yadav and Chauhan, 1998; Bhowmik *et al.,* 2007). With the increasing concentration of Parthenium extracts the seed germination and growth of Eragrostis decreased significantly (Tefera, 2002). Based on a study conducted in the rangelands of Northern Himalayas (India), Kohli *et al.* (2004) found that invasion by Parthenium, significantly decreased species richness in the invaded areas. In similar ways, studies in Australia and India have also demonstrated that Parthenium adversely affects the composition and diversity of species thereby resulting displacement and imbalance in natural and agricultural system (McFadyen, 1992; Chippendale and Panneta, 1994; Sakai *et al.,* 2001; Grice, 2006).

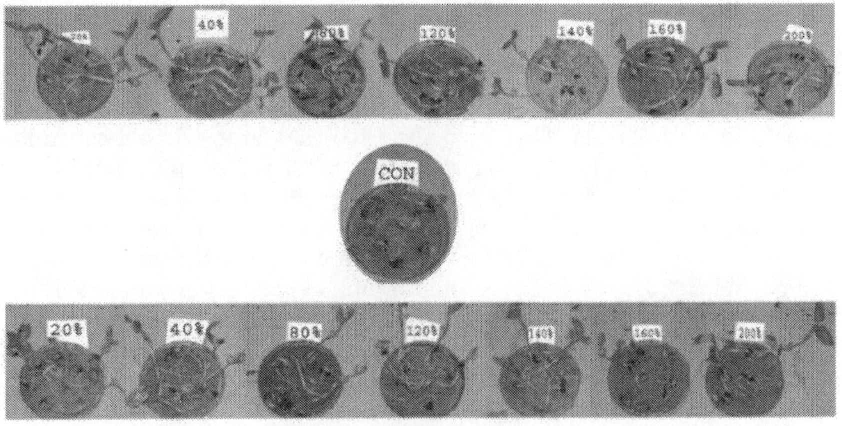

Photo Plate 1. Effect of different parts of Parthenium extracts on Seed germination and seedling growth in Cowpea at 11 days of soaking (Source: Srivastava, 2011).

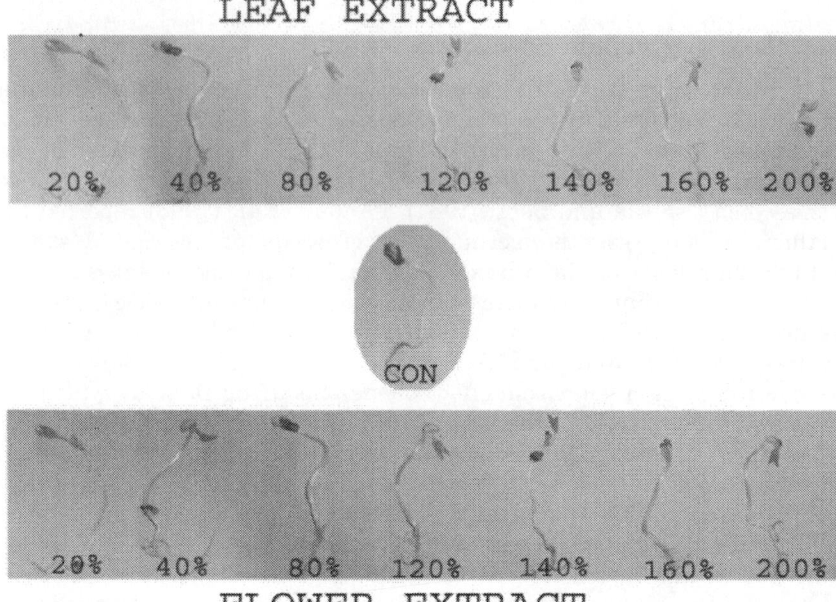

Photo Plate 2. Effect of different parts of Parthenium extracts on seedling growth in Cowpea at 11 days of soaking (Source: Srvastava, 2011)

Germination and seedling growth inhibition was observed in barley and rice (Srivastava *et al.*, 1985; Singh *et al.*, 1992; Rashid *et al.*, 2008; Pandey *et al.*, 2011). Complete failure of seed germination was observed in rice seeds with leaf extract of Parthenium (Singh and Sangeeta, 1991; Maharjan *et al.*, 2007). The lower concentrations upto 80% of RE and SE, upto 60% of LE and FE initiate early germination as compared to control (Srivastava *et al.*, 2011). They also observed that in untreated control set, the maximum germination was recorded (80%) till 9th day, while under treated sets, the maximum germination reached till 3rd, 5th or 7th day. Under the treatment of RE, the maximum germination recorded (100%) with 20% concentration till 7th day and with 40 and 60% concentrations on 3rd day. With 80% of RE, the maximum germination was 80% (non-significant difference with control), but till 3rd day as compare to control (9th day), while 100% of RE showed 80% germination till 5th day. Under SE treatment, the early and maximum germination (100%) was recorded with 20% concentration till 3rd day. Other concentrations i.e. 40 and 60% of SE showed 90 G% and 80% of SE showed 80 G% (non-significant) till 3rd day as compared to control (9th day). Higher concentrations of SE from100 to 200% reduced the G% significantly over control (Photo Plate 1). Under LE treatment only with 20% concentration, G% reached upto 100% till 3rd day, while other concentrations were either non significant or significantly reduced G%, but the maximum germination reached till 5th day. Under FE treatment also 20% concentration showed 100% germination till 3rd day as compared to control (80% G% till 9th day; Photo Plate 2).

Parthenium inhibited the germination and seedling growth of barley, wheat and peas. Kohli *et al.* (1985) suggested that two allelochemicals acting synergistically were responsible for the significant decrease in seed germination and subsequent growth of cabbage, when placed in leaf and inflorescence leachates from Parthenium weed. Patil and Hedge (1988) isolated and purified parthenin from leaves Parthenium and demonstrated that this compound significantly decreased the germination of wheat seeds and adversely affected seedling growth. Nath (1988) studied the effect of Parthenium extract on germination and seedling growth of crops.

Negative allelopathic effects of Parthenium on cultivated plants, like-*Brassica campestris, B. oleracea* and *B. rapa, Glycine max, Lolium multiflorum, Oryza sativa, Phaseolus vulgaris, Raphanus sativus, Cicer arietinum, Triticum aestivum, Vigna radiate, Vigna mungo, Cajanus cajan, Zea mays, Alysicurpus glumaceae*, lettuce, etc., have been well documented (Hussain and Abidi, 1991; Kil and Yun, 1992; Oudhia *et al.*, 1997a; Bajwa *et al.*, 1998; Yadav and Chauhan 1998; Sinha and Deo 1999; Oudhia, 2000a, b, c; Batish *et al.*, 2002 a, b, c, 2005; Singh *et al.*, 2003a; Singh *et al.*, 2005 a, b; Wakjira *et al.*, 2006; Paudel 2007; Maharjan *et al.*, 2007; Gupta, 2008; Patel, 2011; Msafiri *et al.*, 2013; Babu *et al.*, 2014). Effect of Parthenium on radish (Mehta *et al.*, 1995); okra, chilli peppers and clover (Dhawan and Dhawan, 1995), have demonstrated that the germination and yields of traditional Indian pulse crops (guar, black and green gram) were reduced when these were grown in soils previously infested by Parthenium weed. An experiment conducted on allelopathic effect of Parthenium leaf extract on sunflower and sorghum revealed that the germination percentage, shoot and root length, dry weight and vigour index decreased with an increase in the concentration of Parthenium leaf extract from one to 10% (Murthy *et al.*, 1995). Similar results showed with allelopathic effects on germination and seedling vigour of Kodo (Oudhia and Tripathi, 1998).

In agriculture, the inhibitory effect of weed species on germination and growth of crops has been attributed to phytotoxic chemicals released from the leaf litter and roots. In this context, a study was undertaken to elucidate the effects of few common weed species on germination behaviour, root and shoot growth of some of the paddy crops (Krishna *et al.*, 2007).

Swaminathan *et al.* (1989) in their studies also observed the positive effect of leaf extracts on radical growth. The allelopathic effects have been shown with foliar leachates of *P. hysterophorus* on a diverse range of agricultural and tree crops: cowpea, sunflower, *Casuarina, Acacia, Eucalyptus* and *Leucaena* (Swaminathan *et al.,* 1990); rice, wheat, black gram and chickpeas (Singh and Sangeeta 1991); green gram and wheat (Agarwal and Anand, 1992); barley and *Cassia tora* (Singh *et al.,* 1992);

mung beans and guar (Kohli and Rani, 1992); various species of Indian forage crops, pulses and oil seeds (Aggarwal and Kohli, 1992); sorghum (Ayala *et al.,* 1994); maize, ragi (*Eleusine coracana*) and soybeans (Bhatt *et al.,* 1994); sunflower, french beans and cotton(Madhu *et al.,* 1995). Pandey (1994) reported that the relative effects of plant residue on growth of Salvinia and paddy seedlings. The lower concentrations of Parthenium root and stem extract (10 and 20%) enhance seed germination and improves seedling growth parameters in bottle gourd and Brinjal (Singh *et al.,* 2008; Singh, 2010). The studies also revealed that the different extracts of Parthenium were effective with lower concentrations to increase the Emergence Index and the optimum was 40% of RE, SE, LE and FE, which promote the emergence of radicle and plumule. Similarly, Vigour Index of cowpea seedlings also increased with lower concentrations upto 60% of different extracts, while 80% or above it, were either non-significant or inhibitory (Srivastava, 2011). The allelopathic effect of different concentration of solvent extracts of *Parthenium hysterophorus* have been found inhibitory to all parameters viz. Seed germination to metabolism of *Vigna radiata* (mung bean) seeds (Pati and Chowdhury, 2015).

Evaluation of leachates from *Parthenium* plant parts for their effect on growth of agriculturally important plants showed marked increase in growth of some plants. Flower leachate, with high amounts of Auxin hormone extracted in it, showed positive effect on seed germination in *Phaseolus mungo* and metal tolerance of the seeds against Fe, Pb, Hg, and Ni (Lalitha *et al.,* 2012). The higher suppressive action against germination and seedling growth of maize was probably due to higher total phenolic concentrations (6678.2 and 2549.0 mg L^{-1}) and presence of phenolic compounds viz., gallic, caffeic, 4-hydroxy-3-methoxy benzoic, p-coumaric and m-coumaric acids; and ferulic, vanillic, syringic and m-coumaric acids in aqueous leaf extract of parthenium uprooted near the field border (Safdar *et al.,* 2014).

Growth Parameters

Similarly, the effects of different parts of Parthenium extract itself enhanced the nutrients, elevated photosynthesis, biomass and yield in several agricultural crops (Ramaswami, 1997; Khalid, 2000; Park *et al.,* 2003). In India the weed was found to invade the agriculture lands of sugarcane, vegetable crops field and even rice fields (Singh *et al.,* 2001). The presence of Parthenium was reduced pasture production by excluding beneficial forage plants, resulting in a monoculture (Anonymous 2004). Gautam *et al.* (2005) reported a good association between Parthenium and *Euphorbia hirta.* In different studies, it was noticed that species varied considerably in their sensitivity to aqueous extracts of Parthenium for both root growth and germination (Belz *et al.,* 2007; Rashid *et al.,* 2008). Tismina (2007) reported that *Trifolium repens, Imperata* sp., *Chrysopogon aciculatus, Sporobolus* sp. and *Dactyloctenium aegypticum* were affected by Parthenium invasion, but she reported increase in species richness by Parthenium invasion. *Euphorbia hirta* was reported in good proportion in high Parthenium invaded site than the other sites (Ayele, 2007). Wegari (2008) declared that *Chrysopogon aucheri* and *Cynodon dactylon* could out-compete the growth of Parthenium.

Kanchan and Jayachandra (1979; 1981) recorded that Parthenium inhibit nodulation and growth in legumes, ragi bean, and cowpea but have stimulatory affect on bajra (*Pennisetum typhodieum*) Rich C.V.H.B.I. In pot experiment, shoot growth in terms of shoot length and biomass production was markedly suppressed by aqueous extract of Parthenium in different test species (Hsu *et al.*, 1989; Lawrence *et al.*, 1991). Hussain and Abidi (1991) have also reported similar reduction in root growth of *Dicanthium annulatum*, *Chrysopogon montanus* and *Medicago polymorpha*. The reduced root and shoot growth under the allelopathic resulted in declined capitulum's diameter and biomass sunflower test (Bajwa, 2004). Studies showed that the lower concentrations of root and stem extract of Parthenium plant increase the root and stem length of cowpea plants (Srivastava, 2011).

Tamado (2001) found out that 90% of the interviewed farmers rank Parthenium weed as the most serious problem both in rangeland and crop lands. Parthenium exerts strong allelopathic effect and reduces the growth and reproductivity of associated crops. It does these by releasing phytotoxins from its decomposing biomass and root exudates in soil. Bioassay, pot culture and field studies have revealed that all plant parts (shoot, root, inflorescence and seed) are toxic to plants (Yaduraju *et al.*, 2005). The reduced shoot growth under the allelopathic stress may also be attributed to reduced root growth. Root length as well as root fresh and dry weight of test species was significantly suppressed by aqueous leaf extract of Parthenium (Srivastava, 2011).

The leaf growth responses of Parthenium treated Salvinia and paddy plant in relation to number; thickness and weight, etc. were also altered with alternation in doses (Pandey, 1994). Pandey *et al.* (1993) have reported the treatment drastically reduced the number of healthy leaves of *Eichhornia crassipes*. The effect of extract from fresh and dried parts is different for different species and on different stages of plant growth, (Kohli *et al.*, 1996; Belz *et al.*, 2007; Kumar and Kumar, 2010). Application of extract before plant emergence was more effective in promoting growth (Marwat *et al.*, 2008).

Regulation of Productivity with Parthenium Extracts

Parthenium extracts can reduce up to 40% yield (Swaminathan *et al.*, 1990). The competitive ability of Parthenium weed was studied by, Channappagoudar *et al.* (1990), in irrigated sorghum in India. They found that the presence of Parthenium reduced grain yields from 6.47 to 4.25 tons/ hectare and decreased grain weight by almost 30%. Kohli and Batish

(1994) have demonstrated that the yields of traditional Indian pulse crops (guar, black, and green gram) were reduced, when these were grown in soils previously covered by Parthenium. Cryptic factor, which can influence crop yields, is thus even more difficult to quantify than direct competition and much of evidence, such as reduction of grain filling in maize, Mew *et al.* (1982) and Towers and Subba Rao (1992) and is anecdotal Kanchan and Jaychandra (1976).

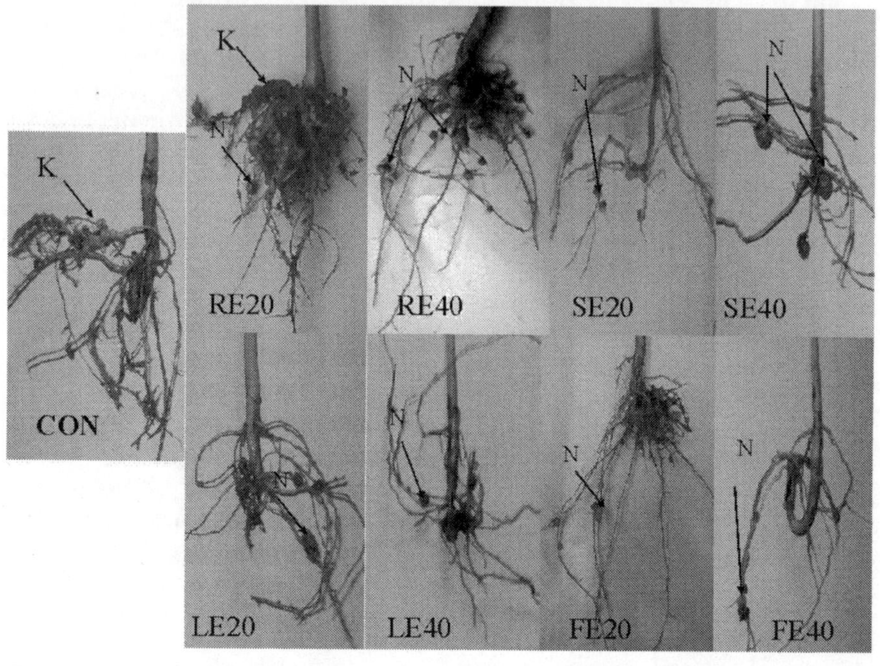

Photo Plate 3. Effect of Different parts of Parthenium extracts on Root Knot Nematodes in Cowpea plants at 45 DAS. [DAS:Days After Sowing; K: Knots; N: Nodules; RE: Root Extract; SE: Stem Extract; LE: Leaf Extract; FE: Flower Extract; 20 & 40: % of extracts] (Source: Srivastava, 2011)

Artificial dusting of pollen on the stigmatic surface of vegetables crops inhibits fruit development in brinjal, tomato and chilies (Paprica). There was a 40% reduction in grain filling in maize by 100-150 clusters of the pollen. Parthenium, which is present in Kenya, Uganda, Tanzania, South Africa, Mozambique and Swaziland is currently considered to be the most important weed both in croplands and grazing areas by 90% of farmers in the lowlands of Ethiopia (Tamado and Milberg, 2000). Sorghum yields being reduced by 97% in experimental fields with high densities of Parthenium (Tamado *et al.,* 2002). Srivastava (2011) worked with Parthenium plant extracts and reported that 40% of root extract, 20 and 40% of stem extarct showed about 4-5 days early initiation of flowers in cowpea. On the other hand, leaf extract showed three days early flowering with 40 and 80% concentrations. The flower extract also showed early flowering by two days with lower concentration i.e. 20%. She also observed increase in cowpea seed yield from 6-25% with 40% concentration of different parts of plant extracts. It has been reported that maize grain yield is reduced between 21 and 50% with Parthenium density ranging from 5-20 plants per m^2 (Safdar *et al.*, 2015).

REGULATION OF BIOCHEMICALS WITH PARTHENIUM EXTRACT

Chloroplastic Pigments

In relation to the reduction of chlorophyll contents in the leaves, Kumari *et al.* (1985) showed that chlorophyll content was markedly reduced, when leaf leachates were sprayed directly sprayed on crop plants. The reduction in chlorophyll content was also observed by Suresh and Rai (1987), in tree plants and Jayakumar *et al.* (1990) in groundnut and corn, under allelochemicals. Above 0.50% (w/v) dry leaf powder treatment caused reduction of chlorophyll contents in the leaves. Pandey (1994) stated total chlorophyll and carotenoid pigments in the leaves of the paddy seedlings grown in the medium were comparable to the amounts in the leaves of seedlings grown in distilled water This demonstrated beneficial effects of the treatments. The reduction in chlorophyll also reported in target species by Einhellig and Rasmussen (1979); Daizy and Kohli (1991); Baziramakenga *et al.* (1994); Dayan *et al.* (1999a); Romagni *et al.* (2000). Singh *et al.* (2002) reported that the content of photosynthetic pigment, chlorophyll was appreciably reduced in the leaves of test plants spray treated with parthenin at 200 μm concentration. However, this inhibitory effect declined with the passage of time and 10 days after treatment, only at 25% reduction in chlorophyll content was observed, compared to 76% on the first day after spray. Likewise, the cellular respiration measured through 2, 3, 5-triphenyl-tetrazolium chloride was found to be less in the treated leaves and inhibitory responses also declined with time. The aqueous extracts of different parts of Parthenium significantly reduced the chlorophyll content in mustard plants (Thakur and Siddiqui, 2003). They observed that even 10% of Leaf extract of Parthenium was inhibitory for chlorophyll content in mustard. Batish *et al.* (2007) suggested that with 500 μm parthenin treatment chlorophyll content was reduced by nearly 70% in *Amaranthus viridis*, where as nearby 80% reduction was observed for *Chenopodium murale* ($P<0.01$). 20 and 40% of aqueous extracts of root of Parthenium increase the total chlorophyll and carotemoin content in brinjal and bottle gourd (Singh *et al.*, 2009; Singh, 2010). Chl 'a' and 'b' content was increased with lower concentrations (20 to 80%) of Parthenium extracts, when sprayed at three stages of growth (Srivastava *et al.*, 2010).

Sugar Content

In relation to sugar content it has been stated that Parthenin (Sesquiterpene lactone) altered this physiological change by reducing the sugar content in Ageratum conyzoides (Singh *et al.,* 2002). Srivastava *et al.* (2010) recorded enhancement in total soluble sugars (mg/100mg dry wt.) in cowpea seeds with different concentrations of Parthenium extracts. The optimum increase was recorded with 40% of root extract (23.35%), 20% of stem extract (12.40%), 20% of leaf extract (15.32%) and 20% of flower extract (10.94%). The promotory effect of 20% and 40% of root extract on total soluble sugars has also been noticed in brinjal and bottle gourd (Singh, 2010).

Other Biochemicals

Pandey *et al.* (1993) have obtained those macromolecules: protein, lipid and nucleic acid resulting in root dysfunction and other inhibitory activities in the root. Picman (1986) pointed out that parthenin, like other sesquiterpenes, may bind to sulfhydryl bonds of amino acids and proteins and cause irreversible alkylation of key enzymes, thereby impairing general metabolism and causing growth inhibition. Reduced activities of amylase, invertase, dehydrogenase, nitrate reductase, peroxidase, polyphenol oxidase in germinating seeds, increase in ATPase activity, decrease in level of reducing and non-reducing sugars (Pawar and Chavan, 2007). Total protein content and leghaemoglobin content in cowpea nodules was also increased with lower concentrations of Parthenium plant extracts (Srivastava, 2011).

EFFECT OF PARTHENIUM COMPOST ON CROPS

It is reported that increasing availability of nitrogen, phosphorus, potassium and sulphur in soil when integrated application composted Parthenium (Gupta *et al.,* 1986). Ramaswami (1997), Tamil Nadu Agricultural University, Coimbatore stated that field experiments were conducted with rice followed by soybean and maize followed by cowpea using the composts prepared out of different organic wastes. All the organic waste composts including Parthenium enhanced organic status of the soil indicating beneficial effect of incorporation of wastes. The highest increase was to the extent of 22.6 and 20.8 percent over NPK alone due to application of composted Parthenium at flowering and harvest stages, respectively. Among the treatments, Parthenium compost recorded higher available soil N at post harvest stage. This could be attributed to the higher N content of Parthenium compost than other composts from organic wastes. All the organic waste composts recorded higher grain yield of rice as main crops over NPK alone. Similarly, all the organic waste composts recorded the highest yield of 1917 Kg per ha and 1285 kg per ha, respectively, as compared to other organic waste composts and application of NPK alone. This residual effect of Parthenium can be exploited for beneficial use. Oudhia *et al.* (1997b) studied the integrated use of 50% recommended dose of N through each of urea and composted Parthenium along with *Azotobacter chroococcum*, which was beneficial to target higher yield of wheat. Inoculation of *A. chroococcum* produced 33-130 % more volume of roots as compared to its corresponding uninoculated treatment indicating synergistic effect of composted Parthenium on activity of organophilic *A. chroococcum*. This clearly showed that integrated use of Parthenium compost and *Azotobacter* increased nitrogen phosphorus, potassium and sulphur acquisition in wheat than urea and Parthenium compost. The maximum uptake N (0.67 g pot^{-1}), P (0.16 g pot^{-1}), K (0.68 g pot^{-1}) and S (0.22 g pot^{-1}) were recorded with treatments T7, where 50% N through each of urea and composted Parthenium were applied with *Azotobacter*. Kishor *et al.* (2010) reported that the application of 100% N through composted Parthenium resulted in significant reduction in plant height, tillers and root volume of plant and ultimately grain and straw yield of wheat.

BIOCONTROL VALUE

The weed population in rice field was influenced by the incorporation of composted organic wastes. Sudhakar (1984) reported the reduction in weed population with Parthenium, in rice fields. Among the treatments, the composted coir pith and Parthenium recorded lower weed population. The application of organic waste composts reduced the weed count from 30.5 to 39.8 percent over NPK at 60 DAT (Days After Treatment). This could be attributed to the role of allelopathic compounds such as phenol present in this two-plant debris even after composting Son (1995). Among different composts, coir pith and Parthenium compost recorded lower weed population in maize. The beneficial effect of organic wastes in reducing the incidence of pests such as stem borer and leaf roller was observed due to the application of organic composts. Generally, under incorporation of organic wastes, the reduction in pest incidence was to the extent of 43.4 to 50 percent at 60 DAT as compared to NPK alone, Son (1995). Incidence of leaf roller in rice crop was the highest with urea application.

SOIL AMENDMENT VALUE

Any organic waste application aids in moisture conservation, which is utilized for better root penetration and crop growth. In general, incorporation of organic wastes enhanced the moisture content of the soil to the tune of 45.5 to 77.4% as compared to application of NPK alone to maize crop (Son, 1995). This enhancement could be attributed to the higher water holding capacity of the soil due to the influence of organic waste application. The moisture in soil due to application of Parthenium compost was 14.5 and 16.5% at 0-15 and 15-30 cm depths as compared to 10.7 and 11.6% at 0-15 and 15-30 cm depths of soil due to application of NPK alone. This may be due to building up of organic carbon status in soil. This behaviour can be well utilized for moisture conservation practices. Parthenium is causing change in vegetation that ultimately results in changing soil characteristics (Daubenmire, 1974; Qureshi *et al.*, 2006). Compost derived from *Parthenium* contains plenty of micronutrients such as Fe, Zn, Mn, and Cu and macronutrients including NPK making it two times richer than farmyard manure (Channappagoudar *et al.*, 2007).

One research study has reported production of improved compost (millicompost), with more nutrients and less allelochemicals, upon introduction of millipede *Harpaphe haydeniana* during composting (Apurva *et al.*, 2010). Organic acids released during composting help in liberation of insoluble K and increase the uptake of P and K (Murthy *et al.*, 2010). Compost also contains abundant enzymes, vitamins, antibiotics, plant growth regulators, and large number of associated useful microorganisms including *Azotobacter* and phosphate solubilizers (Kishor *et al.*, 2010). Chances of weed emergence are reduced greatly if composting is done before flowering in plants as all seeds are not destroyed completely during the process (Kishor *et al.*, 2010; Maurya and Sharma, 2010). Compost formed has shown growth promotion in chilli, *Sorghum* (Murthy *et al.*, 2010), *Vigna radiata* and *Triticum* Khaket *et al.*, 2012), and *Arachis hypogaea* (Rajiv *et al.*, 2013a). Moisture holding capacity of compost increases its utility value (Ambasta and Kumari, 2013). Amendment with other plant materials such as saw dust (Jelin and Dhanarajan, 2013) and poultry manure also gives good quality compost, minimizing the required dose of chemical fertilizers (Khaket *et al.*, 2012). Allelochemicals present in the final compost lessen the chances of infestation by other weeds. Though significant reduction in allelochemicals occurs during composting but better compost is obtained from plants in preinflorescence stage. The compost from Parthenium showed higher N, P and K percentage when compared to farm yard manure (FYM) values (Bhoyar, 2013).

The *Parthenium* weed is with high nutritive value. This nutritive potential of *Parthenium* can be utilized for increasing soil fertility. The widely growing *Parthenium* green bio-mass can be composted by using various techniques and added to soil as organic manure. Inspite of all its ill effects the *Parthenium* plant is having high nutritive value. This nutritive potential of *Parthenium* is utilized for the beneficial use of mankind by converting lots of plants into useful compost. The composting was carried out on large scale using huge amount of *Parthenium* biomass (750 Kg) by Vermi-Compost method. It improves the soil texture and nutrient status. According to the research study conducted by Biradar and Patil (2001) *Parthenium* weed upon composting with *Eudrilus eugeniae* supports growth of worms, indicating potential of weed as good substrate for vermicomposting. Vermicompost produced containsmoderate amounts of useful bacteria, actinomycetes, fungi, phosphate solubilizers, and large number *of Azotobacter* (Nirmalnath *et al.*, 2005).

Vermicomposting has long term effect on crop plants as well as on yield (Yadav and Garg, 2011; Basarkar *et al.,* 2013). However, pathogenic *E. coli* has been found to be absent (Shobha *et al.,* 2012). The manure obtained has low pH and C: N ratio and contains sufficiently high amounts of essential nutrients such as N, P, K, Ca (Chauhan and Joshi, 2010; Sangwan *et al.*, 2011; Basarkar *et al.,* 2013), Zn, Cu, and Mn (Rajiv *et al.*, 2012). Consequently, manure produced increased crop productivity in lady's finger (Vijaya and Seethalakshmi, 2011), *Lampitomauritii* (Kinberg) (Anbalagan and Manivannan, 2012) and sesame (Vijayakumari. and Hiranmai, 2012). *Parthenium* vermicomposted in its vegetative state produced manure with more N content whereas that produced from flowering stage had more P content (Senthilkumari *et al.,* 2013). Also manure obtained shows less toxicity (Chauhan and Joshi, 2010; Basarkar *et al.,* 2013) as allelochemical constituents such as parthenin and phenols are degraded during vermicomposting (Rajiv *et al.*, 2013b). The earthworm has high efficiency to compost the Parthenium weed and the compost was rich in mineral nutrient which help in fast growth and development of the plant *Lycopersicum esculatus* (Rose and Rani, 2014).

Biochar has been formulated successfully from *Parthenium hysterophorus* by its pyrolysis to sequester carbon for negative carbon dioxide emission (Kumar *et al.*, 2013). Addition of this biochar to the soil improved soil quality as evidenced by increased growth of *Zea mays*, increased basal respiration and microbial biomass carbon, increased catalase and dehydrogenase activities, and decreased soil stress and hydrolytic enzymes activities. During charring, ambrosin chemical present in *Parthenium*, having phototoxic effect (Patel, 2011), was lost by degradation at high temperature. Adding large amounts of biochar did not show any negative effect on soil.

GREEN MANURE VALUE

For the main crop of rice, the effect of Parthenium green leaf manure on plant height was comparatively less as compared to other green manures like leucaena and sunnhemp. Whereas, in the ratoon rice crop, Parthenium green leaf manure was superior in influencing the plant height (Sudhakar, 1984). Similarly in the main crop, Parthenium green leaf manure produced less number of filled grains while it produced the highest number of filled grains in the ratoon crop. Among the green leaf manures tried, the residual effect for dry matter production was the highest with Parthenium. Incidence of leaf roller in rice crop was the highest with urea application, whereas it was the lowest with Parthenium as green leaf manure application (Sudhakar, 1984). Composting greatly reduced allelopathic effects of Parthenium compared to fresh Parthenium (Wakjira *et al.*, 2009).

The use of *Parthenium* in agriculture is exploiting its biomass for green manuring. Addition of *Parthenium* leaf manure to rice crop resulted in increased height of plants, increased yield of grains and straw, with no emergence of weed in submerged conditions during rice cultivation (Javaid and Shafique, 2009; Saravanane *et al.*, 2012). Maize growth was also enhanced by green manure from *Parthenium* (Javaid and Shah, 2008; Suryawanshi, 2011). An enhancement has also been noticed in growth of wheat plant when treated with *Parthenium* green manure (Javaid and Shah, (2010). It is advised to utilize this weed for manuring at preflowering stage to avoid spread of weed through dissemination of seeds after seed setting in the plant. *Parthenium* green leaf manure has shown marked increase in number of filled grains in ratoon rice crop and residual effect on biomass of crop (Kishor *et al.,* 2010). Reports have indicated that green manure obtained from *Parthenium hysterophorus* showed high assimilation rate of nitrogen and phosphorus by maize crop (Suryawanshi, 2011). Thus this freely available weed can be utilized for enriching soil with manure, while replacing chemical fertilizers. Allelochemicals are plant secondary metabolites normally released into the environment through volatilization, leaching, root exudation and decomposition of plant residues in the soil (Khalaj *et al.*, 2013).

BIOCIDAL VALUE

The latest thinking includes allelopathy to both harmful and beneficial interactions between the plants (Rizvi *et al.,* 1992). Allelochemicals, which inhibited the growth of some species at certain concentration may stimulate, the growth of same or different species at lower concentration (Narwal, 1994; Narwal *et al.*, 2003). It is used as potential source of herbicide, fungicide and nematicide (Hasan and Jain, 1984; Bano *et al.,* 1986; Ramaswami, 1997; Duke *et al.,* 2000, 2005; Bajwa *et al.*, 2003; Sharma, 2003; Marwat *et al.*, 2008). It has been observed to also have pesticidal and nematicidal properties (Datta and Saxena, 2001).

Species-specific differences in the sensitivity to aqueous extracts of fresh or dry leaf material of Parthenium, were reported by Mersie and Singh (1987); Kohli *et al.* (1993). Marwat *et al.* (2008) reported that pre-emergence application of Parthenium extracts was more effective compared to post-emergence application. These results suggested that higher concentration of Parthenium retard the growth of plants which might be due to inhibition of cell division as allelopathic chemicals have been found to inhibit gibberellin and indole acetic acid function (Tomaszewski and Thimann, 1966). Parthenin is among other inhibitors relevant for residue allelopathy as simulated under laboratory conditions by delaying germination and reducing plant growth (Belz *et al.,* 2007).

Herbicidal Value

Some weeds, which were earlier susceptible, are now herbicide resistant (Ahmad, 1996). The Parthenium become a widespread weed in the zone of U.P. Presently the weed is a major problem in the agriculture filed. The species start their growth before rainy season and cover whole area of agriculture field, which suppressed the growth of other herbaceous vegetation also. Before Kharif crop *P. hysterophorus* remain covered with their peak growth in agricultural land. People through ploughing in agriculture uproot *P. hysterophorus* collect and burned in the agriculture filed. The large-scale burned ash of *P. hysterophorus* remained with the sown crop in the agriculture field and worked as efficient herbicide. Allelochemicals or plant derived chemicals offers a great potential for the pesticides because they are comparatively safer for the environment. In the past two decades, much more work has been done on plant derived compounds as environmentally safe alternatives to herbicides for the weed control (An *et al.*, 1998; Duke *et al.*, 2002).

Extracts from *Parthenium* plant parts have been recorded to show inhibitory effect towards *Eragrostis* (Tefera, 2002) and common aquatic weeds, the water hyacinth (Pandey *et al.*, 1993), and *Salvinia* (Pandey, 1994a, b) due to released phytotoxins from Parthenium leaves (Stephen and Sowerby, 1996). Among various parts, flowers and leaves are the richest in allelochemicals and can kill water hyacinth plant in one month. On the other hand, stem and root parts containing lower phenolics concentrations have been found to show nutritive properties supporting growth of the weed. Therefore, appropriate parts and dose of *Parthenium* plant can be used effectively to control certain weeds. Positive allelopathic effects of many weed on other can be exploited to develop ecofriendly cheap and effective "**green herbicides containing green allelochemicals**" are an integral part of eco or organic farming (Oudhia and Tripathi, 1999; Macias *et al.,* 2001; Batish *et al.* 1997; 2007; Javaid and Anjum, 2006).

Use of natural compounds as herbicides or as chemical basis for the development of new herbicides offers several advantages:

- The wide array of phytotoxic compounds produced by plant provides many complex chemical structures that are unlikely to be discovered in the traditional synthetic strategies used by pesticide companies.
- Degradation of natural compounds in the environment proceeds faster than that of synthetic compound and thus, reduces the environment pollution.

Pesticidal Value

Antifeedent bioassay revealed that lactone was found to be about 2.25 times more active than parthenin against sixth-instar larvae of *Spodoptera litura* and pyrazoline adduct was found to be the most effective as an insecticide against the adults of store grain pest *Callosobruchus maculatus* (Datta and Saxena, 2001). Petroleum ether extracts of leaves, stem and inflorescence of *P. hysterophorus* at 500, 1000, 2000 and

5000 ppm concentrations significantly decreased the life span and progeny production of mustard aphid, *Lipaphis erysimi* (Sohal *et al.,* 2002).

Fungicidal Value

Antifungal potential of different extracts of *P. hysterophorus* against human pathogenic fungi were investigated by Rai and Upadhyay (1990). The dermatophytes and other fungal pathogens have been found to be sensitive to sesquiterpene lactones which are present as active agent in Asteraceous plant *P.hysterophorus* (Rai *et al.*, 2003). The Parthenium extract was also effective against brinjal fruit borer and pathogenic fungi in seeds (Lalitha *et al.,* 2012). The leaf extract prepared with a mixture of methanol proved to be more effective than the other solvent system used in inhibiting the growth of *Candidia albicans* on Potato dextrose agar plates (Malarkodi and Manoharan, 2013). Ethanolic extract of Parthenium can be used as novel fungicide against *Alternaria alternata* (Singh and Srivastava, 2013).

Nematicidal Value

Extract of *P. hysterophorus* show toxicity against root knot nematodes *Meloidogyne incognita* (Kofoid and white), Chitwood, *Helicotylendus dihyslera* (Cobb) sher (Hasan and Jain, 1984; Patel *et al.,* 1993). The study was initiated to survey species of Asteraceae plant (Parthenium) for their susceptibility to *M. incognita* and to evaluate their effect on this nematode (Tiyagi *et al.*, 1985; 1988; Tsay, 2004). However, formulating this plant extract is essential for commercial use of the nematicidal extracts. Cowpea production is beset by an array of pest, root knot nematode and other diseases that can cause serious devastation, thus leading to reduced yield and low profitability (Blade *et al.*, 1997). Crushed leaves admixed into the soil are used to reduced root galling in papaya caused by *M. incognita* (de la Fuente ***et al.***, 2000). It has been observed to also possess pesticidal and nematicidal properties (Datta and Saxena, 2001; Prasad *et al.*, 2002; Randhawa *et al.*, 2002; Sharma and Trivedi, 2002; Shamim and Tiyagi, 2003; Khan *et al.*, 2008). The data collected on number of root knot galls developed on the root surface of cowpea plant showed remarkable reduction with different concentrations of Parthenium extract, when sprayed at different stages of growth (Srivastava, 2011; Photo Plate 3).

Sesquiterpene lactones are one of the very important groups of natural plant products known for their biological activities, including allelopathic properties (Macias *et al.*, 2001). Little has been done to explore their herbicidal potential, although there have been some studies (Batish *et al.* 1997, 2002a; Macias *et al.,* 2001). Among sesquiterpene lactones, parthenin is a natural component of Parthenium hysterophorus known for its phytotoxicity towards other plants including aquatic species (Megharaj *et al.,* 1987; Pandey 1996; Batish *et al.,* 1997, 2002b).

SYNTHESIS OF NANOPARTICLES

The use of noxious weed *Parthenium* has the added advantage that this unwanted plant can be used by nanotechnology processing industries. Applications of such eco-friendly nanoparticles in bactericidal, wound healing and other medical and electronic applications, makes this method potentially exciting for the large-scale synthesis of other inorganic materials (nanomaterials). The bio-reduction of aqueous Ag+ ions by the leaf extract of the plant *Parthenium hysterophorus* L has been demonstrated. The reduction of the metal ions through leaf extracts leading to the formation of silver nanoparticles of fairly well-defined dimensions. This green chemistry approach toward the synthesis of silver nanoparticles has many advantages such as, ease with which the process can be scaled up, economic viability, etc. (Parashar *et al.*, 2009).

Another use of *Parthenium* weed exposed in recent years is use of its extract for synthesising silver nanoparticles (Parashar *et al.*, 2009; Kumar, 2012; Mangrola *et al.,* 2012; Ananda, 2013) and zinc oxide nanoparticles (Rajiv *et al.*, 2013c) having reasonable stability. It is based on the capability of chemicals present in the plant extract to reduce silver ions present in AgNO3 solution (Kumar, 2012). It was found that microwave irradiation could effectively reduce time of synthesis from days to even seconds (Ananda, 2013). Nanoparticles synthesized by this method inhibited the growth of potent bacterial pathogens such as *E. coli, Pseudomonas putida, Klebsiella pneumoniae, Staphylococcus aureus, Salmonella typhi, Pseudomonas aeruginosa, Proteus vulgaris,* and *Bacillus subtilis* (Mangrola *et al.,* 2012; Ananda, 2013), some evenmore than known antibiotics such as gentamicin sulphate (Anwar *et al.,* 2015). Zinc oxide nanoparticles were also found effective against pathogenic fungi.

POTENTIAL ENERGY SOURCE-BIOGAS PRODUCTION

Nallathambi Gunaseelan (1987) used *Parthenium hysterophorus* as an additive (10%) in cattle manure and achieved 60–70% CH4 production, suggesting potential of *Parthenium* weed as a substrate for biogas production. Subsequently Gunaseelan and Lakshmanperumalsamy (1990) successfully produced 75% methane per kg biomass from *Parthenium hysterophorus* alone. Alkali pretreatment of *Parthenium* has been documented to increase biogas production than untreated biomass (Gunaseelan, 1994). During the process of biogas production, degradation of phytotoxic allelochemicals has been seen (Gunaseelan, 1994). Various inocula have been tested in differing ratios for maximizing yield of methane. Readily available sugarcane press mud cake mixed with cow dung augmented methane production from *Parthenium* (Abubacker *et al.,* 1999). Careful monitoring of anaerobic digestion of the weed showed maximum methane formation only after 45 days, with decreasing pH and C :N ratio becoming constant after 45 days (Thakur and Singh, 2000). These results were reproduced by Thakur and Singh (2003), while also using other weeds and agricultural wastes for biogas production. Biogas has also been produced frommixture of cowdung and *Parthenium* (Kannan *et al.*, 2003). Reduction in C : N ratio was suggested due to loss in organic carbon in CH_4 and CO_2 formation and accumulation of hydrolyzed nitrogen in the slurry. Addition of *Parthenium* leachate, obtained by soaking plant in water for one week, to cow dung boosted biogas production fromcowdung (Gitanjali *et al.*, 2009), while slurry left could be used as manure. It can be used as a potential energy source (Ghosh *et al.*, 2013). This situation demands that efforts should be made to develop an alternative technology for weed control. Allelopathy could be an appropriate potential technology for this purpose.

In Phytoremediation

Parthenium hysterophorus has been found to play vital role in accumulation of heavy metals especially in contaminates sites (Mane *et al.*, 2013). Table 3 shows list of some metals extracted by *Parthenium.* Addition of certain chemicals in the soil augments phytoextraction of some metals as observed for Pb with addition of EDTA and GA3 (Hadi and Bano, 2009) and for Zn with addition of EDTA (Sanghamitra *et al.*, 2012). Parthenium has the potential of phytoextraction of Cadmium from the soil (Hadi *et al.*, 2014).

In Pharmaceuticals

Parthenium hysterophorus is known for its various pharmacological activities. Medicinal uses of both the plant extract show remarkable antipyretic, CNS stimulating, skeletal muscle relaxant (Urmilesh *et al.*, 2011) also the anti-oxidant, hypoglycemic activity (Vijay *et al.*, 2008).

Anti-Inflammatory Activity

Oral administration 10, 20, 40 mg/kg of body weight of *P. hysterophorus* extract led to significant antinociceptive and anti-inflammatory effects against acetic acidinduced writhing in mice and carrageenan-induced paw edema in rats, respectively (Jain and Kulkarni, 1999). 200mg/kg of body weight of fresh leaves ethanolic extract exhibited high degree anti-inflammatory in carrageenan induced paw edema rats (Pandey *et al.,* 2012). 1,2 mg/kg of body weight parthenolide administ-ration also produced antinociceptive and antiinflammatory effects (Jain and Kulkarni, 1999). The anti-inflammatory property may be due to an inhibitor of cellular phospholipases, which prevents release of arachidonic acid in response to appropriate physiological stimuli (Makheja and Bailey, 1982).

Antioxidant Activity

The antioxidant phytochemicals protect the cells from oxidative damage caused by free radicals. DPPH (2, 2-diphenyl-1-picrylhydrazyl radical) scavenging assay revealed that Methanolic and ethanolic extract of *Parthenium hysterophorus* showed antioxidant activity. But next time acetone extract was found to have higher anti-oxidant activity than methanol and chloroform extracts (Priya *et al.,* 2011). 200mg/kg of body weight of fresh leaves ethanolic extract showed significant antioxidant activity in rats (Pandey *et al.,* 2012).

Anticancer and Cytotoxic Activity

Anticancer and cytotoxic properties of *P. hysterophorus* are the bright potential in oncology as anticancer therapeutics. The methanolic flower extract exhibited antitumor effects in host mice bearing transplantable lymphocytic leukemia (Mukherjee and Chatterjee, 1993; McFadyen *et al.,* 1995). Crude methanolic extract showed cytotoxic activity against brine shrimp nauplii (Al-mamun *et al.,* 2010). Methanol, Ethanol, Chloroform and Aqueous were found that all extracts were significantly cytotoxic to the investigated human cancerous cell lines Ovary (IGROVI), lung (A-549), prostate (PC-3), breast (MCF-7) and CNS (SF-295) except Lung cancerous cell line (HOP-62) (Haq *et al.,* 2011). Parthenolide, in the presence of α-methylene-glactone moiety, has been reported to have cytotoxic activity in Eagle's 9KB carcinoma of the nasopharynx cell culture system (Berry MI, 1984). *In vitro,* parthenolide functioned as growth inhibitor of mouse fibrosarcoma (MN-11), human lymphoma (TK6), human lung carcinoma (A549), human medulloblastoma (TE671), human colon adenocarcinoma (HT-29) and human umbilical vein endothelial cells (HUVEC) (Ross *et al.,* 1999; Parada-Turska *et al.,* 2007). The methanolic extract of *Parthenium hysterophorus* possesses potent cytotoxic and anthelmintic activities (Durga *et al.*, 2013). Various allelochemicals released from Parthenium includes sesquiterpene lactones and phenolic acids, among these the presence of caffeic, vanillic and ferulic acids is of industrial significance, as they possesses potent free radical scavenging(as anti-oxidant) and anti-cancer activity (Panwar *et al.*, 2015).

a. Antimicrobial Activity

Different extracts of *P. hysterophorus* tested for antimicrobial potential showed varying degree of antimicrobial activities. Dichloromethane extract of leaves was found as the most effective against *E. coli* and methanolic extract of leaves was found nil for *K. pneumoniae* and highest for *S. aureus* (Fazal *et al.,* 2011). All types of organic extracts and aqueous extract of inflorescence were highly effective against *P. aeruginosa* and *C. freundii* (Sharma and Gupta, 2012). Hydroalcoholic extract of *P. hysterophorus* was *in vitro* effective against *Plasmodium falciparum* (Valdes *et al.,* 2010). Crude 50% ethanolic extract of *P. hysterophorus* flowers exhibited trypanocidal *activity against* a flagellate protozoa *Trypanosoma evansi* both *in vitro and in vivo* (Talakal *et al.,* 1995). *In vitro,* this plant demonstrated antiamoebic activity comparable to the standard drug Metronidazole against axenic and polygenic cultures of *Entamoeba histolytica,* responsible for amoebiasis (Sharma and Bhutani, 1988; Khare, 2008). Fusarium wilt, an economically important fungal disease in potato caused by *Fusarium solani* was significantly inhibited by aqueous, methanol and n-hexane extracts (Zaheer *et al.,* 2012). Aqueous extract of inflorescence of this plant was found effective at higher concentrations of 1000 µg/ml and 500µg/ml against *Penicilliusm chrysogenum, Microsporum gypseum and Rhizopus stolonifer* but different organic extracts showed no activity (Sharma and Gupta, 2012). Leaves aqueous extract also showed antifungal activity against *Alternaria alternate* (Ramanujam *et al.,* 2011).

b. Wound healing activity

Externally leaf paste application of *P. hysterophorus* showed wound healing activity (Kumar *et al.,* 2012).

c. Hypoglycemic Activity

Administration of aqueous extract of *P. hysterophorus* flower (100 mg/kg of body weight) significantly decreased the serum glucose level in normal and alloxan induced diabetic rats (Patel *et al.,* 2008). Slightly decreased blood glucose level was found in rats after oral administration of fresh leaves extract (Arya *et al.,* 2012).

d. Thrombolytic Activity

Crude methanolic extract of *P. hysterophorus* showed significant thrombolytic effect comparable to standard thrombolytic agent, streptokinase (Al-mamun *et al.,* 2010). Parthenolide and some other metabolites were determined as the inhibitor of human blood platelet function (Hewlett *et al.,* 1996). The Thrombolytic activity is a possible relevant effect to migraine prophylaxis through the inhibition of serotonin releasing from platelet by platelet aggregating agents: adenosine diphosphate, adrenaline, sodium arachidonate, collagen, and U46619 (Heptinstall *et al.,* 1985; Hewlett *et al.,* 1996).

e. Prevention and Treatment of Migraine

Prevention and treatment of migraine by *P. hysterophorus* is now a matter of debate. However, based on pharmacological studies of the herb and parthenolide, three possible mechanisms are involved: Anti-inflammatory activity, an effect on platelets and inhibition of serotonin binding (Anonymous, 2003). *P. hysterophorus* was reported to inhibit granule secretion in blood platelets, which is related with the etiology of migraine (Heptinstall *et al.,* 1985). A randomised, double-blind, placebo-controlled trial for three months successfully controlled migraine in 17 patients by eating raw feverfew leaves (Johnson *et al.,* 1985). Another randomised, double-blind, placebo-controlled trial demonstrated lowering the migraine pain intensity significantly during powdered leaves (parthenolide 0.2%) 100 mg daily for 60 days administration in 57 patients (Palevitch *et al.*, 1997). 24% reduction of migraine attack in 72 patients was observed after *P. hysterophorus* treatment (70–114 mg *P. hysterophorus* equivalent to 2.19 µg parthenolide; one capsule daily) in another randomised double-blind, placebo-controlled, crossover trial (Murphy ***et al.***, 1988). The above studies suggest that the preparations are admirable in preventing migraine but further well–designed clinical trials are required to establish the beneficial effects of *P. hysterophorus* for migraine prophylaxis.

f. Rheumatoid Arthritis

P. hysterophorus was reported to inhibit granule secretion in blood neutrophils, which is related with the etiology of Rheumatoid arthritis (Heptinstall *et al.,* 1985). But a double-blind, placebocontrolled study demonstrated no beneficial effect when forty-one female patients with inflammatory joint symptoms were given daily either one placebo capsule or one *P. hysterophorus* capsule (70–86 mg equivalent to 2–3 µmol parthenolide), for six weeks (Pattrick *et al.,* 1989).

g. Other Pharmacological Uses

Though *P. hysterophorus* is an obnoxious herb, when exploited by proper means and technology, it can prove to be a valuable asset. In West Indies this weed is used as a remedy against ulcerated sores, certain skin disease, facial neuralgia, fever and anemia (Bhatt *et al.,* 2012). Inflammations, eczema, skin rashes, herpes, rheumatic pain, cold heart trouble, menstrual disorders, difficulty during labour, stomach ache, toothache, diarrhoea, neurologic disorders, urinary infections, dysentery, malaria and insect bites are also treated (Barnes *et al.,* 2007; Patel, 2011). Chemical constituents are useful as insecticide and for curing psoriasis (Kohli and Rani, 1994). A preparation with and ginger is effective for treating migraines during the early pain phase (Kuhn and Winston, 2007). Its broad-spectrum ovicidal, antimicrobial, lervicidal, nematocidal, herbicidal activities designate the improvement of public health and crop production (Bhatt *et al.,* 2012). Greenish yellow dye mordant with chrome produced from stems and leaves is used in wool processing (Kowalchik *et al.,* 1998). It is useable as an additive with cattle manure in biogas production (Patel, 2011).

h. Cautions to Pharmacological Uses

P. hysterophorus should not be taken by individuals who develop various kinds of allergic reactions. Though clinical study showed that the *P. hysterophorus* reduces migraine attack, self-medication should not be undertaken without consulting with doctor. It shows abortifacient activity and affects menstrual cycle (Barnes *et al.,* 2007). This plant is not applicable for the children under two years old (Awang, 1993).

Management of *Parthenium hysterophorus*

It is necessary to control the problematic weed *P. hysterophorus* in time before spreading, because of its negative impact on natural and agro ecosystem. There are several methods available for controlling the parthenium weed. The methods are: manual and mechanical control, chemical and biological control and managing the weed by proper utilization. In a report, Australian Weeds Committee (2012), effective broadscale management of perthenium weeds requires an integrated approach that includes prevention and eradication programs, establishment and implementation of management zones, and the protection of key environmental, social and economic assets in areas where the weeds are already widespread (Fig. 3).

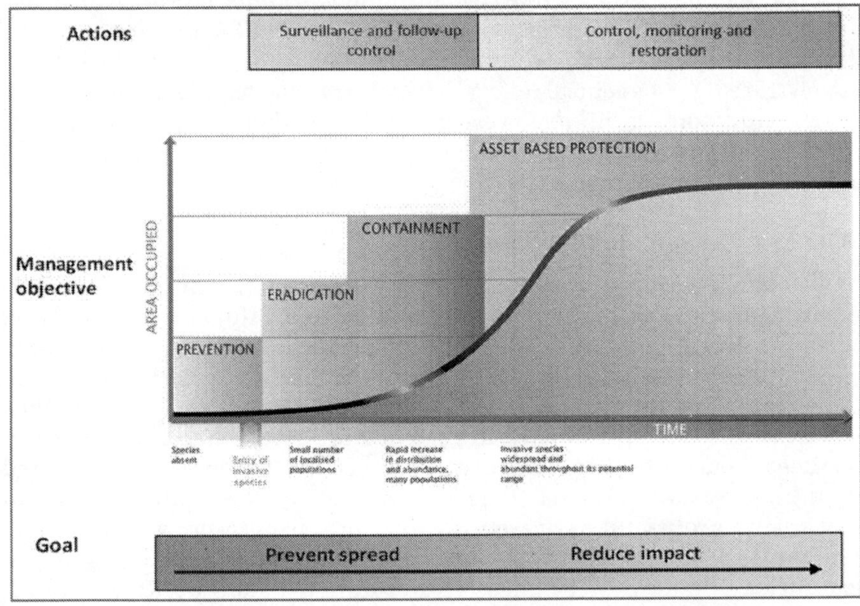

Fig. 3. Stages of weed invasion with corresponding goals, management objectives and actions at each stage (Australian Weeds Committee, 2012)

There are immediate opportunities for management of weed and crop residues and minimize crop losses from allelopathy and also to use allelopathic crops for weed management. Allelopathic-environmental interactions must be considered in efforts to benefit from allelopathy. Allelochemicals may also be adapted to yield stimulants or environmentally sound plant growth regulators. Additionally this enhancement with the nematode pest management and enhanced nodulation can be resulted in the better yield. Thus, incorporating the mechanism of active variety of allelochemicals and using the dual nature of allelochemicals as growth regulating and nematicidal properties (Srivastava, 2011), these can be utilized for promoting the crop yield.

Today with inflation on the rise and inspirations rising, farmers also want a better standard of living. Increasing prices of the agricultural produce can be done only up to a point and therefore increasing yield and quality of produce are better options to increase farmer's income and needs to be focused on. The Indian yields are amongst the lowest in the world and have continued to be stagnant or declined over the last decade. Now is the time, we should consider what kind of innovations could be promoted in agriculture and allelopathy is one of the innovative and cost effective method to be incorporated. Therefore, a comprehensive strategy to do something new is required, as the government needs to work towards increasing the yields significantly.

References

Abhilasha, Dipti, Quintana, Naira, Vivanco, J. and Joshi, J. 2008. Do allelopathic compounds in invasive *Solidago conadensis* s.i. restrain the native European flora? *Journal of Ecology,* **96**: 993-1001.

Abubacker, M. N., Rao, G. R. and Kumaresan, A. 1999. Sugarcane press mud cake: accelerator of biogas production in various weed biomass. *Advances in Plant Sciences*, **12(1)**: 73–78.

Adkins, S.W., Novie, S.C., Graham G.C. and Mcfadyen R.E. 1997. Parthenium weed in Australia: Research underway at the co-operative research center for tropical pest management. In: *Proc. First Internat. Conf. on Parthenium management,* University of Agricultural Science, Dharwad, 6-8 Oct., 1997. pp.13-17.

Adkins, S. and Shabbir, A. 2014, Biology, ecology and management of the invasive Parthenium weed (*Parthenium hysterophorus* L.). *Pest Manage. Sci.,* **70(7)**: 1023-1029.

Afzal, B., Bajwa R. and Javaid A. 2000. Allelopathy and VA mycorrhiza VII. Cultivation of *Vigna radiata* and *Phaseolus vulgaris* under allelopathic stress caused by *Imperata cylindrica. Pakistan J. Biol. Sci.*, **3:** 1926-1928.

Agarwal, C. and Anand, A. 1992. Ecological Studies on allelopathic potential of *Parthenium hysterophorus* in relation to *Phaseolus aureus* and *Triticum aestivum* L. In: *Proc. 1st Nat. Symp. and Allelopathy in Agroecosystem* (Eds. P. Tauro and S.S. Narwal). By Indian Society of Allelopathy, Haryana Agricultural University, Hisar (Haryana), India. Feb. 1992, pp. 64-65.

Aggarwal, A. and Kohli, R.K. 1992. Screening of crops for seed germination against *Parthenium hysterophorus* L. leachates. *Ibid*, pp. 66-68.

Ahmed, M. and Wardle, D.A. 1994. Allelopathic Potential of Vegetative and flowering ragwort (*Senecio jocobaea* L.) plants against associated pasture species. *Plant Soil*, **164:** 61-68.

Ahmad, S. 1996. The potential of allelopathy as a tool for weed management in crops. Presidential Address; 5th Pakistan weed Sci. Conf. NARC, Islamabad, Pakistan.

Ahn, J.K. and Chung, I.M. 2000. Allelopathic Potential of rice hulls on germination and seedling growth of Barnyardgrass. *Agron. J.,* **92** (6): 1162-1167.

Aldrich, J.D. 1984. *Weed Crop Ecology: Principles and Practices.* Breton Publishers, pp. 215-241.

Al-mamun, R., Hamid, A., Islam, M.K., and Chowdhury, J.A. 2010. Cytotoxic and Thrombolytic Activity of Leaves Extract of *Parthenium hysterophorus* (Fam: Asteraceae). *Bangladesh Pharmaceutical Journal*, **13:** 51–54.

Altieri, M.A. and Doll, J.D. 1978. The potential of allelopathy as a tool for weed management in crops. *PANS,* **24:** *495-502*

Ambasta, S.K. and Kumari, S. 2013. A scientific approach of conversion of eco-hazardous *Parthenium* weed into eco-friendly by compost making. *International Journal of Geology, Earth & Environmental Sciences*, **3(1):** 90–94.

Ambiye, N. and Golatkar, V. 2010. Phytochemical analysis of *Lantana camara* and *Parthenium hysterophorus. BIOINFOLET*, **7**(2): 135-136.

An M, Pratley J, and Haig T. 1998. Allelopathy: from concept to reality. In: Proceedings of the 9th Australian Agronomy Conference, Charles Sturt University Convention Centre, http:// www.regional.org.au/au/asa/1998.

Ananda, S. 2013. Exploitation of *Partheniumhysterophorus* L. for the rapid biosynthesis of silver nanoparticles and evaluation of their anti-microbial activity. *Indian Journal of Applied Research*, **3(7):** 79–83.

Anbalagan, M. and Manivannan, S. 2012. Assessment of impact of invasive weed *Parthenium hysterophorus* L. mixed with organic supplements on growth and reproduction preference of *Lampitomauritii* (Kinberg) through vermitechnology. *International Journal of Environmental Biology*, **2(2):** 88–91.

Anonymous 2000. Weeds of national significance: Parthenium weed *(Parthenium hysterophorus*) strategic plan. National weeds strategy executive committee, Launceston, pp.19.

Anonymous 2003. WHO monographs on selected medicinal plants Volume 4 (World Health Organization, Geneva).

Anonymous 2004. Parthenium weed (*P. hysterophorus*). Pest series, Dept. of Natural Resources and Mines, Queensland, Australia. *QNRMO,* **1219**: 2-4.

Anwar, M. F., Yadav, D., Kapoor, S., Chander, J. and Samim, M. 2015. Comparison of antibacterial activity of Ag nanoparticles synthesized from leaf extract of *Parthenium hystrophorus* L. in aqueous media and Gentamicin sulphate: *in-vitro. Drug Development and Industrial Pharmacy*, **41(1):** 43-50.

Apurva, P., Sinha, S. K. and Thakur, P. C. 2010. Composting an obnoxious weed, *Parthenium hysterophorus* L., with the help of a millipede, *Harpaphe haydeniana. Asian Journal of Experimental Biological Sciences*, **1(2):** 337–343.

Arya, A., Abdullah, M.A., Haerian, B.S., and Mohd, M.A. 2012. Screening for Hypoglycemic Activity on the Leaf Extracts of Nine Medicinal Plants: *In-Vivo* Evaluation. *E-Journal of Chemistry, **9***: 1196–1205.

Athanassova, D.P. 1996. Allelopathic effect of *Amaranthus retroflexus* L. on weed and crops. Seisieme Conference du. Journees International Sr. La. Lutter les mauviases herbs, Reims, France, pp 437-442.

Australian Weeds Committee. 2012. Pathenium (*Parthenium hysterophorus* L.) strategic plan 2012–17, Weeds of National Significance, Australian Government Department of Agriculture, Fisheries and Forestry, Canberra.

Awang, D.V.C. 1993. Feverfew Fever: A Headache for the Consumer. *Herbal Gram, 29:* 34–36.

Ayala, J.R., Cruz, A.M. and Miranda 1994. Effects of aqueous extract of *Canavalia ensiformis, Bursera graveolem, Petiveria allicer* and *Parthenium hysterophorus* on Sorghum germination. *Cuban Journal of Agricultural Sciences,* **28:** 371-373

Ayele, S. 2007. The impact of Parthenium (*Parthenium hysterophorus* L.) on the Range Ecosystem Dynamics of the Jijiga Rangeland, Ethiopia [M.A. Dissertation], Department of Animal Sciences, School of Graduate Studies, Haramaya University, pp117.

Babu G. Prasada, Hooda, V., Audiseshamma, K. and Paramageetham, Ch. 2014. Allelopathic effects of some weeds on germination and growth of *Vigna mungo* (L). Hepper. *Int. J. Curr. Microbiol. App. Sci.,* **3(4):** 122-128.

Bais, H.P., Veppachedu, R., Gilory, S., Callaway R.M. and Vivanco J.M. 2003. Allelopathy and exotic plant invasion: from molecules and genes to species interactions. *Science,* **301:** 1377-1380.

Bajwa, R. 2004. Effect of foliar spray of aqueous extract of *Parthenium hysterophorus* on growth of sunflower. *International Journal of Agriculture and Biology,* **3:** 474-478.

Bajwa, R., Hassan, A. and Javaid, A. 1998. Allelopathic potential of *Desmostachya bipinnata* Stapf. *Sci. Inti.* (Lahore), **10***:* 83-86.

Bajwa, R., Khalid, A. and Cheema, T.S. 2003. Antifungal activity of allelopathic plant extracts III. Growth response of some pathogenic fungi to aqueous extract of *Parthenium hysterophorus. Pakistan Journal of Plant Pathology,* **2** (3): 145-156.

Bansal, G.L. 1988. Allelopathic effect of aqueous extract of stem and leaves of three tree species on the germination of some crops and weeds. *Trends inTree* Sciences, **10:** 119-123.

Bano, M, Anwar, S., Tiyagi, S.A. and Alava, M.M. 1986. Evaluation of some nematicidal properties of some members of family Compositae. *Int. Nematol. Network News letter,* **3(1)**: 10

Barnes, J., Anderson, L.A., and Phillipson, J.D. 2007. Herbal Medicines. Published by the Pharmaceutical Press, RPS Publishing, London, UK.

Basarkar, U.G., Kshirsagar, G.S. and Saoji, A.A. 2013. Comparative study of vermicompost using *Parthenium* biodung and usual green biodung. *International Journal of Bioassays,* **02 (05)**: 819-824.

Batish, D.R., Arora, K. and Singh, H.P. 2007. Potential utilization of dried powder of *Tagetes minuta* as a natural herbicide for managing rice weeds. *Crop Protection,* 26 (3): 566-571.

Batish, D.R., Kohli, R.K., Singh, H.P. and Saxena, D.B. 1997. Studies on herbicidal activity of Parthenin, a constituent *of Parthenium hysterophorus* towards billgoat weed (*Ageratum conyzoides*). *Curr. Sci.,* **73***:* 369-371.

Batish, D.R., Singh, H.P., Kohli, R.K. and Kaur, S. 2001. Crop allelopathy and its role in ecological agriculture. In: *Allelopathy in Agroecosystems* (eds. R.K. Kohli, P.S. Harminder, and D.R. Batish). Food Products Press. New York. pp. 121-162.

Batish, D.R., Singh, H.P., Kohli, R.K., Saxena, D.B. and Kaur, S. 2002a. Allelopathic effects of Parthenium against two weedy species, *Avena fatua* and *Bidens pilosa. Environmental and Experimental Botany,* **47**(2): 149-155.

Batish D.R., Singh H.P., Pandher J.K. Arora V. and Kohli, R.K. 2002b. Phytotoxic effect of Parthenium residues on the selected soil properties, growth of radish and chickpea. *Weed Biology and Management,* **2(2):** 73-78.

Batish, D.R., Singh, H.P., Saxena, D.B., Kohli, R.K., and Arora V. 2002c. Effect of parthenin-sesquiterpene lactone from *Parthenium hysterophorus* L. on early growth and physiology of *Ageratum conyzoides. J. Chemical Ecology,* **28:** 2169-2179.

Batish, D.R., Singh, H.P., Pandher, J.K., and Kohli, R.K. 2005. Phytotoxic effect of *Parthenium hysterophorus* L. residues on three Brassica species. *Weed Biology and Management,* **5***:* 105-109.

Batish, D.R., Singh, H.P., Rano, N. and Kohli, R.K. 2006. Assessment of allelopathic interference of *Chenopodium album* through its leachates, debries extracts, rhizosphere and amended soil. *Arch. Argon Soil Sci.,* **53:** 705-715.

Baziramakenga, R., Simard, R.R. and Leroox, G.D. 1994. Effects of Benzoic and cinnamic acids on the growth, mineral composition and chlorophyll content of soybean. *J.Chemical Ecology,* **20***:* 2821-2833.

Bell, D.T. and Koeppe, D.E. 1972. Non-competitive effects of giant foxtail on the growth of Corn. *Agron. J.*, **64:** 321-325.

Belz, R.G., Reinhardtb, C.F., Foxcroftc, C.L. and Hurlea, K. 2007. Residue allelopathy in *Parthenium hysterophorus* L. – Does Parthenium play a leading role? *Crop Protection,* **26(3):** 237-245.

Bennett, R.N. and Wallsgrove, R.M. 1994. Secondary metabolites in plant defence mechanisms. New *Phytol.,* **127:** 617-633

Bergmark, C.L., Jackson, W.A., Volk, R.J. and Blum, U. (1992). Differential inhibition by ferulic acid on nitrate and ammonium uptake in *Zea mays* L. *Plant Physiol.,* **98:** 639-645.

Bertin, C., Weston, L., Huang, T., Jander, G. and Owens, T. 2007. Grass roots chemistry: meta-tyrosine, an herbicidal nonprotein amino acid. *PNAS,* **104(43):** 886-893.

Bezabih, F. and Araya, T. 2002. Spread and status of Parthenium in Tigray Regional State at Woreda level. *Tigray Bureau of Agriculture and Natural Resources*, pp.79.

Bhatt, B.P., Chauhan, D.S. and Todaria, N.P. 1994. Effect of weed leachetes on germination and radical extension of some food crops. *Indian J. Plant Physiol.,* **37:** 177-179.

Bhatt, J.R., Singh, J.S., Singh, S.P., Tripathi, R.S., and Kohli, R.K. 2012. Invasive Alien Plants: An Ecological Appraisal for the Indian Subcontinent (CABI).

Bhowmik, P.C. and Doll, J.D. 1982. Corn and Soybean response to allelopathic effects of weeds and crop residue. *Agron. J.,* **74:** 601-606.

Bhowmik, P.C. and Doll J.D. 1984. Allelopathic effects of annual weeds residues on growth and nutrients uptake of corn and soybean. *Ibid,* ***76*** **(3):** 383 – 388.

Bhowmik, P.C. and Sarkar, D. 2005. *Parthenium hyslerophorus* L.: its world status and potential management. In: *Proc. of the Second Internat. Conf. on*

Parthenium Management, on 5-7, December 2005, University of Agricultural Science, Bangalore, pp 1-6.

Bhowmik, P.C., Sarkar, D. and Yaduraju, N.T. 2007. The status of *Parthenium hypterophorous* L. and its potential management. *Ecoprint,* **14**: 1-17.

Bhullar, M.K., Kalsi, P.S. and Chhabra, B.R. 1997. Methoxy pseudoguaianolides from *Parthenium hysterophorus. Fitoterapia,* **68(1):** 91-92.

Biradar, A. P. and Patil, M. B. 2001. Studies on utilization of prominent weeds for vermiculturing. *Indian Journal of Weed Science,* **33(3-4):** 229–230.

Blade S.F., Shetty V.R., Terao T. and Singh, B.B. 1997. Recent developments in cowpea cropping systems research. In: *Advances in Cowpea Research* (eds. Singh, B.B., Mohan, Raj D.R., Dashiel, K.E. and Jackai, L. EN). IITA, Ibadan and JIRCAS, Japan, pp. 114-128.

Blum, U. 1998. Effects of microbial utilization of phenolic acid and their phenolic acid breakdown products on allelopathic interaction. *J.Chemical Ecology,* **24:** 685-708.

Blum, U. and Shafer, S.R. 1988. Microbial populations and phenolic acids in soil. *Soil Biol. Biochem.,* **20**: 793-800.

Blum, U. Shafer, S.R., and Lehman, M.E. 1999. Evidence for inhibitory allelopathic interaction involving phenolic acid in field soil: Concepts vs on experimental model. *Crit. Rev Plant Sci.,* **18:** 673-693.

Bogatek, R., Gniazdowska, A., Zakezewska, W., Oracz, K. and Gawronski, S.W. 2006. Allelopathic effect of sunflower extracts on mustard seed germination and seedling growth. *Biologia Plantarum,* **650** (1): 156-158.

Bhoyar, M.G. 2013. Possible Uses of Parthenium in Organic Farming. Pop. Kheti, 1(4):136-**141**

Burhan, N. and Shaukat S.S. 1999. Allelopathic potential of *Argemone mexicana* L., a tropical weed. *Pak. J. Biol. Sci.,* **2:** 1268-1273.

Callaway R.A. and Ascjempig, E.T. 2000. Invasive plants verse their new and old neighbors: A mechanism for exotic invasion. *Science,* **290:** 521-523.

Callaway, R.A. and Ridenour, W.M. 2004. Novel weapons, Invasive success and the evolution of increased competitive ability. *Front. Eco. Environ.,* **2**: 436.443.

Channappogoudar, B.B., Panchal, Y.C., Manjunath, S., and Kotia R.V. 1990. Studies on influence of Parthenium on sorghum growth under irrigated conditions. *Farming System,* **6**: 102-104.

Channappagoudar, B. B., Biradar, N. R., Patil, J. B. and Gasimani, C. A. A. 2007. Utilization of weed biomass as an organic source in sorghum. *Karnataka Journal of Agricultural Science,* **20(2):** 245–248.

Chauhan, A. and Joshi, P. C. 2010. Composting of some dangerous and toxic weeds using *Eisenia Foetida. Journal of American Science,* **6(3):** 1–6.

Chhabra, B.R., Kohli, J.C. and Dhillon, R.S. 1999. Three ambrosanolides from *Parthenium hysterophorus. Phytochemistry* (Oxford), **52 (7):** 1331-1334.

Cheng, H.H. 1992. A conceptual framework for assessing allelochemicals in soil environment. In: *Allelopathy: Basic and Applied Aspects.* (Eds, SJH Rizvi and V Rizvi). Chapman and Hall Press, New York, pp. 21-29.

Cheema, Z. A. 1988. *Weed Control in Wheat through Sorghum Allelochemicals.* Ph.D. Thesis. Deptt., of Agronomy, Univ. of Agriculture, Faisalabad.

Cheema, Z. A., Khaliq, A, and Saeed, S. 2004. Weed control in maize (*Zea mays* L.) through Sorghum allelopathy. *J. Sustainable Agric.,* **23** (4): 73-86.

Chippendale, J.F. and Penetta, F.D. 1994. The cost of Parthenium weed in Australia cattle industry. *Plant Protection Quartely,* **9:** 73-76.

Chon, S.U., Kin, R. and Kee, J.C. 2003. Herbicidal potential and quantification of causative allelochemicals from several Compositae weeds. *Weed Res.,* **43***:* 444 – 450.

Chon, S.U. and Nelson, C.J. 2010. Allelopathic in Compositae plants. *A review. Agron.Sustain.Dev.,* **30**: 349-358.

Chung, I.M., Kim, K.H., Ahn, J.K., Lee, S.B., and Hahn, S.J. 2003. Comparision of allelopathic potential of rice leaves, straw and hull extracts on Barnyard grass. *Agron J.,* **95:** 1063-1070.

CPA 2002. http:///www.baa.org.uk/content/knowledgebase/6-prod-pewwer.asp.

Daizy, R., and Kohli, R.K. 1991. Fresh matter is not an appropriate unit for chlorophyll content: experience from experiments on effects of herbicides and allelopathic substances. *PHOTOSYNTHETICA,* **25:** 655-658.

Das, B., Venkataish, B. and Kashinatham, A. 1999. (+)-Syringaresinol from *Parthenium hysterophorus, Phytochemistry* (Amsterdam), **63 (4):** 383-386.

Das, R., Geethangili, M., Majhi, A., Das, B., Rao, Y.K. and Tzeng, Y.M. 2005. A new highly oxygenated pseudoguaianolide from a collection of the flowers of *Parthenium hysterophorus. Chemical and Pharmacological Bulletin* (Tokyo), **53 (7):** 861-862.

Datta, S., and Saxena, D.B. 2001. Pesticidal properties of Parthenin (from *Parthenium hysterophorus*) and related compounds. *Pest Management Science,* **57**: 95-101.

Datta, S.C. and Sinha-Roy, S.P. 1973. Germination and growth inhibitors in *Croton bonaplandium* a perennial weed. In: *Proc. 4th Asian-Pacific Weeds. Sci. Soc. Conf.,* **2:** 485-488.

Datta, S.C. and Sinha-Roy, S.P. 1975. Phytotoxic effects of *Croton bonaplandium* Baill. on weedy associations. *Vegetation,* **30** (3): 157-163.

Daubenmire, R.F. 1974. *Plants and Environment A Textbook of Plant Autecology* 3rd Ed., John Willey and Sons Inc. London, pp.90.

Dayan, F.E., Hernandez, A., Allen, S.N., Morcus, R.M., Vroman, J.A., Avery, M.A., and Duke, S.O. 1999a. Comparative phytotoxicity of artemisinin and several sesquiterpene analogues. *Phytochemistry,* **50:** 607-614.

de Candolle, M.A.P. 1832. *Physiologie Vegetable.* Tome III, Bechet Jenune, Lib, Fac. Med., Paris. pp. 1474-1475.

de la Fuente J.R., Uriburu M.L., Burton G. and Sosa V.S. 2000. Sesquiterpene lactone variability in *Parthenium hysterophorus* L. *Phytochemistry,* **55 (7)**: 769-772.

Dhanger, D.S., Gupta, D.C. and Jain, R.K. 1996. Studies on aqueous extract to different part of *Tagetus erecta* on hatching. Larval mortality larval penetration of M*eloidogyne javanica. Ind. J. Nematol,* **26:** 46-51.

Dhawan, S.R. and Dhawan, P. 1995. Effect of aqueous extracts of some trees on germination and early seedling growth of *Parthenium hysterophorus* L. *World Weeds,* **2:** 217-221.

Dhawan, S.R. and Dhawan, P. 1996. Regeneration in *Parthenium hysterophorus* L. *Ibid*, **3:**181-182.

Dhileepan, K. 2009. Managing Parthenium weeds across diverse landscapes: prospects and limitations. In: *Management of Invasive Weeds* (ed. Inderjit), **5:** 247-259.

Dhileepan, K. and McFadyen, R.E.C. 2001. Effects of gall damage by introduced biocontrol agent *Epiblema strenuana* (Lep., Torticidae) on the weed *Parthenium hysterophorus* (Asteraceae) *Journal of Applied Entomology*, **125** : 1.

Dhima, K.V., Vasilakoglou, I.B., Eleftherohorinos, I.G. and Lithourgidis, A.S. 2006a. Allelopathic potential of winter cereals and their cover crop mulch effect on grass weed suppression and corn development. *Crop Sci.*, **46**: 345-352

Dhima, K.V., Vasilakoglou, I.B., Eleftherohorinos, I.G. and Lithourgidis, A.S. 2006b. Allelopathic potential of winter cereal cover crop mulches on grass weed suppression and sugarbeet development. *Ibid,* **46:** 1682-1691.

Drost, D.C. and Doll, J.D. 1980. The allelopathic effect of yellow nutsedge (*Cyperus esculentus)* on corn and soybeans. *Weed Sci.,* **28:** 229-233.

Duke, S.O., Dayan, F.E., Aliota, G., and Rongani, I.G. 2002. Chemicals form nature for weed management. Weed Sci. **50:** 138-151.

Duke, S.O., Belz, R.G., Baerson, S.R., Pan, Z., Cook, D.D. and Dayan, F.E. 2005. The potential for advance in crop allelopathy. *Outlooks Pest Management,* **16:** 64-68.

Duke, S.O., Dayan, F.E., Romagni, J.G. and Rimando, A.M. 2000. Natural products as soures of herbicides: Current status and future trends. *Weed Res.,* **40:**99-111.

Durga, M. Sesha Sai, Satyavathi, K., Bhojaraju, P., Preeti, P. Mercy, Lovaraju, K. P., Priya, K. Sony, Kumari, Y. Ratna, Prasad, D. Venkatesh and Kanthal, Lakshmi kanta, 2013. Cytotoxic and anthelmintic activities of methanolic extract of *Parthenium hysterophorus* L. *Pharmacophore,* **4 (6):** 275-279.

Ebena, K., Yan, W., Dilday, R.H. Namai, H and Okuno.K. 2001 Variation in the allelopathic effect of rice with water- soluble extracts. *Agron. J.,* **63**:12-16.

Einhellig, F.A. 1987. Interactions among Alletochemicals and other Stress factors of the Plant Environment. In: *Allelochemicals: Role in Agriculture and Forestry* (ed. G.R. Waller). American Chemical Society, Washington DC, pp.343-357.

Einhellig, F.A. 1995. Mechanism of Action of Allelochemical in Allelopathy. In *Allelopathy: Organism,Processes and Application*. American chemical society, Washington, USA., pp.: 96-116.

Einhellig, F.A. and Kaun, L. 1971 Effects of scopoletin and chlorogenic acid on stomatal aperture in tobacco and sunflower. *Bull, Torrey Bot. Club,* **98**:155-162.

Einhelling, F, A., Leather, G.R. and. Hobbs, L.L. 1985. Use of *Lemna minor* L. as a bioassay in allelopayhy. *J. Chemical Ecology,* **11**: 65-72.

Einhelling, F.A. and Rasmussen, J.A. 1979. Effects of three phenolic acids on chlorophyll content and growth of soybean and grain sorghum seedlings. *Ibid,* **5**: 815-824.

Einhelling, F.A., Rice, E.L., Riser, P.G. and Wender, S.H. 1970. Effects of scopoletin on growth, CO_2 exchange rates and concentrations of Scopoletin, Scopolin and chlorogenic acids in tobacco, sunflower and pigweed. *Bull. Torrey Bot. Club.*, **97**: 22-23.

Evans, H.C. 1997. *Parthenium hysterophorus*: A review of its weed status and the possibilities for biological control. *Biocontrol News and Information*, **18 (3):** 89-98.

Eze, S.N.O. and Gill, L.S. 1992. *Chromolaena odorata* a problematic weed. *Compositae Newslett,* **20**: 14-18.

Fazal, H., Ahmad, N., Ullah, I., Inayat, H., Khan, L., and Abbasi, B.H. 2011. Antibacterial potential in *Parthenium hysterophorus*, Stevia rebaudiana and Ginkgo biloba. Pak. J. Bot., **43**: 1307–1313.

Francis, K. 1978. Photochemical activities of Parthenium chloroplasts. *Curr. Sci.,* **47(23):** 899-900.

Furest, E.P. and Putnam, A.R. 1983. Seperating the competitive and allelopathic components of interference: theoritical principles. *J. Chemical Ecology,* **8**: 937-944.

Gahukar, R. T. 2010. Role and perspective of phytochemical in pest managment in India. *Curr. Sci.*, **98:** 897- 899.

Gerald, F.L., Booker, Blum, U. and Fiscus E.L. 1992. Short term effects of ferulic acid on ion uptake and water relations in cucumber seedlings. *Journal Experimental Botany,* **43**: 649-665.

Ghosh. S., Haldar, S., Ganguly, A. and Chatterjee, P.K. 2013. Review on *Parthenium hysterophorus* as a potential energy source. *Renewable and Sustainable Energy Reviews*, **20**: 420-429.

Gitanjali, G., Kumaresan, A., Dharmaraj, M. and Karuppayee, T. 2009. Utilization of parthenium plant leachate for biogas production. *Asian Journal of Microbiology, Biotechnology and Environmental Sciences*, **11(1):** 113–115.

Gniazdowska, A. and Bogatek, R. 2005. Allelopathic interactions between plants: Multisite action of allelochemicals. *Acta Physiol. Plant.*, **27(3B):** 395-407.

Goldsworthy, D. 2005. *Technical Report Parthenium Weed and Health.* University of central Queensland, Rockhampton, Queensland.

Golisz, A., Sugano, M. and Fujii, Y. 2008. Micro array expression profiling of *Arabidopsis thaliana* L. in response to allelochemicals identified in buckwheat. *Journal of Experimental Botany,* **59**(11): 3099-3109.

Goyal, C.P. and Brahma. B.C. 2001. Array of hope against Parthenium in Rajaji National Park. *Indian Forester,* **127**(4): 414.

Goyal, R. and Singh.K.P. 2003. Allelopathic effect of leaf leachates and latex of some plants on seed germination and seedling growth of *Parthenium hysterophorus* L. *Environmental Biology and Conservation*, **8**: 21-33.

Grice, A.C. 2006. The Impact of Invasive plant species on the biodiversity of Australian rangelands. *The Rangeland Journal*, **28**: 1-27.

Grierson, A.J.C. and Long, D.J. 2001. *Flora of Bhutan*, Vol. 2, part 3, Royal Botanical Garden, Edinburgh, pp. 187.

Guin, J.D. 1989. Sesquiterpene-lactone dermatitis. *Immunol. Allergy*, 9: 447-461.

Gunaseelan, V N. 1987. Parthenium as an additive with cattle manure in biogas production. *Biological Wastes,* **21(3):** 195-202.

Gunaseelan, V. N. 1994. Methane production from *Parthenium hysterophorus* L., a terrestrial weed, in semi-continuous fermenters. *Biomass & Bioenergy*, **6(5):** 391–398.

Gunaseelan, V. N. 1998. Impact of anaerobic digestion on inhibition potential of *Parthenium* solids. *Biomass and Bioenergy*, **14(2):** 179–184.

Gunaseelan, V. N. and Lakshmanaperumalsamy, P. 1990. Biogas production potential of Parthenium. *Biological Wastes,* **33(4):**311-314.

Gunaseelan, V. N. 1994. Methane production from *Parthenium hysterophorus* L. a terrestrial weed, in semicontinuous fermenters. *Biomass nd Bioenergy*, **6(5)**: 391-398.

Gunaseelan, V. N. 1998. Impact of anaerobic digestion on inhibition potential of Parthenium solids. *Biomass and Bioenergy,***14(2)**: 179-184.

Gupta, A.P., Antil, R.S. and Gupta, V.K. 1986. Effect of pressmud and zinc on the yield and uptake of zinc and nitrogen by corn. *J. Indian Soc. Soil Sci.*, **34**: 810-814.

Gupta, N., Martin, B.M., Metcalfe, D.D. and Rao, P. V. 1996. Identification of a novel hydroxyproline-rich glycoprotein as the major allergen in Parthenium pollen. *The Journal of Allergy and Clinical Immunology*, **98***:* 903–912.

Gupta, R.K., Dutta, T.R. and Patil, B.D. 1977. Chemical investigation of *Parthenium hysterophorus. Indian Journal of Pharmacy*, **39 (3):** 64-66.

Gupta, S.K. 2008. A system approach to allelopathic responces of Parthenium on germination and seedling vigour of *Cajanus cajan* and control measures for the weed .In: *Proceedings National System Conference, NAC Dec.*17-19, 2008,pp. 645-649.

Guyana, P. 2014. *Parthenium hyterophorus* L. Asteraceae-Parthenium weed, *EPPO Bulletin*, **44(3):** 474-478.

Hadi, F. and Bano, A. 2009. Utilization of *Parthenium hysterophorus* for the remediation of lead-contaminated soil. *Weed Biology and Management*, **9(4):** 307–314.

Hadi, F., Ali, N. and Ahmad, A. 2014. Enhanced phytoremediation of Cadmium-contaminated soil by *Parthenium hysterophorus* plant: Effect of Gibberellic acid (GA_3) and synthetic chelator, alone and in combinations. Bioremediation Journal, **18(1)**: 46-55.

Hall, A.B., Blum, U. and Fites, R.C. 1982. Stress modification of allelopathy of *Helianthus annuus* L. debris on seed germination. *Am. J. Bot.,* **69**: 776-783.

Haq, M.R., Ashraf, S., Malik, C.P., Ganie, A.A. and Shandilya, U. 2011. *In vitro* cytotoxicity of *Parthenium hysterophorus* extracts against human cancerous cell lines. *Journal of Chemical and Pharmaceutical Research*, **3**: 601-608.

Hara, H.Z., Chater, A.O. and Williams, L.H.J. 1982. *Enumeration of Flowering Plants of Nepal*, British Museum (Natural History), London. pp.220.

Harlan, J.R. and deWet, J.M.J. 1965. Some thoughts about weeds. *Economic Botony*, 19(1): 16-24

Harnden, J., MacDougall, A.S. and Sikes, B.A. 2011. Feild based effects of allelopathy in invaded tall grass Prairie. *Botany,* **89**: 227-234.

Harper, J.R. and Balke, N.E. 1981. Charcterization of the Inhibition of K^+ absorption in oat roots by salicylic acid. *Plant Physiol.*, **68**: 1349.

Hasan, N. and Jain, R.K. 1984. Biotoxicity of *Parthenium hysterophorus* extract against *Meloidogyne incognita* and *Helicotylenchus dihystera.* Nematol, Medit., **12**: 239-242.

Haseler, W.H. 1976. *Parthenium hysterophorus* L. in Australlia, *PANS.* 22: 515-517.

Hassan, G. and Marwat, K.B. 2001. Integrated weed management in agricultural crops. *National workshop on Technologies for Sustainable Agriculture*, Sep. 24-26, NIAB, Faisalabad, Pakistan.

Hegab, M.M., Khodary, S.E.A., Hammouda, Ola and Ghareib, H.R. 2008. Autotoxicity of chard and its allelopathic potentiality on germination and some metabolic activties associated with growth of wheat seedling. *African. J. of Biotechnology*, **7(7):** 884-892.

Henderson, L. 2001. *Alian Weeds and Invasive Plants.* Plant Protection Researchn Institute Handbook no. 12. ARC, South Africa, pp. 300.

Heptinstall, S., Williamson, L., White, A., and Mitchell, J.R.A. 1985. Extracts of fever few inhibit granule secretion in blood platelets and polymorphonuclear leucocytes. *The Lancet,* **325:** 1071–1074.

Herz, W. and Watanabe, H. 1959. Parthenin, a new guaianolide. *J. Am. Chem. Soc.*, **81**(22): 6088-6089.

Herz, W., Watanabe, H., Miyazaki, M. and Kishida, Y. 1962. The structures of Parthenin and Ambrosine. *Ibid,* **84**(13): 2601-2610.

Hewlett, M.J., Begley, M.J., Groenewegen, W.A., Heptinstall, S., Knight, D.W., May, J., Salan, U., and Toplis, D. 1996. Sesquiterpene lactones from fever few, *Tanacetum parthenium*: isolation, structural revision, activity against human blood platelet function and implications for migraine therapy. *Journal of the Chemical Society, Perkin Transactions* 1 1979.

Hind, D.J.H., Jeffery, C. and Scott, A.J. 1993. *Composes flora des mascareigenes No.109* 1 Bosser, I.G. and Jeffery (Eds). *The Royal Botanic Garden, Kew,* pp.261.

Hosmani, M.M. 1995. *Integrated Weed Management in Field Crops.* Venkatadri. Printers, Bangalore. pp, 230,

Hsu, F.H., Chin, C.Y. and Chou, C.H. 1989. Action of model allelopathic compounds on seed germination in phytochemical ecology. In: *Allelochemicals, Mycotoxins and lnsects Pheromones and Allomones.* (eds. C.H. Chou, and G.R. Waller) Institute of Botany, Academic Sinica Monograph Series no. **9**, Taipei. pp: 315-328.

Hussain, F. and Abidi, N. 1991. Allelopathy exhibited by *Imperata cylindrical* (L.) P. Beauv. *Pakistan J. Bot.,* **23**: 15-25.

Ilori, O.J., Otusanya, O., Adelusi, A.A. and Sanni, R.O. 2010. Allelopathic activity of some weeds in the Asteraceae family. *International Jouranl of Botany,* **6(2)**: 161-163.

Inderjit 1996. Plant Phenolics in allelopathy. *Botanical Review,* 62: 186-202.

Inderjit 2001. Soil: Enviromental effects on allelochemical activity. *Agron. J.,* 93: 79-84.

Inderjit, Callaway, R.M. and Vivanco, J.M. 2000. Can plant biochemistry contribute to understanding of invasion ecology? *Trends Plant Science,* **11**:574-580.

Inderjit and Dakshini, K.M.M. 1995. Allelpathic potential of an annual weed, *Polypogon monspeliensis*, in crops in India, *Plant Soil*, **173**: 251-257.

Inderjit and Duke, S.O. 2003. Ecophysiological aspects of allelopathy. *Planta,* 217: 529-539.

Inderjit and Keating, K.I. 1999. Allelopathy: Principles, Procedures, Processes, and Promises for biological control. *Adv. Agron.*, **67**: 141-143.

Inderjit, Von Dahl, C.C. and Baldwin, T. 2009. Use of silenced plants in allelopathy bioassays: a novel approach. *Planta,* **229**: 569-575.

Jain, N.K., and Kulkarni, S.K. 1999. Antinociceptive and anti inflammatory effects of *Tanacetum parthenium* L. extract in mice and rats. *Journal of Ethnopharmacology*, **68**: 251–259.

Jalili. A., Abbassia, F., and Bazoobandi, M. 2007. Allelopathic influence of canola on germination of five weeds of canola fields. *Absts. International workshop on allelpathy-current trends and future application.* Univ. of Agri, Faisalabad, Pakistan.

Javaid, A. and Anjum, T. 2006. Control of *Parthenium hysterophorus* L. by aqueous extracts of allelopathic grasses. *Pak. J. Bot.,* **38(1)**: 139-145.

Javaid, A. and Shah, B.M. 2008. Use of parthenium weed as green manure for maize and mungbean production. *The Philippine Agricultural Scientist,* **91(4):** 478–482.

Javaid, A. and Shah, B.M. 2010. Growth and yield response of wheat to EM (effective microorganisms) and parthenium green manure," *African Journal of Biotechnology*, **9(23):** 3373–3381.

Javaid, A. and Shafique, S. 2009. Comparison of *Trifolium alexandrinum* L. and *Parthenium hysterophorus* L. green manures in rice-wheat cropping system. *Philippine Agricultural Scientist*, **92(1):** 110–115.

Javaid, A., Shafique, S. and Shafique, S. 2009. Invasion of noxious alien weed *Parthenium hysterophorus* L. in grazing lands of Lahore, Pakistan. *The Journal of Animal & Plant Sciences,* **19(3):** 149-153.

Javed, K. and Asghari, B. 2008. Allelopathic potential of Sunflower (*Helianthus annus*) on soil metals and leaves extract on physiotgoy of Wheat (*Triticum aestivum* L.) Seedlings. *African J. Biotechnology*, **7:** 3261-3265.

Jayachandra 1971. Parthenium weed in Mysore state and its control. *Curr.Sci.,* 40: 568-9.

Jayakumar, M., Eyini, M and Panner Selvan, A. 1990. Allelopathic effect of *Eucalyptus globulus* Labill. in groundnut and corn. *Comparative Physiol. Ecol.,* **15**: 109-113.

Jayanth, K.P. 1986. Introduction and establishment of *Zygoramma bicolorate* on *Parthenium hystrophorus* at Bangalore, India. *Curr. Sci.*, **56(7):** 310-311.

Jelin, J. and Dhanarajan, M. S. 2013. Comparative physicochemical analysis of degrading Parthenium (*Parthenium hysterophorus*) and saw dust by a new approach to accelerate the composting rate. *International Journal of Chemical, Environmental and Biological Sciences*, **1(3)**: 535–537.

Johnson, E.S., Kadam, N.P., Hylands, D.M., and Hylands, P.J. 1985. Efficacy of feverfew as prophylactic treatment of migraine. *British Medical Journal* (Clinical Research Ed.) **291**: 569–573.

Joshi, S. 1991. Biocontrol of *Parthenium hysterophorus* L. *Crop Protection*, 10(6): 429-431.

Joshi, S. 2006. Interference effects of *Cassia uniflora mill* on *Parthenium hysterophorus* L. *Plant Soil,* **132**:213-218.

Kamal, J. and Bano, A. 2008. Effects of Sunflower (*Helianthus annuus* L.) extracts on wheat (*Triticum aestivum* L.) and physicochemical charecterstics of soil. *African J. of Biotechnology,* **7** (22): 4130-4135.

Kamal, R. and Mathur, N. 1991. Histamine a biogenic amine from *Parthenium hysterophorus* Linn. *Journal of Phytological Research*, **4 (2):** 213-214.

Kannan, N., Guruswamy, T. and Kumar, V. 2003. Design, development and evaluation of biogas plant using donkey-dung and selected biomaterials as feedstock. *Journal of the Institution of Engineers (India): Agricultural Engineering Division*, **84:** 17–23.

Kanchan, S.D. 1975. Growth inhibitors from *Parthenium hysterophorus* L. *Curr. Sci.,* 44: 358-359.

Kanchan, S.D. and Jayachandra 1979. Allelopathic effects of *Parthenium hysterophorus* L. exudation of inhibitors through roots. *Plant Soil,* **53**: 27-35.

Kanchan, S.D. and Jayachandra, K.A. 1981. Effect of *Parthenium hysterophorus on* Nitrogen fixing and nitrifying bacteria. *Canadian Journal of Botany*, **59**: 199-202.

Kassasion, L. 1971. The place of herbicides and weed research in tropical agriculture. *PANS,* **17**: 26-29.

Kaushik, N., Biplas, M. and Singha Ray, A.K. 2005. *Parthenium hysterophorus* L.: A Global view of its weed status and Biodiversity. In: *Proceeding of the Second International Conference on Parthenium management* (Dec 2005), Bangalore, India, pp.5-7

Kesavan, P.C. and Malarvannan, S. 2010. Green to evergreen revolution: ecological and evolutionary perspective in pest management. *Curr. Sci.,* **99(7):** 908-914.

Khaket, T.P., Singh, M., Dhanda, S., Singh, T. and Singh, J. 2012. Biochemical characterization of consortium compost of toxic weeds *Parthenium hysterophorus* and *Eichhornia crassipe. Bioresource Technology*, **123**: 360–365.

Khalaj, M. A., Amiri, M. and Azimi, M. H. 2013. Allelopathy; physiological and sustainable agriculture impact aspects. *International Journal of Agronomy and Plant Production,* **415**: 950 – 962.

Khalid, S. 2000. *Parthenium hysteropharus* L.A. new introduction to Pakistan. *Pak. J. Biol. Sci.,* **3(5):** 846-847.

Khan, A., Sayed, M., Shaukat, S.S. and Handoo, Z.A. 2008. Efficacy of four plant extracts on nematodes associated with papaya in Sind, Pakistan, *Nematol. Meditt,* **36:**93-98.

Khan, A.H. and Vaishya, R.D. 1992. Allelopathic effect of different crop residues on germination and growth of weeds. *Proc. First Nat. Symp.: Allelopathy in Agroecosystems* (agriculture and forestry), CCS Haryana Agric. Univ., Hissar (India), pp.59-60.

Khan, N., George, D., Shabbir, A., Hanif, Z. and Adkins, S.W. 2014. Rising CO_2 can alter fodder-weed interactions and suppression of *Parthenium hysterophorus, Weed Research,* **55(2):** 113-117.

Khare, C.P. 2008. Indian Medicinal Plants: An Illustrated Dictionary (Springer).

Khosla, S.N. and Sobti, S.N. 1981. Effective control of *Parthenium hysterophorus* L. *Pesticides,* **15:**18-19.

Kil, B.S. and Yun, K.W. 1992. Allelopathic effects of water extracts of *Artemisia princeps* var. orientalis on selected plant species. *J. Chemical Ecology,* **18**: 39-51.

Kimber, R.W.L. 1973. Phytotoxicity from plant residues III. The relative effect of toxins and nitrogen immobilization on the germination and growth of wheat. *Plant Soil,* **38:** 543-555.

Kishor, P., Maurya, B.R. and Ghosh, A.K. 2010. Use of uprooted Parthenium before flowering as compost: A way to reduce its hazards worldwide. *Int. J. Soil Sci.,* **5**: 73-81.

Klein, R.R. and Miller, D.A. 1980. Allelopethy and its role in agriculture community. *Soil Sci. Plant Anal.*, **11**: 43-56

Knox, J. 2013. Effect of allelopathic plants on Congress grass (*Parthenium hysterophorus* L.) in a tropical region. *Res. Environ. Life Sci.,* **6(4)**: 113-114.

Kohli, R.K. and Batish, D.R. 1994. Exhibition of allelopathy by *Parthenium hysterophorus* L. in agro ecosystems. *Tropical Ecology,* **35** (2): 295- 307.

Kohli, R.K., Dogra, K.S., Rani, D. and Singh, R.B. 2004. Impact of invasive plants on the structure and composition of natural vegetation of North-Western Indian Himalayas. *Weed Technology,* **18**: 1296-1300.

Kohli, R.K., Kumari, A. and Saxena, D.B. 1985. Auto and teletoxicity of *Parthenium hysterophorus* L. *Acta Universities Agriculture, Bruno (Czechoslovakia),* **33:** 253 -263

Kohli, R.K. and Rani, D. 1992. Indentification and bioefficacy of soil chemics of Parthenium In: Tauro, P. and Narwal, S.S. (eds) *Proc. of the Ist Nat. Symp. on Allelopathy in Agroecosystem.* Feb.1992. Hisar Haryana Agricultural University, Hisar (India). pp 196-198.

Kohli, R.K. and Rani, D. 1994. *Parthenium hysterophorus*-a review. *Res. Punjab Univ.,* **44(1/4):**105-149.

Kohli, R.K., Rani D. and Verma, R.C. 1993. A mathematical model to predict tissue response to parthenin-an allelochemic. *Biologia Plantarum,* **35**: 567 – 576.

Kohli, R. K., Rani, D., Singh, H. P. and Kumar, S. 1996. Response of crop seeds towards the leaf leachates of *Parthenium hysterophorus* L. *Indian Journal of Weed Science,* **28(1-2):** 104-106.

Kololgi, P.D., Kololgi, S. D. and Kololgi, N. P. 1997. Dermatologic hazards of parthenium in human beings. In: *Proc. of the First Internat. Conf. on Parthenium Management.* Mahadevappa M and Patil V C (eds.), 6-8 October 1997, University of Agricultural Sciences, Dharwad, pp.18-19.

Kong, C.H., Hu, F., Cheng, X.H., Chen, Y.P. and Huang, S.S. 2002a. Assessment and utilization of alleopathic crop varietal resources. *Scientia Agriculture Sinca,* **35**: 1159-1164.

Kong, C.H., Xu, X.H., Hu, F., Cheng, X.H., Lin, B. and Tan, Z.W. 2002b. Using specific secondary metabolites as markers to evaluate allelopathic potentials of rice varieties and individual plants. *Chinese Science Bulletin,* **47**: 203-206.

Kowalchik, C., Hylton, W.H., and Carr, A. 1998. Rodale's Illustrated Encyclopedia of Herbs, Rodale Press, Inc.

Krishna A., Ramana, P. and Talekar, M. 2007. Allelopathic effect of weed extract on seed germination of paddy cultivers. *Karnataka J. Agric. Sci.,* **20(3):** 671-673.

Kuhn, M.A. and Winston, D. 2007. Winston & Kuhn's Herbal Therapy and Supplements: A Scientific and Traditional Approach (Lippincott Williams & Wilkins).

Kumar, D. A. 2012. Rapid and green synthesis of silver nanoparticles using the leaf extracts of *Partheniumhystreophorus*: a novel biological approach. *International Research Journal of Pharmacy,* **3(2):** 169–173.

Kumar, M. and Kumar, S. 2010. Effect of *Parthenium hysterophorus* ash on growth and biomass of *Phaseolus mungo. Academia Arena,* **2(1)**: 98-102.

Kumar, S., Khandpu, S., Rao, D.N., Wahaab, S. and Khanna, N. 2012. Immunological response to *Parthenium hysterophorus* in Indian Patients with Parthenium sensitive atopic dermatitis. *Immunological Investigations,* **41**, 75–86.

Kumar, S., Masto, R. E., Ram, L. C., Sarkar, P., George, J. and Selvi, V. A. 2013. Biochar preparation from *Parthenium hysterophorus* and its potential use in soil application. *Ecological Engineering,* **55**: 67–72.

Kumari, A. and Kohli, R.K. 1987. Autotoxicity of ragweed *Parthenium hysterophorus. Weed. Sci.,* **35:** 629-632.

Kumari, A., Kohli, R.K. and Saxena, D.B. 1985. Allelopathic effects of *Parthenium hysterophorus* leachetes and extracts an *Brassica compestris* L. *Annals of Biology,* **1**:189-196

Kushwaha, V.B. and Maurya, S. 2012. Biological utilities of *Parthenium hysterophorus. Journal of Applied and Natural Science,* **4 (1):** 137-143.

Lal, B. and Oudhia, P. 1999. Beneficial effects of allelopathy: I. Crop Production, *Indian J. Weed Sci.,* **31(1&2)**: 103-105.

Lalitha, P., Shivani, K. and Rama, R.R. 2012. *Parthenium hysterophorus*-an economical tool to increase the agricultural productivity. *International Journal of Life Sciences,* **1(1):** 113–127.

Lamb, C. and Collet, F. (Eds.) 1976. *Weed in Australia. Melbourne,* Australia, Inkata Press, pp.332.

Lau, Jennifer, A., Puliafico, Kenneth, P., Kopshever, Joseph, A., Steltzer, Heidi and Jarvis, Edward, P. 2008. Interference is complicated by effects of activate carbon on plant growth. *New Phytologist*, **178**: 412-423.

Lawerence, J.G.A., Colwell, D. and Sexton, D.J. 1991. The Ecological impact of Allelopathy in *Allanthus altissima* (Simaroubaceae). *Am. J. Bot.*, **78**: 948-958.

Lewis, W.H., Dixit, A.B. and Wender, H.J. 1988. Reproductive biology of *Parthenium hysterophorus* L. (Asteraceae). *Journal of Palynology*, **23/24**: 73-82.

Lewis, W.H., Dixit, A.B. and Wender, H.J. 1991. Asteraceae aero pollen of the Western United State Gulf Coast. *Annal of Allergy*, **67:** 37-46.

Li, Z., Wang, Q., Ruan, X. Pan, C. and Jiang, D. 2010. Phenolics and plant allelopathy. *Molecules*, **15:** 8933-8952;

Lieble, R., Simmons, F.W., Wax, L.M. and Stoller, E.W. 1992. Effects of rye (*Secale cereale*) mulch and weed control and soil moisture in soybean (*Glycine max*). *Weed Technol.*, **6:** 838-846.

Lovett, J.V. 1989. Allelochemicals, Mycotoxin and Insect Pheromones and Allomones. In: *Phytochemical Ecology* (eds. C.H. Chou and G.R. Waller). Taipei, ROC. Pp.49-67.

Lovett, J.V. and Ryuntyu, M.Y. 1992. In: *Allelopathic Basic and Applied Aspect* (eds. S.J.H. Rizvi and V. Rizvi). Chapman & Hall, London. pp.**11-20**.

Machado, S. 2007. Allelopathic Potential of various plant species on Downy Brome. *Agron. J.*, **99:** 127-132.

Macias, F.A., Molinillo, J.M.G., Galindo, J.C.G., Varela, R.M., Simond, A.M. and Castellando, D. 2001. The use of allelopathic studies in the search for natural herbicides. *J. Crop. Prod.,* **4:** 237-255.

Mack, R.N. and Lansdale, W.M. 2001. Humans are global plant dispersers: getting more than we have bargained for. *Bioscience*, **51:** 95-102.

Madhu, M., Nanjappa, H.V. and Ramachandrappa, B.K. 1995. Allelopathic effect of weeds on crops. *Mysore Journal of Agricultural Sciences,* **29:** 106-112.

Mahadevappa, N. 1997. Ecology, distribution, menace and management. of Parthenium. In: *Proc. of the 1st Internat. Conf. on Parthenium Mangement.* Oct 6-8, 1997. University of Agricultral Sciences, Dahrwad, Karnataka (India).

Mahajan, V.K., Sharma, N.L. and Sharma, R.C. 2004. Pathenium dermatitis: is it a systemic contact of dermatitis or an airborne contact dermatitis? *Contact Deimatitis*, **51:** 231-234.

Maharajan, S., Shrestha, B.B. and Jha, P.K. 2007. Allelopathic effect of aqueous extract of leaves of *Parthenium hysterophorus* L. on seed germination and seedling growth of some cultivated and wild herbaceous species. *Scientific World*, **5 (5):** 33-39.

Makheja, A.N., and Bailey, J.M. 1982. A platelet phospholipase inhibitor from the medicinal herb feverfew (*Tanacetum parthenium*). *Prostaglandins, Leukotrienes, and Medicine*, **8**: 653–660.

Malarkodi E and Manoharan A. 2013. Antifungal activity of *Parthenium hysterophorus* L. *Journal of Chemical and Pharmaceutical Research,* **5(1)**:137-139

Mane, A. V., Pardeshi, R. G., Gore, V. R., Walave, R. L., Manjrekar, S. S. and Sutar, G. N. 2013. Water quality and sediment analysis at selected locations of Pavana River of Pune district, Maharashtra. *Journal of Chemical and Pharmaceutical Research*, **5(8):** 91–102.

Mangla, A., Vishwanath, B. and Subba Rao, P.V. 1981. Atmospheric survey of pollen of *Parthenium hysterophorus* **L.** *Allergy*, **47:** 192-196.

Mangrola, M. H., Joshi, V. G., Dudhagara, P. R. and Parmar, B. H. 2012. Two step synthesis and biophysical characterization of silver nanoparticles using green approach. *Journal of Environmental Research and Development*, **7(2):** 1021–1025.

Marshall, E.J.P. 2001. Biodiversity, herbicides and non-target plants. In: *Proc. Brighton Crop Protection Conference- Weeds,* Brighton, U.K., pp. 855-862.

Marshall, E.J.P., Brown, V.K., Boatman, N.D., Lutman, P.J.W. and Squire, G.R. 2001. The impact of herbicides on weeds abundance and biodiversity, PNO 940, A report for the U.K. pesticides safely Directorate, IACR-Cong. Ashton Research Station, U.K.

Marshall, E.J.P., Brown, V.K., Boatman, N.D., Lutman, P.J.W., Squire, G.R. and Ward, R.K. 2003. The role of weeds in supporting biological diversity within crop fields. *Weeds Research*, **43**: 77-89.

Marwat, K.B., Khan, M.A., Nawaz, N. and Amin, A. 2008. *Parthenium hysterophorus* L. a potential source of bioherbicide. *Pak. J. Bot.*, **40**: 1933-1942.

Maurya, B. R. and Sharma, P. K. 2010. Studies on the germination and viability of *Parthenium hysterophorus* L. in its compost. *Indian Journal of Weed Science*, **42(3-4):** 244–245.

McClay, A.S. 1987. Observations on the biology and host spacificity of *Epiblema strenuana* (Lepiduptera, Tortricidae) a potential biocontrol agent for *Parthemium hysterophorus* (Compositae). *Entomophaga,* **32(1):** 23-34.

Mcfadyen, R.E. 1992. Biological controls against Parthenium weed in Australia. *Crop Protection,* **11:** 400-407.

McFadyen, RE 1995. Parthenium weed and human health in Queensland. *Aust Fam. Physician*, **24**: 1455.

Medhin, B.G. 1992. *Parthenium hysterophorus* L.: a new weed problem in Ethiopia. *FAO Plant Protection Bulletin*, **40**:49.

Megharaj, M., Rao, A.P., Venkateswarlu, K. and Rao, A.S. 1987. Toxicity of *Parthenium hysterophorus* extract to *Chlorella vulgaris* and *Synechococcus elongates. Plant Soil,* **103:** 292-294.

Mehboob, N., Saleem B. and Qureshi. M.J. 2000. Allelopathic effect of Sunflower (*Helianthus annuus*) on germination and seedling growh of linseed (*Linum usitatissimum.*). *Pak. J. Biol. Sci.,* **3(8)**: 1305-1307.

Mehta, U.C., Mahato, K.C. and Singh, J.N. 1995. Effect of Parthenium extracts on pollen tetrad and pollen sterility in radish (*Raphanus sativus* **L.**). *Cruciferae. Newsletter*, **17:** 48-49.

Meksawat, S. and Pornprom, T. 2010. Allelopathic effect of itchgrass (*Rottobellia cochinchinensis*) on seed germination and plant growth. *Weed Biology and Management,* **10:**16-24.

Mersie, W. and Singh, M. 1987. Allelopathic effect of Parthenium (*Parthenium hysterophorus* L.) extract and residue on some agronomic crops and weeds. *Journal Chemical Ecology,* **13:** 1739-1747.

Mew, D., Balza, F., Towers, G.H.N. and Levey, I.G. 1982. Antitumour effects of the sesquiterpene leactone parthenin. *Planta Medica*, **45**: 23-27.

Miflin-Ben 2000. Crop improvement in the 21st century. *Journal of Experimental Botany*, **51:**1-8.

Miranda, C. A. S. F. de., Cardoso, M. das G., Carvalho, M. L. M. de, Figueiredo, A. C. S., Nelson, D. L., Oliveira, C. M. de, Gomes, M. de S., Andrade, J. de, de Souza, J. A. and de Albuquerque, L. R. M. 2014. Chemical composition and allelopathic activity of *Parthenium hysterophorus* and *Ambrosia polystachya* weeds essential oils. *American Journal of Plant Sciences,* **5**: 1248-1257.

Mishra, K.K. 1991. Parthenium hysterophorus L. A new record for Nepal. Journal of Bombay Natural History Society, **88:** 466-467.

Molisch, H. 1937. Der Enflusseiner Pflanze aufdie andere-allelopathic. Gustav Fischer, Verlag, Jena.

Msafiri, C.J., Tarimo, M.T. and Ndakidemi, P. A. 2013. Allelopathic effects of *Parthenium hysterophorus* on seed germination, seedling growth, fresh and dry mass production of *Alysicurpus glumaceae and Chloris gayana. American Journal of Research Communication,* **1(11):** 190-205

Mukherjee, B., and Chatterjee, M. 1993. Antitumor activity of *Parthenium hysterophorus* and its effect in the modulation of biotransforming enzymes in transplanted murine leukemia. *Planta Medica,* **59**: 513–516.

Mulatu, W., Gezahegn, B. and Belekadu, B. 2006. Allelopathic effect of *Parthenium hysterophorus* extract on seed gerimination and seedling growth of lettuce. *Trop Sci.,* **45:** 159-162.

Muller, C.H. 1966. The role of chemical innibiton (allelopathy) in vegetational composition. *Bull. Torrey Botanical Club,* **93**: 332-351.

Murphy, S.T. and Cheesman, O.D. 2006. The Aid Trade-International Assistance programmes as Pathway for the introduction of Invasive Alian species (a preliminary report). World Bank paper 109. A paper prepared by CABI. Bioscience, UK Center, pp.38.

Murphy, J.J., Heptinstall, S., and Mitchell, J.R. 1988. Randomised double-blind placebo-controlled trial of feverfew in migraine prevention. *Lancet* **2**: 189–192.

Murthy, E.G., Prathibha, N.C. and Thammaiah, N. 1995. Study on allelopathic effect of Parthenium on Sunflower and Sorghum. *World Weeds,* **2:** 161-164.

Murthy, R.K., Raveendra, H. R. and Manjunatha, R.T.B. 2010. Effect of *Chromolaena* and *Parthenium* as green manure and their compost on yield, uptake and nutrient use efficiency on Typic Paleustalf. *European Biological Sciences,* **4(1)**: 41–45.

Nandal, D.P.S., Birla, S.S., Narwal, S.S. and Koushik, J.C. (eds) 1994. Allelopathic interactions in agroferestry systems. In: *Allelopathy in Agricutture and Forestry,* Jodhapur, pp 93-130.

Narwal, S.S. 1994. *Allelopathy in Crop Production.* Scientific Publishers (India), Jodhpur, India. pp. 288

Narwal, S.S. 1999. *Allelopathy Upadate Basic and Applied Aspects*, Science Pub. Inc., New Hampshire, pp. 200-254.

Narwal, S.S., Palaniraj, R., Soti, S.C., Kadiana, H.S. and Dahiya, D.S. 2003. Allelopathic Plants- 8. *Parthenium hysterophorus* L. *Allelopathy J.*, **11**: 151-170.

Naseem, M. 1997. *Allelopathic Effects of Autumn Sunflower Residues on Wheat Productivity and Wheat-weeds*. Ph.D. Thesis, Dept. of Agronomy, Univ. of Agriculture, Faisalabad, Pakistan.

Naseem, M., Cheema, Z.A. and Bazmi, S.A. 2003. Allelopathic effect of sunflower aqueous extracts on germination of wheat and some important wheat weeds. *Pak. J. Scientific Res.,* **55**(3-4): 71-75.

Nath, R. 1988. Note on the effect of Parthenium extract on germination and seedling growth of crops. *Indian J. Agriculture Sci.*, **51**: 601-603.

Navie, S.C., Mcfadyen, R.E., Panetta, F.D. and Adkins, S.W. 1996. The biology of Australian weeds 27. *Parthenium hysterophorus* L. *Plant Protection Quaterly,* **11**:76-88.

Nirmalnath, P. J., Biradar, A. P., Patil, M. B. and Patil, A. B. 2005. Microflora associated with vermicompost obtained from different weeds. *Karnataka Journal of Agricultural Sciences*, **18(1):** 186–187.

Nouguchi, H.K. and Ino, T. 2005. Possible involvement of momilactone B in rice allelopathy. *J. Plant Physiol.,* **162:** 718-722.

O' Donnel, C. and Adkins, S.W. 2005. Management of Parthenium weed through competitive displacement with beneficial plants. *Weed Biology and Management,* **5**: 77-79.

Olofsdotter, M. 1998. *Allelopathy in Rice.* Int. Rice Res. Inst. Makati city, Phillippines.

Olofsdotter, M. 2001. Rice-A step toward use of allalopathy. *Agron. J.,* 93: 3-8.

Olofsdotter, M. Navarez, D. and Moody, K. 1995. Allelopathic potential in rice germplasm. *Ann. Appl. Bot.,* **127:** 543-560.

Oudhia, P. 1996. *Allelopathic effects of selected weeds on crops.* M.Sc (Ag) Thesis, Indira Gandhi, Agricultural University Raipur, (India). pp.250.

Oudhia, P. 2000a. Allelopathic effect of some obnoxious weeds on germination of soybean. *Indian Journal of Plant Physiol.,* **5(2):** 295-296.

Oudhia, P. 2000b. Positive (inhibitory) allelopathic effects of *Parthenium hysterophorus* leaves on germination and seedling vigour of sunflower. *Crop Research,* **20(3):** 560-562.

Oudhia, P. 2000c. Allelopathic effects *Parthenium hysterophorus* L. and *Ageratum conyzoides* on wheat var. Sujata. *Crop Research,* **20**(3): 563-566.

Oudhia, P. 2001. Phyto-sociological studies of rainy season wasteland weeds with special reference to *Parthenium hysterophorus* L. in Raipur district Inida. *Asian Journal of Microbiology, Biotechnology and Environmental Science,* **3**(1-2): 89-92.

Oudhia, P. and Tripathi, R.S. 1998. Allelopathic effects of *Parthenium hysterophorus* L. on Kodo, Mustard and problematic weeds. *In: Proc. of First*

Internat. Conf. on Parthenium Management, 6-8 October 1997, Dharwad, India. **11:** 136-139.

Oudhia, P. and Tripathi, R.S. 1999. Allelopathic effects of *Parthenium hysterophorus, Lantana camara* and *Ageratum conyzoides* on germination and seedling vigour of wheat and selected Rabi weeds. In: *Abstracts II world congress on Allelopathy Critical Analysis and future Prospects* (Ed. A.U. Malik) August 1999.Thunderbay, Ontaria, Canada: Lake head University. pp. 142.

Oudhia, P., Kohli, S.S. and Tripathi, R.S. 1997a. Allelopathic effects of *Parthenium hysterophorus* L. on germination of Linseed. *Indian J. Plant Physiol.*, **2(4):** 327-329.

Oudhia, P., Kohli, S.S. and Tripathi, R.S. 1997b. Allelopathic effect of white top (*Parthenium hysterophorus* L.) on chickpea. *Legume Research*, **20**(2): 117-120.

Palevitch, D., Earon, G., and Carasso, R. 1997. Feverfew (*Tanacetum parthenium*) as a prophylactic treatment for migraine: a double-blind placebo-controlled study. *Phytotherapy Research* **11**: 508–511.

Pandey, D.K 1994a. Inhibition of salvinia (*Salvinia molesta Mitchell*) by parthenium (*Parthenium hysterophorus* L.). I. Effect of leaf residue and allelochemicals. *The Journal of Chemical Ecology*, **20(12):** 3111–3122.

Pandey, D.K 1994b. Inhibition of Salvinia (*Salvinia molesta* Mitchell) By Parthenium (*Parthenium hysterophorus* L.) II. Relative effect of flower, leaf, stem and root residues on Salvinia and paddy. *The Journal of Chemical Ecology,* **20(12):** 3123-3131.

Pandey, D.K. 1996. Phytotoxicity of Sesquiterpene lactone Parthenin on aquatic weeds. *Ibid,* **22:**151-160.

Pandey, H.N. and Dubey, S.K. 1988. Achene germination *Parthenium hysterophorus* L. effects of light, temperature, provenance and achene size. *Weed Research,* **28:**185-190.

Pandey, H.P., Raza, A.K., Chauhan, S.K., 2011. Assessment of allelopathic aggression of *Parthenium hysterophorus* L. on seed germination and seedling growth of some important cereals. *Trends Biosci.* **4:** 68-70.

Pandey, D.K., Kauraw, L.P. and Bhan, V.M. 1993. Inhibitory effect of Parthenium (*Parthenium hysteophorus* L.) residue on growth of water hyacinth (*Eichhornia cyassipes*) relative effect of flower, leaf, stem, and root residues. *Journal of Chemical Ecology*, **19(11):** 2663-2670.

Pandey, D.K., Pani, L.M.S. and Joshi S.C. 2003. Growth, reproduction and photosynthesis of ragweed Parthenin (*Parthenium hysterophorus* L.) *Weed Science*, **51:**191-201.

Pandey, K., Sharma, P.K., and Dudhe, R. 2012. Antioxidant and anti-inflammatory activity of ethanolic extract of *Parthenium hysterophorus* L. *Asian Journal of Pharmaceutical and Clinical Research,* **5**: 28–31.

Panwar, R., Sharma, A.K., Dutt, D. and Pruthi, V. 2015. Phenolic acids from *Parthenium hysterophorus*: Evaluation of bioconversion potential as free radial scavengers and anti-cancer agents. *Advances in Bioscience and Biotechnology*, **6:** 11-17.

Parada-Turska, J., Paduch, R., Majdan, M., Kandefer-Szerszeń, M., and Rzeski, W. 2007. Antiproliferative activity of parthenolide against three human cancer cell lines and human umbilical vein endothelial cells. *Pharmacological Reports*, **59**: 233-237.

Parashar, V., Parashar, R., Sharma, B. and Pandey, A.C. 2009. Parthenium leaf extract mediated synthesis of silver nanoparticles: a novel approach towards Weed utilization. *Digest Journal of Nanomaterials and Biostructures*, **4(1)**:45- 50.

Pareek, A., Suthar, M., Rathore, G.S., and Bansal, V. 2011. Feverfew (*Tanacetum Parthenium* L.): A systematic review. *Pharmacognosy Reviews, 5:* 103-110.

Park, S., Laurence, R.B. and Andrew, R.W. 2003. The theory and application of plant competition models: an agronomic perspective. *Annals of Botany,* **92:** 741-748.

Parsons, W.T., and Cuthbertson, E.G. 2001. Noxious Weeds of Australia (CSIRO Publishing).

Patel, D. J., Patel, H. V., Patel, S. K. and Patel, B. A. 1993. Nematicidal properties of some plant materials for the management of rootknot nematodes in tomato nursery. *Indian Journal of Plant Protection*, **21(**2): 242–244.

Patel, S. 2011. Harmful and beneficial aspects of *Parthenium hysterophorus*: an update. *Biotech.*, **1(1):** 1-9.

Pati, U.K. and Ashim Chowdhury. 2015. A comparison of phytotoxic potential among the crude extracts from *Parthenium hysterophorus* L. extracted with solvents of increasing polarity. *International Letters of Natural Sciences,* **33:** 73-81

Patil, T.M. and Hedge B.A. 1988. Isolation and purification of Sesquiterpene lactones from the leaves of *Pathenium hysterophorus* L.-its allelopathic and cytotoxic effects. *Cur. Sci.,* **57:**1178-1181.

Patterson D.T. 1981. Effects of allelochemicals on growth and physiological responces of soybeans (*Glycine max*). Weeds *Science* **29**, 53-58.

Pattrick, M., Heptinstall, S., and Doherty, M. 1989. Feverfew in rheumatoid arthritis: a double blind, placebo controlled study. *Annals of the Rheumatic Diseases,* **48**: 547-549.

Paudel V.R. 2007. *Allelopathic interaction studies on Parthenium hysterophorus L.* (M.Sc. Dissertation) Central Department of Botany, Tribhuwan University, Kathmandu, Nepal.

Paul, T.K. 2010. The earliest record of *Parthenium hysterophorus* L. (Asteraceae) in India. *Curr. Sci.,* **98**(10): 1272

Paulsen, E. 1992. Compositae dermatitis: a survey. *Contact Dermatitis,* 26(2): 76-86.

Pedrol, N., Gonzalez, L. and Reigosa, M.J. 2006. Allelopathy and abiotic strees. In: *Allelopathy:* A *Physiolgical Process with ecological implication* (eds. M.J, Reigosa, N. Pedrol and L. Gonzalez). Springer, Netherland, pp. 171-209.

Peng, C.I., Hu, L.A. and Kao, M.T. 1988. Unwelcome naturalization of *Parthenium hysterophorus* L. (Asteraceae) in Tiwan. *Journal of Tiwan Museum,* **41:**95-110.

Persons, W.T. and Cuthbertson, E.G. 1992. *Noxious Weed of Australia.* Melbourne: Inkata, pp.692.

Peters, E.J. and Mohammad Zam, A.H.B. 1981. Allelopathic effect of all fescue genotypes. *Agron. J.,* **73:** 56-58.

Picman, A.K. 1986. Biological activities of Sesquiterpene lactones. *Biochem. Syst. Ecol.,* 14: 255-281.

Picman, J. and Picman, A.K. 1984. Autotoxicity in *Parthenium hysterophorus* and its possible role in control of germination. *Ibid,* **12:**287-292.

Picman A.K., Balza F. and Towers G.H.N. 1982. Occurrence of Hysterin and Di-Hydroisoparthenin in *Parthenium hysterophorus, Phytochemistry,* **21(7)**: 1801-1802

Popovici, J., Bertrand, C., Jacquemoud, D., Bellvert, F., Fernandez, M.P., Comte, G. and Piola, F. 2011. An allelochemical from *Myrica gale* with strong phytotoxic activity against highly invasive *Fallopia x bohemica Texa. Molecules,* **16:** 2323-2333.

Prasad, D., Ram, D. and Imtiyaz, A. 2002. Management of plant parasitic nematodes by the use of botanicals. *Annals of Plant Protection Sciences,* **10(2):** 360-364.

Priya, V., Radhika, and Srinivasa 2011. Evaluation of invitro free radical scavenging sctivity of different organic extracts of *Parthenium hysterophorus* leaves. *International Journal of Pharmacy & Pharmaceutical Sciences,* **3**: 135–138.

Putnam, A.R. 1985. Weed allelopathy. In: *Weed Physiology, Reprodution and Ecophysiology* (*ed.* S.O. Duke). *CRC Press,* **1**:131-155.

Putnam, A.R. and Defrank, J. 1979. Use of Cover crops to inhibit weeds. In: *Proc. IX Intl' Cong. Plant Port.* pp. 580-582.

Putnam, A.R. and Duke, W.B. 1974. Biological suppression of weeds: evidence for allelopathy in accession of cucumber. *Science,* **18**:370-372.

Putnam, A.R. and Tang, C.S. 1986. Allelopathy: State of the Science. In: *The Science of Allelopathy* (eds. A.R. Putnam & C.S. Tang), Wiley, New York, pp. 1-19.

Purivs, C.E. Jessop, R.S. and Lovelt, J.V. 1985. Selective Regulation of germination and growth of annual weeds by crop residues. *Weed Res.,* **25(**6): 415-421.

Qasem J.R. 1993. Allelopathic effects of some common weed species in cereal crops. *Dirasat.,* **20B**: 7-28.

Qasem, J.R. 1995. Allelopathic effects of *Amaranathus retroflexus* and *Chenopodium murale* on vegetable crops. *Allealopathy J.,* **2**:49-66.

Qasem, J.R. 2001. Allelopathic potential of white top and Syrian sage on vegetable crops. *Agron. J.,* **93:**67-71

Qureshi, R.A., Gilani, S.S., Ghufran, M.A. and Sultana, K.N. 2006. Effect of *Parthenium hysterophorus* L. on soil characterstics initially inhibited to *Cannabis sativa* L. *Pakistan J. Biol. Sci.,* **9 (15**): 2794-2797.

Rai M.K. and Upadhyay S.K. 1990. In vitro efficacy of different extract of *Parthenium hysterophorus* Linn. Against human pathogenic fungi using different techniques. *Indian J. Pathol. and Microbiol.,* **33(2)** : 179-181.

Rai M.K., Deepak Acharya and Wadegaonkar P. 2003. Plant derived antimycotics: Potential of Asteraceous plants. In: *Plant derived antimycotics :Current Trends and Future prospects.* Howarth Press, London, pp. 165-185.

Rainhardt C., Kraus S., Walker F., Foxcroft L., Robbertse P. and Hurle K. 2005. The allelochemical parthenin is sequestered at high level in capitate sessile trichomes on leaf surfaces of *Parthenium hysterophorus. Allelopathy Journal,* 15:2.

Rajan, L. 1973. Growth inhibitors from *Parthenium hysterophorus. Curr. Sci.,* 42:729-730.

Rajendrudu, G. and Rama Das, V.S. 1981. *Parthenium hysterophorus* L. (Asteraceae) exhibitory low photorespiration. *Ibid,* **50**(3): 592-593.

Rajiv, P., Narendhran, S. and Sivaraj, R. 2012. Bioconversion of *Parthenium hysterophorus* using earthworms and assessment of its macro and micronutrients. In: *Proceedings of the International Bio Conference & Event, Leonia International Conventional Centre,* Hydrabad, India, December 2012.

Rajiv, P., Narendhran, S., Subhash, K.M., Sankar, A., Sivaraj, R. and Venckatesh, R. 2013a. *Parthenium hysterophorus* L. compost: assessment of its physical properties and allelopathic effect on germination and growth of *Arachis hypogeae* L. *International Research Journal of Environmental Science,* **2(2):** 1-5.

Rajiv, P., Rajeshwari, S., Hiranmai Yadav, R. and Rajendran, V. 2013b. Vermi remediation: detoxification of parthenin toxin from *Parthenium* weeds. *Journal of Hazardous Materials,* **262**: 489-495.

Rajiv, P., Rajeshwari, S. and Venckatesh, R. 2013c. Bio-Fabrication of zinc oxide nanoparticles using leaf extract of *Parthenium hysterophorus* L. and its size-dependent antifungal activity against plant fungal pathogens. *Spectrochimica Acta. Part A. Molecular and Biomolecular Spectroscopy,* **112**: 384–387.

Rajulu, G.S., Gowri, N. and Perumal, S.S. 1976. Biological control for the pernicious weed *Parthenium hysterophorus* Linn. *Ibid,* **45(17):** 624-625.

Ramanujam, J.R., Kulothungan, S., Anitha, S., and Deepa, K. 2011. A Study on Compatibility of *Pseudomonas fluorescenes* L. and *Parthenium hysterophorus* L . as a Biocontrol agent to leaf spot by *Alternaria alternata* f . sp . *lycopercisi* in Tomato. *South As. J. Biol. Sci.,* **1:** 71–86.

Ramaswami, P.P. 1997. Potential use of Parthenium. In: *Proc. First International Conference on Parthenium Management.* Vol. **I**: 77-80.

Ramesh C., Ravindranah N., Das B, Prabhakar A, Bharatam J., Ravikumar K., Kashinatham A. and Mc Morris T.C. 2003. Pseudoguaianolides from the flowers of *Parthenium hysterophorus. Phytochemistry* (Amsterdam), 64 (4): 841-844

Randhawa, N., Sakhuja, P.K. and Singh, I. 2002. Management of Root-Knot nematode *Meliodogyne incognita* in *Abelmoschus esculentus* through botanicals extract and organic amendments. *Indian J. Nematol.,* **32**(2): 129-131.

Rao, R.S. 1956. Parthenium: A new record for India. *J. Bombay Nat. Hist. Soc.,* 54: 218-220.

Rashid, H., Khan, M.A., Amin, A., Nawab, K., Hussain N. and Bhowmik, P.K. 2008. Effect of *Parthenium hysteropherus* L. root extracts on seed germination and growth of maize and barley. *Am. J. Plant Sci. Biotechnol.,* **2:**51-55.

Ratwat, L.S., Ratwat, D.S.K., Narwal, S.S., Palaniraj, R. and Sati, S.C. 2002. Allelopathic effect of aqueous extracts of sunflower (*Helianthus annuus* L.) root on some winter oil seed crops. *Geobios,* **29**:225-228.

Rebaz, Z., Shaukat, S.S. and Siddiqui, I.A. 2001. Allelopathic potential of *Anagallis arvesis* L.: a cosmopolitan weed. *Pak. J. Biol. Sci.,***4:** 446-450.

Rice, E.L. 1974. *Allelopathy*. Academic Press. New York

Rice E.L. 1984. *Allelopathy*. 2nd edition, Academic Press. New York.

Rizvi, S.J.H., Haque, H., Singh, V.K. and Rizvi, V. 1992. A discipline called allelopathy. In: *Allelopathy: Basic and Aapplied Aspects* (eds. S.J.H. Rizvi, and V. Rizvi). Chapmann & Hall Publishers. Pp.1-8.

Rodosevich, S.R. and Holt, S.T. 1984. *Weed Ecology Implication for Management*, 2nd Ed. J. Willey, New York.

Rodriguez, E. 1975. Chemistry and Distribution of Sesquiterpene lactone and Flavonoids in Parthenium (Comositae): Systematic and Ecological Implicatons. Ph.D. Thesis, University of Texas.

Romo de Vivar, A., Bratoeff, E.A. and Rios, T. 1996. Structure of hysterin, a new sesquiterpene lactone. *J. Org. Chem.,* **31**: 673-677.

Romagni, J.G., Allen, S.N. and Dayan, F.E. 2000. Allelopathic effect of volatile cineoles on two weedy plant species. *Journal of Chemical Ecology,* **26**: 303-313.

Rose, J.Caroline and Rani, R. Helen Niruba 2014. Vermicomposting of *Parthenium histeroporous* by *Eudrilis eugenia* and effect of compost on *Lycopersicum esculatus. Int. J. Inv. Pharm. Sci.,* 2(2);650-652

Ross, J.J., Arnason, J.T. and Birnboim, H.C. 1999. Low concentrations of the feverfew component parthenolide inhibit *in vitro* growth of tumor lines in a cytostatic fashion. Planta Medica *65*, 126–129.

Safdar, M.E., Tanveer, A., Khaliq, A. and Naeem, M.S. 2014. Allelopathic action of *Parthenium* and its rhizospheric soil on Maize as influenced by growing conditions 1 Planta Daninha, Viçosa-MG, **32(2)**: 243-253.

Safdar, M.E., Tanveer, A., Khaliq, A. and Riaz, M.A. 2015. Yield losses in maize (*Zea mays*) infested with Parthenium weed (*Parthenium hysterophorus* L.). Crop Protection, **70**: 77-82.

Saghir, A.R. and Bhatti, S.M. 1970. The influence of herbicides on the chemical composition soybean seeds. *Proc. 10th British Weed Control Conf.*, **1**: 384-388.

Saini, A., Aggarwal, N.K., Sharma, A., Kaur, M. and Yadav, A. 2014. Utility Potential of *Parthenium hysterophorus* for its Strategic Management. *Advances in Agriculture*, 1-16 pp.

Sakai, A.K., Allendrof, F.W., Holt, J.S., Lodge, D.M., Molofsky, J., Baughman, K.A., Cabin, S., Cohen, R.J., Allstrand, J.E., McCauley, N.C., O'Nell, D.E., Parker, I.M., Thompson, J.N. and Waller, S.G. 2001. The population biology of invasive species. *Annual Review of Ecology and Systems,* **32:** 305-332.

Sampietro, D.A. and Vattuone, M.A. 2006. Sugercane Straw and its phytochemicals and growth regulators of weed and crop plants. *Plant Growth Regul.*, **48:** 21-27.

Sangeetha, C. and Baskar, P. 2015. Allelopathy in weed management: A critical review. *African Journal of Agricultural Research*, 10(9): 1004-1015.

Sanghamitra, K., Prasada Rao, P. V. V. and Naidu, G. R. K. 2012. Uptake of Zn (II) by an invasive weed species *Parthenium hysterophorus* L. *Applied Ecology and Environmental Research*, **10(3):** 267–290.

Sangwan, P., Kaushik, C. P. and Garg, V. K. 2011. Nutrient recycling and management of press mud, parthenium and biogas plant slurry employing earthworms. *International Journal of Environment and Waste Management*, **7(3-4):** 382–394.

Saravanane, P., Poonguzhalan, R. and Chellamuthu, V. 2012. Parthenium (*Parthenium hysterophorus* L.) distribution and its bioresource potential for rice production in Puduchery, India. *Pakistan Journal of Weed Science Research*, 18: 551–555.

SasiKumar, K., Vijayalakshmi, C. and Prathiban, K.T. 2001. Allelopathic effects of four *Eucalyptus* species on redgram *(Cajanus cajan* **L.**). *Journal of Tropical Agriculture,* **39**:134-138

Schon, M.K. and Einhellig F.A. 1982. Allelopathic effect of cultivated sunflower on grain sorghum. *Bot. Gaz.*, **143:** 505-510.

Seigler, D.S. 1996. Chemistry and mechanisms of allelopathic interactions. *Agron J.,* 88: 876-885.

Senthilkumari, M., Vasanthi, K., Saradha, T. and Bharathi, R. 2013. Studies on the Vermiconversion of different leaf wastes by using *Eudrilus Eugeniae* (Kinberg). *International Journal of Advanced Research,* **1(3):** 96–101.

Shabbir, A. and Bajwa.R. 2006. Distribution of Parthenium weed (*Parthenium hysterophorus* L.) and alien invasive weed species threatening the biodiversity of Islamabad. *Weed Biology and Management,* **6:** 89-95.

Shad R.A 1987. Status of weed science activities in Pakistan. *Prog. Farm.*, 7(1): 10-16.

Shamim, A. and Tiyagi, S.A. 2003. Ulilization of noxious weeds for menagement of plant parasitic nematodes infesting some vegetable crops. *Plant Pathology and Nematodogy,* **36**: 211-219.

Sharma, R. 2003. Performance of different herbicides for control of congress grass (*Parthenium hysterophorus* L.) in some cropping areas. *Indian. J. Weed Sci.,* **35:** 242-245.

Sharma, G.L. and Bhutani K.K. 1988. Plant based antiamoebic drugs: Part II. Amoebicidal activity of Parthenin isolalated from *Parthenium hysterophorus. Planta Medica,* **54:** 20-22.

Sharma, S., and Gupta, N. 2012. Antimicrobial potential of a weed plant *Partherium hysterphorus*: an invivo study. *International Journal of Pharmaceutical Research and Development,* **4:** 112–118.

Sharma, K., Rao, P.B. and Pandey N. 2009. Effect of seven-weed extract on seed germination in different varieties of wheat (*Triticum aestivum* L.). *J. Indian bot. Soc.,* **88**(1 & 2): 35-40.

Sharma, N. and Trivedi, P.C. 2002. Screening of leaf extracts of some plants for their nematicidal and fungicidal properties against *M. incognita* and *Fusarium oxyspoum*. *Asian J. Exp. Sci.,* **16(1 &2):** 21-28.

Shaukat, S.S., Ghazala, P., Khan, D. and Ahmed, M. 1985. Phytotoxic effect of *Citrullus colocynthies* L., Schardon certain crop plants. *Pak. J. Bot.*, **17:** 235-246.

Shaukat, S.S. and Siddique, I.A. 2002. Allelopathic and antifungal potential of *Lantana camara* root leachates in soil. *Pak. J. Biol. Sci.,* **5:** 51-53.

Shaukat, S.S., Tajuddin, Z. and Siddiqui, I.A. 2003. Allelopathic potential of *Launaea procumbens* (Roxb.) Rammya and Rajgopal: A tropical weed. *Pak. J. Biol. Sci.,* **6**(3) 225-230.

Shen, M.C., Rodriquez, E., Ken, K. and Mabry, T.J. 1976. Flavanoid of four species of Parthenium (compositae). *Phytochemistry,* **15**(6): 1045-1051.

Sheng-peng, Z. and Young-qing, M. 2007. A preliminary study on systems engineering based method for evolution of allelopathic potential in crops and its application. *Agricutural Sciences in China,* **6(1):** 68-77.

Shobha, A., Reddy, S. A., Akila, A. and Kale, R.D. 2012. Management of secondary sewage sludge by vermicomposting for use as soil amendment. *Global Journal of Biochemistry and Biotechnology,* **7(1)**: 13–18.

Shrestha, B.B., Shabbir, A. and Adkins, A.W. 2014. *Parthenium hysterophorus* in Nepal: A Review of its weed status and possibilities for managemenr. *Weed Research,* **55(2):** 132-144.

Shrivastava, J.N., Shukla, J.P. and Shrivastava, R.C. 1985. Effect of *Parthenium hysterophorus* L. extract on seed germination and seedling growth of barley, pea and wheat. *Acta Botanica Indica,* **13:** 194-197.

Singh, H.P., Batish, D.R., Kohli, R.K., Saxena, D.B. and Arora, V. 2002. Effect on parthenin-a sesquiterpene lactone from *Parthenium hysterophorus* L. on early growth and physiology of *Ageratum conyzoides. Journal of Chemical Ecology,* **28:** 2169-2179.

Singh, H.P., Batish, D.R., Pandher, J.K. and Kohli R.K. 2003a. Assesment of allelopathic properties of *Parthenium hysterophorus* residues. *Agric Ecosyst Environ.,* **95**: 537-541.

Singh, H.P., Batish, D.R., Kaur, S. and Kohli, R.K. 2003b. Phytotoxic interference of *Ageratum conyzoides* with wheat (*Triticum aestivum) J. Agron, Crop Sci.,* **189**: 341-346.

Singh, H.P. Batish D.R., Setia, N. and Kohli, R.K. 2005a. Phytotoxic effects of *Parthenium hysterophorus* L. residues on three *Brassica* species, *Weed Biology and Management.* **5**; 105-109.

Singh, H.P., Batish, D.R., Setia, N. and Kohli, R.K. 2005b. Herbicidal activity of volatile oils from *Eucalyptus citriodora* against *Parthenium hysterophorus* L. *Annals of Applied Biology,* **146:** 89-94.

Singh, H.P., Kohli, R.K. and Batish, D.R. 2001. Allelopathy in agroecosystems: an overview. *J.Crop Prod.,* **4**:1-41.

Singh, K., Shaki, A.K., Pal, S. and Bahyan, S.S., 1992. Phyto-allelopathic influence of Parthenium hysterophorus L. In: Tauro, P., Nawal, S.S. (Eds.), Proceedings of the 1st National Symposium on Allelopathy in Agroecosystems,

Hisar, India, February 1992. Hisar, Haryana Agricultural University, pp. 61-63.

Singh, L., Raghava. N., Raghava R.P. and Jayanti 2008. Allelopathic effect of Parthenium on *Lagenaria siceraria* (Mol) Standl. *Indian J. Applied and Pure Bio.,* **23**(2): 285-294.

Singh, L. Raghava, N. and Raghava, R.P. 2009. Allelopathic substances of Parthenium weed and interaction of *Solanum melongena. Bioscience, Biotechnology, Research Asia,* **6 (2):** 855- 862.

Singh, L. 2010. *Studies on Allelopathic effect of Parthenium on some Vegetables.* Ph.D. Thesis, VBS Purvanchal University, Jaunpur, India.

Singh, N.B. and Thaper, R. 2002. Allelopathic effect of *Croton bonplandianum* on *Parthenium hysterophours. Allelopathy Journal,* **10:**163-170.

Singh N.P. 1983. Potential Biological control of *Parthenium hysterophorus* L. *Curr. Sci.,* **95**: 537-541.

Singh, S.P. and Sangeeta 1991. Allelopathic potential of *Parthenium hysterophorus* L. *Journal of Agronomy and Crop Science,* **167**: 201-206.

Singh, S., Yadav A., Balyan R.S., Malik, R.K. and Singh, M. 2004. Control of Ragweed Parthenium (*Parthenium hysterophorus)* and associated weeds. *Weed Technology,* **18**: 658-664.

Singh, V.K. and Rizvi, V. 1992. *A discipline called allelopathy.* In: *Allelopathy: Basic and An applied Aspects* (eds. S.J.H. Rizvi and V. Rizvi). Chapman and Hall, London, pp:**9.**

Singh, B, and Srivastava, P. 2013. Phytochemical screening and *in vitro* antifungal investigation of *Parthenium hysterophorus* extracts against *Alternaria alternata. Int. Res, J. Pharma.*, **4(7):** 190-193.

Sinha, P. and Deo, P.M. 1999. Allelopathic effect of Parthenium root extracts on germination and early seedling growth of kidney bean (*Phaseolus vulgaris* L.). *Geobios,* **26**(2-3): 127-130.

Sisodia, S. and Siddique M.B. 2009. Allelopathic potential of rhizosphere soil of *Croton bonaplandianum* on growth and establishment of some crop and weed plants. *Afr. J. Agric. Res.,* **4(5):** 461-467.

Smith, A.E. and Martin, L.D. 1994. Allelopathic characterstics of three cool season grass species in the forage ecosystem. *Agron J.,* **6;** 243-246.

Sohal, S.K., Rup, P.J., Kaur, H., Kumari, N., and Kaur, J. 2002. Evaluation of the pesticidal potential of the congress grass, *Parthenium hysterophorus* Linn. on the mustard aphid, Lipaphis erysimi (Kalt.). *Journal of Environmental Biology/Academy of Environmental Biology*, India, **23**: 15–18.

Son, T.T.N. 1995. *Bioconversion of organic wastes for sustainable agriculture* Ph.D. Thesis. Tamil Nadu Agricultural University, Coimbatore.

Sonkaran, S., Kandasamy, O.S., Swamippan, M., Doraiswamy, S. and Vidively, S. 1996. Biological control approaches in the management of *Parthenium hystrophorus, Solanum elaeaguifalium* and *Eupatorium glandulosum,* Report, pp.71.

Srirama, Rao, S., Naggpal, B.S., Rao, O., Prakash and Rao, P.V. 1991. Immediate hypersensitivity of *Parthenium hysterophorus,* clinical studies on the prevalence of parthenium rhinitis. *Clinical Experiment Alleregy,* **21:** 55-62.

Srivastava, J. 2011. Potential of Parthenium to Control Root Knot Nematode in Cowpea. The Ph.D Thesis, VBS Urvanchal Univrsity, Jaunpur, India, pp 122-126.

Srivastava, J., Raghava, N., Raghava, R. P. and Singh, L. 2010. Potential use of Parthenium extract on growth parameters, chlorophyll content and total soluble sugar of cowpea. *Biomedical & Pharmacology Journal*, **3(2):** 357-363.

Srivastava, J., Raghava, N. and Raghava, R. P. 2011. Allelopathic potential of Parthenium to reduce water absorption in germinating cowpea seeds. *Indian Journal Scientific Research*, **2(3):**58-64.

Stachon, W.J. and Zindel, R.L. 1980. Allelopathic Activity of Canada thistle, *Cirsium arvense* in Colorado. *Weed Sci.,* **28:** 83-86.

Stephen, W.A. and Sowerby, M.S. 1996. Allelopathic potential of the weed, *Parthenium hysterophorus* L. in Australia. *Plant Potection Quarterly*, **11**: 20-23.

Stoller, E.W., Wax, L.M. and Slife F.W. 1979. Yellow nutsedge (*Cyperus esculentus*) competion and control in corn (*Zea maize*). *Weed Science,* **27**: 32-37.

Subba Rao, P.V., Mangla, B.S. Subba Rao and Prakash, K.M. 1976. Parthenium allergic weed pp 17-18. In proceedings of seminar on Parthenium a positive Danger, Spt. 4 Banglore Internatioanl cities relationship organization and University of Agricultural Science Banglore, Karnataka, India.

Sudhakar, P. 1984. *Substitute of fertilizer nitrogen through green manure in lowland rice*. M.Sc, Thesis, Tamil Nadu Agricultural University, Coimbatore.

Suresh, K.K. and Rai, R.S.V. 1987. Studies on the allelopathic effects of some agroforestry tree crops. *International Tree Crops Journal,* **4:** 109-115.

Suryawanshi, D. S. 2011. Utilization of weed biomass as an organic source in maize. *Life Science Bulletin*, 8(1): 10–12.

Swaminathan, C., Vinayrai, R.S. and Suresh, K.K. 1989. Allelpathic proactivities of *Acacia nilotica. Journal of Tropical Forest Science,* **2**: 56-60.

Swaminathan, C., Vinayrai,, R.S. and Suresh, K.K. 1990. Allelopathic effect of *Parthenium hysterophorus* L. on germination and seedling growth of few multipurpose trees and arable. *Int. Tree Crops. J.,* **6(2-3):** 143-150.

Swaminathan, M. S. 2006. An Evergreen revolution. *Crop Sci.*, **46**(5): 2293 2303.

Swaminathan, M. S. 2007. India's tryst with destiny in agriculture. *Yojana,* **51** (8): 61-65.

Swaminathan, M.S. 2008. Growth in agriculture. *Yojana,* **52(2):** 40-48.

Syamasundar, J. and Mahadevappa, M. 1986. *Cassia sericia,* to fight *Parthenium hysterophorus* L. *Curr. Sci.,* **55(5):** 261-262.

Tajuddin, Z., Shaukat, S.S. and Siddique, I.A. 2002. Allelopathic potential of *Solanum forskalii* Dunal.: A tropical weed. *Pak. J. Biol. Sci.*, **5:** 866-868.

Talakal, T.S., Dwivedi, S.K., and Sharma, S.R. 1995. In vitro and in vivo therapeutic activity of *Parthenium hysterophorus* against *Trypanosoma evansi. Indian Journal of Experimental Biology,* **33***:* 894-896.

Tamado, T. 2001. Performance and field Stability of medium duration groundnut genotypes under erratic rainfall condition in eastern Ethiopia. *Trop Sci.*, **41:**192-198.

Tamado, T. and Milberg, P. 2000. Weed flora in arable fields of eastern Ethiopia with emphasis on the occurance of *Parthenium hysterophorus*. *J. Weed. Res.*, **40**: 507-521.

Tamado, T. and Milberg, P. 2004. Control of Parthenium (*Parthenium hysterophorus* L.) in grain Sorghum (*Sorghum bicolor*). In the small holder farming system in eastern Ethopia. *Weed Technology*, **18**: 100-105.

Tamado, T., Schutz, W. and Milberg, P. 2002. Germination ecology of the weed *Parthenium hysterophorus* L. in eastern Ethopia. *Annual Applied Biology*, **140**: 263-270.

Tanveer, A., Khaliq, A., Ali, H.H., Mahajan, G. and Chauhan, B.S. 2015. Interference and managements of Parthenium: the world's most important invasive weed. *Crop Protection*, **68**: 49-59.

Tawaha, A.M. and Turk, M.A. 2003. Allelopathic effect of black mustard (*Brassica nigra*) on germination and growth of wild barley (*Hordeum sponataneum*). *J. Agron. Crop. Sci.*, **189:** 298-303.

Taye, T. 2002. *Investigation of pathogens for Biological Control of Parthenium hysterophorus in Ethiopia*. Ph.D Thesis presented to Humboldt Unversitat Zu Berlian, Landwirts Chaztlich Gartnerrischen Fakultat, Barlia.

Tefera, T. 2002. Allelopathic effects of *Parthenium hysterophorus* L. extracts on seed germination and seedling growth of *Eragrostis tef*. *Journal of Agronomy & Crop Science.*, **188:** 306-310.

Thakur, R. and Siddique, N. 2003. Allelopathic effect of aqueous extract of three common weeds on photosynthetic content of mustard. *Indian. J. Applied and Pure Bio.*, **18 (1):** 99-102.

Thakur, S.K. and Singh, K.D.N. 2000. Efficiency of agricultural wastes and weeds for biogas production. *Journal of Research, Birsa Agricultural University*, **12(1):** 11–15.

Thakur, S.K. and Singh, K.D.N. 2003. Anaerobic digestion of agricultural wastes and weeds for biogas production. *Annals of Biology*, **19(2):** 245–249.

Thapar, R. and Singh, N.B. 2003. Allelopathic effects of *Amaranthus viridis* on *Parthenium hysterophorus*. *J. Indian bot Soc.*, **82:** 93-96.

Theophrastus, 300 BC. *Enquiry into Plants*. Quoted from Hort, A.F. Transl. and Edite (1916). Book I-V, G.P.Putnam's Sons, New York.

Thijs, H., Shann, J.R. and Weidenhamer, J.D. 1994. The effect of phytotoxins on competitive outcome in model system. *Ecology*, **75**: 1959-1964.

Tismina, B. 2007. Impact of *Parthenium hysterophorus* L. invasion on Soil and plant spesies composition of Grasslands of central Nepal. [M.Sc. Dissers-tation] Central Department of Botany. Tribhuvan University, Kathmandu Nepal, pp. 61.

Tiwari, S., Adhikari, B., Siwakoti, M. and Subedi, K. 2005. An Inventory and Assesment of Invasive Alien Plant Species of Nepal IUCN- The world conservation Union Nepal, pp.115.

Tiyagi, S.A., Mukhtar, J. and Alam, M.M. 1985. Preliminary studies on nematicidal nature of two plants of the family Compositae. *Int. Nematol. Network Newsl.*, **2(3):**16-17.

Tiyagi, S.A., Bano, M.L. and Alam. M.M. 1988. Evaluation of nematocidal potential in plant species belonging to family compositae. *Indian J. of Nematology,* **18(2):** 228-231.

Tomaszewski, M. and Thimann, K.V. 1966. Interaction of phenolic acid metallic ions and chelating agents on auxin induced growth. *Plant Physiol.,* **41**: 1443-1454.

Towers, G.H.N. 1981. Allergic eczematous contact dermatitis from Parthenium weed (*Parthenium hysterophorus*). Proc. of the 6th Australian Weeds Conference, Gold coast, Queensland, Queensland Weed Society, Broadbeach, pp. 143-150.

Towers, G.H.N., Mitchell, J.C., Rodriguez, E., Bennett, F.D. and Subba Rao, P.V. 1977. Biology and chemistry of *Parthenium hysterophorus* L. a problem weeds in India. *Journal of Scientific and Industrial Research,.* **36:** 672-684.

Towers, G.H.N. and Subba Rao, P.V. 1992. Impact of pan tropical weed *Parthenium hysterophorus* L. on human. *Weed Science Society of Victoria.* pp. 134-138.

Tsay, T.T., Wu, S.T. and Lin, Y.Y. 2004. Evaluation of Asteraceae plants for control of *Meloidogyne incognita. Journal of Nematology,* **36**(1): 36–41.

Tudor, G.D., Ford, A.L., Armstrong, T.R. and Bromage, E.K. 1982. Taints in meat from sheep grazing Parthenium weed. *Australian Journal of Experinmental Agriculture and Animal Husbandry.,* **22(115)** :43-46.

Urmilesh J, Prites J, Chhajed, Rajesh J, Oswal, *et al.* 2011. Skeletal muscle relaxant activity of methanolic extract of *Parthenium hysterophorus* L. leaves in Swiss albino mice. *Inter J Pharm Life Sci.,* 2: 1211-1213.

Vaitor, R.P., Johnson, R.M., Grimm C.C. and Richard, J.E.P. 2006. Allelopathic autotoxic and hermetic effect of post harvest sugercane residue. *Agron. J.,* **98:** 1526-1531.

Vasilakoglou, I., Dhima, K. and Eleltherohorinos, I. 2005. Allelopathic potential of Bermudagrass and Jhonsongrass and their interference with cotton and corn. *Agron. J.,* **97**: 303-313.

Vasilakoglou, I., Dhima, K., Eleftherohorinos, I. and Lithourgidis 2006. Winter cereal cover crop mulches and inter-row cultivation: Effect on cotton development and grass weed suppression. *Agron. J.,* **98:** 1290-1297.

Valdes, A.F.C., Martínez, J.M., Lizama, R.S., Gaiten, Y.G., Rodríguez, D.A., and Payrol, J.A. 2010. *In vitro* antimalarial activity and cytotoxicity of some selected Cuban medicinal plants. *Revista Do Instituto De Medicina Tropical De Sao Paulo,* **52**: 197–201.

Venkataiah, B., Ramesh, C., Ravindranath, N. and Das, B. 2003. Charminarone, a seco-pseudoguaianoide from *Parthenium hysterophorus. Phytochemistry* (Amsterdam), **63 (4):** 383-386.

Verma, M. and Rao, P.B. 2006. Allelopathic effect of four weeds species extracts on germination, growth and protein in different varieties of *Glycine max* (L.) Merill. *J. Env Bio.,* **127**(3): 571-577.

Verma, K.K., Sirka, C.S. and Ramam, M. 2001. Contact dermatitis due to plants challenges and prospects. *The Second Practice,* **54**(11): 791-796.

Vijay, S.P., Chitra, V., Lakshmi, P., and Krishnaraju, V. 2008. Hypoglycemic effect of aqueous extract *Parthenium hysterophorus* L. in normal and alloxan induced diabetic rats. *Indian J Pharmacol,* **40:** 183-185.

Vijaya K. S. and Seethalakshmi, S. 2011. Contribution of *Parthenium* vermicompost in altering the growth yield and quality of *Abelmoschus esculents* (I) Moench. *Advanced Biotechnology,* **11(2):** 44–47, 2011.

Vijayakumari, B. and Hiranmai, Y. R. 2012. Influence of fresh, composted and vermicomposted *Parthenium* and poultry manure on growth characters of sesame (*Sesamum indicum*). *Journal of Organic Systems,* **7(1):** 14–19.

Wahab, S. 2005. Management of Parthenium through an integrated approach initiatives, achievements and research apportunities in India. In: *Proc. of the second International Conference on Parthenium Management,* 5-7 December 2005, Banglore, India, pp. 55-59.

Wakjira, M., Berecha, G. and Bulati, B. 2006. Allelopathic effects of *Parthenium hysterophorus* extracts on seed germination and seedling growth of lettuce . *J. Tropical Science,* **45(4):** 159-162.

Wakjira, M. Berecha, G. and Tulu, S. 2009. Allelopathic effects of an invasive alien weed *Parthenium hysterophorus* L. compost on lettuce germination and growth. *African Journal of Agricultural Research,* **4 (11):** 1325-1330.

Waller, G.R. 1987. Allelochemical action of some natural products. In: Phytochemical Ecology, Allelochemicals and Insect Pheremones and Allomones (eds. C.H. Chou and G.R. Waller). *Institute of Botany, Academia Sinca, Taipei,* pp.129-154.

Wegari, K.E. 2008. Perceived Socio-Ecomimic Impact of Parthenium weed and its effect on Crop Production in Babile Haramaya and Tulo Districts, East and west Hararge Zone. Ethiopia M.A. dessertation). Development and Agriculture Extention, School of Graduate Studies, Haramaya University, pp.88.

Weston, L.A. 1996. Utilization of allelopathy for weed management in agroecosystem. *Agron. J.,* **88:** 860-866.

Weston, L.A. 2005. History and current Trends in the use of Allelopathy for weed Management. *Weed Management,* **13**:529-534.

Weston, L.A. and Duke, S.O. 2003. Weed and crop allelopathy. *Crit. Rev. Plant Sci.,* 22: 367-389.

Weidenhammer, J.D. 1996. Distinguishing resource competition and chemical interference overcoming the methodological impose. *Agron. J.,* **88:** 866-875.

Whittaker, R.H. and Feeny, P.P. 1971. Allelochemicals: Chemical enteraction between species. *Science,* **171**: 757-770

Wills, R.J. 1985 .The historical basis of the concept of allelopathy. *Journal of the History of Biology,* **18:** 71-102.

Wise, R.M., Vanwilgen, W.B., Hill, P.M., Schulthese, F., Tweddle, D., Chabi oley, A., and Zimmeran G.H. 2007. The Economic Impect and Apporoprate Management of Selected Alien Invasive species on the African Continent, CSIR report No. CSIR/ NRI/ RBSD/ ER/ 2007/44.

Yadav, A. and Chauhan, S.V.S. 1998. Studies on allelopalthic effect of some weed. *Journal of Physiological Research,* **11(1)**: 15-18.

Yadav, A and Garg, V.K. 2011. Vermicomposting – An effective tool for the management of invasive weed *Parthenium hysterophorus. Bioresource Technology,* **102**: 5891-5895.

Yaduraju, N.T., Sushil Kumar, M.B.B., Prasad Babu and Gogoi, A.K. 2005. *Parthnium hysterophorus* L. Distribution, problem and management strategy in India. In: *Proc. of the Second International Conf. on Parthenium Managemnt,* 5-7 December 2005. University of Agricultual Science, Banglore India. pp 6-11

Yongqing, M.A. 2005. Allelopathic studies of common wheat (*Triticum aestivum* L.). *Weed Biology and Management,* **5:** 93-104.

Yu, J.Q., Ye, S.F., Zhang, M.F. and Hu, W.H. 2003. Effects of root exudates and aqueous root extracts of cucumber (*Cucumis sativus*) and allelochemicals, on photosynthesis and antioxidant enzymes in cucumber. *Biochem. Syst. Ecol.,* **31**: 129-139.

Zaheer, Z., Shafique, S., Shafique, S., and Mehmood, T. 2012. Antifungal potential of *Parthenium hysterophorus* L. plant extracts against *Fusarium solani. Scientific Research and Essays,* **7**: 2049–2054.

Zhou, J., Xie, G., and Yan, X. 2011. Encyclopedia of Traditional Chinese Medicines - Molecular Structures, Pharmacological Activities, Natural Sources and Applications. Vol. 1- 5: Isolated Compounds A—Z, References, TCM Plants and Congeners, Berlin, Heidelberg: Springer Berlin Heidelberg.

SUBJECT INDEX